James Glimm
Arthur Jaffe

Quantum Physics —
A Functional Integral Point of View

With 43 Illustrations

Springer-Verlag
New York Heidelberg Berlin

James Glimm
The Rockefeller University
New York, NY 10021
USA

Arthur Jaffe
Harvard University
Cambridge, MA 02138
USA

Library of Congress Cataloging in Publication Data

Glimm, James,
    Quantum physics.

    Bibliography: p.
    Includes index.
    1. Quantum field theory. 2. Quantum theory.
3. Statistical physics. I. Jaffe, Arthur, joint
author. II. Title.
QC174.45.J33   530.1'43   81-17

Printed in the United States of America.

9 8 7 6 5 4 3 2 1

ISBN 0-387-90551-0 Springer-Verlag New York Heidelberg Berlin (hard cover)
ISBN 3-540-90551-0 Springer-Verlag Berlin Heidelberg New York (hard cover)
ISBN 0-387-90562-6 Springer-Verlag New York Heidelberg Berlin (soft cover)
ISBN 3-540-90562-6 Springer-Verlag New York Heidelberg Berlin (soft cover)

Contents

Contents

PART III   The Physics of Quantum Fields

13   Scattering Theory: Time-Dependent Methods   237

14   Scattering Theory: Time-Independent Methods   256

15   The Magnetic Moment of the Electron   270

16   Phase Transitions   280

17   The ϕ^4 Critical Point   304

Introduction

This book is addressed to one problem and to three audiences.

The problem is the mathematical structure of modern physics: statistical physics, quantum mechanics, and quantum fields. The unity of mathematical structure for problems of diverse origin in physics should be no surprise. For classical physics it is provided, for example, by a common mathematical formalism based on the wave equation and Laplace's equation. The unity transcends mathematical structure and encompasses basic phenomena as well. Thus particle physicists, nuclear physicists, and condensed matter physicists have considered similar scientific problems from complementary points of view.

The mathematical structure presented here can be described in various terms: partial differential equations in an infinite number of independent variables, linear operators on infinite dimensional spaces, or probability theory and analysis over function spaces. This mathematical structure of quantization is a generalization of the theory of partial differential equations, very much as the latter generalizes the theory of ordinary differential equations. Our central theme is the quantization of a nonlinear partial differential equation and the physics of systems with an infinite number of degrees of freedom.

Mathematicians, theoretical physicists, and specialists in mathematical physics are the three audiences to which the book is addressed.

Each of the three parts is written with a different scientific perspective. Part I is an introduction to modern physics. It is designed to make the treatment of physics self-contained for a mathematical audience; it covers quantum theory, statistical mechanics, and quantum fields. Since it is addressed primarily to mathematicians, it emphasizes conceptual structure—the definition and formulation of the problem and the meaning of the answer—rather than techniques of solution. Because the emphasis differs from that of conventional physics texts, physics students may find this part a useful supplement to their normal texts. In particular, the development of quantum mechanics through the Feynman–Kac formula and the use of function space integration may appeal to physicists who want an introduction to these methods.

Part II presents quantum fields. Boson fields with polynomial self-interaction in two space-time dimensions—$P(\phi)_2$ fields—are constructed. This treatment is mathematically complete and self-contained, assuming some knowledge of Hilbert space operators and of function space integrals. The original construction of the authors has been replaced by successive improvements and simplifications accumulated for more than a decade. This development is due to the efforts of a small and dedicated group of some thirty constructive field theorists including Fröhlich, Guerra, Nelson, Osterwalder, Rosen, Schrader, Simon, Spencer, and Symanzik, as well as the authors. Physicists may find Part II useful as a supplement to a conventional quantum field text, since the mathematical structure (normally omitted from such texts) is developed here.

Part II contains the resolution of a scientific controversy. For years physicists and mathematicians questioned whether nonlinear field theory is compatible with relativistic quantum mechanics. Could quantization defined by renormalized perturbation theory be implemented mathematically? The mathematically complete construction of $P(\phi)_2$ fields presented here and the construction of Yukawa$_{2,3}$, ϕ_3^4, sine-Gordon$_2$, Higgs$_2$, etc., fields in the literature provide the proof. Central among the issues resolved by this work is the meaning of renormalization outside perturbation theory. The mathematical framework for this analysis includes the theory of renormalization of function space integrals. From the viewpoint of mathematics the implementation of these ideas has involved essentially the creation of a new branch of mathematics.

Whether the equations are mathematically consistent in four space-time dimensions has not been resolved. There is speculation, for example, that the equations for coupled photons and electrons (in isolation from other particles) may be inconsistent, but that the inclusion of coupling to the quark field may give a consistent set of equations. A proper discussion of this issue is beyond the scope of this book, but is alluded to in Chapters 6 and 17.

Particle interaction, scattering, bound states, phase transitions, and critical point theory form the subject of Part III. Here we develop the consequences of the Part II existence theory and make contact with issues of broad concern to physics. This part of the book is written at a more advanced level, and is addressed mainly to theoretical and mathematical physicists. It is neither self-contained nor complete, but is intended to develop central ideas, explain main results of a mathematical nature, and provide an introduction to the literature.

Condensed matter physicists may find interesting the discussion of phase transitions and critical phenomena. The central matters are series expansions and correlation bounds. These methods find application in diverse areas. We give detailed justification of the connection (by analytic continuation) between quantum fields and classical statistical mechanics. Professional physicists could well start directly in Part III, returning to earlier material only as necessary.

Readers interested in the historical development of constructive quantum field theory are referred to the various survey articles of the authors and others. In this book the specific, detailed references are minimized, especially in the self-contained Parts I and II. A large bibliography has been included; we apologize for the inevitable omissions.

Numerous colleagues, students, and friends helped make this book possible. Of particular importance were R. D'Arcangelo, R. Brandenberger, B. Drauschke, J.-P. Eckmann, J. Gonzalez, W. Minty, K. Peterson, P. Petti, the staff at Springer Verlag, and especially our wives Adele and Nora. We are also grateful to the ETH, the IHES, the University of Marseilles and the CEN Saclay for hospitality as well as to the Guggenheim Foundation and the NSF for support.

Conventions and Formulas

Fourier transforms

$$f(x) = (2\pi)^{-d/2} \int e^{ipx} f^{\sim}(p) \, dp,$$

$$f^{\sim}(p) = (2\pi)^{-d/2} \int e^{-ipx} f(x) \, dx,$$

$$f(\theta) = (2\pi)^{-d/2} \sum e^{in\theta} f^{\sim}(n),$$

$$f^{\sim}(n) = (2\pi)^{-d/2} \int_0^{2\pi} e^{-in\theta} f(\theta) \, d\theta.$$

Minkowski vectors

$$x = (x_0, \mathbf{x}) = (x_0, \ldots, x_{d-1}),$$

$$x^2 = x \cdot x = -x_0^2 + \mathbf{x}^2, \qquad p^2 = p \cdot p = -p_0^2 + \mathbf{p}^2,$$

$$x \cdot p = \sum x_i p^i = -x_0 p_0 + \mathbf{x} \cdot \mathbf{p},$$

$$\Box = -\partial_t^2 + \Delta = -\partial x_0^2 + \sum_{i=1}^{d-1} \partial x_i^2.$$

Euclidean vectors

$$x_d = ix_0,$$

$$x^2 = x \cdot x = \sum_{i=1}^{d} x_i^2,$$

$$\Delta = \sum_{i=1}^{d} \partial x_i^2.$$

Schrödinger's equation

$$\hbar = h/2\pi,$$

$$i\hbar\dot{\theta} = H\theta, \qquad \theta(t) = e^{-itH/\hbar}\theta(0),$$

$$p = -i\hbar\frac{\partial}{\partial q}, \qquad [p(x), q(y)] = -i\hbar\,\delta(x - y)$$

Covariance operators $C_m \in \mathscr{C}_m$ satisfy

$$(-\Delta + m^2)C_m = \delta.$$

σ and γ matrices

$$\sigma_0 = \begin{pmatrix} 1 & 0 \\ 0 & 1 \end{pmatrix}, \qquad \sigma_1 = \begin{pmatrix} 0 & 1 \\ 1 & 0 \end{pmatrix},$$

$$\sigma_2 = \begin{pmatrix} 0 & -i \\ i & 0 \end{pmatrix}, \qquad \sigma_3 = \begin{pmatrix} 1 & 0 \\ 0 & -1 \end{pmatrix},$$

$$\gamma_i = \begin{pmatrix} 0 & \sigma_i \\ \sigma_i & 0 \end{pmatrix}, \qquad i = 1, 2, 3,$$

$$\gamma_0 = \begin{pmatrix} I & 0 \\ 0 & -I \end{pmatrix}, \qquad \gamma_5 = \gamma_0\gamma_1\gamma_2\gamma_3 = \begin{pmatrix} 0 & I \\ I & 0 \end{pmatrix},$$

$$\not{a} = \sum a_\mu \gamma_\mu,$$

$$\not{a}^2 = \sum a_\mu^2 = a^2.$$

Dirac equation (zero field)

$$(\hbar\not{\partial} - mc)\psi = 0.$$

Dirac equation in external field A

$$\left(\hbar\not{\partial} + i\frac{e}{c}\not{A} - mc\right)\psi = 0.$$

List of Symbols

a, a^*, A, A^*	annihilation and creation operators
a, A	free energy
A	antisymmetrization operator
$\mathscr{A}$	action
$\mathfrak{A}, \mathscr{A}$	algebra of operators
b	bond
B	observable; region in space-time
$\mathscr{B}$	set of bonds
c	diagonal values of C, $c(x) = C(x, x)$; critical (as a subscript); constant
cr	critical
cl	classical
C	covariance; (Chap. 7) complex numbers
$\mathscr{C}, \mathscr{C}_m$	a class of covariance operators (Sec. 7.9)
d	dimension of space-time
D	Dirichlet boundary conditions
$\mathscr{D}$	domain of an operator; C_0^∞ test function space
$\mathscr{D}'$	Schwartz distribution space
$\mathscr{D}^{(j)}$	domain for irreducible spin j representation of $SU(2, C)$
E	energy level; eigenvalue for H; Euclidean transformation; Euclidean group

$\mathscr{E}$	Euclidean group; Euclidean Hilbert space; time strip (Section 10.5)
f	test function; free energy
$\mathscr{F}$	Fock space
g	test function
$\mathscr{G}$	group
$h, \hbar$	Planck's constant
h	external field
H	Hamiltonian
HS	Hilbert–Schmidt
$\mathscr{H}(\mathbf{x})$	Hamiltonian density
$\mathscr{H}$	Hilbert space of quantum states
I	identity operator
J	interaction strength for Ising ferromagnet
$j, \mathbf{J}$	angular momentum
k	Boltzmann's constant
$\mathscr{K}$	kernel of semigroup
K	kernel of Bethe–Salpeter equation
L	angular momentum (Section 15.1)
L_s, L_i, L	lines in Feynman graphs (self-interacting, interacting)
$\mathscr{L}$	Lagrangian; lattice; multiple reflection norm (Section 10.5); Lorentz group
m, M	mass; magnetization; multiple reflection norm (Section 10.5)
n	number of field components; degree of polynomial P
nn	nearest neighbor
N	Neumann boundary condition; $N(f) =$ norm of f.
$\mathscr{N}$	null space for inner product
p	period boundary conditions; pressure; Lebesgue index; degree of polynomial P
p, P	momenta; momentum operator; momentum space
P	polynomial interaction; projection operator
P_n	Hermite polynomial
q, Q	configuration; configuration space; Lebesgue index
R	real numbers; multiple reflection norm (Section 10.5)
R^d	Euclidean d-space
s	time

s, S	entropy
S	generating function; Schwinger function; sphere; symmetrization operator
$\lvert S^n \rvert$	volume of n-sphere
$\mathscr{S}$	Schwartz space of rapidly decreasing test functions
$\mathscr{S}'$	Schwartz space of tempered distributions
$\mathfrak{S}_n$	symmetric group on n elements (permutation group)
t	Euclidean time $(=x_d)$; Minkowski time $(=x_0)$
T	time ordering; truncation
U, V	unitary operator on Hilbert space
V	potential
W	Wightman function
dW	Wiener measure
$\mathscr{W}$	Wiener path space
$\mathscr{X}$	phase space
x	point in space time
$\mathbf{x}$	point in space
z	fugacity; activity
Z	partition function; field strength renormalization constant; integers
Z_+	nonnegative integers; partition function
β	$(1/kT)$ inverse temperature
γ	critical exponent
γ, Γ	boundary; phase boundary
Γ	Dirichlet boundary conditions on Γ; inverse to propagator or two point function
$\lvert \Gamma \rvert$	length or area of Γ
δ	Dirac δ function; Kronecker δ function; lattice spacing; critical exponent
Δ	Laplacian; special solution of wave or Laplace equation (propagator) also unit square
ε	$2\theta - 1$ (a type of Heaviside function); lattice spacing; reduced temperature $(T - T_c)/T_c$
ζ, η	critical exponents
θ	reflection operator; Heaviside function; state in $\mathscr{H}$
κ	momentum cutoff
λ	coupling constant
Λ	bounded region of space

$\lvert \Lambda \rvert$	area or volume of Λ
μ	$(-\Delta + m^2)^{1/2} = (p^2 + m^2)^{1/2}$; chemical potential; external field
$d\mu$	statistical weight or ensemble
ν	frequency; critical exponent
$d\nu$	statistical weight or ensemble
ξ	random variable
Ξ	partition function
π	3.14159; momentum conjugate to field ϕ
Π	projection operator; hyperplane
$\Pi_{\pm}$	half spaces of $R^d \backslash \Pi$
ρ	density
σ	mass^2; Ising spin variable; time
Σ	proper self-energy
ϕ, Φ	quantum field; configuration of classical field
$d\Phi_C$	Gaussian measure, covariance C
χ	susceptibility; random variable; state in $\mathscr{H}$; characteristic function
ψ	quantum field; state in $\mathscr{H}$
ω	frequency; Wiener path; angular integration variable
Ω	vacuum state; ground state; equilibrium state
∂	derivative; boundary operator
∇	gradient; divergence
$\lvert \cdot \rvert$	absolute value; area, volume or number of $\cdot$; norm
$^\wedge$	projection operator from the Euclidean path space to the Hilbert space of quantum states
$\sim$	Fourier transform
$\langle\ ,\ \rangle$	inner product
$\langle\ \rangle$	expectation; integral with respect to $d\mu$
$[\ ,\ ]$	commutator: $[a, b] = ab - ba$
$\{\ ,\ \}$	anticommutator: $\{a, b\} = ab + ba$
$\varnothing$	free boundary conditions; empty set
$\times$	vector product
$\cdot$	time derivative; position for missing variable as in $f(\cdot) = f$ for a function f.
$\backslash$	set theoretic difference: $A \backslash B = \{x : x \in A,\ x \notin B\}$
$-$	complex conjugation; closure

PART I

AN INTRODUCTION TO MODERN PHYSICS

CHAPTER 1

Quantum Theory

1.1 Overview

Classical mechanics is the limit $\hbar \to 0$ of quantum mechanics. Non-relativistic (Newtonian) mechanics is the limit $c \to \infty$ of special relativity. Here $\hbar$ is Planck's constant and c is the speed of light. Quantum field theory is the combination of quantum mechanics with special relativity. It contains both parameters c and $\hbar$, and it has two distinct degenerate limits: $c \to \infty$ and $\hbar \to 0$. The classical limit of a quantum field $(\hbar \to 0)$ actually gives rise to two distinct limits: a classical field limit and a classical particle limit, with choice between these two limits depending on the sequence of states in which the limit is taken. Only the second (particle) limit has the further nonrelativistic limit of classical particle mechanics.

1.2 Classical Mechanics

We present some basic definitions in classical mechanics (and classical statistical mechanics) for comparison with quantum mechanics. We consider n point particles of mass m_j moving in the field of a time-independent potential energy V. The phase space of the system, $\mathscr{X} = R^{6n}$, is defined as the sum of a configuration space $Q = R^{3n}$ and a conjugate momentum space $P = R^{3n}$. A classical *observable* is a function $B(q, p)$ on phase space, $(q, p) \in (Q, P) = \mathscr{X}$.

The time development of a point in phase space is the subject of classical dynamics. The time development of a probability distribution on phase

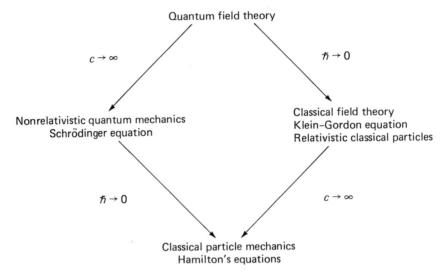

Figure 1.1 The classical and nonrelativistic limits of a quantum field.

space is the subject of nonequilibrium classical statistical mechanics. In either case, the Hamiltonian or total energy observable determines the dynamical laws. In Cartesian coordinates,

$$H(q, p) = \sum_{i=1}^{n} \frac{p_i^2}{2m_i} + V(q). \tag{1.2.1}$$

We consider a curve $(q(t), p(t))$ in phase space, with initial value at time t_0. The curve $(q(t), p(t))$ is obtained by integrating Hamilton's (Newton's) equations

$$\frac{dq_i(t)}{dt} = \nabla_{p_i} H = \frac{p_i}{m_i},$$

$$\frac{dp_i(t)}{dt} = -\nabla_{q_i} H = -\nabla_{q_i} V = F_i(q) \tag{1.2.2}$$

with initial conditions $q(t_0) = q_0$, $p(t_0) = p_0$.

More generally, we may consider the time development of an arbitrary observable B:

$$B_t(q, p) = B(q(t), p(t))$$

with initial conditions $B_{t_0}(q, p) = B(q, p)$. By (1.2.2), we obtain

$$\frac{\partial B_t(q, p)}{\partial t} = D_H B_t(q, p), \tag{1.2.3}$$

where D_H is the vector field

$$D_H = \sum_{i=1}^{n} \left\{ \frac{1}{m_i} p_i \cdot \nabla_{q_i} + F_i(q) \cdot \nabla_{p_i} \right\} \tag{1.2.4}$$

on $\mathscr{X}$. Assuming integrability of (1.2.3),

$$B_t(q, p) = (e^{(t-t_0)D_H}B)(q, p) = B(q(t), p(t)).\tag{1.2.5}$$

Proposition 1.2.1 (Liouville's theorem). *Define the Liouville measure by*

$$d\mu = \prod_{i=1}^{n} dp_i\, dq_i,$$

and let

$$D_H = \sum_{i=1}^{n} \left(\frac{\partial H}{\partial p_i}\frac{\partial}{\partial q_i} - \frac{\partial H}{\partial q_i}\frac{\partial}{\partial p_i} \right).$$

Then D_H is formally skew adjoint on $L_2(\mathscr{X}, d\mu)$, and e^{tD_H} is formally unitary.

Remark. By *formally* skew adjoint we mean that for functions F and G in $C_0^\infty(\mathscr{X})$,

$$\langle D_H F, G\rangle_{L_2(d\mu)} = -\langle F, D_H G\rangle_{L_2(d\mu)}.$$

Thus on the domain $C_0^\infty(\mathscr{X})$, we have $D_H^* = -D_H$. The extension of this result to establish unitarity of e^{tD_H} involves a technical question of the integrability of (1.2.3) and the existence and uniqueness of the exponential e^{tD_H}. There exists an extensive literature about this question, and we do not discuss it further. We only mention that restrictions are required on $V(q)$ to ensure the classical result.

PROOF. For F, G in C_0^∞, integration by parts yields

$$\int F^-(D_H G)\, d\mu = - \int (D_H F)^- G\, d\mu - \sum_i (F^- G)\left(\frac{\partial^2 H}{\partial q_i\, \partial p_i} - \frac{\partial^2 H}{\partial p_i\, \partial q_i} \right).$$

The last term vanishes, so $D^* = -D$ on C_0^∞ as claimed.

An alternative statement of Liouville's theorem is that the measure $d\mu$ is invariant under the classical dynamics determined by H. In terms of the Jacobian of the transformation $(q, p) \to (q(t), p(t))$ we state:

Proposition 1.2.2. *Assuming the unique integrability of Hamilton's equations, a solution satisfies*

$$\frac{\partial(q(t), p(t))}{\partial(q, p)} \equiv 1.$$

Thus the volume of any subset of phase space is left invariant by the flow given by the solution to Hamilton's equations. Namely, if $\mathscr{X}_0 \subset \mathscr{X}$ and $\mathscr{X}_t = e^{tD_H}\mathscr{X}_0$, then

$$\mu(\mathscr{X}_0) = \int_{\mathscr{X}_0} d\mu = \int_{\mathscr{X}_t} d\mu = \mu(\mathscr{X}_t).$$

An equivalent formulation of the dynamics of observables is given by the Poisson bracket invariant of two observables,

$$\{B, C\} = \sum_{i=1}^{n} (\nabla_{q_i} B \cdot \nabla_{p_i} C - \nabla_{p_i} B \cdot \nabla_{q_i} C). \tag{1.2.6}$$

The bracket $\{\,,\,\}$ gives the algebra of observables a Lie algebra structure. We note that the *canonical* Poisson brackets are

$$\{p_i, p_j\} = 0 = \{q_i, q_j\}, \qquad \{q_i, p_j\} = \delta_{ij} I. \tag{1.2.7}$$

Since these brackets are constant functions, they are time-independent by (1.2.3). Furthermore,

$$\{H, B\} = -D_H B.$$

Thus the equation (1.2.3) can be written

$$\frac{\partial B_t}{\partial t} = -\{H, B\}, \tag{1.2.8}$$

so assuming convergence

$$B_t = e^{(t-t_0)D_H} B = \sum_{n=0}^{\infty} \frac{(t_0 - t)^n}{n!} \{H, \{H, \ldots, \{H, B\}, \ldots\}\}. \tag{1.2.9}$$

The automorphism $B \to B_t$ corresponds to the Heisenberg picture in quantum mechanics. The $\hbar \to 0$ limit of Heisenberg quantum dynamics yields (1.2.9).

The other basic quantity, aside from the observables, is the *state*. The state of a classical system is determined by a single point (a, b) in phase space, which specifies the observables $B_{t_0}(a, b)$ at time t_0. In classical statistical mechanics, one considers a more general type of state, namely a probability density $d\rho(q, p)$ on phase space,

$$\int d\rho(q, p) = 1, \qquad d\rho(q, p) \geq 0. \tag{1.2.10}$$

The classical states with probability concentrated at one point have the form

$$d\rho(q, p) = \delta(q - a)\, \delta(p - b)\, dq\, dp,$$

and are called *pure* states. The state ρ assigns a number, the expected value $\rho(B)$, to each observable B:

$$\rho(B) = \int B(q, p)\, d\rho(q, p). \tag{1.2.11}$$

For the classical state above, $\rho(B) = B(a, b)$.

We may consider the dynamics as a mapping of states, rather than observables. We consider a state ρ at time t_0. We then define a state ρ_t such that for each observable B,

$$\rho_t(B) = \rho(B_t). \tag{1.2.12}$$

Since D_H is formally skew adjoint,

$$dp_t(q, p) = (e^{-(t-t_0)D_H} \, dp)(q, p)$$
$$= dp(q(-t), p(-t)). \tag{1.2.13}$$

This point of view of dynamics corresponds in quantum mechanics to the Schrödinger picture. The $\hbar \to 0$ limit of the Schrödinger equation yields (1.2.13).

1.3 Quantum Mechanics

Classical physics breaks down at the level of atoms and molecules. Consider the hydrogen atom, for instance, which is composed of two particles: the nucleus, a proton of charge $+e$ and mass m_p, and an electron of charge $-e$ and mass m_e. The nucleus is small and heavy, with $m_p/m_e \approx 2000$ and the radius of the proton $\approx 10^{-3}$ times the atomic radius. According to classical ideas, the attractive Coulomb potential

$$V(r) = -\frac{e^2}{r}$$

would cause the electron to orbit around the proton, much like the earth–moon gravitational system. However, the accelerated, charged electron, according to classical theory, would continuously radiate energy, causing the atom to collapse.

The original task of quantum mechanics was to explain the stability of atoms and molecules, as well as to explain the discrete frequencies of light radiated by excited atoms. The quantitative success of quantum mechanics in predicting observed atomic and molecular spectra is a major triumph of twentieth century science. There is no doubt that quantum mechanics is a correct description of nature. In this chapter, we present the basic principles of quantum mechanics as postulates. We do not attempt to justify or derive them. We prefer to regard the classical mechanics of Section 1.2 as a limit of quantum mechanics as Planck's constant $\hbar \to 0$. We divide the postulates into general ones (labeled P) and others valid in a restricted class of theories.

Postulate P1.   The pure states of a quantum mechanical system are rays in a Hilbert space $\mathscr{H}$ (i.e., unit vectors, with an arbitrary phase).

Specifying a pure state in quantum mechanics is the most that can be said about a physical system. In this respect, it is analogous to a classical pure state.

The concept of a state as a ray in Hilbert space leads to the probability interpretation in quantum mechanics. Given a physical system in the state θ, the probability that it is in the state χ is $|\langle \theta, \chi \rangle|^2$. Clearly

$$0 \le |\langle \theta, \chi \rangle|^2 \le 1.$$

We note that while the phase of a vector θ has no physical significance, the relative phase of two vectors does, i.e., for $|\alpha| = 1$, $|\langle \alpha\theta, \chi \rangle|$ is independent of α, but $|\langle \theta_1 + \alpha\theta_2, \chi \rangle|$ is not. It is most convenient to regard pure states θ simply as vectors in $\mathscr{H}$, and to normalize them in an appropriate calculation.

Postulate P2. Quantum mechanical observables are self-adjoint operators on $\mathscr{H}$. The expected (average) value of the observable B in the state θ is

$$E_\theta(B) = \frac{\langle \theta, B\theta \rangle}{\langle \theta, \theta \rangle}. \tag{1.3.1}$$

Examples of observables are the Hamiltonian (energy) observable, the momentum observable, and the position observable.

We remark that statistical mixtures in quantum mechanics lead to quantum statistical mechanics. The usual statistical mixture is described by a positive trace class operator ρ, yielding the expectation

$$\rho(B) = \frac{\operatorname{tr}(\rho B)}{\operatorname{tr} \rho}. \tag{1.3.2}$$

(Trace class means $\operatorname{tr} \rho < \infty$.) If ρ has rank 1, then $\rho(B)$ is the pure state (1.3.1) with $\rho/\operatorname{tr} \rho$ the projection onto $\theta/\|\theta\|$. Otherwise, $\rho(B)$ is a convex linear combination of pure states,

$$\rho(B) = \sum \alpha_i \langle \theta_i, B\theta_i \rangle, \tag{1.3.3}$$

where the θ_i are the (orthonormal) eigenvectors of ρ and $\sum \alpha_i = 1$.

Postulate P3. The Hamiltonian H is the infinitesimal generator of the unitary group $U(t) = e^{-itH/\hbar}$ of time translations. The momentum $\mathbf{P}$ is the infinitesimal generator of the unitary space translation group $e^{i\mathbf{x} \cdot \mathbf{P}/\hbar}$. The angular momentum $\mathbf{J}$ is the infinitesimal generator for the unitary space rotation group.

The time translation group $U(t)$ determines the dynamics. There are two standard descriptions: the Schrödinger picture and the Heisenberg picture. In the Schrödinger picture, the states $\theta \in \mathscr{H}$ evolve in time,

$$\theta(t) = e^{-itH/\hbar}\theta, \tag{S1}$$

while the observables do not evolve. The vectors satisfy the Schrödinger equation

$$i\hbar \frac{d\theta(t)}{dt} = H\theta(t). \tag{S2}$$

The time-dependent state $\theta(t)$ yields the expectation $E_{\theta(t)}(B)$.

The second description of dynamics is the Heisenberg picture, in which the states remain fixed, and the observables evolve in time according to the automorphism group

$$B \to B(t) = e^{itH/\hbar} B e^{-itH/\hbar}$$

$$= U(t)^* B U(t). \tag{H1}$$

Thus the observables B satisfy the dynamical equation

$$\hbar \frac{dB(t)}{dt} = [iH, B(t)], \tag{H2}$$

with the formal solution

$$B(t) = \sum_{n=0}^{\infty} \frac{(it/\hbar)^n}{n!} [H, [H, \ldots, [H, B], \ldots]]. \tag{H3}$$

We remark the similarity between (H2–3) and (1.2.8–9) with $\{\ ,\ \}$ replaced by $[\ ,\ ] (i\hbar)^{-1}$. The unit of action $\hbar$ has the same dimension as pq.

The relation between the Heisenberg and Schrödinger pictures is given by

$$E_{\theta(t)}(B) = E_\theta(B(t)). \tag{1.3.4}$$

We remark that Postulate P3 ensures that the results of an experiment, i.e., inner products $\langle \theta, \chi \rangle$, are independent of the time at which the experiment is performed:

$$|\langle \theta, \chi \rangle| = |\langle \theta(t), \chi(t) \rangle|.$$

Conversely, let θ, θ' be a symmetry of $\mathscr{H}$, i.e., a 1-1 transformation of $\mathscr{H}$ onto itself which conserves probabilities: $|\langle \theta, \chi \rangle| = |\langle \theta', \chi \rangle|$.

Theorem 1.3.1 (Wigner). *Every symmetry of $\mathscr{H}$ can be implemented either by a unitary transformation U on $\mathscr{H}$,*

$$\theta' = U\theta,$$

or by an antiunitary operator A on $\mathscr{H}$,

$$\theta' = A\theta.$$

The interpretation of this result is that every symmetry of $\mathscr{H}$ can be regarded as a coordinate transformation. In particular, the group of time translations is implemented by a unitary group of operators $U(t)$. Only certain discrete symmetries (e.g., time inversion in nonrelativistic quantum mechanics) are implemented by antiunitary transformations.

We now pass to the second part of this section, in which we specialize to nonrelativistic quantum mechanics. One usual representation for a system of n particles moving in a potential field V is

$$\mathscr{H} = L_2(Q), \tag{1.3.5}$$

where $Q = R^{3n}$ is configuration space. The choice (1.3.5) is called the Schrö-dinger representation (as distinct from the Schrödinger picture). The function $\psi(q) \in \mathscr{H}$ has the interpretation of giving the probability distribution $\rho(q) = |\psi(q)|^2$ for the position of the particles in Q.

By P3, we have

$$p_i = \frac{\hbar}{i} \nabla_{q_i}, \tag{1.3.6}$$

and a Hamiltonian of the form

$$H = \sum \frac{p_j^2}{2m_j} + V(q) \tag{1.3.7}$$

becomes the elliptic differential operator

$$H = -\sum_j \frac{\hbar^2}{2m_j} \Delta_{q_j} + V(q). \tag{1.3.8}$$

Note that the (canonical) commutation relations

$$[q_i, q_j] = 0 = [p_i, p_j],$$
$$[p_i, q_j] = -i\hbar\, \delta_{ij} I \tag{1.3.9}$$

have the form (1.2.7), with $\{\ ,\ \}$ replaced by $[\ ,\ ] (i\hbar)^{-1}$, and are preserved by the Heisenberg picture dynamics.

The representation (1.3.5) does not account for the spin of the basic particles, e.g., spin 0 for π mesons (pions), spin $\frac{1}{2}$ for electrons, muons, protons, or neutrons, spin 1 for photons, and higher spins for other particles or nuclei. In order to consider spin-dependent forces (e.g., the coupling of the spin magnetic moment to a magnetic field), we generalize (1.3.5) to the n-fold tensor product

$$\mathscr{H} = \otimes L_2(Q, S). \tag{1.3.10}$$

Here $Q = R^3$ and L_2 denotes functions defined on Q with values in the finite dimensional spin space S. For spin zero particles, $S = C$, and we are reduced to (1.3.5). For nonzero spin, $S = C^{2s+1}$. We write $\theta(q)$ as a vector with components $\theta(q, \zeta)$. A space rotation (generated by the angular momentum observable $\mathbf{J}$) will rotate both q and ζ, the latter by a linear transformation of the ζ coordinates according to an n-fold tensor product of a representation of the spin group SU(2, R) (the universal covering group of the rotation group SO(3)).

It is a physically verifiable fact that particles of a given type are indistinguishable. In other words, in a 5-particle theory, if we find 3 particles in a region B, we cannot tell which of the 5 particles these are. To obtain indistinguishable particles, we restrict ourselves to a subset of (1.3.10) invariant under an irreducible representation of the symmetric group (permutation group) of the n particle coordinates (q_i, ζ_i), $i = 1, 2, \ldots, n$. The standard choices are the totally symmetric representation for integer spin particles, e.g., π mesons or photons, and the totally anti-symmetric representation for half-integer spin particles, e.g., electrons, protons, or neutrons.

The choice of antisymmetry for atomic and molecular problems with spin $\frac{1}{2}$ is known as the Pauli exclusion principle. In quantum field theory, one can prove that integer spin particles cannot be antisymmetrized and half-integer spin particles cannot be symmetrized. Other, more complicated representations (called parastatistics) cannot be mathematically excluded. There is, however, no experimental evidence for their occurrence. Particles with integer spin are called *bosons*; those with half-integer spin are called *fermions*.

Postulate P4. A quantum mechanical state is symmetric under the permutation of identical bosons, and antisymmetric under the permutation of identical fermions.

1.4 Interpretation

The most important physical aspects of quantum mechanics are the interpretation of the Hamiltonian H and the predictions of particle scattering. Unlike the classical Hamiltonian, which always takes a continuum of values, a quantum mechanical Hamiltonian can have both discrete eigenvalues and a continuous spectrum. The eigenvalues are associated with bound states of the system, which roughly are states in which the particles remain in a bounded region of space. The continuous spectrum is associated with scattering states, in which some of the particles move off to infinite distances.

Let us consider a simple case (discussed in more detail in Sections 1.6 and 1.7), a single particle in a Coulomb potential, $V = -1/r$. Classically the particle can be bound in an elliptic (negative energy) orbit, or it can lie in a hyperbolic (positive energy) orbit. Quantum mechanically, the spectrum of H consists of eigenvalues $E_n = -\alpha/n^2$, $n = 1, 2, \ldots$, where α is some constant, and a continuum $[0, \infty)$.

The eigenvalues give the possible quantum mechanical bound state energies E_n. The eigenvector for E_1 is called the ground state, while the eigenvectors for $E_2, E_3, \ldots$, are called the first, second, $\ldots$, excited states. In isolation (e.g., from the electromagnetic field) these excited states are stable, but under interaction with some external system, they are in general not stable. As a simplification, we regard interaction with the external system as changing the eigenvector and eigenvalue from E_m to E_n. In the case of an atom interacting in this fashion with the electromagnetic field, the atom emits light, a photon of frequency

$$v = \frac{E_m - E_n}{h},$$

for $E_n < E_m$. If $E_n > E_m$, the atom has absorbed a photon of this frequency. The observed frequencies of emission or absorption are always differences of energies in the spectrum of H. Before the discovery of quantum mechanics, the fact that observed spectral lines could be written as differences of energies was known as the *Ritz combination principle*.

The positive energy states, or scattering states, describe the scattering of a quantum mechanical particle by the attractive potential. Under an external perturbation a particle can be *captured*, with a transition from a state of positive energy to an eigenstate $E_n < 0$. This capture is accompanied by the emission of photons with a continuous range of observed frequencies and energies.

1.5 The Simple Harmonic Oscillator

This elementary example appears in every book on quantum mechanics. Unlike most quantum mechanics problems, the harmonic oscillator can be solved explicitly in terms of elementary (Hermite) functions and Gaussian integrals. Thus we can illustrate certain general principles by explicit calculation. Many of these properties carry over to other quantum mechanical problems whose solution cannot be given in terms of Gaussian states. The simple harmonic oscillator also plays an important role in field theory, because a free quantum field can be formulated as an assembly of an infinite number of harmonic oscillators. With this picture an interacting field becomes an assembly of anharmonic oscillators, and can be regarded as an anharmonic perturbation of the harmonic oscillators.

The Hamiltonian for the simple harmonic oscillator has the same form

$$H_{osc} = \frac{1}{2m}p^2 + \tfrac{1}{2}kq^2$$

as the classical oscillator energy. The classical oscillator frequency is $\mu = (k/m)^{1/2}$, and it is convenient to introduce dimensionless variables

$$Q = (m\mu/\hbar)^{1/2}q, \qquad P = (m\mu\hbar)^{-1/2}p, \qquad H = (\hbar\mu)^{-1}H_{osc} \qquad (1.5.1)$$

so that

$$H = \tfrac{1}{2}(P^2 + Q^2) \qquad (1.5.2)$$

and

$$[P, Q] = \hbar^{-1}[p, q] = -i.$$

We realize P and Q in the Schrödinger representation, so that $P = -id/dy$ and Q is multiplication by y, as operators on $\mathcal{H} = L_2(dy)$. We take for the domain of H, and other operators in this chapter, the Schwartz space $\mathcal{S}$. By definition, $\theta \in \mathcal{S}$ if θ and each of its derivatives have rapid decrease at infinity. The operators we consider, e.g., H, P, Q, all map $\mathscr{S}$ into $\mathscr{S}$, so that $\mathscr{S}$ is an "invariant domain".

In this chapter we study two fundamental properties of H: the completeness of its eigenfunctions and the positivity-preserving property of e^{-tH}. We verify these properties by direct computation for H. We can then use more abstract methods to establish these same features for a wide variety of Hamiltonians. The positivity of e^{-tH} is an especially important property, related to the uniqueness of the ground state and to the Feynman–Kac integral representation of Chapter 3.

Theorem 1.5.1. *The operator H_{osc} is essentially self-adjoint and has spectrum $\hbar\mu(n + \frac{1}{2})$. The resolvent of H_{osc} is compact.*

PROOF. We define "creation" and "annihilation" operators A^* and A by

$$A^* = \frac{1}{\sqrt{2}}(Q - iP)$$

$$A = \frac{1}{\sqrt{2}}(Q + iP),$$ (1.5.3)

so

$$[A, A^*] = 1.$$ (1.5.4)

Then elementary algebra shows that

$$H = \tfrac{1}{2}(P^2 + Q^2) = A^*A + \tfrac{1}{2}.$$ (1.5.5)

Furthermore

$$[H, A] = -A, \qquad [H, A^*] = A^*.$$ (1.5.6)

From (1.5.5) we see that a vector Ω_0 satisfying $A\Omega_0 = 0$ is also an eigenvector of H, namely $H\Omega_0 = \tfrac{1}{2}\Omega_0$. The equation $A\Omega_0 = 0$ can be written in the Schrödinger representation as

$$\frac{d\Omega_0}{dy} = -y\Omega_0,$$ (1.5.7)

from which we infer that Ω_0 is the Gaussian function

$$\Omega_0(y) = \text{const } e^{-y^2/2} = \pi^{-1/4}e^{-y^2/2},$$ (1.5.8)

where we choose the constant so that $\|\Omega_0\| = 1$.

By (1.5.6), we see that since Ω_0 is an eigenvector of H, so is $A^{*n}\Omega_0$ and

$$HA^{*n}\Omega_0 = A^{*n}H\Omega_0 + [H, A^{*n}]\Omega_0$$

$$= (\tfrac{1}{2} + n)A^{*n}\Omega_0.$$ (1.5.9)

Hence the spectrum of H contains the points $(\tfrac{1}{2} + n)$, $n = 0, 1, 2, \dots$.

To complete the proof of the theorem, we need to know that the eigenfunctions $\{A^{*n}\Omega_0\}$ form a complete set in L_2. We give an elementary proof of this well-known fact in Proposition 1.5.7 below. We remark that the eigenfunctions are elements of $\mathscr{S}$, so H is essentially self-adjoint, on the domain $\mathscr{S}$.[1] In other words, there is a unique self-adjoint operator (which we also denote by H) which agrees with H on the original domain $\mathscr{S}$.

We first obtain some explicit properties of the eigenfunctions. Their normalization follows from

$$\langle \Omega_0, A^n A^{*n}\Omega_0 \rangle = \langle \Omega_0, A^{n-1}[A, A^{*n}]\Omega_0 \rangle = n\langle \Omega_0, A^{n-1}A^{*n-1}\Omega_0 \rangle$$

$$= \cdots = n!,$$

so

$$\Omega_n = n!^{-1/2}A^{*n}\Omega_0 \tag{1.5.10}$$

is a normalized eigenvector. The relations

$$A^*\Omega_n = \sqrt{n+1}\,\Omega_{n+1},$$

$$A\Omega_n = \sqrt{n}\,\Omega_{n-1}, \tag{1.5.11}$$

$$A^*A\Omega_n = n\Omega_n$$

can be interpreted by regarding the state Ω_n of the oscillator as an n particle or n quantum state, with each quantum of energy equal to $\hbar\mu$. The operator A^* adds one quantum, or particle, to a state, and increases its energy by $\hbar\mu$. The adjoint operator A absorbs, or annihilates, a quantum. The total energy is $\hbar\mu/2$ plus $\hbar\mu$ times the number n of quanta. For any state $\theta = \sum c_n\Omega_n$, $|c_n|^2$ is the probability of having n quanta, and

$$\langle \theta, A^*A\theta \rangle = \sum n|c_n|^2$$

is the expected number of quanta. Thus A^*A is the operator which measures the number of quanta.

In terms of the Schrödinger representation, the wave functions $\Omega_n(y)$ are normalized Hermite functions, which we see as follows. Let $P_n(x)$ be the nth Hermite polynomial,

$$P_n(x) = \sum_{j=0}^{[n/2]}(-1)^jc_{n,\,j}x^{n-2j}, \tag{1.5.12}$$

with $[n]$ the integer part of n and

$$c_{n,\,j} = \frac{n!}{(n-2j)!\,2^jj!}. \tag{1.5.13}$$

---

[1] Essential self-adjointness of a symmetric operator H defined on a dense domain $\mathscr{D}$ means $H^{**} = H$. This is equivalent to the statement that range $(H \pm i)$ is dense, or equivalently to the statement that H^* does not have $\pm i$ as an eigenvalue. In case H has a complete set of eigenvectors $\Omega_n \in \mathscr{D}$, then H is essentially self-adjoint on $\mathscr{D}$. In fact, let χ in the domain of H^* be a solution to the eigenvalue equation $H^*\chi = i\chi$. Then

$$0 = \langle (H^* - i)\chi, \Omega_n \rangle = \langle \chi, (H + i)\Omega_n \rangle = \langle \chi, \Omega_n \rangle(E_n + i),$$

Thus $\langle \chi, \Omega_n \rangle = 0$ for all n, and by completeness, $\chi = 0$.

Lemma 1.5.2. $(x - d/dx)P_n = P_{n+1}.$

PROOF. The left hand side has the expansion

$$\sum_{j=0}^{[n/2]} (-1)^j c_{n,j} [x^{n+1-2j} - (n-2j)x^{n+1-2(j+1)}]. \tag{1.5.14}$$

We note that

$$c_{n,j} = \left(1 - \frac{2j}{n+1}\right) c_{n+1,j} = \frac{2(j+1)}{(n+1)(n-2j)} c_{n+1,j+1}.$$

We substitute these identities in the expansion for $(x - d/dx)P_n$, and after re-labeling the indices, we obtain the expansion (1.5.12) for P_{n+1}.

Lemma 1.5.3. *The inversion of* (1.5.12) *is*

$$x^n = \sum_{j=0}^{[n/2]} c_{n,j} P_{n-2j}(x). \tag{1.5.15}$$

PROOF This lemma is true for $n = 0$. We use induction on n, and suppose that it has been proved for $n \le r$. By Lemma 1.5.2,

$$x^{r+1} = \sum_{j=0}^{[r/2]} c_{r,j} x P_{r-2j}(x)$$

$$= \sum_{j=0}^{[r/2]} c_{r,j} \left(P_{r+1-2j}(x) + \frac{d}{dx} P_{r-2j}(x) \right).$$

By the induction hypothesis, the derivative terms sum to rx^{r-1}, which can be reexpanded in Hermite polynomials, again using the induction hypothesis. This yields

$$x^{r+1} = \sum_{j=0}^{[r/2]} c_{r,j} P_{r+1-2j}(x) + \sum_{j=1}^{[(r+1)/2]} r c_{r-1,j-1} P_{r+1-2j}(x).$$

We note that for $j \le [r/2]$,

$$c_{r+1,j} = c_{r,j} + r c_{r-1,j-1},$$

and if r is odd and $j = (r+1)/2$, then

$$c_{r+1,j} = r c_{r-1,j-1}.$$

Thus,

$$x^{r+1} = \sum_{j=0}^{[(r+1)/2]} c_{r+1,j} P_{r+1-2j}(x).$$

Proposition 1.5.4.

$$(A^{*n}\Omega_0)(y) = P_n(\sqrt{2}\,y)\Omega_0(y),$$
$$\Omega_n(y) = n!^{-1/2} P_n(\sqrt{2}\,y)\Omega_0(y).$$

PROOF. The proposition is true for $n = 0$. We proceed by induction, and suppose it has been proved for n. Using (1.5.10),

$$\Omega_{n+1} = (n+1)^{-1/2} A^* \Omega_n = (n+1)!^{-1/2} A^* P_n(\sqrt{2}\, y)\Omega_0(y).$$

By (1.5.3), $A^* = 2^{-1/2}(y - d/dy)$, so

$$\Omega_{n+1} = (n+1)!^{-1/2}\left[2^{1/2}\, yP_n(\sqrt{2}\, y) - 2^{-1/2}\frac{dP_n(\sqrt{2}\, y)}{dy}\right]\Omega_0(y).$$

By Lemma 1.5.2, $\Omega_{n+1} = (n+1)!^{-1/2} P_{n+1}(\sqrt{2}\, y)\Omega_0$, completing the induction and the proof.

The operator $P_n(\sqrt{2}\, Q)$ is just $\sum (-1)^j c_{n,j}(\sqrt{2}\, Q)^{n-2j}$, and in the Schrödinger representation on $\mathscr{H} = L_2(dy)$, $P_n(\sqrt{2}\, Q)$ is the multiplication operator by the function $P_n(\sqrt{2}\, y)$. We define the "Wick ordered monomials" of Q by

$$:Q^n: = 2^{-n/2} P_n(\sqrt{2}\, Q) = Q^n + \text{degree}(n-2)$$

$$= \sum_{j=0}^{[n/2]} (-1)^j c_{n,j} 2^{-j} Q^{n-2j}, \qquad (1.5.16)$$

and extend this definition to polynomials by linearity. The Wick monomials $:Q^n:$ are just degree n polynomials, orthogonal when integrated with respect to the measure

$$d\phi = \Omega_0^2 \, dy = \pi^{-1/2} e^{-y^2} \, dy. \qquad (1.5.17)$$

In other words,

$$\int :Q^n: \, :Q^m: \, d\phi = \langle \Omega_0, \, :Q^n: :Q^m: \Omega_0\rangle$$

$$= 2^{-(n+m)/2}\langle A^{*n}\Omega_0, A^{*m}\Omega_0\rangle$$

$$= 2^{-n} n! \, \delta_{nm}.$$

Proposition 1.5.5. *On the domain given by finite linear combinations of the eigenvectors Ω_n,*

$$:Q^n: = 2^{-n/2} \sum_{j=0}^{n} \binom{n}{j} A^{*j} A^{n-j}. \qquad (1.5.18)$$

Remark. This formula is just the binomial expansion for $Q^n = 2^{-n/2}(A^* + A)^n$, rearranged so that each creator A^* occurs to the left of each annihilator A (i.e., in Wick order). There is an analysis of these formulas in terms of graphs, according to which the coefficients $c_{n,j}$ can be identified by inspection as the number of ways of selecting j unordered pairs from n objects.

PROOF. Let L denote the right hand side of this equation. On the domain $\mathscr{S}$,

$$[Q, L] = 2^{-(n+1)/2} \sum_{j=0}^{n} \binom{n}{j} \{j A^{*j-1} A^{n-j} - (n-j) A^{*j} A^{n-j-1}\}$$

$$= 0.$$

Thus $[P_r(\sqrt{2}Q), L] = 0$. Using Proposition 1.5.4 and (1.5.16),

$$0 = \{:Q^n: -L\}\Omega_0 = r!^{-1/2} P_r(\sqrt{2}Q)\{:Q^n: -L\}\Omega_0$$

$$= \{:Q^n: -L\}\Omega_r,$$

as claimed.

Proposition 1.5.6. The vectors

$$\chi_v = e^{ivQ}\Omega_0, \qquad v \text{ real},$$

span L_2.

PROOF. Suppose $\theta \in L_2$ is orthogonal to the χ_v. Then

$$0 = \langle \theta, \chi_v \rangle = (\theta e^{-Q^2/2})\tilde{\ }(v).$$

Since Fourier transform is unitary on L_2, $\theta e^{-Q^2/2} = 0$. We conclude that $\theta = 0$ and the χ_v span L_2.

Proposition 1.5.7. *The normalized Hermite functions $\Omega_n(y)$ are a complete orthonormal set in L_2.*

PROOF. We show that the vectors χ_v lie in the span of the Ω_n. The proposition then follows from Proposition 1.5.6. The inversion of (1.5.16) is given by (1.5.15):

$$Q^n = \sum_{j=0}^{[n/2]} c_{n,j} 2^{-j} :Q^{n-2j}:, \tag{1.5.19}$$

so $Q^n \Omega_0$ lies in the span of the Ω_n, with

$$Q^n \Omega_0 = \sum_{j=0}^{[n/2]} c_{n,j} 2^{-n/2} (n-2j)!^{1/2} \Omega_{n-2j}. \tag{1.5.20}$$

Thus

$$\|Q^n \Omega_0\|^2 = \sum_{j=0}^{[n/2]} c_{n,j}^2 \, 2^{-n} (n-2j)!$$

$$= \sum_{j=0}^{[n/2]} \frac{(n!)^2}{(n-2j)!(j!)^2 2^{2j+n}} \leq O(1)n!,$$

and the series

$$\chi_v = \sum_{n=0}^{\infty} (ivQ)^n (n!)^{-1} \Omega_0$$

converges in L_2, as desired.

Proposition 1.5.8. *For any complex v, the defining series converge and*

$$e^{vQ}\Omega_r = e^{v^2/4}e^{vA^*/\sqrt{2}}e^{vA/\sqrt{2}}\Omega_r. \tag{1.5.21}$$

PROOF. Convergence of the series with $r = 0$ follows by (1.5.20). By (1.5.16), (1.5.19),

$$\sum_{n=0}^{\infty}(vQ)^n n!^{-1}\Omega_0 = \sum_{n=0}^{\infty}\sum_{j=0}^{[n/2]}v^n n!^{-1}c_{n,j}2^{-n/2}A^{*n-2j}\Omega_0$$

$$= \sum_{n=0}^{\infty}\sum_{j=0}^{[n/2]}\frac{v^{2j}}{2^{2j}j!}\frac{v^{n-2j}}{(n-2j)!\,2^{(n-2j)/2}}A^{*n-2j}\Omega_0$$

$$= e^{v^2/4}e^{vA^*/\sqrt{2}}\Omega_0 = e^{v^2/4}e^{vA^*/\sqrt{2}}e^{vA/\sqrt{2}}\Omega_0.$$

However,

$$f(v) \equiv e^{vA^*}Ae^{-vA^*} = f(0) + vf'(0) = A - v$$

since $f^{(n)}(v) = 0$, $n \geq 2$. Hence $e^{vA^*}A = (A - v)e^{vA^*}$ and $[Q, e^{vA^*}e^{vA}]\Omega_0 = 0$. Thus multiplication by $(r!)^{-1/2}P_r(\sqrt{2}Q)$ yields (1.5.21), and a converging series for $r \neq 0$.

We note that (1.5.21) has the form

$$e^{R+S} = e^R e^S e^{-\frac{1}{2}[R,S]}, \qquad [R, S] = \alpha I,$$

valid when R, S are bounded and $[R, S]$ is a multiple of the identity. To avoid domain questions, we have given a direct proof of (1.5.21). Abstractly, the identity (1.5.21) follows from $f(\lambda) = I$ being the unique solution to the first order equation $f'(\lambda) = 0, f(0) = I$, where

$$f(\lambda) = e^{\lambda(R+S)}e^{-\lambda S}e^{-\lambda R}e^{\lambda^2[R,S]/2}.$$

We have already remarked that once we have verified the completeness of a given set of eigenfunctions, we can then use an abstract argument to prove completeness for a wide class of Hamiltonians. Only discrete eigenstates occur for any Hamiltonian H_1 larger than the harmonic oscillator Hamiltonian H_{osc}. This is the case for the anharmonic oscillator Hamiltonian

$$H_1 = H_{\text{osc}} + \lambda q^4 - \mu q + \text{const},$$

as we now show.

Theorem 1.5.9. *Let H and H_1 be self-adjoint operators, and let*

$$0 \leq H \leq \text{const } H_1. \tag{1.5.22}$$

Then H_1 has a compact resolvent and a complete set of eigenfunctions if H does.

PROOF. From (1.5.22), we infer that $B = H^{1/2}H_1^{-1/2}$ is bounded. Then $H_1^{-1/2} = H^{-1/2}B$ is the product of a bounded and a compact operator, and hence is compact. Completeness of the eigenfunctions follows.

This theorem illustrates the power of an *a priori* estimate, (1.5.22), in comparing two objects of interest. In this case we might compare the complicated unknown H_1 with the simple, concrete H_{osc}.

We now pass to the second main feature of H, the fact that in the Schrödinger representation e^{-tH} has a positive integral kernel, $p_t(y, y')$, defined by

$$(e^{-tH}\theta)(y) = \int p_t(y, y')\theta(y') \, dy'. \tag{1.5.23}$$

In order to obtain conservation of probability ((1.5.25) below), it is necessary to renormalize H and replace (1.5.2) by $H = \frac{1}{2}(P^2 + Q^2) - \frac{1}{2} = A^*A$.

Theorem 1.5.10. *The kernel for e^{-tH} satisfies*

$$p_t(y, y') = p_t(y', y) > 0. \tag{1.5.24}$$

$$\int p_t(y, y')e^{(y^2 - y'^2)/2} \, dy' = 1. \tag{1.5.25}$$

In particular, the kernel is given by Mehler's formula

$$p_t(y, y') = \pi^{-1/2}(1 - e^{-2t})^{-1/2} \exp\left(-\frac{y^2 - y'^2}{2} - \frac{(e^{-t}y - y')^2}{1 - e^{-2t}}\right). \tag{1.5.26}$$

PROOF. The symmetry follows from (1.5.23) and the symmetry of H. (It also follows from (1.5.26).) (1.5.25) states that $e^{-tH}\Omega_0 = \Omega_0$, and follows from $H\Omega_0 = 0$, using our new normalization for H. The positivity of p_t follows from Mehler's formula, which we now verify. Using (1.5.21), we compute

$$e^{-tH}\chi_v = e^{-tH}e^{ivQ}\Omega_0 = e^{-tH}e^{-v^2/4}e^{ivA^*/\sqrt{2}}\Omega_0$$

$$= e^{-v^2/4}e^{ive^{-t}A^*/\sqrt{2}}\Omega_0 = \exp[-v^2(1 - e^{-2t})/4] \exp[ive^{-t}Q]\Omega_0.$$

Thus by the definition (1.5.23) of the kernel,

$$\int p_t(y, y')e^{-y'^2/2}e^{ivy'} \, dy' = \exp\left(-v^2 \frac{1 - e^{-2t}}{4} + ive^{-t}y - \frac{y^2}{2}\right).$$

The proof is completed by taking the inverse Fourier transform in the variable v. After multiplication by $(2\pi)^{-1}e^{-ivy}$, the preceding line can be written as

$$\frac{1}{2\pi} \exp\left[\left(\frac{e^{-t}y - y'}{\sqrt{1 - e^{-2t}}} + iv\frac{\sqrt{1 - e^{-2t}}}{2}\right)^2 - \frac{(e^{-t}y - y')^2}{1 - e^{-2t}}\right] \exp\left(-\frac{y^2}{2}\right).$$

After a complex translation in the path of integration, we are reduced to the integral $\int e^{-av^2/2} \, dv = (2\pi)^{1/2}a^{-1/2}$, yielding (1.5.26).

A common transformation is to use the representation $\mathscr{H} = L_2(d\phi(y))$ in which $d\phi(y) = \Omega_0^2 \, dy$, as in (1.5.17) and in which the ground state is represented by the function 1. In this representation, other states are represented by $\Omega_0(y)^{-1} = \pi^{1/4}e^{y^2/2}$ times their Schrödinger representation. The kernel of e^{-tH} is defined in this representation by

$$(e^{-tH}\theta)(y) = \int \mathcal{K}_t(y, y')\theta(y') \, d\phi(y')$$

where

$$\mathcal{K}_t(y, y') = \Omega_0(y)^{-1}p_t(y, y')\Omega_0(y')^{-1}$$

$$= (1 - e^{-2t})^{-1/2} \exp\left(y'^2 - \frac{(e^{-t}y - y')^2}{1 - e^{-2t}}\right).$$

1.6 Coulomb Potentials

The Coulomb potential is the basic potential of atomic physics. It describes the electromagnetic forces which account for the binding of electrons and nuclei to form atoms and molecules.

In the case of two particles (the hydrogen atom) the Schrödinger equation can be solved in terms of elementary functions. In the many particle case, only estimates and approximate calculations can be made.

Consider a Hamiltonian H_N for N particles of mass m_i, charge e_i, momentum p_i, and position q_i respectively, $i = 1, 2, \ldots, N$. Then

$$H_N = \sum_{i=1}^{N} \frac{1}{2m_i} p_i^2 + \sum_{1 \le i < j \le N} \frac{e_i e_j}{|q_i - q_j|}. \qquad (1.6.1)$$

The potential

$$V = \sum_{i < j} \frac{e_i e_j}{|q_i - q_j|}$$

is a sum of Coulomb pair potentials.

Theorem 1.6.1. *The Hamiltonian H_N of* (1.6.1) *is essentially self-adjoint on the N-fold tensor product $L^2(R^3)$, and in the case of equal particle masses, on the antisymmetric subspace.*

Remark. This basic result of Kato easily reduces to estimates on the two particle system. To avoid complication, we only consider this case $N = 2$. We rewrite H_2 in terms of center of mass and relative coordinates,

$$H_2 = \frac{1}{2M} P^2 + \frac{1}{2\mu} p^2 + \frac{e_1 e_2}{|q|}, \qquad (1.6.2)$$

where the canonical center of mass coordinates are

$$Q = \frac{m_1 q_1 + m_2 q_2}{m_1 + m_2}, \qquad P = p_1 + p_2 \qquad (1.6.3)$$

and the (canonical) relative coordinates are

$$q = q_1 - q_2, \qquad p = \mu\dot{q} = \frac{\mu}{m_1} p_1 - \frac{\mu}{m_2} p_2, \qquad (1.6.4)$$

and $\mu = m_1 m_2/(m_1 + m_2)$ is the reduced mass. The self-adjointness of H_2 then follows by writing $1/|q| \in L_\infty + L_2$, i.e.

$$\frac{1}{|q|} = \frac{1}{|q|}(1 - \chi(q)) + \frac{1}{|q|}\chi(q),$$

where χ is the characteristic function of the set $|q| \le 1$. We use

Theorem 1.6.2. *Let $V(q)$, $q \in R^3$, be the sum of L_∞ and L_2 functions. Then $-\Delta + V$ is essentially self-adjoint on $\mathscr{S}(R^3)$.*

Lemma 1.6.3. *Let A be an essentially self-adjoint operator with domain $D(A)$, and let B be a symmetric operator with domain $D(B) \supset D(A)$. Suppose*

$$\|B\theta\| \le a\|A\theta\| + b\|\theta\| \qquad (1.6.5)$$

for certain constants $a < 1$ and b. Then $A + B$ is essentially self-adjoint on the domain $D(A)$.

PROOF. For y sufficiently large, we show that the range of $A + B \pm iy$ is dense in the Hilbert space $\mathscr{H}$. A Neumann series establishes the existence of $(I + B(A \pm iy)^{-1})^{-1}$, as a bounded operator on $\mathscr{H}$. Since

$$(A + B \pm iy)(A \pm iy)^{-1} = I + B(A \pm iy)^{-1},$$

the statement on $\text{Range}(A + B \pm iy)$ follows on multiplication by $(I + B(A \pm iy)^{-1})^{-1}$ on the right. To prove convergence of the Neumann series

$$(I + B(A \pm iy)^{-1})^{-1} = \sum_{n=0}^{\infty} [-B(A \pm iy)^{-1}]^n,$$

we must establish the bound

$$\|B(A \pm iy)^{-1}\| < 1. \qquad (1.6.6)$$

In fact

$$\|B(A \pm iy)^{-1}\theta\| \le a\|A(A \pm iy)^{-1}\theta\| + b\|(A \pm iy)^{-1}\theta\|$$

$$\le (a + b/y)\|\theta\|.$$

The desired bound holds for $a + b/y < 1$.

Remark. If A is bounded from below, we may take y large negative above, thereby showing that $A + B$ is semibounded also.

Lemma 1.6.4. *Let $f \in L_2(R^3)$ and let $\varepsilon > 0$. Then there exists $b < \infty$ such that*

$$\|f\theta\|_2 \le \varepsilon\|\Delta\theta\|_2 + b\|\theta\|_2.$$

PROOF.   The bound is proved by Fourier transformation. Let $\|\cdot\|_p$ denote an L_p norm. For $0 < \delta < \frac{1}{2}$,

$$\|f\theta\|_2 \le \|f\|_2 \|\theta\|_\infty \le \|f\|_2 \|\theta^\sim\|_1$$
$$\le \|f\|_2 \|(1 + p^2)^{-3/4 - \delta}\|_2 \|(1 + p^2)^{3/4 + \delta}\theta^\sim\|_2$$
$$\le \varepsilon\|\Delta\theta\|_2 + b\|\theta\|_2.$$

In the last inequality, use $x^\alpha \le \varepsilon x + b(\varepsilon)$ for $\alpha < 1$, $0 \le x$.

PROOF OF THEOREM 1.6.2.   Let $V = f_1 + f_2$, with $f_1 \in L_2$, $f_2 \in L_\infty$. By Lemma 1.6.4,

$$\|V\theta\|_2 \le \|f_1\theta\|_2 + \|f_2\theta\|_2 \le \varepsilon\|\Delta\theta\|_2 + (b + \|f_2\|_\infty)\|\theta\|_2.$$

Thus by Lemma 1.6.3, $-\Delta + V$ is essentially self-adjoint.

We remark that a real function $V \in L_{3/2+\varepsilon}$, $\varepsilon > 0$, does not define an operator on the domain $\mathscr{S}$. It does, however, define a bilinear form, and a related Neumann series argument shows that the bilinear form $-\Delta + V$ uniquely determines a self-adjoint operator. A more important application of the perturbation theory of bilinear forms is to the Dirac equation with a Coulomb potential. The Dirac equation describes a relativistic electron and, in calculations, provides corrections to the eigenvalues determined by (1.6.1). The bilinear form perturbation theory of the Dirac equation breaks down for large N. A second, and more serious defect of the Dirac equation is that the energy spectrum is not bounded from below.

The fact that H_N is bounded from below follows from the remark after Lemma 1.6.3. The N dependence of this bound is a deep result of [Dyson and Lenard, 1967–8]; it is related to the $N \to \infty$ limit (the stability of matter in the thermodynamic limit), established in [Lieb and Lebowitz, 1972]. See also [Lieb and Thirring, 1975] and [Lieb and Simon, 1977a,b].

Theorem 1.6.5.   *Let* $|e_i|$, $m_i \le M$. *Then there is a constant* B *such that*

$$0 \le H_N + NB.$$

The spectrum of H_N is most easily understood on a subspace of definite total momentum P. In the language of physics, we fix the motion of the center of mass. On a definite momentum subspace, H_N has discrete eigenvalues and continuous spectrum. The eigenvalues and eigenvectors refer to N-particle bound states. For suitable choice of e_i and m_i, these states describe the $N - 1$ electron atom in its ground state or one of its excited states. The states in the continuous spectrum are called scattering states. In these states, one or more electrons are free (not bound), while the remaining particles constitute a singly or multiply ionized atom. The continuous parameter in the spectrum comes from the relative momentum of the free electron(s) and of the ionized atom.

$N = 2$ is the familiar case of the hydrogen atom. The eigenfunctions can be found explicitly, and in the Schrödinger representation have the form

$$\psi = L_n(r/a)P_n^j(\theta, \phi)e^{-r/a}, \tag{1.6.7}$$

where L_n is a Laguerre polynomial, P_n^j is a spherical harmonic and a is a constant. The generalized eigenfunctions in the continuous spectrum can also be found explicitly, and also involve Laguerre functions. There is an extensive mathematical literature on the two body scattering problem. This theory concerns potentials $V + 1/|q|$. (V generally has more rapid decrease at infinity than $1/|q|$.) These problems cannot be solved explicitly. Among the potentials which arise naturally in physics problems, we mention the Yukawa potential, $V(q) \sim e^{-m|q|}/|q|$, which occurs in a nonrelativistic approximation to nuclear forces, and the Lennard–Jones potential, $\alpha r^{-12} - \beta r^{-6}$, $\alpha, \beta > 0$, which is sometimes used to approximate molecular forces. The molecular forces are electromagnetic in origin, and can in principle be traced back to Coulomb interactions. The nuclear forces govern the interactions of mesons, neutrons, and protons. The nonrelativistic approximation, $V \sim e^{-m|q|}/|q|$, is derived from relativistic quantum field theory, with, for example, a $\lambda\phi^4 + g\psi^\dagger\psi\phi$ interaction.

$N = 3$ is the helium atom. The existence of a large number of bound states has been established by [Kato, 1951b]. The theory of [Faddeev, 1963], which allows sufficiently regular potentials, but not the Coulomb potential, gives a construction of the three-body scattering states and proves asymptotic completeness. Asymptotic completeness is the statement that the bound states together with the scattering (ionized) states are complete in $\mathscr{H} = L_2(R^{3N})$, or (in the case of identical particles) in some symmetric or antisymmetric subspace of $L_2(R^{3N})$. This theory was extended to arbitrary N in [Hepp, 1969a], [Sigal, 1978], and [Hagedorn, 1980]. The problem of asymptotic completeness has been considered by a number of authors, for various classes of potentials, but for the Coulomb potential the mathematical theory is still incomplete. Nonetheless, the spectrum of H_N can be described on the basis of formal analysis, perturbation calculations, and experimental evidence. The bound states of an $N - 1$ electron atom can be described approximately, leading to an explanation of the periodic table of chemical elements, as well as the absorption and emission of spectral lines. As an approximation, one omits the repulsive electron–electron forces. Then H_N, acting on $L_2(R^{3N})$, is a direct sum of hydrogen atoms, and hence exactly solvable. The effect of antisymmetrization is the Pauli exclusion principle: an eigenstate of H_N is a tensor product of $N - 1$ hydrogen atom bound states, one for each of the $N - 1$ electrons, and each electron must be in a different state. If N is not too large, the effects of the electron–electron forces can then be treated approximately in perturbation calculations.

The $N \geq 3$ scattering states are described in terms of clusters. A cluster is just a subset of the N particles, and a cluster decomposition is a partition of the N particles into clusters. A scattering state corresponding to a given cluster decomposition is a state with the following large time behavior: the particles within a given cluster remain together, and form a bound state of these particles. The distinct clusters separate and move independently. For the case of one atom, the only possible clusters correspond to singly or multiply ionized atoms, together with the one or more free electrons, since there are no bound states formed by two or more electrons, because of the repulsive electron-electron force. For the case of several atoms (or a molecule), there will be in general more allowed clusters. For example, water allows the decomposition $H_2O \rightleftharpoons H_2 + O$, but $H + H + O$ are also allowed clusters, as well as the ionized states, in which the individual constituents lose or gain one or more electrons, e.g. $H^+ + OH^-$.

1.7 The Hydrogen Atom

The simplest atom is hydrogen, composed of an electron and a proton. The Schrödinger Hamiltonian H which describes the quantum hydrogen atom is

$$H = -\frac{\hbar^2}{2m_e}\Delta_{x_1} - \frac{\hbar^2}{2m_p}\Delta_{x_2} - \frac{e^2}{|x_1 - x_2|}. \tag{1.7.1}$$

Here m_e, m_p are the masses of the electron and proton with coordinates x_1 and x_2 respectively and electric charge $-e$ and e respectively. The charge often occurs in the dimensionless combination called the fine structure constant,

$$\alpha = \frac{e^2}{\hbar c} = (137.035963 \pm 0.000015)^{-1}. \tag{1.7.2}$$

The rest mass m_e of the electron has the rest energy

$$\mu_e = m_e c^2 = 0.5110034 \pm 0.0000014 \text{ MeV}, \tag{1.7.3}$$

where MeV denotes 10^6 electron volts; similarly the reduced mass $m_r = m_e m_p/(m_e + m_p)$ has the rest energy

$$\mu_r = (1 + m_e/m_p)^{-1}\mu_e$$
$$= 0.999449819\mu_e. \tag{1.7.4}$$

After separating the center of mass motion from (1.7.1) (see also Section 13.2 for a general discussion), we are left with a Hamiltonian describing the relative motion of the electron and proton. It has the form of a single body of mass m_r moving in the potential $-e^2/|x|$,

$$H = -\frac{\hbar^2}{2m_r}\Delta - \frac{e^2}{|x|}. \tag{1.7.5}$$

H has eigenvalues

$$E_n = \frac{\mu\alpha^2}{2n^2}, \qquad n \in Z_+, \tag{1.7.6}$$

and E_n has multiplicity n^2. Numerically, for $n = 1$, the ground state energy is

$$E_1 = -13.5983 \text{ electron volts.} \tag{1.7.7}$$

This is the negative of the ionization energy of hydrogen, i.e., 13.6 electron volts energy is sufficient to unbind the electron from the proton.

As explained in Section 1.4, energy differences $E_n - E_m = hv$ determine a frequency v of light emitted. Equivalently, spectroscopic measurements are sometimes quoted in wave numbers i.e., inverse wavelength,

$$\lambda^{-1} = \frac{v}{c} = \frac{E_n - E_m}{hc}. \tag{1.7.8}$$

Since $hc = 12.39852 \times 10^{-5}$ eV cm, it follows that in these units the calculated value of E_1 is

$$-E_1/hc = 109677 \text{ cm}^{-1}. \tag{1.7.9}$$

In fact, because of the accuracy of spectroscopic measurements, E_1 can be experimentally determined to a greater accuracy, namely, to an error of one part in 10^8.

The observed spectra from hydrogen were classified by various methods. The transitions from $n > 1$ to $m = 1$ levels are known as the Lyman series, from $n > 2$ to $m = 2$ the Balmer series, etc.; cf. Figure 1.2.

The multiplicity n^2 of the eigenvalue E_n can be understood by analyzing the unitary irreducible representations of the group SO(3) of rotations in 3-space, which acting on $L_2(R^3)$ leave H invariant. These representations $\mathscr{D}^{(j)}$ are $2j + 1$ dimensional, with $j = 0, 1, 2, \ldots$, a nonnegative integer (the angular momentum quantum number). For a given n, the eigenspace $\mathscr{H}_n$ of E_n can be decomposed according to its irreducible components under SO(3), and we find the isomorphism

$$\mathscr{H}_n \cong \bigoplus_{j=0}^{n-1} \mathscr{D}^{(j)}. \tag{1.7.10}$$

Thus

$$\text{dimension } \mathscr{H}_n = \sum_{j=0}^{n-1} (2j + 1) = n^2.$$

The interpretation of j in terms of angular momentum is a consequence of Section 1.3, Postulate P3. In fact, the classical angular momentum $\mathbf{J}$ of a single particle is the vector product of its position $\mathbf{x}$ with momentum $\mathbf{p}$,

$$\mathbf{J} = \mathbf{x} \times \mathbf{p}.$$

Thus using the Schrödinger representation $p = -i\hbar\nabla_x$, we find that the components of $\mathbf{J}$ satisfy the commutation relations of the SO(3) Lie algebra,

$$[J_i, J_j] = i\hbar J_k$$

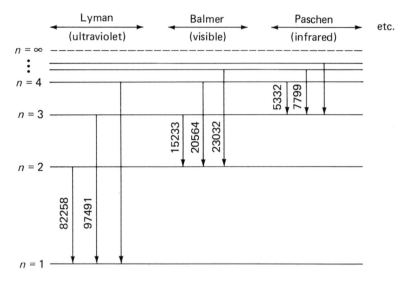

Figure 1.2. Observed transitions in atomic hydrogen, with some wave numbers in cm^{-1}. Frequencies are obtained by multiplication by $c = 2.99792458 \times 10^{10}$ cm sec^{-1}.

where (i, j, k) is a cyclic permutation of $(1, 2, 3)$. Exponentiation yields the representation of SO(3) on $\mathcal{H} = L_2(R^3)$ given by rotation on R^3. Thus

$$V(\mathbf{n}, \theta) = \exp(i\mathbf{J} \cdot \mathbf{n}\theta/\hbar)$$

is the group element representing a rotation by angle θ about the axis in the direction $\mathbf{n}$, with $\mathbf{n}$ a unit 3-vector. The Casimir operator

$$J^2 \equiv J_1^2 + J_2^2 + J_3^2$$

commutes with the representation, and has eigenvalues $\hbar^2 j(j+1), j = 0, 1, 2, \ldots$, each with multiplicity one. The eigenspace corresponding to a particular j is $2j + 1$ dimensional, and V restricted to this subspace is the representation $\mathcal{D}^{(j)}$ of angular momentum j. Since SO(3) is compact, every unitary representation can be reduced to a sum of finite dimensional representations $\mathcal{D}^{(j)}$. The space $\mathcal{D}^{(j)}$ has a natural basis consisting of spherical harmonics P_n^j, which is the origin of the P_n^j in (1.6.7).

In the case of several particles, the total angular momentum

$$\mathbf{J}_{\text{tot}} = \sum_{i=1}^{N} \mathbf{x}_i \times \mathbf{p}_i$$

has the properties of $\mathbf{J}$ above, and yields a representation of the rotation group on $L_2(R^{3N})$. As above, the irreducible components are $\mathcal{D}^{(j)}$. For a central force problem such as (1.7.1), H and $\mathbf{J}_{\text{tot}}$ commute. Thus an eigenspace of H can be reduced to a sum of $\mathcal{D}^{(j)}$'s.

1.8 The Need for Quantum Fields

The ability to perform accurate atomic spectroscopy measurements and accurate perturbative calculations means that quantum theory has been very accurately tested. In these tests, physical effects which deviate from the theory of Sections 1.6 and 1.7 can be detected. The effects are consequences of spin and relativity. Moreover, for nuclear physics, new interparticle force laws arise. Spin can be incorporated easily into the nonrelativistic quantum mechanics discussed here. (See also Section 1.3 and (1.3.10).)

A first effort to make a relativistic theory of the electron leads to the Dirac equation (see Section 15.3). Spin and the Dirac electron greatly improve the agreement between theory and experiment in atomic spectroscopy. A more fundamental combination of special relativity and quantum mechanics is given by quantum field theory. Field theory gives further corrections to atomic spectroscopy, including the celebrated Lamb shift and the anomalous magnetic moment of the electron. Within the accuracy of the measurements and the calculations, the inclusion of field theory effects allows complete agreement between theory and experiment in atomic spectroscopy. Chapter 15 is devoted to a discussion of these issues.

In the case of nuclear and elementary particle physics, the short distances and high energies which occur mean relativistic effects are large. Because particles can be created or destroyed in such interactions, field theory effects are important. The proton radius and the interparticle distance in the nucleus are comparable numbers, so a theory of point objects is not correct for protons and neutrons in nuclear matter. Thus nuclear and elementary particle physics are intrinsically field theoretic, and in contrast to atomic physics, approximation by N body quantum mechanics does not provide a coherent theory of the dominant effects. In summary, quantum field theory is needed for

1. intellectual coherence as the combination of quantum mechanics and special relativity,
2. small but accurately determined corrections to atomic spectroscopy,
3. the leading effects for elementary particle physics and the language in which the fundamental force laws of nuclear physics are formulated.

REFERENCES

[Baym, 1969], [Kato, 1966], [Jauch and Rohrlich, 1976].

CHAPTER 2

Classical Statistical Mechanics

2.1 Introduction

Statistical mechanics is the bridge between molecular science and continuum mechanics. The input to statistical mechanics is a force law between particles. The particles can be atoms in a crystal, molecules in a gas or liquid, electrons in a plasma, amino acid units in a protein, elementary constituents in a complex polymer, etc. The forces between particles originate from Coulomb forces between electric charges and from magnetic dipole forces between magnetic moments. The classical force laws may be modified by the quantum mechanics (especially the Pauli exclusion principle) describing the particles. Normally the force laws are quite complicated and are not given by a simple analytical expression. Rather, they are given in one or more of the following four ways: (1) In terms of an unknown function, for which qualitative properties are postulated as laws of physics. (2) As a specific function, such as the Lennard–Jones potential or hard sphere potential. Such functions are chosen because they have representative features in common with the true force laws; for example, they may be asymptotically exact in some limiting region. (3) As a result of numerical calculation, based e.g. on the complicated exact force law and a Hartree–Fock approximation. (4) As a result of experimental measurement.

The task of statistical mechanics is to take such a force law and deduce from it bulk properties of matter. The bulk properties are known as thermodynamic functions and transport coefficients, including density, pressure, temperature, thermal and electrical conductivity, magnetization, tensile strength, fluid viscosity, specific heat, chemical reaction rates, etc. These properties are not in general independent, but are subject to relation(s) defined by an equation of state. An example is the ideal gas law

$$p = \frac{NkT}{V}.$$ (2.1.1)

The equation of state is the part of the interparticle force law which survives in the continuum mechanics description of matter. It is the starting point for thermodynamics and continuum mechanics, because it gives the linear or nonlinear "response functions" which are manipulated by the laws of thermodynamics and are required by the equations of continuum mechanics. Statistical mechanics is the exact science from which these laws and equations are derived.

Statistical mechanics can also be defined by its relation to mathematics. It is a subfield of probability theory, or more generally of analysis. Statistical mechanics concerns infinite systems, and the relevant analysis is analysis over infinite dimensional spaces. The physical state of each of the component subsystems is described mathematically as a point in a finite dimensional space X_i, for example

$$X_i = R^1, R^n, S^0 = Z_2 = \{-1, +1\}, Z_n, S^n, \text{ or } SU(n).$$ (2.1.2)

Then statistical mechanics is concerned with probability measures defined on the product space of individual components:

$$X = \overset{\infty}{\underset{i=1}{\mathsf{X}}} X_i.$$ (2.1.3)

Usually we are given a measure $d\mu_i$ on X_i, and then the simplest measure on X is the infinite product measure

$$d\mu = \overset{\infty}{\underset{i=1}{\mathsf{X}}} d\mu_i.$$ (2.1.4)

This measure is not very interesting, because it corresponds to a situation in which there is no interaction between the particles or degrees of freedom of the problem. However, $d\mu$ is not completely misleading, because the measures of statistical mechanics are typically close to tensor product measures.

The probability distribution of the ith particle will in general depend on the state of the jth particle, $j \neq i$. In order to have statistical behavior, it is necessary to have this dependence primarily limited to a finite number of particles—the neighbors of the ith particle. In physics terminology this amounts to short range, stable forces. To accommodate this situation, we consider measures $d\mu^{(n)}$ defined on

$$X^{(n)} = \overset{n}{\underset{i=1}{\mathsf{X}}} X_i.$$

The measures $d\mu^{(n)}$ are not tensor product measures, but generally have the form

$$d\mu^{(n)} = e^{-U} \bigotimes_{i=1}^{n} d\mu_i, \qquad (2.1.5)$$

where U is the interaction energy. For example, the energy of the form

$$U = \sum_{i<j} V(x_i - x_j) \qquad (2.1.6)$$

results from a two particle interaction V, and V is determined as in (1)–(4) above. [See also (2.3.1).] We assume that

$$d\mu = \lim_{n \to \infty} d\mu^{(n)} \qquad (2.1.7)$$

exists. The limit (2.1.7) is a proper generalization of (2.1.4) and includes measures of interest to statistical mechanics.

There is an essential difference between analysis over finite and over infinite dimensional spaces. The difference is that in the infinite dimensional case, the measure $d\mu$ and hence the physical solution may depend discontinuously on the parameters of the problem. The discontinuity, when it occurs, is not an irritating pathology, but is a central and dominant feature. In fact one expects on physical grounds that the solution will be a piecewise real analytic function of the parameters (temperature, external field, etc.). The discontinuities form a set of codimension one in the parameter space, and correspond to the physical phenomenon of phase transitions. The boundary points of codimension two at the edge of discontinuities are called critical surfaces. In the common but special case in which two parameters describe the solution, the critical surfaces of codimension two become points; hence the term critical points. The behavior of physical systems in the neighborhood of critical points plays a fundamental role in experimental investigation, and one goal of the theory is to give quantitative predictions in this domain.

2.2 The Classical Ensembles

To make the above discussion more specific with regard to $d\mu$, let us consider the case of N identical classical particles of mass m in a domain $\Lambda \subset R^3$, interacting via a pair potential $V = V(q_i - q_j)$. Assume reflecting boundary conditions on the walls $\partial \Lambda$ of Λ. Furthermore suppose each particle is subject to an external potential $\Phi(q_i)$. Then the N particle Hamiltonian is

$$H_N(q, p) = H(q, p)$$

$$= \sum_{i=1}^{N} \left\{ \frac{1}{2m} p_i^2 + \Phi(q_i) \right\} + \frac{1}{2} \sum_{i \neq j} V(q_i - q_j). \qquad (2.2.1)$$

The classical one particle state space is $X_i = \Lambda \otimes R^3$, namely the phase space of a single particle. On the state space $X^{(N)} = X_{i=1}^N X_i$, the Liouville measure defined as

$$d\mu_{\text{Liouville}} = d\mu_N = \prod_{i=1}^N dp_i \, dq_i \qquad (2.2.2)$$

plays a distinguished role: it is invariant under the n-particle classical dynamics defined by H, as was established in Propositions 1.2.1–2.

The fundamental postulate of Gibbs concerning the equilibrium distribution of the states of an isolated mechanical system is: On a constant energy surface $H = E$, the *a priori* probability distribution is uniform with respect to Liouville measure, restricted to $H = E$. The consideration of an isolated system is called the Gibbs microcanonical ensemble. The assumption is that the *a priori* probability of a classical state with energy E is proportional to

$$\delta(H - E) \, d\mu_N. \qquad (2.2.3)$$

We can normalize (2.2.3) by dividing by the invariant surface area $V(E) = \int \delta(H - E) \, d\mu_N$ of the constant energy surface.[1]

How can this hypothesis be justified? In case there are other integrals of motion, such as the total angular momentum, corresponding δ functions can be included in (2.2.3), leading to a new (micro-microcanonical) ensemble. Having taken into account the obvious integrals of motion, we make the mathematical postulate that there are no further (measure theoretic) integrals, or in other words that the micro-microcanonical ensemble is ergodic. Consider now a classical trajectory $(q(t), p(t))$, which necessarily is a curve in a surface of constant energy E, angular momentum j, etc. For $F(q, p)$ an observable function on X, we measure $F(q(t), p(t))$ at equilibrium by a time average, defined by the formula

$$\lim_{T \to \infty} \frac{1}{T} \int_0^T F(q(t), p(t)) \, dt = \langle F \rangle_{\text{av}}. \qquad (2.2.4)$$

Assuming ergodicity (i.e., that no proper subset, with respect to $d\mu_N$ measure, is left invariant under time translation), the time average equals the expectation in $d\mu_N$,

$$\langle F \rangle_{\text{av}} = V(E, j, \ldots)^{-1} \int F(q, p) \, \delta(H - E) \, \delta(J - j) \cdots d\mu_N, \qquad (2.2.5)$$

[1] Note that, as in the proof of Liouville's theorem (Proposition 1.2.1), and using $\int \delta(H - E) \, dp = \|\text{grad } H\|^{-1}$ on the $H = E$ surface, we see that (2.2.3) is the invariant area under motions on $H = E$. However, the example $H = p^2 + \alpha q^2$ shows that the invariant arc length

$$\int_p \delta(H - E) \, dp \, dq = \frac{1}{2\sqrt{E - \alpha q^2}} \, dq$$

does not integrate to the length of the ellipse $H = E$. (For $\alpha = 1$, $V(E) = \pi$, while $H = E$ is a circle of length $2\pi E^{1/2}$.) In case $\|\text{grad } H\|$ is not constant on the $H = E$ surface, these measures differ by more than normalization. The invariance of $\delta(H - E) \, d\mu_N$ remains fundamental for the choice (2.2.3).

for almost all initial conditions, where V is the volume of the micro-micro-canonical ensemble. For simplicity, we consider below the case in which H is the only integral of motion. In spite of serious efforts and successes in special cases, the ergodicity problem has proved elusive. It may be that other aspects of the important question of approach to equilibrium are more promising scientifically.

In terms of the microcanonical distribution, we define the measure

$$d\mu_{\text{microcanonical}} = \frac{1}{N!} \frac{1}{V(E)} \delta(H - E) \, d\mu_N. \tag{2.2.6}$$

A fundamental derivation of classical statistical mechanics must start from quantum statistical mechanics. If this is done, the effect of the symmetrization or anti-symmetrization in the boson or fermion statistics is the origin of the $1/N!$.

The entropy, or statistical weight for different energies and particle numbers, will be defined as

$$S(E) = \ln\left(\frac{1}{N!} \int \delta(H - E) \, d\mu_N\right). \tag{2.2.7}$$

This entropy is an *extensive* quantity, in the sense that $S(E)$ diverges as the number of particles N and the volume $|\Lambda|$ tend to ∞. (The thermodynamic limit for the grand canonical ensemble defined below is the limit $|\Lambda| \to \infty$ while the density $\rho = N/|\Lambda|$ remains fixed.) Entropy, like energy, may be changed by an additive constant without affecting the physics, so the definition (2.2.7) involves an arbitrary choice. The corresponding *intensive* quantity is

$$s(E) = \frac{S(E)}{|\Lambda|} = \text{entropy per unit volume},$$

and $s(E)$ normally has a well-behaved limit as $N \to \infty$, with $\rho \equiv N/|\Lambda|$ fixed.

While the microcanonical distribution is appropriate for an isolated system, many physical systems are not isolated. One particular kind of lack of isolation is the case of a system with a fixed number of particles in equilibrium with a constant temperature heat bath. Hence energy is exchanged between the system and the heat bath until both achieve temperature T. Microscopically the temperature is a measure of the average kinetic energy per particle.

The Gibbs postulate for such a distribution function is called the canonical ensemble:

$$d\mu_{\text{canonical}} = \frac{1}{N!} e^{-\beta H} \, d\mu_N$$

$$= \frac{1}{N!} e^{-(1/2) \sum_{i \neq j} |V(q_i - q_j)|}$$

$$\times \prod_{i=1}^{N} \left(e^{-[(p_i^2/(2m_i) + \Phi(q_i)]} \, dp_i \, dq_i\right), \tag{2.2.8}$$

where $\beta = (kT)^{-1}$ and k is Boltzmann's constant. Note that (2.2.8) has the general form (2.1.5) of e^{-V} times a tensor product measure, with $V = \frac{1}{2} \sum_{i \neq j} V(q_i - q_j)$. For the canonical ensemble, the Helmholtz free energy is defined as

$$A_N = A = -\beta^{-1} \ln \int d\mu_{\text{canonical}} . \tag{2.2.9}$$

The free energy A is an extensive function, and the corresponding intensive quantity is

$$a = \frac{A}{|\Lambda|} = \text{Helmholtz free energy per unit volume.} \tag{2.2.10}$$

The canonical partition function is

$$Z = Z_N = \int d\mu_{\text{canonical}} = e^{-\beta A_N}. \tag{2.2.11}$$

The canonical ensemble describes an experimental situation in which temperature is known exactly, being determined by the heat bath. With certain idealized assumptions, one can derive the canonical ensemble from the microcanonical one. Consider system A in a heat bath B and assume the combined system $A \cup B$ is isolated, so that it obeys the microcanonical distribution. The canonical distribution for system A could be defined as the conditional distribution obtained by averaging over the B degrees of freedom,

$$\frac{\int_B d\mu_{A \cup B, \text{microcanonical}}}{\int_{A \cup B} d\mu_{A \cup B, \text{microcanonical}}}, \tag{2.2.12}$$

and taking some idealized limit including

$$|\text{volume } B| \to \infty, \qquad |\text{energy per } B \text{ particle}| \to e_B. \tag{2.2.13}$$

We outline the arguments here, using additional simplifying assumptions:

$$B \text{ is a perfect gas: } H_B = \frac{1}{2m} \sum_{i=1}^{|B|} p_i^2. \tag{2.2.14i}$$

$$A \text{ and } B \text{ are weakly coupled: } H_{A \cup B} = H_A + H_B. \tag{2.2.14ii}$$

$$\text{The average energy per particle in the heat bath is } \tfrac{3}{2}kT. \tag{2.2.14iii}$$

Proposition 2.2.1. *Under the assumptions* (2.2.14i–iii),

$$\lim_{|B| \to \infty} \frac{\int_B d\mu_{A \cup B, \text{microcanonical}}}{\int_{A \cup B} d\mu_{A \cup B, \text{microcanonical}}} = Z_A^{-1} e^{-\beta H_A} \, d\mu_A$$

$$= Z_A^{-1} \, d\mu_{A, \text{canonical}}, \tag{2.2.15}$$

where $\beta = (kT)^{-1}$ is defined by the condition (2.2.14iii).

PROOF. Under these assumptions, (2.2.12) equals

$$\frac{\int_B \delta(H_A + H_B - E) \, d\mu_B \, d\mu_A}{\int_{A \cup B} \delta(H_A + H_B - E) \, d\mu_B \, d\mu_A}$$

$$= \frac{V_B(E - H_A)}{V_{A \cup B}(E)} \, d\mu_A = \left(\frac{V_B(E)}{V_{A \cup B}(E)}\right)\left(\frac{V_B(E - H_A)}{V_B(E)}\right) d\mu_A, \quad (2.2.16)$$

and by definition has total integral 1. Using (2.2.14), and defining $n = |B|$,

$$V_B(E) = \int \delta(H_B - E) \, d\mu_B = (2m)^{3n/2} |\Lambda|^n \int \delta\left(\sum_{i=1}^{n} p_i^2 - E\right) dp_1 \cdots dp_n$$

$$= (2m)^{3n/2} |\Lambda|^n |S^{3n-1}| \int_0^\infty \delta(k^2 - E)k^{3n-1} \, dk = cE^{(3n-2)/2},$$

where $c = 2^{-1}(2m)^{3n/2} |\Lambda|^n |S^{3n-1}|$ and $|S^n|$ is the volume of the n-sphere. Assumption (2.2.14iii) implies that as $n \to \infty$, the average energy of a B particle converges to constant e_B. For $H_A = E_A$ fixed, also $E/|B| \to e_B$. In this limit,

$$\frac{V_B(E - H_A)}{V_B(E)} = \left(1 - \frac{H_A}{E}\right)^{(3n-2)/2} \to \exp\left[-\frac{3}{2}\frac{H_A}{e_B}\right]. \quad (2.2.17)$$

The classical value of e_B is at this point a parameter. The standard definition of temperature arises from choosing the average energy per particle e_B in a heat bath to equal $3kT/2 = \frac{3}{2}\beta^{-1}$. Then the limit of (2.2.17) is $e^{-\beta H_A}$. The factor $V_B(E)/V_{A \cup B}(E)$ in (2.2.16) normalizes (2.2.16). Since Z_A^{-1} normalizes $d\mu_{A, \text{canonical}}$, we have therefore established the proposition.

Although for finite systems the microcanonical and canonical distributions differ, it is believed that in the thermodynamic limit (infinite volume $|\Lambda|$, fixed density $\rho = N/|\Lambda|$) the two distributions yield the same thermodynamic functions.

Another connection between the microcanonical and canonical points of view is the relation between free energy and entropy:

$$a(\beta) = \inf_E \{\beta E - s(E)\} \beta^{-1}. \quad (2.2.18)$$

The transformation (2.2.18) is known as a Legendre transformation, with inverse

$$s(E) = \sup_\beta \{\beta E - \beta a(\beta)\}. \quad (2.2.19)$$

We do not derive these relations, but note that if $s(E)$ and $a(\beta)$ are convex, then the transformations (2.2.18–19) are inverses, i.e., one formula holds if the second does. While we do not use the equivalence of these ensembles, we remark on its fundamental character.

A third experimental situation is one in which both energy and matter may be exchanged between the system under consideration and the surrounding medium. In other words, we relax the condition that N is fixed (for fixed Λ). For this situation Gibbs generalized the canonical ensemble to the grand canonical ensemble. This ensemble arises, for example, in a chemical reaction or in considering diffusion though a permeable membrane. In these cases we have a particle reservoir as well as a heat reservoir. Define

$$d\mu_{\text{grand canonical}} = \sum_{N=0}^{\infty} z^N \, d\mu_{\text{canonical, }N} \qquad (2.2.20)$$

Here $d\mu_{\text{canonical, }N}$ is the canonical ensemble

$$d\mu_{\text{canonical, }N} = \frac{1}{N!} e^{-\beta H_N} \, d\mu_N, \qquad (2.2.21)$$

and z is the activity or fugacity. We relate z to another parameter h by $z = e^{\beta h}$, with h called the chemical potential. In other words the number of particles, N, is replaced as a parameter by the potential per particle, $-h$, which can be achieved by adding a constant to Φ in the example (2.2.1). Furthermore, in the example, if we restrict attention to averages of functions $F(q)$ which are independent of p, the p integrals can be performed, yielding an effective $z = (2\pi m/\beta)^{3/2} e^{\beta h}$. In the grand canonical ensemble,

$$\Xi = \sum_{N=0}^{\infty} z^N \int d\mu_{\text{canonical, }N} = \sum_{N=0}^{\infty} z^N Z_N$$

$$= \sum_{N=0}^{\infty} \frac{z^N}{N!} \int e^{-\beta H_N} \, d\mu_N$$

$$= \sum_{N=0}^{\infty} \frac{1}{N!} e^{-\beta(A_N - hN)} \qquad (2.2.22)$$

is the grand canonical partition function. The expected number of particles is

$$\langle N \rangle = \Xi^{-1} \sum_{N=0}^{\infty} \frac{z^N}{(N-1)!} \int d\mu_{\text{canonical, }N} \cdot \qquad (2.2.23)$$

In the grand canonical ensemble, the density

$$\rho = \lim_{|\Lambda| \to R^3} \frac{\langle N \rangle}{|\Lambda|} \qquad (2.2.24)$$

and the pressure

$$p = \beta^{-1} \lim_{|\Lambda| \to R^3} \left(|\Lambda|^{-1} \ln \Xi \right) \qquad (2.2.25)$$

are functions of β and h. This fundamental relation between ρ, p, β, and h is known as the equation of state. For the ideal gas (but not in general) $p\beta = \rho$, and a single equation $\rho = \rho(\beta, h)$ completes the equation of state.

The grand canonical ensemble differs from the canonical ensemble for finite $|\Lambda|$, but both are expected to give the same thermodynamic functions in the infinite volume $|\Lambda| \to R^3$ limit. In particular the canonical and grand canonical thermodynamic functions are related by a Legendre transformation in the variable ρ or its conjugate variable h:

$$p(\beta, h) = \sup_{\rho} \{\rho h - a(\beta, \rho)\},$$

$$a(\beta, \rho) = \inf_{h} \{\rho h - p(\beta, h)\}. \tag{2.2.26}$$

2.3 The Ising Model and Lattice Fields

In principle, solids and liquids, as well as gases, are described by the statistical ensembles of Section 2.2 and an interparticle potential $V(q_i - q_j)$. The occurrence of a local lattice structure dominated by defects (for liquids) or a global crystal lattice with isolated defects (crystalline solid) should follow from these data at suitable temperatures. However, the existence of a lattice which is derived mathematically from first principles and which assumes only the existence of a potential V has yet to be established; moreover, if it were, it would still be an inconvenient starting point for the study of solids. Rather we introduce a lattice into the formulation of the problem; the points of the lattice can be thought of as equilibrium locations of the atoms or atomic groups in a crystal. For application to quantum fields, the lattice Z^d is just an approximation to the continuum R^d. In either case, for $i \in Z^d$, the variable $\xi_i \in X_i$ defines the state at the location i. For example, ξ_i may denote a field strength, the orientation of a magnetic moment, or a discrete variable such as the presence or absence of an impurity at the lattice site $i \in Z^d$.

To be more precise, we choose in (2.1.2), (2.1.5),

$$X_i = R^1, \qquad d\mu_i = \frac{e^{-P(\xi_i)} \, d\xi_i}{\int e^{-P} \, d\xi}, \tag{2.3.1}$$

with $\xi_i \in R^1$, $d\xi$ Lebesgue measure, and P a semibounded polynomial. We also allow the limiting case

$$d\mu_i = e^{h\xi_i} \frac{\delta_{-1}(\xi_i) + \delta_{+1}(\xi_i)}{2 \cosh h} \, d\xi_i, \tag{2.3.2}$$

where $\delta_{\pm 1}(\xi) = \delta(\xi \mp 1)$. We are interested in cooperative phenomena (or occasionally in anti-cooperative phenomena) in which the occurrence of some event at the site i—i.e., some value of ξ_i—encourages similar values of ξ_j, for j near i. The most common way to do this is the following. Let Δ be the second difference operator on Z^d. After partial summation, we have

$$\langle \xi, \Delta\xi \rangle \equiv - \sum_{i \in Z^d} \sum_{v=1}^{d} (\xi_{i+e_v} - \xi_i)^2, \tag{2.3.3}$$

where e_v is the unit vector in the vth coordinate direction and the bracket $\langle \, , \, \rangle$ is the inner product in $l_2(Z^d)$. Also let $\Delta_{\partial\Lambda}$ be the second difference operator on Z^d with Dirichlet boundary conditions on the boundary $\partial \Lambda$ of a region $\Lambda \subset R^d$. (See Section 9.5 for the definition and properties of $\Delta_{\partial \Lambda}$.) Let

$$d\mu_\Lambda = e^{\frac{1}{2}\beta \langle \xi, \Delta_{\partial \Lambda} \xi \rangle} \prod_{i \in \Lambda} d\mu_i \qquad (2.3.4)$$

and

$$d\mu = \lim_{\Lambda \uparrow R^d} \frac{d\mu_\Lambda}{\int d\mu_\Lambda}. \qquad (2.3.5)$$

The limit measure $d\mu$ depends on β ($\sim 1/T = $ (temperature)$^{-1}$) and on the coefficients of P. For (2.3.1) we get a lattice field, while for (2.3.2) we get the Ising model with external field h. The limit (2.3.5) always exists, and in general is a discontinuous function of β and P. The Ising model is used to describe metallic impurities (alloys). It is also used as a qualitative description of magnetism, although the classical Heisenberg ferromagnet ($X_i = S^2$ in (2.1.2)) is more realistic. A great variety of lattices, state spaces X_i, and measures $d\mu_i$ arise from examples in solid state physics.

It should be noted that the lattice models (2.3.5) describe perfect crystals, in that no allowance is made for defects in the crystal lattice structure. Perfect crystals are rare in nature. They have very different physical properties from imperfect crystals. In fact the crystal defects in ordinary real materials govern their mechanical properties (strength, brittleness, fatigue, etc.); see [Ashcroft and Mermin, 1976]. There are speculations in quantum field theory that a related phenomenon may occur, namely that field configurations far from minimum action or equilibrium configuration (bags, vortices, instantons, etc.) may dominate the measure $d\mu$ and thus the solution of the quantum theory. In the case of quantum field theory, all configurations (near or far from equilibrium) are contained in the ensemble (2.3.5). The effect of configurations close to equilibrium can be analyzed by perturbation theory; the effect of other configurations is not contained in perturbation theory and requires a more fundamental analysis.

2.4 Series Expansion Methods

The cluster expansion of statistical mechanics provides a convergent series expansion to study the measure $d\mu$ in (2.1.5) or (2.3.5). The leading term in this expansion is the decoupled case, or product measure, and successive expansion terms introduce the interaction. The expansion converges in a region where the interaction is small. These methods are very useful in the analysis of lattice fields, Ising models, and (Euclidean) quantum fields (cf. Part III). Here we illustrate the methods with the grand canonical ensemble gas dynamics, (2.2.20). The leading term is the ideal gas.

We want to show that the pressure p, defined by (2.2.25), is a smooth function of z and β for small z and small β. Small β means high temperature, and small activity z corresponds to low densities. Thus we conclude that a gas at low density and/or high temperature does not undergo a phase transition (e.g., does not liquify). The proof reveals all the properties of the gas in this parameter region. We show that

$$\beta p = \sum_{n=1}^{\infty} b_n z^n, \tag{2.4.1}$$

where the b_n are given explicitly in terms of integrals over $R^{3(n-1)}$. The series (2.4.1) is called the Mayer series. We show that it is convergent, and it follows that p is analytic for small β and z.

We introduce these two hypotheses:

$$\|V\|_{L_1} = \int |V(\xi)| \, d\xi < \infty \qquad \text{(short range potential)}, \tag{2.4.2a}$$

and there is a positive constant B such that

$$\inf_{\xi_i \in R^3} \sum_{1 \le i < j \le n} V(\xi_i - \xi_j) \ge -nB \qquad \text{(stability)} \tag{2.4.2b}$$

Proposition 2.4.1 Ξ_Λ is an entire function of z.

This follows by substituting (2.4.2b) in (2.2.20). It reflects the fact that finite systems do not have phase transitions. The proof is completely useless for the limit $\Lambda \uparrow R^3$. To control this limit, we must use the approximate independence of the distinct particles, and we must take advantage of the ln and the division by $|\Lambda|$ in (2.2.25). Let

$$Z_\Lambda^{(n)} = \frac{1}{n!} \int_{\Lambda^n} \exp\left(-\beta \sum_{1 \le i < j \le n} V(\xi_i - \xi_j)\right) \prod_{i=1}^{n} d\xi_i \tag{2.4.3}$$

so that

$$\Xi_\Lambda = \sum z^n Z_\xi^{(n)}. \tag{2.4.4}$$

We will construct quantities $K_\Lambda^{(i)}$ which satisfy the equation

$$Z_\Lambda^{(n)} = \sum_{1 \le i \le n} \frac{i}{n} K_\Lambda^{(i)} Z_\Lambda^{(n-i)}. \tag{2.4.5}$$

Proposition 2.4.2.   Given (2.4.5), with $|K_\Lambda^{(n)}| \le e^{O(n)}$,

$$\Xi_\Lambda = \exp\left(\sum_n z^n K_\Lambda^{(n)}\right). \tag{2.4.6}$$

Remark.   From (2.4.6), we see that

$$b_n = \lim_{\Lambda \uparrow R^3} |\Lambda|^{-1} K_\Lambda^{(n)} \tag{2.4.7a}$$

is the desired coefficient in (2.4.1). We will find that

$$b_1 = 1, \qquad b_2 = -\frac{\beta}{2} \int_0^1 ds \int d\xi \, V(\xi) e^{-\beta s V(\xi)}. \tag{2.4.7b}$$

PROOF. We multiply (2.4.5) by nz^n and sum over n to obtain

$$z \frac{d}{dz} \Xi_\Lambda = \left(\sum_{i=1}^{\infty} iz^i K_\Lambda^{(i)} \right) \Xi_\Lambda .$$

Since the same equation is satisfied with Ξ_Λ replaced by $\exp(\sum_n z^n K_\Lambda^{(n)})$, we obtain (2.4.6).

The equation (2.4.5) expresses the partial decoupling of the first particle from the other particles. This equation and in particular the $K_\Lambda^{(i)}$ are of central importance. To derive this equation and to obtain an explicit expression for $K_\Lambda^{(i)}$, we introduce parameters s_j, $0 \leq s_j \leq 1$, which interpolate linearly between zero interaction ($s_j = 0$) and full interaction ($s_j = 1$). The parameter s_j refers to the interactions $V(\xi_i - \xi_k)$ between all the ith and kth particles, $i \leq j < k$. Thus the parameter s_j modifies the interaction as follows:

$$V(\xi_i - \xi_k) \rightarrow s_j V(\xi_i - \xi_k) = (1 - s_j) \cdot 0 + s_j V(\xi_i - \xi_k) \quad (2.4.8a)$$

for $i \leq j < k$, and

$$V(\xi_i - \xi_k) \rightarrow V(\xi_i - \xi_k) = (1 - s_j)V(\xi_i - \xi_k) + s_j V(\xi_i - \xi_k) \quad (2.4.8b)$$

for i, $k \leq j$ and for $j < i$, k. Note that (2.4.8) maps V into a convex sum of potentials V_{ik}. Each summand, and thus the convex sum, satisfies the stability condition (2.4.2b):

$$\inf_{\xi} \sum_{1 \leq i < j \leq n} V_{ij}(\xi_i - \xi_j) \geq -nB.$$

Performing this modification for each s_j, $1 \leq j \leq n - 1$, leads to the n particle potential energy

$$W^{(n)}(\sigma_{n-1}) = \sum_{1 \leq i < j \leq n} s_i \cdots s_{j-1} V(\xi_i - \xi_j),$$

where

$$\sigma_{n-1} = \{s_1, \ldots, s_{n-1}\}.$$

The next result is evident.

Proposition 2.4.3. $W^{(n)}(\sigma_{n-1}) \geq -nB.$

With these definitions, we proceed to derive the equation (2.4.5). For a function $Z^{(n)}(\sigma_{n-1})$, we have the fundamental theorem of calculus:

$$Z^{(n)}(\sigma_{n-1} = 1) = Z^{(n)}(s_1 = 0, s_2 = \cdots = s_{n-1} = 1)$$

$$+ \int_0^1 ds_1 \frac{d}{ds_1} Z^{(n)}(s_1, s_2 = \cdots = s_{n-1} = 1). \quad (2.4.9)$$

In particular let

$$Z^{(n)}(\sigma_{n-1}) = \frac{1}{n!} \int_{\Lambda^n} \exp(-\beta W^{(n)}(\sigma_{n-1})) \prod_{i=1}^{n} d\xi_i. \quad (2.4.10)$$

For $s_1 = 0$, $Z^{(n)}$ factors, and for $s_2 = \cdots = s_{n-1} = 1$, one of the factors is $Z^{(n-1)}$. Thus with

$$K_\Lambda^{(1)} = |\Lambda|$$

the first term in (2.4.9) yields the $i = 1$ term in (2.4.5). The second term can be written as

$$\frac{1}{n} \int_0^1 ds_1 \frac{-\beta}{(n-2)!} \int_{\Lambda^n} V(\xi_1 - \xi_2)$$

$$\times \exp\left(-\beta \sum_{2 \le i < j \le n} V(\xi_i - \xi_j) - \beta \sum_{2 \le j \le n} s_1 V(\xi_1 - \xi_j)\right) \prod_{i=1}^n d\xi_i$$

Here we use the symmetry among the $n-1$ final particles to replace

$$\sum_{2 \le j \le n} V(\xi_1 - \xi_j) \quad \text{by} \quad (n-1)V(\xi_1 - \xi_2).$$

As above, we apply the fundamental theorem of calculus in the variable s_2. The term with $s_2 = 0$ is

$$\frac{1}{n} \int_0^1 ds_1 \left(-\beta \int_{\Lambda^2} V(\xi_1 - \xi_2)e^{-\beta s_1 V(\xi_1 - \xi_2)} \, d\xi_1 \, d\xi_2\right) Z^{(n-2)}, \quad (2.4.11)$$

and so with

$$K_\Lambda^{(2)} = \frac{-\beta}{2} \int_0^1 ds_1 \int_{\Lambda^2} V(\xi_1 - \xi_2)e^{-\beta s_1 V(\xi_1 - \xi_2)} \, d\xi_1 \, d\xi_2, \quad (2.4.12)$$

we see that (2.4.12) yields the $i = 2$ term in (2.4.5). Continuing in this fashion, the $s_i = 0$ term gives $(i/n)K_\Lambda^{(i)} Z_\Lambda^{(n-i)}$ in (2.4.5) and a definition of $K_\Lambda^{(i)}$.

To get an explicit definition of $K_\Lambda^{(i)}$, we introduce the notion of a tree graph, defined by an integer valued function $\eta(l)$, satisfying $1 \le \eta(l) < l$. The graph is then formed by bonds joining the pairs $(\eta(l), l)$, $l = 1, \ldots, i-1$, as illustrated in Figure 2.1. We also define

$$f(\eta, \sigma_{i-1}) = \prod_{l=1}^{i-1} s_{l-1} s_{l-2} \cdots s_{\eta(l+1)}$$

with the (empty) product $s_{l-1} s_{l-2} \cdots = 1$ if $\eta(l+1) = l$.

Proposition 2.4.4. *With*

$$K_\Lambda^{(i)} = \sum_\eta K_\Lambda^{(i)}(\eta), \quad (2.4.13a)$$

$$K_\Lambda^{(i)}(\eta) = \frac{(-\beta)^{i-1}}{i} \int d\sigma_{i-1} \int_{\Lambda^i} d\xi$$

$$\times f(\eta, \sigma_{i-1}) \prod_{l=1}^{i-1} V(\xi_{l+1} - \xi_{\eta(l+1)})e^{-\beta W^{(i)}(\sigma_{i-1})}, \quad (2.4.13b)$$

(2.4.5) *is valid.*

Figure 2.1

The proof follows the patterns used in defining $K_\Lambda^{(1)}$ and $K_\Lambda^{(2)}$, and is omitted.

Proposition 2.4.5. *Assume* (2.4.2). *Then*

$$|K_\Lambda^{(i)}| \le \beta^{i-1} \|V\|_{L_1}^{i-1} e^{i\beta B} \frac{e^{i-1}}{i} |\Lambda|.$$

Remark. Our main result, convergence of the limit $\Lambda \uparrow R^3$ in p and of the Mayer series (2.4.1), follows for

$$|z| \le (e^{\beta B + 1} \beta \|V\|_{L_1})^{-1}.$$

PROOF. We show that

$$\sum_\eta \int d\sigma_{i-1} f(\eta, \sigma_{i-1}) \le e^{i-1}. \tag{2.4.14}$$

This bound and stability (Proposition 2.4.3) complete the proof. The prototype for this inequality is

$$\int_0^1 ds \, v e^{sv} \le e^v. \tag{2.4.15}$$

The left hand side of (2.4.14) is bounded by

$$\sum_\eta \int_0^1 ds_1 \cdots \int_0^1 ds_{i-1} f(\eta, \sigma_{i-1}) \exp\left(\sum_{l=1}^{i-1} s_{i-1} \cdots s_l\right).$$

Using (2.4.15) successively in the iterated integral completes the proof.

REFERENCES

[Friedman, 1962], [Huang, 1963], [Uhlenbeck and Ford, 1963], [Ruelle, 1969], [Lanford, 1973], [Thompson, 1980].

CHAPTER 3

The Feynman–Kac Formula

3.1 Wiener Measure

We have presented the equations of quantum mechanics in Chapter 1. In only a handful of cases, however, do these equations have solutions equal to well-known special functions, or can the spectra be written in closed form. Thus calculations in quantum mechanics are made by some approximate method, such as computing the first few terms in a formal power series. For example, series in coupling constants are known as perturbation theory; the series in Planck's constant is known as the classical approximation.

In order to have a qualitative picture of the solutions, as well as to establish error estimates, it is extremely useful to have an integral representation for the solution. The Feynman–Kac formula provides this. This formula for general potentials gives the kernel $\mathscr{K}_t(q, q')$ of e^{-tH}, i.e.,

$$(e^{-tH}\theta)(q) = \int \mathscr{K}_t(q, q')\theta(q')\, dq',$$

even though $\mathscr{K}_t$ cannot be expressed in terms of elementary functions. Feynman's idea was to give such a representation for the kernel of the unitary group e^{-itH}. Consider the action

$$\mathscr{A}\left(-\frac{t}{2}, \frac{t}{2}\right) = \int_{-t/2}^{t/2} \mathscr{L}(q(s), \dot{q}(s))\, ds \qquad (3.1.1)$$

for a classical path $q(s)$, depending on the time parameter s. Here $\mathscr{L}$ is the Lagrangian, obtained from the Hamiltonian

$$H(p, q) = \tfrac{1}{2}p^2 + V(q) \tag{3.1.2}$$

by the Legendre transformation

$$\mathscr{L}(q, \dot{q}) = \sup_p \left[\dot{q}p - H(p, q)\right] = \tfrac{1}{2}\dot{q}^2 - V(q). \tag{3.1.3}$$

Let $\mathscr{W}(q, q', t)$ be the set of continuous paths $q(s)$ which take the values $q(-t/2) = q$ and $q(t/2) = q'$ at their endpoints. Feynman's formula is

$$(\text{kernel } e^{-itH})(q, q') = \text{const} \int_{\mathscr{W}(q, q', t)} e^{i\mathscr{A}(-t/2, t/2)} \prod_{-t/2 < s < t/2} dq(s). \tag{3.1.4}$$

This formula has been extensively used by physicists, because its transformation properties make it amenable to formal manipulation. However, the complex measure $e^{i\mathscr{A}} \prod dq(s)$ has not been given a satisfactory mathematical meaning, and for this reason Feynman's formula has not played a large role in the mathematically rigorous treatment of quantum mechanics.

The Feynman–Kac formula is a similar path space integral representation for the kernel of e^{-tH}. In this case the path space measure is positive and has a rigorous mathematical basis, as we now explain. Since e^{-tH} is obtained from e^{-itH} by an analytic continuation which replaces t by $-it$, we motivate our discussion by making the same analytic continuation in (3.1.3). Using the substitutions

$$ds \to -i\, ds, \qquad \dot{q}^2 = \left(\frac{dq}{ds}\right)^2 \to -\dot{q}^2,$$

we are led to the formal expression

$$(\text{kernel } e^{-tH})(q, q') = \int_{\mathscr{W}(q, q', t)} \exp\left(-\int_{-t/2}^{t/2} [\tfrac{1}{2}\dot{q}(s)^2 + V(q(s))]\, ds\right)$$

$$\times \prod_{-t/2 < s < t/2} dq(s). \tag{3.1.5}$$

We first consider the case $V = 0$,

$$H = H_0 = \tfrac{1}{2}p^2. \tag{3.1.6}$$

This case leads to the definition of conditional Wiener measure on the set $\mathscr{W}(q, q', t)$ of continuous paths $q(s)$ running from q to q' in the time interval $[-t/2, t/2]$. For simplicity, all formulas are written in three dimensions.

The kernel $\mathscr{K}_t^0(q, q')$ of the operator e^{-tH_0} is the fundamental solution for the heat equation

$$\frac{\partial}{\partial t} u(q, t) = H_0 u = \tfrac{1}{2}\, \Delta u(q, t). \tag{3.1.7}$$

Here $\Delta = \sum_{i=1}^{3} \partial^2/\partial q_i^2$. Thus if $u(q, 0) = f(q)$, the solution to (3.1.7) is

$$u(q, t) = (e^{-tH_0}f)(q) = \int \mathscr{K}_t^0(q, q')f(q')\, dq'. \tag{3.1.8}$$

The kernel is the well-known Gaussian distribution

$$(2\pi t)^{-3/2} e^{-(q-q')^2/2t} = \mathscr{K}_t^0(q, q') \qquad (3.1.9)$$

obtained by Fourier transformation of $e^{-p^2 t/2}$. It has the basic properties

   i. $0 < \mathscr{K}_t^0(q, q')$,

   ii. $\int \mathscr{K}_t^0(q, q')\, dq' = 1$,

   iii. $\mathscr{K}_{t+s}^0(q, q') = \int \mathscr{K}_t^0(q, r)\mathscr{K}_s^0(r, q')\, dr$.

Properties i, ii allow interpretation of $\mathscr{K}_t^0(q, q')$ as a probability density, while iii is the semigroup property $e^{-(t+s)H_0} = e^{-tH_0}e^{-sH_0}$. These properties i–iii also allow the definition of conditional Wiener measure on $\mathscr{W}(q, q', t)$. (The "condition" is the fact that the two endpoints of the path are fixed, while the usual Wiener measure fixes only the starting point of the path.) We define the total mass of $\mathscr{W}(q, q', t)$ to be $\mathscr{K}_t^0(q, q')$. Furthermore the measure of the set

$$\{q(s) \in \mathscr{W}(q, q', t) : q(t_1) \in I_1\},$$

namely those paths restricted to lie in a subset $I_1 \subset R^3$ at a time t_1, is defined to be

$$\int_{I_1} \mathscr{K}_{(t/2)+t_1}^0(q, q_1)\mathscr{K}_{(t/2)-t_1}^0(q_1, q')\, dq_1. \qquad (3.1.10)$$

This set of paths is illustrated in Fig. 3.1. By property iii this definition is consistent, because if we choose $I_1 = R^3$, then (3.1.10) equals the total mass $\mathscr{K}_t^0(q, q')$ of $\mathscr{W}(q, q', t)$.

More generally, consider the subset of $\mathscr{W}(q, q', t)$ for which the path $q(s)$ is specified at n intermediate times t_j to lie in Borel subsets I_j of R^3. Here $j = 1, 2, \ldots, n$ and we take $-t/2 < t_1 < t_2 < \cdots < t_n < t/2$. These subsets of

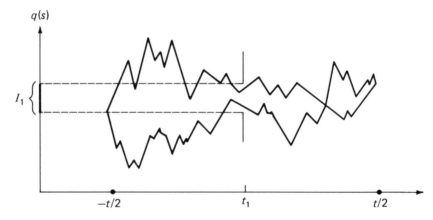

Figure 3.1. Illustration of two paths $q(s)$ in $\mathscr{W}(q, q', t)$ with $q(t_1) \in I_1$.

$\mathscr{W}(q, q', t)$, namely

$$\{q(s) : q(-t/2) = q, q(t/2) = q', q(t_j) \in I_j, j = 1, 2, \ldots, n\}, \quad (3.1.11)$$

are called cylinder sets, and the measure of (3.1.11) is defined to be

$$\int_{I_1} dq_1 \int_{I_2} dq_2 \cdots \int_{I_n} dq_n \ \mathscr{K}^0_{t_1 + t/2}(q, q_1) \mathscr{K}^0_{t_2 - t_1}(q_1, q_2) \cdots \mathscr{K}^0_{t/2 - t_n}(q_n, q')$$

$$= \int_{I_1} dq_1 \cdots \int_{I_n} dq_n \prod_{j=1}^{n+1} [2\pi(t_j - t_{j-1})]^{-3/2} \, e^{-(q_i - q_{i-1})^2/2(t_j - t_{j-1})}, \quad (3.1.12)$$

where

$$t_0 \equiv -t/2, \quad q_0 \equiv q, \quad t_{n+1} \equiv t/2, \quad q_{n+1} \equiv q'.$$

We state without proof the basic existence theorem for Wiener measure:

Theorem 3.1.1. *The conditional Wiener measure (3.1.12) is countably additive on the cylinder subsets of $\mathscr{W}(q, q', t)$ and it has a unique extension to the Borel subsets of $\mathscr{W}(q, q', t)$.*

Let $dW^t_{q, q'}$ denote integration with respect to conditional Wiener measure. A limiting case of (3.1.12), with I_i given the weight $A_i(q)$, follows from the definition (3.1.8) of $\mathscr{K}$:

Corollary 3.1.2. *Let A_i be the operator of multiplication on $L_2(q, dq)$ by a bounded function $A_i(q)$, $i = 1, 2, \ldots, n$, and let $-t/2 \le t_1 \le \cdots \le t_n \le t/2$. Then*

$$\int \prod_{i=1}^n A_i(q(t_i)) \, dW^t_{q, q'}$$

$$= (\text{kernel}(e^{-(t_1 + t/2)H_0} A_1 e^{-(t_2 - t_1)H_0} A_2 \cdots A_n e^{-(t/2 - t_n)H_0}))(q, q').$$
$$(3.1.13)$$

In the special case for which each A_i is the identity, (3.1.13) reduces to

$$\int dW^t_{q, q'} = (\text{kernel } e^{-tH_0})(q, q') = \mathscr{K}^0_t(q, q'). \quad (3.1.14)$$

We end this section with the remark that, on a formal level, these definitions are consistent with the expression (3.1.5) in the case $V = 0$. In fact $e^{-tH_0} = (e^{-tH_0/n})^n$, so as in (3.1.12), with

$$I_j = R^3 \text{ and } t_j = -t/2 + jt/n, \quad j = 0, 1, 2, \ldots, n,$$

we have

(kernel $e^{-tH_0})(q, q')$

$$= \left(\frac{2\pi t}{n}\right)^{-3n/2} \int \cdots \int \exp\left[-\frac{1}{2}\sum_{i=1}^n \left(\frac{\Delta q_i}{\Delta t_i}\right)^2 \Delta t_i\right] \prod_{i=1}^{n-1} dq_i. \quad (3.1.15)$$

Here $\Delta q_i = q(t_i) - q(t_{i-1})$, $q_0 = q$, $q_n = q'$, $\Delta t_i = t/n$. Formally, the $n \to \infty$ limit yields $\exp[-\frac{1}{2}\int_{-t/2}^{t/2} \dot{q}(s)^2 \, ds]$. Neither the constant $(2\pi t/n)^{-3n/2}$ nor the measure $\prod_i dq_i$ has an $n \to \infty$ limit. It can be proved that the set of paths $q(s)$ which are Hölder continuous with exponent at least $\frac{1}{2}$ has Wiener measure zero. Similarly, $\exp[-\int \frac{1}{2}\dot{q}(s)^2 \, ds] = 0$ for a.e. Wiener path, and so none of the three factors in (3.1.15) has a limit as $n \to \infty$. However, the product of these three factors in the integrand (3.1.15) has a limit, which is just conditional Wiener measure $dW_{q,q'}^t$, as defined before.

3.2 The Feynman–Kac Formula

Let e^{tA}, e^{tB} be one parameter groups on a Hilbert space $\mathscr{H}$ with infinitesimal generators A, B. Assuming that A, B are bounded, the Lie product formula composes the two given groups to form the group $e^{t(A+B)}$.

Theorem 3.2.1. *For bounded operators A, B,*

$$e^{A+B} = \lim_{n \to \infty} \left(e^{A/n}e^{B/n}\right)^n. \tag{3.2.1}$$

PROOF. Let $C = e^{(A+B)/n}$ and $D = e^{A/n}e^{B/n}$. We show that as $n \to \infty$, $\|C^n - D^n\| \to 0$. In fact,

$$\|C^n - D^n\| = \left\| \sum_{m=0}^{n-1} C^m(C-D)D^{n-m-1} \right\| \leq \text{const } n\|C - D\|, \tag{3.2.2}$$

where we have used

$$\|C\|, \|D\| \leq \exp\left[\frac{\|A\| + \|B\|}{n}\right] \leq \text{const}^{1/n}.$$

Furthermore

$$\|C - D\| \leq \text{const } n^{-2}, \tag{3.2.3}$$

by the cancellation of the order zero and order one terms in the power series expansion of

$$C - D = e^{(A+B)/n} - e^{A/n}e^{B/n}.$$

Substituting (3.2.3) in (3.2.2) completes the proof.

In case A and B are not bounded, the proof of (3.2.1) requires further assumptions. We state one form of the theorem.

Theorem 3.2.2. *Let H_0, V be bounded below and essentially self-adjoint. Let $H = H_0 + V$ be essentially self-adjoint. Then*

$$e^{-H} = \text{strong } \lim_{n \to \infty}(e^{-H_0/n}e^{-V/n})^n. \tag{3.2.4}$$

We establish the Feynman–Kac formula

$$\mathcal{K}_t(q, q') \equiv (\text{kernel } e^{-tH})(q, q')$$

$$= \int \exp\left(-\int_{-t/2}^{t/2} V(q(s))\, ds\right) dW_{q,\,q'}^t, \qquad (3.2.5)$$

for $H = H_0 + V = -\tfrac{1}{2}\Delta + V$. We assume sufficient regularity for the potential $V(q)$ to simplify the proof, and more general potentials may then be treated as limits. We remark that for V real, the Feynman–Kac formula shows that $\mathcal{K}_t(q, q') > 0$.

Theorem 3.2.3. *Let $V(q)$ be a continuous, real valued function on R^d, bounded from below, and let $H = -\tfrac{1}{2}\Delta + V$ be essentially self-adjoint. Then the kernel $\mathcal{K}_t(q, q')$ of e^{-tH} is given by* (3.2.5).

Remark. For $V \in L_p(R^d)$, $p > d/2$, a Sobolev inequality shows that V is a Kato perturbation of Δ and that H is essentially self-adjoint. If V is a polynomial, bounded from below, then H is also essentially self-adjoint.

PROOF.   By Theorem 3.2.2,

$$e^{-tH} = \lim_{n \to \infty} (e^{-tH_0/n} e^{-tV/n})^n, \qquad (3.2.6)$$

where $H_0 = -\tfrac{1}{2}\Delta$. By Corollary 3.1.2,

$$(\text{kernel}(e^{-tH_0/n} e^{-tV/n})^n(q, q')) = \int \exp\left[-\frac{t}{n}\sum_{j=1}^{n} V\left(q\left(-\frac{t}{2}+\frac{jt}{n}\right)\right)\right] dW_{q,\,q'}^t.$$
$$(3.2.7)$$

By (3.2.6), the kernel on the left of (3.2.7) converges in the sense of Schwartz distributions as $n \to \infty$ to $\mathcal{K}_t(q, q')$. Pointwise on path space,

$$\frac{t}{n}\sum_{j=1}^{n} V\left(q\left(-\frac{t}{2}+\frac{jt}{n}\right)\right) \to \int_{-t/2}^{t/2} V(q(s))\, ds,$$

since $V(q(s))$ is Riemann integrable. Hence the integrand in (3.2.7) converges pointwise to $\exp[-\int_{-t/2}^{t/2} V(q(s))\, ds]$. Since $V(q)$ is bounded below, the integrand is bounded above, uniformly in n, by a constant, which is integrable with respect to $dW_{q,\,q'}^t$. Thus by the Lebesque dominated convergence theorem, (3.2.7) converges as $n \to \infty$ to (3.2.5).

We consider two modifications of the Feynman–Kac formula (3.2.5). Here we take a more general H_0, giving a more general $\mathcal{K}_t^0(q, q')$ and a path space measure more general than Wiener measure. In Section 3.4, we establish a path space formula for expectations in the ground state Ω of H.

The first generalization involves replacing $H_0 = -\tfrac{1}{2}\Delta$ by the second order elliptic operator $H_0 = -\tfrac{1}{2}\Delta + V$. (The fact that H_0 is second order elliptic results in the positivity of the kernel e^{-tH_0}.) We could, for example, replace $\mathcal{K}_t^0(q, q')$ by $\mathcal{K}_t(q, q')$ of (3.2.5). Often, however, one wants the H_0 part of the problem to be explicitly solvable and to generate a Gaussian measure, amenable to computation. A common example is $H_0 = -\frac{1}{2}\Delta + \frac{1}{2}q^2 - \frac{1}{2}$, which we studied in detail in Section 1.3. The associated diffusion process is the Ornstein–Uhlenbeck velocity process, for which $\mathscr{K}^0_t(q, q')$ is given by the Mehler formula (1.5.26). The analogs of Theorem 3.1.1 and Corollary 3.1.2 remain valid for the Ornstein–Uhlenbeck velocity process. Let $dU^t_{q,q'}$ denote the measure on $\mathscr{W}(q, q', t)$ constructed in this way from the Mehler kernel. Then the Feynman–Kac formula for $H_0 = -\frac{1}{2}\Delta + \frac{1}{2}q^2 - \frac{1}{2}$, $H = H_0 + V$, becomes

$$\mathscr{K}_t(q, q') = (\text{kernel } e^{-tH}(q, q')) = \int \exp\left(-\int_{-t/2}^{t/2} V(q(s))\, ds\right) dU^t_{q,q'}. \quad (3.2.8)$$

We now introduce the second generalization of (3.2.5). Let $\Omega_0(q)$ be the ground state of $H_0 = -\frac{1}{2}\Delta + \frac{1}{2}q^2 - \frac{1}{2}$, given by (1.5.8) for $d = 1$, or its appropriate generalization for $q \in R^d$, namely $\Omega_0(q) = \prod_{i=1}^d \Omega_0(q_i)$. Let

$$d\phi_0 = \int_{R^d \times R^d} \Omega_0(q)\Omega_0(q')\, dU^t_{q,q'}. \quad (3.2.9)$$

The measure $d\phi_0$ is defined on functions of continuous paths $q(s)$, $s \in (-t/2, t/2)$. Thus if $-t/2 < t_1 \le t_2 \cdots \le t_n < t/2$ and $A_i(q)$ are bounded functions of q,

$$\int \prod_{i=1}^n A_i(q(t_i))\, d\phi_0 = \langle \Omega_0, A_1 e^{-(t_2 - t_1)H_0} A_2 \cdots A_n \Omega_0 \rangle. \quad (3.2.10)$$

Since the integrals (3.2.10) do not depend on t, we can extend the definition of $d\phi_0$ to functions of continuous paths $q(s)$ for s belonging to any bounded interval. The measure $d\phi_0$ is Gaussian, as we see by elementary calculation. By the properties of Section 1.5,

$$\int q(t_1)\, d\phi_0 = \langle \Omega_0, q\Omega_0 \rangle = 0,$$

$$\int q(t_1)q(t_2)\, d\phi_0 = \langle \Omega_0, qe^{-|t_1 - t_2|H_0} q\Omega_0 \rangle$$

$$= e^{-|t_1 - t_2|}\langle \Omega_0, q^2\Omega_0 \rangle = \tfrac{1}{2}e^{-|t_1 - t_2|}.$$

We denote the real inner product $\langle q, f \rangle$ by $q(f) = \int q(s)f(s)\, ds$. Consider f smooth, real, and with compact support. By taking Riemann sum approximations, and using the fact that the Fourier transform of $e^{-|t|}$ is $\text{const}(1 + E^2)^{-1}$, we find that

$$\int q(f)^2\, d\phi_0 = \left\langle f, \left(1 - \frac{d^2}{ds^2}\right)^{-1} f \right\rangle_{L_2}$$

$$= \int |f^\sim(E)|^2 (1 + E^2)^{-1}\, dE$$

$$= \|f\|^2_{-1}.$$

In other words the second moment of the measure $d\phi_0$ is determined by the positive operator $(1 + E^2)^{-1}$, which defines the Sobolev inner product space H_{-1} with norm $\|\cdot\|_{-1}$. Also calculation gives

$$\int e^{iq(f)} \, d\phi_0 = e^{-\frac{1}{2}\langle f, f\rangle_{-1}}, \qquad (3.2.11)$$

the full Gaussian property. In other words,

$$\int q(f)^{2n} \, d\phi_0 = (2n - 1)!! \, \|f\|^{2n}_{-1}$$

$$= (2n - 1)(2n - 3) \cdots 1 \|f\|^{2n}_{-1}, \qquad (3.2.12)$$

$$\int q(f)^{2n+1} \, d\phi_0 = 0.$$

The derivation of (3.2.12) follows most easily by expanding $q = (1/\sqrt{2}) \times (A + A^*)$, as in (1.5.3), and using (3.2.10) along with the property (1.5.21). Choosing $f = \sum z_j f_j$ and differentiating (3.2.11), we obtain

$$\int q(f_1) \cdots q(f_{2n}) \, d\phi_0 = \sum_{\text{pairings}} \langle f_{i_1}, f_{i_2}\rangle_{-1} \cdots \langle f_{i_{2n-1}}, f_{i_{2n}}\rangle_{-1}, \qquad (3.2.13)$$

where the sum extends over the $(2n -\!\!- 1)!!$ pairings of the $2n$ functions f_j.

3.3 Uniqueness of the Ground State

It is general experience that the ground states of Hamiltonians of the form $H = -\Delta + V$ are unique. This uniqueness property of the ground state is related to the Perron–Frobenius theorem for matrices with strictly positive entries, or to integral operators A with strictly positive kernels and is proved in this section. For the eigenvalue problem in a region $\mathscr{B}$, the ground state Ω of H can be chosen to be a positive function in the interior of $\mathscr{B}$, i.e., it is "nodeless". The existence of a ground state, for $\mathscr{B}$ unbounded (e.g., $\mathscr{B} = R^3$), requires further restrictions on V; cf. Theorem 1.5.9.

Definition 3.3.1. Let A be an operator on a Hilbert space $\mathscr{H}$, and let $\mathscr{H}$ be represented as $L_2(X, dv)$ for some measure space X and some measure dv on X. Then A has a strictly positive kernel if for each choice of non-negative function $\theta \in \mathscr{H}$, $\|\theta\| \neq 0$,

$$0 < A\theta \quad \text{almost everywhere.} \qquad (3.3.1)$$

Such an operator has the property that if θ, ψ are positive vectors, not identically zero, then $0 < \langle \theta, A\psi\rangle$.

Remark. The operator $A = e^{-tH_0}$ studied in Theorem 1.5.10 has a strictly positive kernel. We see later that the same is true for $A = e^{-tH}$, with suitable restrictions on V. The ground state of H is the eigenvector of e^{-tH} corresponding to its maximum eigenvalue.

Theorem 3.3.2. *Let A have a strictly positive kernel, and let $\|A\| = \lambda$ be an eigenvalue of A. Then λ has multiplicity 1, and the corresponding eigenvector Ω can be chosen to be a strictly positive function.*

PROOF. Since A maps real functions into real functions, we may assume that Ω is real. Also, by the positivity of the kernel of A,

$$\lambda\|\Omega\|^2 = \langle A\Omega, \Omega\rangle \le \langle A|\Omega|, |\Omega|\rangle \le \|A\|\ \|\Omega\|^2 = \lambda\|\Omega\|^2.$$

Thus

$$\langle A\Omega, \Omega\rangle = \langle A|\Omega|, |\Omega|\rangle. \tag{3.3.2}$$

Write $\Omega = \Omega_+ - \Omega_-$, where $\Omega_\pm$ are the positive and negative parts of Ω. Then by (3.3.2),

$$\langle \Omega_+, A\Omega_-\rangle + \langle \Omega_-, A\Omega_+\rangle = 0, \tag{3.3.3}$$

contradicting positivity (3.3.1) unless $\Omega_+ = 0$ or $\Omega_- = 0$. Thus we may choose $\Omega \ge 0$. If $\|\theta\| \neq 0$, and $\theta \ge 0$, then by (3.3.1),

$$0 < \langle \theta, A\Omega\rangle = \lambda\langle \theta, \Omega\rangle.$$

Since θ is arbitrary, $\Omega > 0$ almost everywhere.

Finally, if ψ and Ω were linearly independent eigenvectors of A with eigenvalue λ, we could repeat the above argument with the component of ψ orthogonal to Ω. This would yield two strictly positive, orthogonal vectors, which is impossible. Hence λ has multiplicity 1 and Ω is unique.

In order to apply this theorem to Hamiltonians in quantum mechanics, we give sufficient conditions to ensure that the kernel of e^{-tH} is strictly positive, using the Feynman–Kac representation of Section 3.1.

Theorem 3.3.3. *Suppose V is continuous and bounded from below, and let $H = H_0 + V$ be essentially self-adjoint. Then*

$$0 < (\text{kernel } e^{-tH})(q, q') = \int \exp\left(-\int_{-t/2}^{t/2} V(q(s))\, ds\right) dW_{q, q'}^t. \tag{3.3.4}$$

PROOF. Let $M(R) = \sup\{V(q): |q| \le R\}$, and let $I(R)$ be the measure (with respect to $dW_{q, q'}^t$) of the set $\mathscr{W}(R)$ of paths $q(s)$ for which $\sup\{|q(s)|: -t/2 \le s \le t/2\} \le R$. Then

$$(\text{kernel } e^{-tH})(q, q') \ge e^{-M(R)t} \int_{\mathscr{W}(R)} dW_{q, q'}^t \ge I(R)e^{-M(R)t}.$$

Since $I(R) \to 1$ as $R \to \infty$, $I(R)$ is nonzero for R sufficiently large, which completes the proof.

Corollary 3.3.4. *Let H satisfy the hypotheses of Theorem 3.3.3 and have a ground state Ω. Then Ω is unique up to a phase $e^{i\theta}$, θ real, and Ω can be chosen strictly positive.*

Remark. In the theorem and its corollary, we may replace $dW_{q,\,q'}^{t}$ by the Ornstein–Uhlenbeck measure $dU_{q,\,q'}^{t}$.

3.4 The Renormalized Feynman–Kac Formula

We now return to the second generalization of (3.2.5), namely the "renormalized" Feynman–Kac formula. With $H = H_0 + V$,

$$H_0 = -\tfrac{1}{2}\Delta + \tfrac{1}{2}q^2 - \tfrac{1}{2},$$

and V continuous, we define

$$\hat{H} = H - E_0 \tag{3.4.1}$$

to be the "renormalized" Hamiltonian. We assume that H is essentially self-adjoint, with a ground state Ω, and then E_0 is chosen so that $\hat{H}\Omega = 0$. By Theorems 3.3.2, 3.3.3, Ω is unique and $\langle \Omega, \Omega_0 \rangle \neq 0$. Thus using the spectral representation of H,

$$e^{-tH} = \int e^{-t\lambda}\, dE(\lambda),$$

we obtain

$$\Omega\langle \Omega, \Omega_0 \rangle = \lim_{t \to \infty} e^{-t\hat{H}}\Omega_0. \tag{3.4.2}$$

We now translate (3.4.2) into path space language. We define a probability measure

$$d\mu_t = Z_t^{-1} \exp\left(-\int_{-t/2}^{t/2} V(q(s))\, ds \right) d\phi_0 \tag{3.4.3}$$

where $d\phi_0$ is defined in (3.2.9) and where the normalization constant Z_t is given by

$$Z_t = \int \exp\left(-\int_{-t/2}^{t/2} V(q(s))\, ds \right) d\phi_0. \tag{3.4.4}$$

By the Feynman–Kac formula, for times $-t/2 < t_1 \leq t_2 \leq \cdots \leq t_n < t/2$, and A_i bounded functions of q,

$$\frac{\langle e^{-(t_1+t/2)H_0}\Omega_0,\, A_1 e^{-(t_2-t_1)H} A_2 \cdots A_n e^{-(t/2-t_n)H}\Omega_0 \rangle}{\|e^{-tH/2}\Omega_0\|^2}$$

$$= \int \prod_{i=1}^{n} A_i(q(t_i))\, d\mu_t. \tag{3.4.5}$$

By (3.4.2), the left side converges as $t \to \infty$, and $|\langle \Omega_0, \Omega \rangle|^2$ factors from the numerator and denominator. Thus we have established

Theorem 3.4.1. *Let H satisfy the hypotheses above. Then*

$$\langle \Omega, A_1 e^{-(t_2-t_1)\hat{H}} A_2 e^{-(t_3-t_2)\hat{H}} A_3 \cdots A_n \Omega \rangle$$

$$= \lim_{t \to \infty} \int \prod_{i=1}^{n} A_i(q(t_i)) \, d\mu_t(q(\cdot)). \quad (3.4.6)$$

We can now reformulate this result in terms of a measure $d\mu$ which is countably additive on the space $\mathscr{D}'(R^1)$ of real distributions[1] on the line. The convergence as $t \to \infty$ in (3.4.6) also extends to convergence in the moments of $d\mu_t$ and the Fourier transform. We define the (inverse) Fourier transform $S_t\{f\}$ of the measure $d\mu_t$ by

$$S_t\{f\} = \int e^{iq(f)} \, d\mu_t, \qquad f \in \mathscr{D}(R^1). \quad (3.4.7)$$

The function $S_t\{f\}$ maps $\mathscr{D}$ into scalars C, and hence is a functional. It is sometimes called the generating functional or characteristic functional of $d\mu_t$. It is easy to check that $S_t\{f\}$ satisfies the following three criteria:

1. *Continuity:* $S\{f_n\} \to S\{f\}$ if $f_n \to f$ in $\mathscr{D}$. (Uniform convergence $f_n \to f$ and $D^j f_n \to D^j f$ on compact sets.)
2. *Positive definiteness:* $0 \le \sum_{i,j=1}^{N} \bar{c}_i c_j S\{f_i - f_j\}$ for all sequences $f_i \in \mathscr{D}$, $c_i \in C$, $i = 1, 2, \ldots, N$.
3. *Normalization:* $S\{0\} = 1$.

Note that defining $A(q) = \sum_{j=1}^{N} c_j \exp[iq(f_j)]$, condition 2 is ensured by $0 \le \int |A|^2 \, d\mu_t$, namely the positivity of $d\mu_t$.

Conversely, a functional $S\{f\}$ satisfying conditions 1–3 is always the Fourier transform of a measure on $\mathscr{D}'$, and this property allows us to ensure the existence of measures for limiting functionals $S\{f\}$. We state the result for R^d replacing R^1.

Theorem 3.4.2 (Minlos). *Let $S\{f\}$ be a functional on $\mathscr{D}(R^d)$ satisfying conditions 1–3 above. Then there exists a unique Borel probability measure $d\mu(q)$ on $\mathscr{D}'(R^d)$ related to $S\{f\}$ by Fourier transformation:*

$$S\{f\} = \int e^{iq(f)} \, d\mu(q). \quad (3.4.8)$$

---

[1] Distributions are continuous linear functionals on the space $\mathscr{D}$ of C^∞ functions with compact support. We write formally $q(f) = \int q(t) f(t) \, dt$, $q \in \mathscr{D}'$, $f \in \mathscr{D}$. Convergence in $\mathscr{D}$ is defined as uniform convergence of each derivative $D^n f$ on compact subsets of R. A distribution $q \in \mathscr{D}'$ is defined by the two properties of linearity and convergence: $q(f_n) \to 0$ when $f_n \to 0$ in $\mathscr{D}$. See [Schwartz, 1950–1] or [Gelfand and Shilov, Vol. I, 1964–8.] for further details.

Remark. This theorem generalizes the standard result, Bochner's theorem, which characterizes the Fourier transform of a Borel probability measure on R^N as a continuous, positive definite, normalized function $S\{f\}$ on R^N (the dual space of R^N). A proof of Minlos' theorem is given in [Gelfand and Vilenkin, 1964]. Note that if $S\{f\}$ extends continuously to $\mathscr{S}(R^d)$, the Schwartz space of well-behaved functions

$$\mathscr{S} = \{f : x^r D^s f \in L_\infty, \text{ all } r, s \geq 0\},$$

then $d\mu$ is concentrated on $\mathscr{S}'(R^d)$, the space of tempered distributions.

We now illustrate how to use this theorem in order to construct a measure $d\mu(q(\cdot))$ on $\mathscr{D}'(R)$ which is *time-translation-invariant*. Integrating with this measure yields the renormalized Feynman–Kac formula: for $t_1 \leq t_2 \leq \cdots \leq t_N$,

$$\langle \Omega, A_1 e^{-(t_2 - t_1)\hat{H}} A_2 \cdots A_N \Omega \rangle = \int \prod_{i=1}^{N} A(q(t_i))\, d\mu. \tag{3.4.9}$$

We call this formula renormalized because it expresses expectations in the ground state Ω of $\hat{H}$ rather than in the ground state Ω_0 of H_0. The construction of $d\mu$ involves taking the limit $t \to \infty$ in $d\mu_t$. In this limit the measures $d\mu_t$ are not proved to converge weakly, but the Fourier transforms $S_t\{f\}$ do converge to define $d\mu$.

In order to avoid technical complications, and to ensure that our results are a special case of the results of Part II, we now restrict $V(q)$ to be an even + linear polynomial. The following holds, however, for a much wider class of potentials $V(q)$.

Theorem 3.4.3. *Under the above hypotheses, and with $f \in \mathscr{D}(R)$,*

$$\lim_{t \to \infty} S_t\{f\} = S\{f\} \tag{3.4.10}$$

exists and satisfies conditions 1–3 above for Theorem 3.4.2. Hence a measure $d\mu$ exists on $\mathscr{D}'(R)$ which satisfies (3.4.8–9).

Remark. For a proof, we appeal to Part II. A direct proof, however, could be given along these lines: define, for supp $f \subset [\tau, T]$,

$$q(f_N) = N^{-1} \sum_{j=1}^{N} q(t_j) f(t_j), \qquad t_j = \tau + (T - \tau)\frac{j}{N}.$$

Then by Theorem 3.4.1, $S\{f_N\} = \lim_{t \to \infty} S_t\{f_N\}$ exists. The limit $N \to \infty$ can be justified by generalizing Theorem 3.2.2.

We point out two properties of $S\{f\}$. It is invariant under time translation or reflection, and it satisfies a further positivity property. Let $f_t(s) = f(s - t)$ denote translation, and let $(\theta f)(s) = f(-s)$ be the action on $\mathscr{D}$ of translation and reflection about time zero.

Corollary 3.4.4.   *The generating function $S\{f\}$ in* (3.4.10) *satisfies:*
(1) *Invariance:* $S\{f\} = S\{f_t\} = S\{\theta f\}$.
(2) *Reflection positivity: If $f_i(s)$ is real and vanishes for $s < 0$, for $i = 1, 2$,*
..., N, then the matrix $M_{ij} = S\{\theta f_i - f_j\}$ has positive eigenvalues.

PROOF.   The invariance is obvious; the positivity follows by using the representation (3.4.9) for $\sum_{ij} \bar{\xi}_i \xi_j M_{ij}$.

REFERENCES

[Kac, 1959], [Gelfand and Vilenkin, 1964].

CHAPTER 4

Correlation Inequalities and the Lee–Yang Theorem

Correlation inequalities are general inequalities relating expectations of correlations of statistical mechanical systems. The Lee–Yang theorem is included here because its proof is related to that of the correlation inequalities.

The correlation inequalities are used in quantum field theory in the proof of convergence of the infinite volume limit (Chapter 11), in the study of phase transitions (Chapter 16) and in the approach to the critical point (Chapter 17). The proof of these inequalities for continuum quantum fields is based on a lattice approximation. Here we give proofs for the lattice case only. The lattice case is of interest in its own right, as a description of crystalline solids. Some simple applications are included here.

4.1 Griffiths Inequalities

The simplest inequalities are the Griffiths inequalities, which state that expectations and pair correlations are positive for general ferromagnetic interactions, i.e.,

$$0 \le \langle \xi^A \rangle, \qquad \text{first Griffiths inequality (Theorem 4.1.1),}$$

$$0 \le \langle \xi^A \xi^B \rangle - \langle \xi^A \rangle \langle \xi^B \rangle, \quad \text{second Griffiths inequality (Theorem 4.1.3).} \tag{4.1.1}$$

Here $A = \{a_i\}$

$$\xi^A \equiv \prod_i \xi_i^{a_i} \tag{4.1.2}$$

denotes a product of spin variables (with possible duplications in A). Given the polynomial Hamiltonian

$$H = -\sum_A J_A \xi^A, \tag{4.1.3}$$

we define expectations which generalize those considered in Chapter 2. The Hamiltonian (4.1.3) is said to be *ferromagnetic* if $0 \le J_A$ for all A. It is *nearest neighbor* if $J_A = 0$ unless A contains either one lattice site or two nearest neighbor sites. Let $d\mu_i(\xi_i)$ be the single spin distribution, i.e., a measure on R. We assume the property

$$\int |\xi|^N e^{|H(\xi)|} \prod_{i=1}^n d\mu_i(\xi_i) < \infty \tag{4.1.4}$$

for all N. We define the partition function

$$Z = \int e^{-H(\xi)} \, d\mu(\xi), \tag{4.1.5}$$

where $d\mu(\xi) = \prod_i d\mu_i(\xi_i)$. The expectation $\langle \cdot \rangle$ of a function $F(\xi)$ is defined by

$$\langle F \rangle = \frac{1}{Z} \int F(\xi) e^{-H(\xi)} \, d\mu(\xi). \tag{4.1.6}$$

(This notation differs from that of Chapter 2 by separating the factor $e^{-H(\xi)}$ from the measure $d\mu$.) It is in terms of the expectation (4.1.6) that Griffiths inequalities hold for ferromagnetic interactions.

To study these inequalities, introduce a duplicate lattice labeled by duplicate variables

$$\xi = (\xi_1, \ldots, \xi_n) \quad \text{and} \quad \chi = (\chi_1, \ldots, \chi_n),$$

coordinate functions on $R^n \oplus R^n$. We also introduce the rotated coordinate system

$$t_i = \frac{1}{\sqrt{2}} (\xi_i + \chi_i), \qquad q_i = \frac{1}{\sqrt{2}} (\xi_i - \chi_i). \tag{4.1.7}$$

The inverse transformation is

$$\xi_i = \frac{1}{\sqrt{2}} (t_i + q_i), \qquad \chi_i = \frac{1}{\sqrt{2}} (t_i - q_i). \tag{4.1.7'}$$

As above, we define χ^A, t^A, q^A. Furthermore, for a function $F = F(\xi, \chi)$ we define the expectation

$$\langle F \rangle = Z^{-2} \int F(\xi, \chi) e^{-(H(\xi) + H(\chi))} \, d\mu(\xi) \, d\mu(\chi). \tag{4.1.8}$$

For a function F of ξ alone, or of χ alone, (4.1.8) reduces to the expectation (4.1.6).

Theorem 4.1.1. *Let H be ferromagnetic, let $d\mu_i(\xi_i)$ be symmetric under $\xi_i \to -\xi_i$, and let (4.1.4) be satisfied. Then all moments are nonnegative:*

$$0 \le \langle \xi^A \rangle. \tag{4.1.9}$$

PROOF. We expand the exponent in (4.1.6) in a Taylor's series. By (4.1.4), the sum and integral can be interchanged, so that

$$\langle \xi^A \rangle = Z^{-1} \sum_{j=0}^{\infty} \frac{1}{j!} \int \left(\sum_B J_B \xi^B \right)^j \xi^A \prod_{i=1}^{n} d\mu_i(\xi_i).$$

Since $Z = \int e^{-H} \prod_{i=1}^{n} d\mu_i(\xi_i) \ge 0$, it is sufficient to show that each term above is nonnegative. Expanding $\left(\sum_B \cdots \right)^j$ and using the fact that $0 \le J_B$, we are reduced to showing that

$$0 \le \prod_{i=1}^{n} \int \xi_i^{C_i} d\mu_i(\xi_i) = \int \xi^C \prod_{i=1}^{n} d\mu_i(\xi_i)$$

However, by the symmetry of the measure, $d\mu$, $\int \xi_i^{C_i} d\mu_i(\xi)$ is either zero (for C_i an odd integer) or positive (for C_i an even integer), and so the proof is complete.

Lemma 4.1.2. *For any exponent A, $2^{|A|/2}(\xi^A \pm \chi^A) = (q + t)^A \pm (q - t)^A$ is ferromagnetic (has positive coefficients) as a polynomial in q and t.*

PROOF. By the binomial theorem,

$$(q + t)^A = \sum_{0 \le b_i \le a_i} \prod_{i=1}^{n} \binom{a_i}{b_i} q^{a_i - b_i} t_i^{b_i},$$

$$(q - t)^A = \sum_{0 \le b_i \le a_i} (-1)^{\sum_{i=1}^{n} b_i} \prod_{i=1}^{n} \binom{a_i}{b_i} q_i^{a_i - b_i} t_i^{b_i}.$$

The terms with opposite sign cancel, and the terms with the same sign are ferromagnetic.

Theorem 4.1.3. *With the same assumption as the previous theorem,*

$$0 \le \langle q^A t^B \rangle, \tag{4.1.10}$$

$$0 \le \langle \xi^A \xi^B \rangle - \langle \xi^A \rangle \langle \xi^B \rangle. \tag{4.1.11}$$

PROOF. By the lemma, $H(\xi) + H(\chi)$ is ferromagnetic, as a polynomial in q and t. As in Theorem 4.1.1, we expand the exponent $e^{-(H(\xi) + H(\chi))}$, and thus reduce (4.1.10) to the inequality

$$0 \le \int q_i^{a_i} t_i^{b_i} d\mu_i(\xi_i) d\mu_i(\chi_i). \tag{4.1.12}$$

for all integers $0 \le a_i, b_i$. However,

$$d\mu(\xi) d\mu(\chi) = d\mu(2^{-1/2}(q + t)) d\mu(2^{-1/2}(q - t))$$

$$= d\mu(2^{-1/2}(-q - t)) d\mu(2^{-1/2}(-q + t)) \tag{4.1.13}$$

by the symmetry of the measure $d\mu$. By inspection, (4.1.13) is symmetric under $(q, t) \to (-q, t)$ and $(q, t) \to (-q, -t)$, and so (4.1.12) is either zero (if a_i or b_i is odd) or positive (if a_i and b_i are both even).

To prove (4.1.11), we use the duplicate variable χ to write

$$\langle \zeta^A \zeta^B \rangle - \langle \zeta^A \rangle \langle \zeta^B \rangle = \langle \zeta^A (\zeta^B - \chi^B) \rangle$$
$$= 2^{-(|A|+|B|)/2} \langle (t+q)^A [(t+q)^B - (t-q)^B] \rangle.$$

Here $|A| = \sum_{i=1}^n a_i$. The bracket $[\cdots]$ is ferromagnetic by Lemma 4.1.2, and so the expectation is nonnegative by (4.1.10).

4.2 The Infinite Volume Limit

As a simple application of Griffiths inequalities, we show that the Ising model correlation functions converge in the limit of infinite volume. The correlation functions

$$\langle \zeta^B \rangle = \langle \zeta_1^{b_1} \cdots \zeta_n^{b_n} \rangle \tag{4.2.1}$$

are the moments of the integral defined by (4.1.6).

Proposition 4.2.1. *Let H be ferromagnetic. Then $\langle \zeta^B \rangle$, considered as a function of the couplings J_A in H, is monotone increasing.*

PROOF. By Theorem 4.1.3,

$$0 \le \langle \zeta^B \zeta^A \rangle - \langle \zeta^A \rangle \langle \zeta^B \rangle = \frac{d}{dJ_A} \langle \zeta^B \rangle.$$

Proposition 4.2.2. *The Ising model has ferromagnetic couplings, and $\langle \zeta^A \rangle \le 1$ for any correlation function.*

PROOF. For the Ising model, $\zeta_i = \pm 1$, and so $\zeta^A = \pm 1$. Since $\langle \cdot \rangle$ defines a normalized probability measure, $|\langle \zeta^A \rangle| \le 1$, also.

To show that the Ising model is ferromagnetic, recall that $\zeta_i^2 = 1$, and expand (2.3.3):

$$\langle \zeta, \Delta_{\partial \Lambda} \zeta \rangle = 2 \sum_{v=1}^d \sum_{i \in \Lambda, \, i + e_v \in \Lambda} (\zeta_{i+e_v} \zeta_i - 1).$$

The second term above, -1, contributes equally to the numerator and denominator in (2.3.5) and may be omitted. The first term is ferromagnetic.

We remark that with a positive external field, $0 \le h$, the Ising model measure

$$d\mu_{h, \Lambda} = \frac{e^{h \sum_{i \in \Lambda} \zeta_i} d\mu_\Lambda}{\int e^{h \sum_{i \in \Lambda} \zeta_i} d\mu_\Lambda} \tag{4.2.2}$$

is still ferromagnetic.

Theorem 4.2.3. *Let* $0 \le h$ *in* (4.2.2). *As* $\Lambda \uparrow R^d$, *the correlation functions* (4.2.1) *of the Ising model converge.*

PROOF. In the measure (2.3.4) and expectation (4.1.6), $J_A = \beta$ if $A = \{a_i, a_{i+e_v}\}$ with $a_i = a_{i+e_v} = 1$ and both points i, $i + e_v$ in Λ, and $J_A = 0$ otherwise. In (4.2.2), a nonzero value $J_A = h$ occurs for $A = \{a_i\}$, $a_i = 1$, $i \in \Lambda$ also. In both cases, increasing Λ amounts to increasing certain J_A, and so $\langle \xi^B \rangle$ is monotone increasing in Λ, by Proposition 4.2.1. Then the upper bound (Proposition 4.2.2) guarantees convergence.

In the case of lattice fields with a single spin distribution of the form $e^{-P(\xi)} d\xi$ where P is a semibounded polynomial of the form $P = $ even $+$ linear, the same convergence proof applies. In this case the uniform upper bound $\langle \xi^A \rangle \le$ const requires a separate argument. This uniform bound is easily proved using the multiple reflection methods developed in Chapter 10 for continuum fields. Their use on the lattice is simpler, but we postpone the discussion to Part II.

4.3 ξ^4 Inequalities

For the single spin distribution $d\mu_i = e^{-P_i(\xi_i)} d\xi_i$, we choose

$$P_i(\xi_i) = \lambda_i \xi_i^4 + \sigma_i \xi_i^2, \qquad 0 < \lambda_i, \text{ or } 0 = \lambda_i \text{ and } 0 < \sigma_i. \qquad (4.3.1)$$

Hence we have a quartic interaction and speak of ξ^4 inequalities. In this case further correlation inequalities are valid.

In addition to the duplicate variables ξ, χ introduced in Section 4.1, we introduce new variables ξ', χ', and

$$\begin{aligned}
t_i' &= 2^{-1/2}(\xi_i' + \chi_i'), & q_i' &= 2^{-1/2}(\xi_i' - \chi_i'), \\
\alpha_i &= 2^{-1/2}(t_i + t_i'), & \beta_i &= 2^{-1/2}(t_i - t_i'), & (4.3.2) \\
\gamma_i &= 2^{-1/2}(q_i + q_i'), & \delta_i &= 2^{-1/2}(q_1' - q_i).
\end{aligned}$$

Note $(\alpha, \beta, \gamma, \delta)$ is related to (ξ, χ, ξ', χ') by an orthogonal transformation of R^4.

Theorem 4.3.1. *Let* P *satisfy* (4.3.1), *and let*

$$H = -\sum_{i,j} J_{ij}\xi_i\xi_j - \sum_i h_i\xi_i,$$

where $J_{ij}, h_i \ge 0$, *satisfy* (4.1.4). *Then*

$$0 \le \langle \alpha^A \beta^B \gamma^C \delta^D \rangle \qquad (4.3.3)$$

in the fourfold duplicate variable (tensor product) expectation constructed as in (4.1.8).

The point to correlation inequalities in the duplicate variables, such as (4.3.3), is to obtain twice subtracted inequalities in the original variables. As an intermediate step, we obtain once subtracted inequalities in the t, q variables.

Corollary 4.3.2 (Lebowitz Inequalities)

$$0 \leq \langle t^A t^B \rangle - \langle t^A \rangle \langle t^B \rangle,$$

$$0 \leq \langle q^A q^B \rangle - \langle q^A \rangle \langle q^B \rangle,$$

$$0 \geq \langle t^A q^B \rangle - \langle t^A \rangle \langle q^B \rangle.$$

PROOF. In the inequalities below, we use Lemma 4.1.2 to show that each quantity in brackets $[\cdots]$ is ferromagnetic in the variables $\alpha, \ldots, \delta$, and then we apply (4.3.3):

$$\langle t^A t^B \rangle - \langle t^A \rangle \langle t^B \rangle = \langle t^A (t^B - t'^B) \rangle$$
$$= 2^{-(|A|+|B|)/2} \langle (\alpha + \beta)^A [(\alpha + \beta)^B - (\alpha - \beta)^B] \rangle \geq 0,$$

$$\langle q^A q^B \rangle - \langle q^A \rangle \langle q^B \rangle = \langle q'^A (q'^B - q^B) \rangle$$
$$= 2^{-(|A|+|B|)/2} \langle (\gamma + \delta)^A [(\gamma + \delta)^B - (\gamma - \delta)^B] \rangle \geq 0,$$

$$\langle t^A \rangle \langle q^B \rangle - \langle t^A q^B \rangle = \langle t_A (q'^B - q^B) \rangle$$
$$= 2^{-(|A|+|B|)/2} \langle (\alpha + \beta)^A [(\gamma + \delta)^B - (\gamma - \delta)^B] \rangle \geq 0.$$

PROOF OF THEOREM 4.3.1.   We assume $0 < \lambda_i$ for $i \in \Lambda$. The general case follows by limits. Up to a factor Z^{-4}, the expectation in (4.3.3) has the form

$$\int \alpha^A \cdots \delta^D e^{-[H(\xi) + H(\chi) + H(\xi') + H(\chi')]} \prod_i (d\mu_i(\xi_i) \cdots d\mu_i(\chi_i')). \qquad (4.3.4)$$

Here

$$H(\xi) + \cdots + H(\chi') = - \sum_{i,j} J_{ij} [\xi_i \xi_j + \cdots + \chi_i' \chi_j']$$
$$- \sum_i h_i [\xi_i + \chi_i + \xi_i' + \chi_i'].$$

We reexpress this sum in the variables $\alpha, \ldots, \delta$. Since the transformation

$$(\xi_i, \chi_i, \xi_i', \chi_i') \to (\alpha_i, \beta_i, \gamma_i, \delta_i)$$

is orthogonal, and since the coefficient $[\cdots]$ of J_{ij} has the form of an inner product in the $\xi, \ldots, \chi'$ variables, this coefficient is equal to the same inner product, expressed in $\alpha, \ldots, \delta$ variables. That is,

$$H(\xi) + \cdots + H(\chi') = - \sum J_{ij} [\alpha_i \alpha_j + \cdots + \delta_i \delta_j] - 2 \sum h_i \alpha_i, \qquad (4.3.5)$$

using also the fact that $2\alpha_i = 2^{1/2}(t_i + t_i') = \xi_i + \chi_i + \xi_i' + \chi_i'$. Since $0 \leq J_{ij}$ and $0 \leq h_i$, the Hamiltonian (4.3.5) is ferromagnetic. Hence we can expand the exponent in (4.3.4) and factor over lattice sites to reduce the proof to the inequality

$$0 \leq \int \alpha_i^k \beta_i^l \gamma_i^m \delta_i^n \, d\mu_i(\xi_i) \cdots d\mu_i(\chi_i') \qquad (4.3.6)$$

for all $i, k, \ldots, n$.

For simplicity we drop the index i. We use the detailed form of the measure

$$d\mu(\xi) = e^{-\lambda \xi^4 - \sigma \xi^2} \, d\xi,$$

where $\lambda > 0$ and σ equals the sum of the σ in (4.3.1) plus a term proportional to $\sum_j J_{ij}$ from the expansion of $(\xi_i - \xi_j)^2$. Because the change of variables ξ, $\ldots \leftrightarrow \alpha, \ldots$ is orthogonal, $d\xi \cdots d\chi' = d\alpha \, d\beta \, d\gamma \, d\delta$ and

$$\sigma(\xi^2 + \chi^2 + \xi'^2 + \chi'^2) = \sigma(\alpha^2 + \beta^2 + \gamma^2 + \delta^2).$$

An explicit calculation shows that

$$\begin{aligned}
2^4(\xi^4 + \chi^4 + \xi'^4 + \chi'^4) &= (\alpha + \beta + \gamma - \delta)^4 + (\alpha + \beta - \gamma + \delta)^4 \\
&\quad + (\alpha - \beta + \gamma + \delta)^4 + (-\alpha + \beta + \gamma + \delta)^4 \\
&= 4(\alpha^4 + \beta^4 + \gamma^4 + \delta^4) + 12(\alpha^2\beta^2 + \alpha^2\gamma^2 + \cdots) \\
&\quad - 4! \, 4\alpha\beta\gamma\delta.
\end{aligned}$$

Thus we conclude that $P(\xi) + P(\chi) + P(\xi') + P(\chi')$ equals a ferromagnetic term, $-c\alpha\beta\gamma\delta$, plus an even function of α, β, γ, δ, and we write (4.3.6) as

$$0 \le \int \alpha^k \beta^l \gamma^m \delta^n e^{c\alpha\beta\gamma\delta} e^{-Q} \, d\alpha \, d\beta \, d\gamma \, d\delta.$$

Here c is a positive constant and Q is an even function of α, β, γ, and δ. Expand $e^{c\alpha\beta\gamma\delta}$ in a power series. Hence we are reduced to proving that

$$0 \le \int \alpha^k \beta^l \gamma^m \delta^n e^{-Q} \, d\alpha \, d\beta \, d\gamma \, d\delta \qquad (4.3.7)$$

for all k, l, m, n. But Q is even, so (4.3.7) vanishes unless k, l, m, n are all even. This completes the proof.

Remark. Although the proof extends to polynomials $P(x) = \sum_{j=2}^n c_j x^{2j} + \sigma x^2$ with all $c_j \ge 0$, only the ξ^4 polynomials (4.3.1) survive renormalization in dimensions $d \ge 2$.

Corollary 4.3.3. *Let $h_i \equiv 0$, and let $|A|$ and $|B|$ be even. Then*

$$0 \le \langle \xi^A \xi^B \rangle - \langle \xi^A \rangle \langle \xi^B \rangle \le \sum \langle \xi^{A_1} \xi^{B_1} \rangle \langle \xi^{A_2} \xi^{B_2} \rangle,$$

where the sum runs over partitions $A = (A_1, A_2)$, $B = (B_1, B_2)$ with $|A_1|$, $|B_1|$ odd.

PROOF. The first inequality follows from Theorem 4.1.3. To prove the second, we use the third inequality from Corollary 4.3.2, as follows:

$$\begin{aligned}
2^{(|A| + |B|)/2} \langle t^A q^B \rangle &= \sum_{\substack{A = (A_1, A_2) \\ B = (B_1, B_2)}} (-1)^{|B_2|} \langle \xi^{A_1} \chi^{A_2} \xi^{B_1} \chi^{B_2} \rangle \\
&= \sum (-1)^{|B_2|} \langle \xi^{A_1} \xi^{B_1} \rangle \langle \xi^{A_2} \xi^{B_2} \rangle \\
&\le 2^{(|A| + |B|)/2} \langle t^A \rangle \langle q^B \rangle \\
&= \sum (-1)^{|B_2|} \langle \xi^{A_1} \rangle \langle \xi^{A_2} \rangle \langle \xi^{B_1} \rangle \langle \xi^{B_2} \rangle.
\end{aligned}$$

We assert that terms with a nontrivial partition and with $|B_2|$ even can be omitted without affecting the inequality. In fact, let

$$\langle \xi^A \xi^B \rangle^T = \langle \xi^A \xi^B \rangle - \langle \xi^A \rangle \langle \xi^B \rangle.$$

By Theorem 4.1.3 $\langle \cdots \rangle^T \geq 0$. However,

$$\langle \xi^{A_1} \xi^{B_1} \rangle \langle \xi^{A_2} \xi^{B_2} \rangle - \langle \xi^{A_1} \rangle \langle \xi^{B_1} \rangle \langle \xi^{A_2} \rangle \langle \xi^{B_2} \rangle$$

$$= \langle \xi^{A_1} \xi^{B_1} \rangle^T \langle \xi^{A_2} \xi^{B_2} \rangle^T + \langle \xi^{A_1} \rangle \langle \xi^{B_1} \rangle \langle \xi^{A_2} \xi^{B_2} \rangle^T + \langle \xi^{A_1} \xi^{B_1} \rangle^T \langle \xi^{A_2} \rangle \langle \xi^{B_2} \rangle$$

$$\geq 0,$$

which proves the assertion. Terms on the right side with $|B_2|$ odd are zero. In fact $\langle \xi^{B_2} \rangle = 0$, because the expectation is invariant under the symmetry $\xi \to -\xi$. For the same reason terms on the left with $|B_2|$ odd and $|A_2|$ even are zero. The remaining terms give the desired inequality.

Corollary 4.3.4. *Let $0 \leq h_i$ for all i. Then under the hypotheses of the theorem,*

$$0 \leq \langle \xi_i \rangle, \qquad 0 \leq \langle \xi_i \xi_j \rangle - \langle \xi_i \rangle \langle \xi_j \rangle,$$

and

$$0 \geq \langle \xi_i \xi_j \xi_k \rangle - \langle \xi_i \rangle \langle \xi_j \xi_k \rangle - \langle \xi_j \rangle \langle \xi_i \xi_k \rangle - \langle \xi_k \rangle \langle \xi_i \xi_j \rangle$$
$$+ 2 \langle \xi_i \rangle \langle \xi_j \rangle \langle \xi_k \rangle.$$

PROOF.   The last inequality follows from Corollary 4.3.2, since

$$2^{3/2} \langle t_i q_j q_k \rangle = 2 (\langle \xi_i \xi_j \xi_k \rangle - \langle \xi_i \xi_j \rangle \langle \xi_k \rangle - \langle \xi_i \xi_k \rangle \langle \xi_j \rangle + \langle \xi_i \rangle \langle \xi_j \xi_k \rangle)$$

$$\leq 2^{3/2} \langle t_i \rangle \langle q_j q_k \rangle$$

$$= 2 (2 \langle \xi_i \rangle \langle \xi_j \xi_k \rangle - 2 \langle \xi_i \rangle \langle \xi_j \rangle \langle \xi_k \rangle).$$

With $h_i \geq 0$, $\langle \xi_j \rangle \geq 0$, since $\langle \xi_j \rangle = 0$ if $h_i = 0$ for all i, and

$$\frac{d \langle \xi_j \rangle}{d h_i} = \langle \xi_i \xi_j \rangle - \langle \xi_i \rangle \langle \xi_j \rangle \geq 0$$

by Theorem 4.1.3. This completes the proof.

Remark. $\langle \xi_i \rangle = \partial \ln Z / \partial h_i$, and the truncated correlations (Ursell functions or connected n point functions)

$$U(i_1, \ldots, i_v) \equiv \frac{\partial^v \ln Z}{\partial h_{i_1} \cdots \partial h_{i_v}}$$

are exactly the combinations of correlations considered in the corollary, $v = 1, 2, 3$. For all $h_i = 0$, and v odd, $U(i_1, \ldots, i_v) = 0$, by the $\xi \to -\xi$ symmetry. For even v and all $h_i = 0$, we have $U(i_1, i_2) \geq 0$ (Theorem 4.1.3), $U(i_1, i_2, i_3, i_4) \leq 0$ (a special case of Corollary 4.3.3). Also $U(i_1, \ldots, i_6) \geq 0$ [Cartier, 1974; Percus, 1975; Sylvester, 1975]. It has been conjectured that

$$(-1)^{v/2} U(i_1, \ldots, i_v) \leq 0$$

for all even v.

Corollary 4.3.5.   *Let $0 \le h_i$ for all i. Then*

$$\langle \xi_{i_1} \cdots \xi_{i_n} \rangle \le (n-1)! \sum_{j=0}^{[n/2]} \sum \langle \xi_{i_1} \xi_{i_2} \rangle \cdots \langle \xi_{i_{2j-1}} \xi_{i_{2j}} \rangle \langle \xi_{i_{2j+1}} \rangle \cdots \langle \xi_{i_n} \rangle,$$

where the inner sum runs over all ways of selecting j pairs from n objects.

Proof.   We use Corollary 4.3.2:

$$2^{(n+2)/2} \langle t_1 \cdots t_n q_{n+1} q_{n+2} \rangle = \sum_{\substack{\{1,\ldots,n\} = A_1 \cup A_2 \\ \{n+1,\,n+2\} = B_1 \cup B_2}} (-1)^{|B_2|} \langle \xi^{A_1} \chi^{A_2} \xi^{B_1} \chi^{B_2} \rangle$$

$$= \sum (-1)^{|B_2|} \langle \xi^{A_1} \xi^{B_1} \rangle \langle \xi^{A_2} \xi^{B_2} \rangle$$

$$\le \sum (-1)^{|B_2|} \langle \xi^{A_1} \rangle \langle \xi^{A_2} \rangle \langle \xi^{B_1} \rangle \langle \xi^{B_2} \rangle.$$

Dropping negative terms from the right (B_2 odd) and all terms with B_2 even and the $A = A_1 \cup A_2$ partition nontrivial (as in the proof of Corollary 4.3.3), we have

$$\langle \xi_1 \cdots \xi_{n+2} \rangle \le \langle \xi_1 \cdots \xi_n \rangle \langle \xi_{n+1} \xi_{n+2} \rangle$$

$$+ \sum_{A = A_1 \cup A_2} \langle \xi^{A_1} \xi_{n+1} \rangle \langle \xi^{A_2} \xi_{n+2} \rangle.$$

From this inequality, the corollary follows by induction on n.

4.4 The FKG Inequality

The FKG inequality arrives at the same conclusion as the Griffiths second inequality, but the hypotheses are different, in terms of both the allowed interactions and the allowed observables. We define an order in R^n by the relation

$$\xi = (\xi_1, \ldots, \xi_n) \le \chi = (\chi_1, \ldots, \chi_n) \iff \xi_i \le \chi_i \quad \text{for all } i. \quad (4.4.1)$$

A function $F(\xi)$ is called *monotone* if it is monotone with respect to this order.

Theorem 4.4.1.   *Let F and G be monotone increasing functions of ξ, and let the expectation $\langle \cdot \rangle$ be defined by (2.3.1) and (2.3.4). Then*

$$\langle F \rangle \langle G \rangle \le \langle FG \rangle. \quad (4.4.2)$$

Remark.   In the quadratic terms defining the Gaussian part of the expectation, an arbitrary number of boundary terms of the form

$$\alpha_j (\xi_j - \gamma_j)^2$$

$\alpha_j \ge 0$ are allowed, because these terms can be included in the interaction polynomials P_j. The P_j are not required to have the form even + linear.

Proof.   With duplicate variables ξ and χ, we must show that

$$0 \le \langle [F(\xi) - F(\chi)][G(\xi) - G(\chi)] \rangle. \quad (4.4.3)$$

Let $n = |\Lambda|$ be the number of sites in Λ. For $n = 1$, $F(\xi) - F(\chi)$ and $G(\xi) - G(\chi)$ have the same sign by monotonicity, and the proof is complete in this case.

Now suppose by induction that the theorem has been proved for $n - 1$ sites, and write

$$\xi = (\check{\xi}, \xi_n), \qquad \check{\xi} = \xi_1, \ldots, \xi_{n-1},$$

and similarly for χ. Writing (4.4.3) as an iterated integral, we show that the $d\check{\xi}\, d\check{\chi}$ integral is nonnegative for each value of ξ_n, χ_n. Explicitly, let

$$Z(\alpha) = \langle \delta(\xi_n - \alpha) \rangle,$$

$$\langle F \rangle_{\xi_n = \alpha} = Z(\alpha)^{-1} \langle \delta(\xi_n - \alpha) F(\xi) \rangle, \tag{4.4.4}$$

and let $\langle \; \rangle_{\xi_n = \alpha, \chi_n = \gamma} \equiv \langle \; \rangle_{\alpha\gamma}$ be the corresponding duplicate variable expectation.

We use the identity

$$\langle [F(\xi) - F(\chi)][G(\xi) - G(\chi)] \rangle_{\alpha\gamma}$$

$$= \langle FG \rangle_\alpha + \langle FG \rangle_\gamma - \langle F \rangle_\alpha \langle G \rangle_\gamma - \langle F \rangle_\gamma \langle G \rangle_\alpha$$

$$= [\langle FG \rangle_\alpha - \langle F \rangle_\alpha \langle G \rangle_\alpha] + [\langle FG \rangle_\gamma - \langle F \rangle_\gamma \langle G \rangle_\gamma]$$

$$+ [\langle F \rangle_\alpha - \langle F \rangle_\gamma][\langle G \rangle_\alpha - \langle G \rangle_\gamma], \tag{4.4.5}$$

true by the normalization (4.4.4). We must show that (4.4.5) is nonnegative. The first two terms are nonnegative by the induction hypothesis, and so we complete the proof by showing that each factor in the third term has the same sign. In fact, using the dependence of Z on α given by (4.4.4) and the definitions (2.3.3–4),

$$\frac{d}{d\alpha} \langle F \rangle_\alpha = \beta \sum_l \{ \langle (\alpha - \xi_l) F(\xi) \rangle_\alpha - \langle \alpha - \xi_l \rangle_\alpha \langle F \rangle_\alpha \} + \left\langle \frac{dF}{d\xi_n} \right\rangle_\alpha$$

where the sum over l extends over the nearest neighbors to the nth site. Note that terms proportional to $P'(\xi_n)$ cancel. Since the linear function $\xi \to \alpha - \xi_l$ is monotone, the inductive hypothesis applies again, and it follows that $\langle F \rangle_\alpha$ is monotone increasing in α. The same is true for $\langle G \rangle_\alpha$, and so the third term in (4.4.5) is nonnegative.

4.5 The Lee–Yang Theorem

It is a general fact that thermodynamic functions such as free energy, pressure, etc., are piecewise analytic with jump discontinuities in the function or its derivative. These jumps occur across phase transition surfaces. In special cases, it is possible to establish the absence of phase transitions for nonvanishing magnetic field h (i.e., for a nonzero linear interaction in the Hamiltonian). The Lee–Yang theorem is such a result, which establishes analyticity in h for Re $h \neq 0$, or in terms of the fugacity $z = e^h$, for $|z| \neq 1$.

We consider a single spin distribution $d\mu_i(\xi) = e^{-P_i(\xi)}\, d\xi$ and Hamiltonian $H(\xi)$ with

$$P_j(\xi_j) = a_j \xi_j^4 + b_j \xi_j^2, \qquad H(\xi) = -\sum_{i,j} J_{ij} \xi_i \xi_j - \sum_i h_i \xi_i, \tag{4.5.1}$$

in (4.1.6) with $a_j > 0$, $J_{ij} \geq 0$, and b real. The Lee–Yang theorem shows an absence of phase transition for Re $h_i \neq 0$.

The free energy f is defined as

$$f = f_\Lambda = \frac{1}{|\Lambda|} \ln Z_\Lambda, \tag{4.5.2}$$

where the partition function Z_Λ for volume Λ is

$$Z_\Lambda(\{h_i\}) = \int e^{-H(\xi)} \, d\mu(\xi)$$

$$= \int e^{-H(\xi)} \prod_{i \in \Lambda} e^{-P_i(\xi_i)} \, d\xi_i. \tag{4.5.3}$$

For a finite volume Λ, Z_Λ is manifestly analytic in h. Thus f_Λ is analytic in h in any domain for which Z_Λ has no zeros.

Theorem 4.5.1. *Assume* (4.5.1) *and ferromagnetic pair interactions,*

$$0 \leq J_{ij}. \tag{4.5.4}$$

Let $h_i = h$. *If* Re $h \neq 0$, *then* $Z \neq 0$.

Remark 1. The proof [Simon and Griffiths, 1973] is somewhat involved; it first establishes the Lee–Yang theorem for the Ising model and then approximates the $e^{-P(\xi)}$ distribution by a superposition of Ising models. We present a simpler proof due to [Dunlop, 1977] which is valid in the smaller region $|\text{Im } h_i| \leq \text{Re } h_i$; this proof gives, moreover, a lower bound for $|Z|$.

Theorem 4.5.1'. *Assume* (4.5.1) *and ferromagnetic pair interactions,* (4.5.4). *Let* $|\text{Im } h_i| \leq \text{Re } h_i$ *for all* i. *Then*

$$0 < Z_\Lambda(h_i = 0) \leq Z_\Lambda(\text{Re } h_i - |\text{Im } h_i|) \leq |Z_\Lambda(h_i)|. \tag{4.5.5}$$

Remark 1. Since $Z_\Lambda(h) = Z_\Lambda(-h)$, it follows that in the reflected sector, $|\text{Im } h_i| \leq -\text{Re } h_i$ for all i,

$$Z_\Lambda(h_i = 0) \leq Z_\Lambda(-\text{Re } h_i - |\text{Im } h_i|) \leq |Z_\Lambda(h_i)|. \tag{4.5.6}$$

Remark 2. For most applications, $h_i = h$ (i.e., a spatially independent magnetic field) is sufficient.

In proving the theorem, we use the notion of a positive definite function $f(\theta)$, $\theta \in [0, 2\pi]^N$, with Fourier series

$$f(\theta) = (2\pi)^{-N/2} \sum_{n=-\infty}^{\infty} f_n e^{in\theta}, \qquad n\theta \equiv \sum_{j=1}^{N} n_j \theta_j. \tag{4.5.7}$$

Definition 4.5.2. The function f is positive definite if its Fourier coefficients are all nonnegative, $f_n \geq 0$. Let $\mathscr{P}$ denote the set of positive definite functions f. $\mathscr{P}$ defines an order: $f \leq g$ if $g - f \in \mathscr{P}$, i.e. if $f_n \leq g_n$ for all n.

Proposition 4.5.3. *The set $\mathscr{P}$ is closed under sums, products, complex conjugates, exponentials, and multiplication by positive real constants. In other words, $\mathscr{P}$ is a multiplicative convex cone. These operations preserve the order defined by $\mathscr{P}$.*

PROOF. Addition and complex conjugation clearly preserve $\mathscr{P}$. Multiplication becomes convolution of Fourier coefficients, so it also preserves positivity. Exponentials can be obtained by their power series.

PROOF OF THEOREM 4.5.1′. We use a duplicate variable representation for $Z(h)$. In terms of the spins ξ, χ we define duplicate variables t, q and polar coordinates ρ, θ:

$$t_j = \frac{1}{\sqrt{2}}(\xi_j + \chi_j) = \rho_j \cos \theta_j,$$

$$q_j = \frac{1}{\sqrt{2}}(\xi_j - \chi_j) = \rho_j \sin \theta_j. \tag{4.5.8}$$

The plan of the proof is to show that

$$|Z(h)|^2 = \int e^{-[H(\xi) + H(\chi) - 1]} \prod_{i \in \Lambda} \left(e^{-[P(\xi_i) + P(\chi_i)]} \, d\xi_i \, d\chi_i \right), \tag{4.5.9}$$

when expressed in polar coordinates, is the integral of a product of positive definite functions. The Fourier coefficients of these functions depend in an explicit monotonic fashion on Re $h \pm$ Im h.

Thus we start the proof by writing

$$\exp[-P_i(\xi_i) - P_i(\chi_i)] \, d\xi_i \, d\chi_i = v_i(\rho_i, \theta_i) \, d\rho_i \, d\theta_i, \tag{4.5.10}$$

and showing that $v_i(\rho, \theta)$ is a positive definite function of θ. Since $\prod_i (v_i(\rho_i, \theta_i) \, d\rho_i \, d\theta_i)$ is a product of positive definite measures, it is also positive definite as a function of $(\theta_1, \ldots, \theta_N)$.

Now we verify the positive definiteness of v_i. Note that $d\xi \, d\chi = \rho \, d\rho \, d\theta$, and

$$v_i(\rho, \theta) = \rho \exp[-P_i(\xi) - P_i(\chi)]$$

$$= \rho \exp[-a_i(\xi^4 + \chi^4) - b_i(\xi^2 + \chi^2)]. \tag{4.5.11}$$

Since (4.5.8) is orthogonal,

$$\xi^2 + \chi^2 = t^2 + q^2 = \rho^2 \tag{4.5.12a}$$

and

$$\xi^2 - \chi^2 = \tfrac{1}{2}[(t + q)^2 - (t - q)^2] = 2tq = \rho^2 \sin 2\theta. \tag{4.5.12b}$$

Thus

$$\xi^4 + \chi^4 = \tfrac{1}{2}[(\xi^2 + \chi^2)^2 + (\xi^2 - \chi^2)^2] = \tfrac{1}{2}\rho^4[1 + \sin^2 2\theta]$$

$$= \tfrac{1}{4}\rho^4[3 - \cos 4\theta], \tag{4.5.13}$$

and

$$v_i(\rho, \theta) = \rho \exp[-\tfrac{3}{4}a_i\rho^4 - b_i\rho^2] \exp[\tfrac{1}{4}a_i\rho^4 \cos 4\theta]. \tag{4.5.14}$$

Since $0 < a_i$, and $\cos 4\theta$ is a positive definite function, it follows by Proposition 4.5.3 that v_i is positive definite, as desired.

The second step in the proof is to write $-[H(\xi) + H(\chi)^-]$ in polar coordinates, and to verify that it is a sum of two positive definite functions. In fact

$$\xi_i \xi_j + \chi_i \chi_j = t_i t_j + q_i q_j = \rho_i \rho_j \cos(\theta_i - \theta_j),$$

so

$$\sum_{i,j} J_{ij}(\xi_i \xi_j + \chi_i \chi_j) = \frac{1}{2} \sum_{i,j} J_{ij} \rho_i \rho_j [e^{i(\theta_i - \theta_j)} + e^{-i(\theta_i - \theta_j)}] \qquad (4.5.15)$$

is positive definite. Also

$$h_j \xi_j + \bar{h}_j \chi_j = (\text{Re } h_j)(\xi_j + \chi_j) + i(\text{Im } h_j)(\xi_j - \chi_j)$$

$$= \sqrt{2}\,(\text{Re } h_j)t_j + i\sqrt{2}\,(\text{Im } h_j)q_j$$

$$= \frac{1}{\sqrt{2}} \rho_j (\text{Re } h_j + \text{Im } h_j)e^{i\theta_j}$$

$$+ \frac{1}{\sqrt{2}} \rho_j (\text{Re } h_j - \text{Im } h_j)e^{-i\theta_j}, \qquad (4.5.16)$$

which is positive definite as long as $0 \le \text{Re } h_j \pm \text{Im } h_j$, i.e., $|\text{Im } h_j| \le \text{Re } h_j$. Since $-H(\xi) - H(\chi)^-$ is the sum of (4.5.15) with the sum over j of (4.5.16), it follows from Proposition 4.5.3 that

$$\exp[-H(\xi) - H(\chi)^-] = \exp\left[\sum_{ij} J_{ij}(\xi_i \xi_j + \chi_i \chi_j)\right] \exp\left[\sum_j (h_j \xi_j + \bar{h}_j \chi_j)\right]$$

is the product of two positive definite functions. Furthermore by (4.5.16), it follows that $|Z(h)|^2$ given by (4.5.9) is monotonic as a function of $\text{Re } h_j + \text{Im } h_j$ and of $\text{Re } h_j - \text{Im } h_j$. Since

$$0 \le \text{Re } h_j - |\text{Im } h_j| = \min_{\pm} (\text{Re } h_j \pm \text{Im } h_j),$$

the inequality (4.5.5) follows from this monotonicity and Proposition 4.5.3.

4.6 Analyticity of the Free Energy

In this section we establish the analyticity of the free energy for an infinite volume lattice field model to which the Lee–Yang theorem applies. We also discuss the cluster property of the pair correlation function for such models and for their Ising limits.

Let Z_Λ denote the partition function in volume Λ. The free energy in volume Λ is

$$f_\Lambda = \frac{1}{|\Lambda|} \ln Z_\Lambda. \qquad (4.6.1)$$

For the ξ^4 models discussed in the previous section, the Lee–Yang theorem shows that $f_\Lambda(h_j)$ is analytic for $\text{Re } h_j \ne 0$. Here we study limits of these models, either as $\Lambda \to \infty$, or as we vary the single spin distribution to approximate a continuum theory or an Ising model.

We first remark that the convergence of f_Λ as $\Lambda \to \infty$ is a special case of the conditioning inequalities to be established in Chapter 10. For β small, the convergence also follows by the series expansion methods of Chapter 2. We do not prove convergence here, but state a special case of Proposition 10.3.3, adapted to a lattice field model.

Proposition 4.6.1. *Let Z_Λ denote the partition function for a lattice field with nearest neighbor, translation-invariant, ferromagnetic pair interaction*

$$H = -J \sum_{nn} \xi_i \xi_j - \sum_j h_j \xi_j, \qquad 0 \le J, h_i,$$

with single spin distribution $d\mu_i(\xi_i) = e^{-P(\xi_i)} d\xi_i$ satisfying (4.1.4). As $\Lambda \uparrow \infty$, f_Λ converges,

$$f_\Lambda \to f. \tag{4.6.2}$$

Theorem 4.6.2. *Let Z_Λ be as in the previous proposition. If the Lee-Yang bound (4.5.5) holds for $Z_\Lambda(h)$, then $f(h)$ is analytic for $|\text{Im } h_j| < \text{Re } h_j$.*

Remark 1. In order to avoid the technicalities of a function of infinitely many complex variables h_j, we set $h_j = h$ and prove that $f(h)$ is analytic for $|\text{Im } h| < \text{Re } h$. In fact, using reflection and the Lee-Yang theorem 4.5.1 (rather than 4.5.1', which we proved in the previous section), the analyticity domain extends to $\text{Re } h \ne 0$.

PROOF.   Consider the function

$$g_\Lambda(h) = Z_\Lambda(h)^{1/|\Lambda|} = e^{f_\Lambda(h)}. \tag{4.6.3}$$

Then

$$g_\Lambda(\text{Re } h - |\text{Im } h|) \le |g_\Lambda(h)| \le g_\Lambda(\text{Re } h), \tag{4.6.4}$$

where the upper bound follows by inserting an absolute value in the definition of Z_Λ, and the lower bound follows by Theorem 4.5.1'. By Proposition 4.6.1, $g_\Lambda(h) \to g(h)$ for h real, so (4.6.4) shows that $g_\Lambda(h)$ is bounded uniformly in Λ for h lying in a compact subset K of $|\text{Im } h| < \text{Re } h$. Furthermore, since $Z_\Lambda(h) \ne 0$ for $h \in K$ by Theorem 4.5.1', $g_\Lambda(h)$ is analytic in h. By Proposition 4.6.1, $f_\Lambda(h)$ converges for h real. Thus we choose K to meet the real h axis, and on their intersection $g_\Lambda(h) \to g(h)$. We therefore infer from the Vitali convergence theorem that $g_\Lambda(h) \to g(h)$ for all h satisfying $|\text{Im } h| < \text{Re } h$. Furthermore $g(h)$ is analytic on this domain and is uniformly bounded on any compact subset K of the domain. Also (4.6.4) shows that for any such h,

$$|g(h)| = \lim_{\Lambda \uparrow \infty} |g_\Lambda(h)| \ge \lim_{\Lambda \uparrow \infty} g_\Lambda(\text{Re } h - |\text{Im } h|)$$

$$= \exp[f(\text{Re } h - |\text{Im } h|)]. \tag{4.6.5}$$

By Proposition 4.6.1, f_Λ is increasing in Λ. Hence $|g(h)| \neq 0$ for $|\operatorname{Im} h| < \operatorname{Re} h$. It follows that $\ln g(h)$ exists and is analytic in h, which completes the proof.

Remark 2. A similar limiting argument can be used for the sequence of single spin distributions

$$d\mu(\xi) = \frac{e^{-P_\lambda(\xi)\, d\xi}}{\int e^{-P_\lambda(\xi)}\, d\xi}, \qquad P_\lambda(\xi) = \lambda(\xi^2 - 1)^2. \tag{4.6.6}$$

The limit $\lambda \to \infty$ yields the Ising model [J. Rosen, 1977] and for $\Lambda < \infty$, as $\lambda \to \infty$, $Z_\Lambda(\lambda, h) \to Z_\Lambda^{\text{Ising}}(h)$. This convergence follows for h real by the Lebesgue dominated convergence theorem. Thus we conclude as above that the Lee–Yang bound holds for the Ising model, and hence that $f^{\text{Ising}}(h)$ is analytic for $|\operatorname{Im} h| < \operatorname{Re} h$.

Remark 3. We can also repeat this argument in case the continuum limit of the lattice field exists, for example with appropriate choices for J, a, b as functions of the lattice spacing ε. In space-time dimensions $d = 1, 2, 3$ the $\varepsilon \to 0$ limit (continuum field theory limit) is known to exist. We establish the existence of the $d = 2$ limit in Part II. Hence we conclude that the free energy $f(h)$ for the ϕ^4 quantum field model is analytic for $|\operatorname{Im} h| < \operatorname{Re} h$.

Remark 4. The Lee–Yang theorem is also known to hold for 2 and 3 component spins invariant under O(2) or O(3) rotations, respectively. In other words, $\xi_i \xi_j$ is replaced by $\boldsymbol{\xi}_i \cdot \boldsymbol{\xi}_j$, and ξ_i^4 is replaced by $(\boldsymbol{\xi}_i^2)^2$, where $\boldsymbol{\xi}$ denotes a 2 or 3 component vector [Dunlop and Newman, 1975]. It is not known whether the Lee–Yang theorem holds for four or more spin components.

Remark 5. The Lee–Yang theorem and the proof given here are known to extend to Z_2 lattice gauge theories [Dunlop, 1980]; cf. Section 20.9. It is not known whether other gauge groups lead to a Lee–Yang theorem.

Remark 6. For a general even polynomial $P(\xi)$ of degree 6 or more, we cannot expect analyticity for all $\operatorname{Re} h \neq 0$. This is also the case for a "spin 1" Ising model where $d\mu_i(\xi) = (\frac{1}{2} - \alpha)\delta_{-1}(\xi) + 2\alpha\delta(\xi) + (\frac{1}{2} - \alpha)\delta_1(\xi)$, with $0 < \alpha < \frac{1}{2}$; such a $d\mu_i$ can be obtained as a limit of measures, defined for a sequence of P's of degree 6. In the case of the $d\mu_i$ given, or for $P = \xi^2(\xi^2 - 1)^2 + \alpha\xi^2$, we have a phase diagram of the form shown in Figure 4.1. The fact that there are phase transition lines for $h \neq 0$ shows that certain degree 6 polynomials cannot obey the Lee–Yang theorem. No general criterion has been given for polynomials P (e.g., with positive coefficients) which satisfy the Lee–Yang theorem. Analyticity for general polynomials P has only been studied by series expansion methods. The mean field picture, which is the leading term of the series expansions, provides a qualitative analysis of phase diagrams such as Figure 4.1. The mean field picture will be discussed systematically in the next chapter.

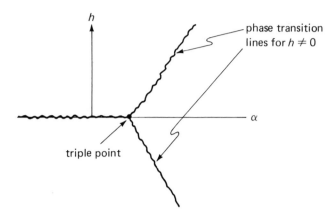

Figure 4.1. Phase transition lines for $P(\xi) = \xi^2(\xi^2 - 1)^2 + \alpha\xi^2$ or for the spin 1 Ising model with $d\mu_i = (\frac{1}{2} - \alpha)(\delta_{-1} + \delta_1) + 2\alpha\delta_0$. See also Section 20.5.

4.7 Two Component Spins

The correlation inequalities and the Lee–Yang theorem have generalizations to vector valued spins. We mention here the simplest results for two component spins $\boldsymbol{\xi}_i = (\xi_i^1, \xi_i^2)$. Let $\boldsymbol{\xi}_i \cdot \boldsymbol{\xi}_j = \sum_{\alpha=1}^2 \xi_i^\alpha \xi_j^\alpha$, and consider

$$H = -\sum_{i,j} J_{ij} \boldsymbol{\xi}_i \cdot \boldsymbol{\xi}_j - \sum_i \mathbf{h}_i \cdot \boldsymbol{\xi}_i \tag{4.7.1}$$

where $0 \le J_{ij}$, h_i^α. In case $\mathbf{h}_i = 0$, the interaction (4.7.1) is invariant under simultaneous SO(2) rotation of all spin vectors $\boldsymbol{\xi}_i$. As single spin distribution, choose the SO(2) invariant measure

$$d\mu_i(\boldsymbol{\xi}_i) = e^{-P_i(\xi_i)} \, d\xi_i^1 \, d\xi_i^2, \tag{4.7.2}$$

where

$$P_i(\xi_i) = \lambda_i(\boldsymbol{\xi}_i \cdot \boldsymbol{\xi}_i)^2 + \sigma_i(\boldsymbol{\xi}_i \cdot \boldsymbol{\xi}_i), \qquad 0 < \lambda_i, \text{ or } 0 = \lambda_i, 0 < \sigma_i. \tag{4.7.3}$$

As a limiting case, also consider the rotator:

$$d\mu_i(\boldsymbol{\xi}_i) = \delta(|\xi_i|^2 - 1) \, d\xi_i^1 \, d\xi_i^2. \tag{4.7.2'}$$

Introduce the variables

$$t_i = \xi_i^1, \qquad q_i = \xi_i^2, \tag{4.7.4}$$

and as before $t^A = \prod_{i \in A} t_i$, etc.

Theorem 4.7.1. *With the Hamiltonian and single spin distribution* (4.7.1–3),

$$0 \le \langle t^A q^B \rangle, \tag{4.7.5}$$

$$0 \le \langle t^A t^B \rangle - \langle t^A \rangle \langle t^B \rangle, \tag{4.7.6}$$

$$0 \le \langle q^A q^B \rangle - \langle q^A \rangle \langle q^B \rangle, \tag{4.7.7}$$

$$0 \le \langle t^A \rangle \langle q^B \rangle - \langle t^A q^B \rangle. \tag{4.7.8}$$

PROOF. In terms of t, q variables,

$$H = - \sum_{i, j} J_{ij}(t_i t_j + q_i q_j) - \sum_i (h_i^1 t_i + h_i^2 q_i), \tag{4.7.9}$$

and

$$\begin{aligned} P(\xi_i) &= \lambda_i(t_i^2 + q_i^2)^2 + \sigma(t_i^2 + q_i^2) \\ &= \lambda_i(t_i^4 + q_i^4 + 2t_i^2 q_i^2) + \sigma(t_i^2 + q_i^2). \end{aligned} \tag{4.7.10}$$

The proof of (4.7.5) now follows the proof of the Griffiths inequality, expanding e^{-H} in a power series. To establish (4.7.6–8), introduce duplicate variables ξ_i' to ξ_i, or in other words duplicate variables (t_i', q_i') to (t_i, q_i). Define α_i, β_i, γ_i, δ_i by the formulas (4.3.2). Then as in (4.3.5),

$$H(\xi) + H(\xi') = - \sum J_{ij}(\alpha_i \alpha_j + \cdots + \delta_i \delta_j) - 2^{1/2} \sum (h_i^1 \alpha_i + h_i^2 \gamma_i)$$

is ferromagnetic in α, β, γ, δ, and using (4.7.10),

$$P(\xi) + P(\xi') = \text{even} - 4\lambda\alpha\beta\gamma\delta$$

is even plus ferromagnetic. The proof now follows the proof of Theorem 4.3.1 and Corollary 4.3.2.

Two special cases of these inequalities follow.

Corollary 4.7.2. *With the above Hamiltonian and single spin distribution,*

$$0 \le \left\langle \left(\prod_{i \in A} \xi_i^1 \right) \left(\prod_{j \in B} \xi_j^2 \right) \right\rangle, \tag{4.7.11}$$

$$0 \le \langle \boldsymbol{\xi}_i \cdot \boldsymbol{\xi}_j \rangle - \langle \boldsymbol{\xi}_i \rangle \cdot \langle \boldsymbol{\xi}_j \rangle. \tag{4.7.12}$$

REFERENCE

[Ruelle, 1969].

CHAPTER 5

Phase Transitions and Critical Points

5.1 Pure and Mixed Phases

For statistical behavior in a family ξ_i of random variables, we required in Chapter 2 the property of almost independence: ξ_i should be almost independent of all but a finite number of ξ_j. This property, in the sense of short range stable forces, was satisfied by the examples of Chapter 2. We now draw a further distinction of weak vs. strong interactions. Weak means that ξ_i should be almost independent of all ξ_j, $j \neq i$, while strong means that ξ_i is strongly correlated to a finite number of ξ_j, $j \neq i$. The weak coupling situations are handled by expansions, as in Chapter 2, and should be regarded as perturbations of an infinite tensor product, zero interaction model.

The Ising and lattice field models of Chapter 2 describe cooperative phenomena. Each ξ_i influences its neighbors to follow suit: $(\xi_i - \xi_{i+\varepsilon_v})^2 \approx 0$. If this influence is strong enough (large β—low temperature in (2.3.5)), a global tendency may develop for all ξ_i to coincide. If we further suppose that $\xi_i = \pm 1$ as in (2.3.2) and that neither $+1$ nor -1 is favored, as in (2.3.5), then the global tendency $\xi_i \approx 1$ vs. $\xi_i \approx -1$ is not unique, and is in general resolved by some additional feature of the problem, such as the nature of the limit process $\Lambda \uparrow R^d$. In this case we have a phase transition, with

$$d\mu = \alpha\,d\mu^+ + (1 - \alpha)\,d\mu^-, \qquad 0 \leq \alpha \leq 1, \tag{5.1.1}$$

and the measures $d\mu^{\pm}$ are pure phases in the sense that in the measure $d\mu^{+}$, the tendency $\xi_i \approx +1$ is realized for most i, except for a set of $d\mu^{+}$ measure zero. Mathematically, the measure $d\mu$ is decomposed into its ergodic components $d\mu^{+}$ and $d\mu^{-}$, the pure phases. In other words

$$\langle \xi_i \rangle_+ \equiv \int \xi_i \, d\mu^+ > 0, \tag{5.1.2}$$

and similarly $\langle \xi_i \rangle_- < 0$. Moreover, $d\mu^{\pm}$ are extreme points in some convex set, and thus are in some sense indecomposable. (The decomposition into pure phases is discussed in Chapter 18.)

To continue the discussion, we note two distinct questions. The simpler question, which we consider in this section, is whether a given infinite volume measure can be decomposed. This is the question of pure vs. mixed phases. A related, but distinct question is whether a phase transition occurs for a given set of parameters (i.e., β, P in (2.3.1–5); see Sections 5.2 and 16.1).

We give three criteria which characterize pure phases. The first is that the lattice translation group, acting on the measure space defined by $d\mu$, should be ergodic. The second criterion is that the transfer matrix have the function 1 as a simple eigenfunction for the eigenvalue 1. The transfer matrix is constructed in Chapter 6 and the equivalence of these two properties (for continuum fields) is established in Section 19.7.

A much simpler criterion is effective in many cases. It is often sufficient to study the pair correlation function

$$\langle \xi_i \xi_j \rangle^T \equiv \int \xi_i \xi_j \, d\mu - \int \xi_i \, d\mu \int \xi_j \, d\mu$$

$$= \langle \xi_i \xi_j \rangle - \langle \xi_i \rangle \langle \xi_j \rangle \tag{5.1.3}$$

$$= \langle (\xi_i - \langle \xi_i \rangle)(\xi_j - \langle \xi_j \rangle) \rangle.$$

The physical interpretation of (5.1.3) can be understood in terms of the fluctuation δ of ξ from its mean value: $\delta_i \equiv \xi_i - \langle \xi_i \rangle$. Thus

$$\langle \xi_i \xi_j \rangle^T = \langle \delta_i \delta_j \rangle, \tag{5.1.4}$$

and the pair correlation function is a measure of the joint fluctuation at i and j. For example consider the measure (5.1.1) with

$$0 < \pm \langle \xi_i \rangle_\pm = \int \xi_i \, d\mu^\pm = M_\pm = \pm M$$

and

$$\lim_{|i-j| \to \infty} (\langle \xi_i \xi_j \rangle_\pm - \langle \xi_i \rangle_\pm \langle \xi_j \rangle_\pm) = 0.$$

Then $\langle \xi_i \rangle = -M(1 - 2\alpha)$ and

$$\lim_{|i-j| \to \infty} \langle \xi_i \xi_j \rangle = \alpha M_+^2 + (1 - \alpha)M_-^2 = M^2,$$

so that

$$\lim_{|i-j|\to\infty} \langle \xi_i \xi_j \rangle^T = 4\alpha(1-\alpha)M^2. \qquad (5.1.5)$$

Thus (5.1.5) takes on its maximum value at $\alpha = \frac{1}{2}$ and vanishes only for pure phases $\alpha = 0, 1$.

In Section 16.1, we show that the vanishing of $\langle \xi_i \xi_j \rangle^T$ at $i - j = \infty$ is a necessary and often sufficient condition for $d\mu$ to be a pure phase.

In statistical mechanics, the mixed states are physically meaningful (as in a mixture of ice and water), but in quantum field theory the experimental facts suggest a unique vacuum.

5.2 The Mean Field Picture

The simplest intuitive discussion of phase transitions uses the lattice field (2.3.1–5), or better yet its continuum limit, given by the formal expression

$$d\mu = Z^{-1} \exp\left[-\int_{R^d} [\tfrac{1}{2}(\nabla\phi)^2 + P(\phi)] \, dx \right] \prod_x d\phi(x), \qquad (5.2.1)$$

where the continuum field $\phi(x)$, $x \in R^d$, has replaced ξ_i, $i \in Z^d$. The measure (5.2.1) is to be regarded as a measure on $\mathcal{S}'(R^d)$, the space of tempered distributions. The mathematics of defining this measure will be discussed in Chapters 6–12.

We expect the measure $d\mu$ in (2.3.5) or (5.2.1) to be concentrated near maximum values of the exponent, i.e. near minimum values of

$$\sum_{i \in Z^d} [\tfrac{1}{2}(\nabla\xi_i)^2 + P(\xi_i)]. \qquad (5.2.2)$$

Clearly this minimum is achieved if

$$\xi_i = \xi^c \quad \text{for all } i, \quad \text{and} \quad \xi^c = \text{global minimum of } P(\cdot). \qquad (5.2.3)$$

The global minima ξ^c are called classical values of ξ. As a simple approximation (which will be modified below), a unique minimum indicates no phase transition, while several minima $\xi^c, \xi^{c'}, \dots$ indicate occurrence of a phase transition, with distinct infinite volume measures (pure phases)

$$d\mu_{\xi^c}, d\mu_{\xi^{c'}}, \dots \qquad (5.2.4)$$

and a general measure

$$d\mu = \alpha_{\xi^c} \, d\mu_{\xi^c} + \alpha_{\xi^{c'}} \, d\mu_{\xi^{c'}} + \cdots \qquad (5.2.5)$$

which is a convex sum of pure phase measures $d\mu_{\xi^c}$, etc.

This picture must be modified on two levels. Sometimes it is approximately correct and sometimes it is grossly incorrect. To test these two possibilities, we return to our original premise, that fluctuations about ξ^c should behave statistically. Let

$$\chi_i = \xi_i - \xi^c \qquad (5.2.6)$$

be the fluctuation field, and rewrite $P(\xi)$ as a polynomial in χ. Since ξ^c is a global minimum,

$$P(\xi) = P(\xi^c) + \frac{1}{2} P''(\xi^c)\chi^2 + \frac{1}{3!} P'''(\xi^c)\chi^3 + \cdots. \qquad (5.2.7)$$

For such a polynomial, the simplest criterion is that if (a) P'' dominates all higher derivatives,

$$P''(\xi^c) \gg P^{(j)}(\xi^c) \quad \text{for all } j \geq 3, \qquad (5.2.8)$$

and if (b) the highest order terms in P are not approximately degenerate, in the sense that $|P^{(j)}(\xi_c)| \leq \text{const } P^{(n)}(\xi_c)$ for $3 \leq j < n = \deg P$, then the above classical picture is approximately correct. Roughly speaking the condition (5.2.8) means that the potential barrier in P which separates ξ^c from other minima should be sufficiently high and wide. Let us suppose that this is the case, and consider the (semiclassical) approximation

$$P_{sc} = P(\xi^c) + \tfrac{1}{2}P''(\xi^c)\chi^2. \qquad (5.2.9)$$

Then P_{sc} is quadratic, and the measure

$$d\mu_{sc} = (\text{normalization})^{-1} e^{-\Sigma_i[1/2(\nabla\chi_i)^2 + P_{sc}(\chi_i)]} \prod_{i \in Z^d} d\chi_i \qquad (5.2.10)$$

is Gaussian, with mean $\langle\chi_i\rangle_{sc} = 0$, i.e., $\langle\xi\rangle_{sc} = \xi^c$. Here $d\mu_{sc}$ is an intermediate approximation between the exact measure $d\mu$ and the classical approximation ξ^c. (Written as a measure, the classical approximation is

$$d\mu_c = \delta_0(\xi - \xi^c),$$

where δ_0 is the delta function measure on function space.) The quadratic energy term $(\sim \chi^2)$ in P_{sc} leads to linear force laws and a linear equation of motion. Thus (5.2.10) is a linearization of the statistical problem around the classical value ξ^c. In statistical mechanics, the higher order corrections in (5.2.7) are often ignored.

The statistical properties of Gaussian measures are easily understood. (See also Chapters 3 and 6.) The covariance of the measure (5.2.10) is the kernel of the operator

$$(-\Delta + P''(\xi^c))^{-1}.$$

At long distances, the asymptotic behavior is

$$\langle\chi_i\chi_j\rangle_{sc} = \text{kernel } (-\Delta + P''(\xi^c))^{-1}$$
$$\sim |i - j|^{-(d-1)/2} e^{-P''(\xi^c)^{1/2}|i-j|} \qquad (5.2.11)$$

(up to corrections for the effect of the lattice). In particular, with $P'' > 0$, the fluctuations χ_i have exponentially small correlations, as anticipated above, and are short ranged events.

The semiclassical approximation (5.2.9–10) may be taken as the starting point of an expansion which is similar to the Mayer expansion of Section 2.4, but more complicated. Just as the Mayer expansion $p = \beta^{-1}z + b_2z^2 + \cdots$

shows that the interaction terms modify the zero interaction (ideal gas) value $p = \beta^{-1}z$ of the pressure, so also the higher terms $(1/3!)P'''(\xi^c)\chi^3 + \cdots$ in (5.2.7) modify the theory. One modification which may occur is to the exact identity of $P(\xi)$ at the distinct minima

$$P(\xi^c) = P(\xi^{c'}).$$

In other words, the interaction may eliminate the phase transition entirely. In this case, however, P and $d\mu$ are close to a phase transition, in the sense that for some polynomial $P_{\text{eff}} = P + \delta P$, with δP small, a phase transition does occur (exactly). For one of the pure phase measures $\langle \xi_i \rangle = \xi^c$, for another $\langle \xi_i \rangle = \xi^{c'}$, etc. Here δP can be regarded as a renormalization effect, and can be computed perturbatively by methods similar to those of Sections 9.4 and 14.3. The fact that it is small means that the coefficients of δP are small, and δP effectively takes on values near $\chi = 0$.

The above description is stronger than presently proven theorems, but typical special cases have been established. It seems likely that a number of linearizations in statistical physics can be justified by these methods. Establishing these results as theorems boils down to showing that the statistical (entropy) factors are dominated by the small probabilities associated with fluctuations away from the mean field values.

This completes the discussion of the case in which the semiclassical picture is approximately correct—namely the case in which the potential barrier in P is large. Then P has deep wells with well-separated minima, so e^{-P} is an approximate product of Gaussians. In this case $P + \delta P$ gives the structure of the phases, with δP small. If, however, the distinct minima are not well separated by a high barrier in P, or if the higher order terms in (5.2.7) are large, then the approximations (5.2.9–10) are not good and may be grossly misleading.

A critical point is by definition a boundary point of the space of phase transitions. In the classical picture above, it occurs as in Figure 5.1, when two minima coalesce, to give in place of (5.2.7)

$$P(\xi) = P(\xi^c) + \frac{1}{4!} P^{(iv)}(\xi^c)\chi^4 + \cdots. \tag{5.2.12}$$

Because $P''(\xi^c) = 0$, and distinct minima in P are not separated, but in fact coincide, the classical description of the critical point is not supposed to be

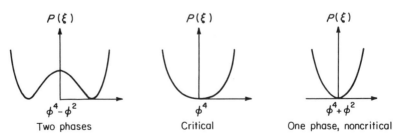

Figure 5.1

very accurate. Indeed, it is not accurate, although it may be used as a rough guide. At a classical critical point, the semiclassical measure (5.2.10) is Gaussian, with covariance

$$\langle \xi_i \xi_j \rangle = \text{kernel } (-\Delta)^{-1}(i, j) \sim |i - j|^{-d+2}. \tag{5.2.13}$$

This type of long range (polynomial) correlation is typical of critical theories, although the exponent, $-d + 2$, is sensitive to the omitted terms in (5.2.12). (For $d = 2$, the long range behavior of (5.2.13) is $-(1/2\pi) \ln |i - j|$.)

Correlation inequalities give one sided bounds on relations between the corrections to (5.2.13). For example in a critical field theory, as $|x - y| \to \infty$,

$$\langle \phi(x)\phi(y) \rangle \leq O(|x - y|^{-d+2-\eta}), \tag{5.2.14}$$

where η is called the anomalous dimension. For the d dimensional Ising model, at its critical temperature, similarly

$$\langle \xi_i \xi_j \rangle \sim |i - j|^{-d+2-\eta}. \tag{5.2.15}$$

For $d = 1$, it follows that $\eta = 1$. For $d = 2$, $\eta = \frac{1}{4}$. For $d = 3$, $\eta = 0.041$, by approximate calculation.

For $d = 1$, 2 and presumably for $d = 3$, these critical points are not Gaussian. Since (5.2.13) does not give the correct long distance behavior, the higher order terms in (5.2.12) are at least as important as the semi-classical terms $\frac{1}{2}(\nabla \xi)^2 + P_{sc}(\xi) = \frac{1}{2}(\nabla \xi)^2 + \text{const. } \xi^2$. Thus the semiclassical theory is not a promising starting point for an expansion. A variety of methods is used by physicists to study the critical point. Most of these methods, including the renormalization group, do not have a well-developed mathematical foundation, but are nonetheless successful in finding approximate numerical values for critical exponents such as above. An introduction to critical phenomena may be found in [Stanley, 1971]. A number of two dimensional models, including notably the two dimensional Ising model, are soluble in closed form. For these cases, the critical behavior is known exactly and in considerable detail; see [McCoy, Tracy and Wu, 1977], [McCoy and Wu, 1973]. For the hierarchical model [Dyson, 1969a], there is a mathematical theory giving some control over critical behavior [Bleher and Sinai, 1973, 1975; Collet and Eckmann 1978], although the model is not exactly soluble.

5.3 Symmetry Breaking

Phase transitions and symmetry breaking are distinct concepts, although they often occur together—in fact, so often that it is worthwhile to discuss the special structures which result from their combination.

In quantum mechanics we found that a Hamiltonian of the form $-\Delta + V(q) = H$ always has a unique ground state Ω, and hence every symmetry group of H must leave Ω invariant. Furthermore, it is the case that for statistical mechanics of a finite number of degrees of freedom, the measure $d\mu_\Lambda$ of (2.3.4) is invariant under any symmetry which leaves $d\mu(\xi_i)$ and $\sum (\xi_i - \xi_{i'})^2$ invariant.

On the other hand, for systems with an infinite number of degrees of freedom (both the statistical systems considered here and the quantum fields considered later), the measure $d\mu$ need not reflect the symmetries of the action which defines it. In that case, we say that symmetry breaking occurs. The Ising model with $h = 0$ is an example in which the transformation $\xi_i \rightarrow -\xi_i$ for all ξ_i leaves the action invariant. However, for $\beta > \beta_c$ there are two pure phases $d\mu_\pm$ for which $M_+ = \int \xi_i \, d\mu_+ = - \int \xi_i \, d\xi_- \neq 0$ breaks the symmetry. In fact $d\mu_+(\xi) = d\mu_-(-\xi)$, so the symmetry maps the two ground states into each other.

In this case the breaking of symmetry explains the occurrence of a phase transition. It is possible, however, for phase transitions to occur in the absence of symmetry (or its breaking). Such an example occurs for a $P(\phi)$ lattice theory of the form

$$P(\phi) = \lambda\phi^2(\phi^2 - a)^2 + \varepsilon\phi^5 - \mu\phi. \tag{5.3.1}$$

By choosing λ, a, $\varepsilon^{-1} \geqslant 0$ and fixed, one can find $\mu = \mu(\lambda, a, \varepsilon)$ near zero such that $P(\phi)$ has two global minima (one near $\phi = 0$, one for $\phi \gg 0$). Thus by the mean field analysis of Section 5.2, we expect a phase transition for $P + \delta P$ with δP small. This phase transition occurs with no change in the symmetry group.

Let us give some other examples. Another phase transition without symmetry breaking is the liquid–gas transition. The symmetry group of each of the two phases is the Euclidean group $\mathscr{E}(R^3)$ of symmetries of space R^3. On the other hand, the liquid–solid transition breaks the symmetry group $\mathscr{E}(R^3)$. The solid, thought of as a perfect crystal defined by a lattice $\mathscr{L}$, has a symmetry group

$$\mathscr{E}(\mathscr{L}) = \{g \in \mathscr{E}(R^3) : g\mathscr{L} = \mathscr{L}\},$$

which is a discrete subgroup of $\mathscr{E}(R^3)$. In this example, the lattice $\mathscr{L}$ is specified by the crystal state. (Individual atoms in the crystal may move from—i.e., vibrate about—their equilibrium position specified by $\mathscr{L}$, but $\mathscr{L}$ does not move as a whole.) The equivalent, but distinct lattices

$$g\mathscr{L} \neq \mathscr{L}$$

do not arise from statistical fluctuations of the atoms in the state specified by $\mathscr{L}$. In other words, $g\mathscr{L}$ and $\mathscr{L}$ specify distinct states or phases. It is often the case that $\mathscr{L}$—or $g\mathscr{L}$—completely specifies the statistical state. By this we mean that the variables (pressure, temperature, etc.) which specify the state of the liquid must be supplemented by a new variable, $\zeta = \xi^c$, to describe the state of a solid. The new variable

$$\zeta \in \mathscr{E}(R^2)/\mathscr{E}(\mathscr{L}) \tag{5.3.2}$$

runs over the coset space, which is the space of inequivalent translated lattices. Any theory specified only by fixing the values of the original variables (pressure, temperature, etc.) would in general be a mixture of various ζ theories. As such, it could be expressed as a direct integral over its constituent phases, labeled by ζ. In this language, we can restate the common occurrence noted above. In the simplest case, the ζ theories are pure phases and (5.3.2) labels all possible pure phases.

Another example of a symmetry breaking phase transition occurs with a $P(\phi)$ lattice field for an even polynomial P. With P even, $\phi \leftrightarrow -\phi$ is a symmetry of the theory. As we should expect from the considerations of Section 5.1, the polynomial

$$P(\phi) + \sigma\phi^2 \quad \text{(for } \beta \text{ fixed)} \tag{5.3.3}$$

substituted for P in (2.3.1–5) gives rise to a unique theory (no phase transition) for $1 \ll \sigma$ and β fixed, while the same polynomial for $\sigma \ll -1$ and β fixed gives rise to two distinct theories (two pure phases) interchanged by $\phi \leftrightarrow -\phi$. In both cases, the symmetry group is Z_2 and the set of pure phases is the coset space Z_2/H, where $H \subset Z_2$ is the symmetry subgroup of a single pure phase. ($H = Z_2$ and $Z_2/H = \{I\}$ for $1 \ll \sigma$, while $H = \{I\}$ and $Z_2/H = Z_2$ for $\sigma \ll -1$.)

Another way to parameterize symmetry breaking phase transitions is to introduce, as an explicit perturbation, an operator breaking the symmetry. For example ϕ is not invariant under the symmetry $\phi \leftrightarrow -\phi$, and we consider

$$P(\phi) + \sigma\phi^2 - h\phi \tag{5.3.4}$$

as a perturbation of (5.3.3).

In the language of magnetism, h represents an external magnetic field. This external field h breaks the $\phi \leftrightarrow -\phi$ (up–down) symmetry of the magnet, i.e., it magnetizes the magnet. Also

$$M \equiv \langle \phi(x) \rangle \tag{5.3.5}$$

is the magnetization. Here $\langle \ \rangle$ is the expectation in the measure $d\mu$ defined by (2.3.5). We write

$$d\mu = d\mu_{\sigma, h} = d\mu_{\sigma, h, P, \beta}, \tag{5.3.6}$$

and $M = M(\sigma, h)$. M is called an order parameter, and it serves to characterize the phase transition. In Figure 5.2, we show the locus of phase transitions for the case $P(\phi) = \phi^4$.

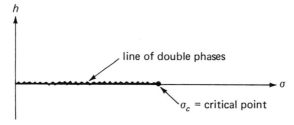

Figure 5.2. Locus of phase transitions for ϕ^4.

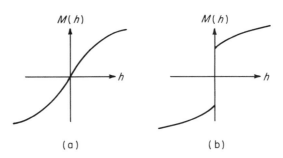

Figure 5.3. $M(h)$ vs. h, (a) for $\sigma \gg 1$, (b) for $\sigma \ll -1$.

The Lee–Yang theorem shows that for $P = \phi^4$ there is a unique pure phase (no transition) for $h \neq 0$, and the cluster expansion shows the same for $\sigma \gg 1$. A different (two phase) cluster expansion is applicable for $\sigma \ll -1$ and shows the existence of at least two phases. For $\sigma \gg 1$, M is a smooth function of h, while for $\sigma \ll -1$, M has a jump discontinuity at $h = 0$. (See Figure 5.3.) We note that $0 \leq dM/dh \leq \text{const}$, and for $0 \leq h$, $d^2M/d^2h \leq 0$ by the correlation inequalities of Chapter 4. Also $dM/d\sigma|_{h=0} \leq 0$. The concavity results from Corollary 4.3.4 and reflects the saturation of the magnetization $M(h)$ by the external field h. In the case of cooperative phenomena, large h leads asymptotically to the alignment of a large fraction of the spins. In case the spins are bounded, as in the Ising model, $M(h)$ tends to the asymptotic value given by total alignment. Much is known about the behavior near $\sigma = \sigma_c$; see Chapter 17.

The phase structure for the Ising model is qualitatively the same as for the ϕ^4 interaction. A variety of interactions, with spin spaces X_i (2.1.2), with nearest neighbor or longer range interaction, and various lattices $\mathscr{L}$, are used as simplified models of the exact molecular and atomic configurations in solid state physics and physical chemistry. In general the equation of state is a piecewise analytic function of the thermodynamic variables—h and σ in Figure 5.2. The phase transitions are the points of nonanalyticity, and the multiple phases arise from points of nonsingle valuedness. Note that in (5.3.5), $M = d \ln Z/dh$, where Z is the partition function. For this problem $M = M(h, \sigma)$ is the equation of state. When, as in this case, the nonanalyticity is reflected in a jump discontinuity in the order parameter, the phase transition is called first order.

5.4 The Droplet Model and Peierls' Argument

For the Ising model at low temperature, the analysis of Sections 5.1–5.2 can be completed mathematically without great difficulty. The proofs are instructive, because they exhibit the distinct configurations which contribute to the two ($\pm$) phases. The proofs are based on the droplet model, which we discuss first.

In the d dimensional Ising model, configurations are specified by a function

$$\sigma_i: Z^d \to Z_2$$

which assigns $+$ or $-$ to each site $i \in Z^d$. At low temperature, there is a strong tendency for neighboring sites to align, and correspondingly an equivalent but distinct description of the configuration becomes more useful. Instead of focusing on the regions

$$X_\pm = \{i \in Z^d: \sigma_i = \pm 1\},$$

we focus on $\partial X \equiv \partial X_+ = \partial X_-$, which is the set of boundary elements across which σ_i changes sign. To be more precise, consider a bond (i.e., a line segment joining nearest neighbor sites i, i' for which $\sigma_i \sigma_{i'} = -1$). Then ∂X contains a lattice element of hypersurface, which is a normal bisector to this bond. (See Figure 5.4.) Each set of phase boundaries ∂X corresponds to exactly two configurations σ_i (interchanged by a global $\pm$ sign change).

At low temperature, the phase boundaries are a better description of the configuration, because there are few of them, and the effect of the phase boundaries can be understood as a perturbative correction to zero temperature, where there are only two allowed configurations ($\sigma_i \equiv +1$ and $\sigma_i \equiv -1$), and no phase boundaries. According to this picture, a typical configuration at low temperature is a sea of single sign (say $+$) in which one finds occasional isolated islands of the opposite sign. The islands are mainly small, and purely of the opposite sign, but with still smaller probability, there will be occasional larger islands, which may contain sub-islands ("lakes") of the original sign, etc. The configurations in which the dominant sign is $-$ contribute to the $-$ phase. Any infinite volume measure, constructed as a limit of finite volume measures, must, in this picture, be a mixture of the $+$ and $-$ phases as in (5.1.1). The choice of the mixture, i.e., the α in (5.1.1), depends on the details of the $V \uparrow \infty$ limit, especially the boundary conditions. For $\alpha = 0$ or $\alpha = 1$, the mixture is a pure phase. The case $d = 1$ is exceptional; for $d = 1$ the complement of a phase island is disconnected, and moreover there is no phase transition for temperatures $T > 0$ $(\beta < \infty)$.

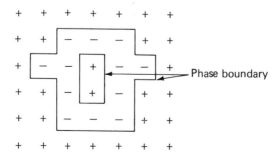

Figure 5.4.

To be quantitative, recall that the Ising model Hamiltonian

$$H_\Lambda = -\beta \sum_{nn} \sigma_i \sigma_{i'}$$

in the bounded region Λ defines the measure

$$d\mu_\Lambda = Z_\Lambda^{-1} e^{-H_\Lambda} \prod_{i \in \Lambda} \delta(\sigma_i^2 - 1) \, d\sigma_i,$$

$$Z_\Lambda = \int e^{-H_\Lambda} \prod_{i \in \Lambda} \delta(\sigma_i^2 - 1) \, d\sigma_i,$$

and the infinite volume measure $d\mu$ is constructed as in Section 4.2. Writing H as a function of a phase boundary γ, and its hypersurface area $|\gamma|$ (length for $d = 2$), we have

$$H = H(\gamma) = 2\beta |\gamma|,$$

where a constant

$$H_{\min} = \min_\sigma H(\sigma) = -\beta \sum_{nn} 1$$

has been subtracted from H. The constant cancels between numerator and denominator in $d\mu_\Lambda$, and thus the subtraction does not change the measure $d\mu_\Lambda$, nor the physical consequences of H.

Next consider other configurations, with phase boundaries ∂X, and let γ also denote the event (i.e. the set of configurations) for which $\gamma \subset \partial X$. Then

$$\Pr(\gamma) = \frac{\sum_{\partial X \supset \gamma} e^{-H(\partial X)}}{\sum_{\partial X} e^{-H(\partial X)}},$$

where ∂X runs over all phase boundaries, or in other words over all spin configurations.

Proposition 5.4.1. $\Pr(\gamma) \leq e^{-2\beta|\gamma|}$.

PROOF. If $\partial X \supset \gamma$, let $(\partial X)^*$ denote the configuration obtained by flipping all spins inside γ. The resulting phase boundary removes γ, or in other words

$$(\partial X)^* = \partial X \setminus \gamma.$$

Thus

$$e^{-H(\partial X)} / e^{-H((\partial X)^*)} = e^{-2\beta|\gamma|}.$$

The mapping $\partial X \leftrightarrow (\partial X)^*$ is a 1-1 map from phase boundaries ∂X containing γ to a subset of those not containing γ. In fact the image is exactly the phase boundaries ∂Y such that

$$\partial Y \cap \gamma = \emptyset,$$

since for such a ∂Y, we can again flip all spins inside γ to obtain $\partial X = \partial Y \cup \gamma$ and then $\partial Y = (\partial X)^*$. Accordingly in the inequalities

$$\Pr(\gamma) = \frac{\sum_{\partial X \supset \gamma} e^{-H(\partial X)}}{\sum_{\partial Y} e^{-H(\partial Y)}} \leq \frac{\sum_{\partial X \supset \gamma} e^{-H(\partial X)}}{\sum_{\partial Y \cap \gamma = \varnothing} e^{-H(\partial Y)}}$$

$$= \frac{\sum_{\partial X \supset \gamma} e^{-H(\partial X)}}{\sum_{(\partial X)^*} e^{-H((\partial X)^*)}} = e^{-2\beta|\gamma|},$$

we have majorized a ratio by omitting positive terms from the denominator.

Next we examine the effect of $+$ boundary conditions. Let $d\mu_\Lambda^+$ be the measure which results from the boundary condition $\sigma_i \equiv +1$ for $i \notin \Lambda$, and let $\langle \ \rangle_{+,\Lambda}$ be the expectation in this measure.

Proposition 5.4.2. *For β sufficiently large and $2 \leq d$,*

$$0 \leq 1 - \langle \sigma_i \rangle_{+,\Lambda} \leq e^{-\beta}.$$

PROOF. $1 - \langle \sigma_i \rangle = \langle 1 - \sigma_i \rangle$, and only configurations with $\sigma_i = -1$ contribute to this expectation. For any such configuration with phase boundary ∂X, there must be a $\gamma \subset \partial X$ enclosing the site i. Let $\gamma(\partial X)$ be the smallest such γ. Then

$$1 - \langle \sigma_i \rangle = 2 \frac{\sum_\gamma \sum_{\{\partial X: \, \gamma(\partial X) = \gamma\}} e^{-H(\partial X)}}{\sum_{\partial X} e^{-H(\partial X)}}$$

In the numerator we have summed first over ∂X's with a fixed smallest γ, and then over all such γ. We majorize the ratio by adding positive terms to the numerator, so that the first sum is over $\{\partial X: \partial X \supset \gamma\}$. This sum is the numerator in $\Pr(\gamma)$, and so

$$1 - \langle \sigma_i \rangle \leq 2 \frac{\sum_\gamma \sum_{\partial X \supset \gamma} e^{-H(\partial X)}}{\sum_{\partial X} e^{-H(\partial X)}}$$

$$= 2 \sum_\gamma \Pr(\gamma) \leq 2 \sum_\gamma e^{-2\beta|\gamma|}.$$

Let $N(|\gamma|)$ be the number of phase boundaries γ of area $|\gamma|$ (length in $d = 2$) which enclose the site i. There is a translational degree of freedom bounded by $|\gamma|^d$ in γ, since γ must be contained in a d dimensional cube of length $|\gamma|$ centered at i. Starting from any element of γ, and adding hypersurface elements one at a time, we construct γ in at most $c^{|\gamma|}$ distinct ways, where $c = c(d)$. Thus $N(|\gamma|) \leq |\gamma|^d c^{|\gamma|}$ and

$$1 - \langle \sigma_i \rangle \leq 2 \sum_{r=4}^{\infty} r^d c^r e^{-2\beta r} \leq e^{-\beta}$$

for β sufficiently large. This completes the proof.

Note that $d = 1$ is exceptional: for $d = 1$, $|\gamma|$ is the number of points in γ. In this case γ is disconnected and $N(|\gamma|)$ is unbounded for $|\gamma| = 2$.

The circle of ideas developed in this section also has consequences for continuum field theory, as we illustrate in Figure 5.5. Consider a $\phi^4 - \phi^2$ interaction as illustrated in Figure 5.1. The classical (zero temperature) ground state is defined by a constant field configuration. We illustrate typical field configurations contributing to the quantum (i.e. positive temperature) ground state. The dominant configurations are close to classical because of the $e^{-\mathscr{A}}$ factor in the statistical weight, but they contain various fluctuations. In Figure 5.5(a) we illustrate a configuration containing two types of fluctuations. The small scale fluctuations are the fluctuations within a single well of the W-shaped potential of Figure 5.1, and are described by the mean field theory of Section 5.2. The large scale fluctuations are the tunneling transitions between distinct wells, and are described by the Ising fluctuations and droplet model of this section. In Figure 5.5(b) we show a configuration belonging to a soliton (one particle) excitation of

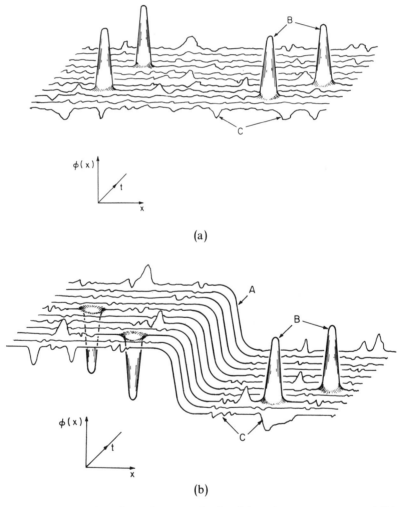

(a)

(b)

Figure 5.5. A typical configuration contributing (a) to the vacuum state and (b) to the one soliton state. Shown here is the soliton transition A, fluctuations to the other vacuum B, and fluctuations within a single vacuum C.

the vacuum. Here we start from a classical, time-independent (soliton) solution of the equation

$$\phi_{tt} - \phi_{xx} + P'(\phi) = 0,$$

namely $\phi(x) = a \tanh bx$. The illustrated configuration deviates from the classical solution by both small and large scale fluctuations, as in the case of the vacuum configuration. These figures provide the rationale for the quantum field phase transitions of Chapter 16 and for the low and high temperature cluster expansion of Chapters 18 and 20.5.

5.5 An Example

We end this chapter by showing that the mean field picture can describe other mechanisms for phase transitions than degeneracy of the ground state. In thermodynamics, a phase transition is said to occur whenever any thermodynamic function is nonanalytic. Such nonanalyticity may occur even though the ground state is unique. For example, in $d = 2$ systems with a continuous group of symmetries, the ground state is known to be symmetric (the Mermin–Wagner theorem). The simplest example of this situation is the rotator, or XY model, described by an Ising type measure of the form

$$d\mu_\Lambda = Z(\Lambda)^{-1} \exp\left[-\frac{\beta}{2} \sum_{\substack{i \in \Lambda \\ |i-i'|=1}} (\xi_i - \xi_{i'})^2\right] \prod_{i \in \Lambda} dv(\xi_i), \qquad (5.5.1)$$

where $\xi_i \in S^{(1)}$ is a unit vector in 2-space and $dv(\xi)$ is the uniform distribution on $S^{(1)}$. The group of symmetry transformations $U(1) = SO(2)$ is given by the simultaneous rotation of every spin vector ξ_i. The Mermin–Wagner theorem states that for all boundary conditions, $\lim_\Lambda d\mu_\Lambda = d\mu$ remains invariant under the symmetry group and ergodic under translations. See Section 16.3.

In spite of the uniqueness of the ground state, Kosterlitz and Thouless [1973] have given a simple and appealing picture which predicts a phase transition in this model for all $T < T_c$, with T_c some positive temperature. This phase transition can be interpreted on the one hand as arising from the degeneracy of the particle states for $T < T_c$. On the other hand, it has a mean field interpretation. In particular the configurational states of the XY model can be regarded as a superposition of two independent configurations. In angular variables, with $\xi = (\cos\theta, \sin\theta)$, $\theta = \theta_{sw} + \theta_v$ are the spin wave and vortex parts. The assumption that θ_{sw} and θ_v obey independent distribution laws is the mean field approximation. The energy of θ_{sw} is the kinetic energy of a $d = 2$, zero mass excitation, while the energy of the vortex part θ_v is the energy of a $d = 2$ dipole gas with Coulomb interaction. (The vortices always arise in pairs, to yield finite energy configurations, and each pair forms an elementary dipole.) We can then easily understand the pair correlation function for the rotator model. In the mean field approximation the energies add, so the expectations multiply:

$$\langle \xi_i \xi_j \rangle = \langle e^{i(\theta_i - \theta_j)} \rangle_{sw} \langle e^{i(\theta_i - \theta_j)} \rangle_v$$

$$= |i - j|^{-1/4\pi\beta} \langle e^{i(\theta_i - \theta_j)} \rangle_v. \tag{5.5.2}$$

Since the vortex expectation is bounded by 1, the mean field approximation yields vanishing at infinity of the two point function for all T, in agreement with the Mermin–Wagner theorem. However, the vortex expectation (Coulomb dipole gas) decays exponentially for high temperature (small β), yielding exponential decay of $\langle \xi_i \xi_j \rangle$. This is the disordered phase. At low temperature, $\beta \sim T^{-1} \gg 0$, one expects the dipoles to condense, yielding long range order for the vortex–vortex correlations. In this regime, $\langle \xi_i \xi_j \rangle$ decays polynomially, rather than exponentially. The correlation length $m(T)^{-1}$ describing the exponential decay: $\langle \xi_i \xi_j \rangle \sim \exp(-m|i-j|)$ thus diverges for $T < T_c$. One therefore expects the inverse correlation to behave as in Figure 5.6a. The figure is drawn to scale to represent the believed asymptotic behavior $m(T) \sim \exp(-c(T-T_c)^{-1/2})$ for $T \searrow T_c$ [Kosterlitz, 1974]. Compare this with the behavior $m(T) \sim |T - T_c|$ for the dimension 2 Ising model in Figure 5.6b. See Section 17.7 for a discussion of critical exponents and asymptotic behavior near T_c.

In summary, the picture of the rotator phase transition involves the inverse correlation length or mass, which vanishes for all $T \leq T_c$. In this region the long range order is zero, the ground state is unique, but there is no gap in the spectrum of the transfer matrix. The phase transition can be interpreted as a dipole condensation in the space of rotator states. At this time no mathematical argument has substantiated this picture due to Kosterlitz and Thouless, but it appears quite convincing.

Finally we mention that in certain gauge theories one conjectures that phase transitions are described by decay of the parallel transport operators.

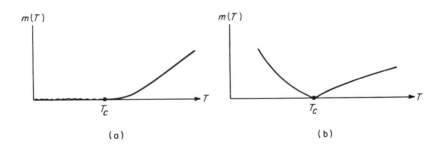

Figure 5.6. (a) The supposed behavior of the inverse correlation length for the dimension 2 rotator model, illustrating a line of critical points for $T < T_c$. (b) The inverse correlation length for the dimension 2 Ising model.

These phase transitions again occur in theories with a unique ground state, but have mean field interpretations in terms of vortex configurations of the theory.

REFERENCES

[Domb and Green, 1972-], [Ruelle, 1969], [Stanley, 1971].

CHAPTER 6

Field Theory

In this chapter we introduce the notion of quantum field on a technical level. This chapter is the end of the introductory material; it also can be regarded as the basic background for Part II. Readers already familiar with quantum theory and statistical mechanics may start here.

6.1 Axioms

(i) Euclidean Axioms

We define quantum fields in terms of their analytic continuation to imaginary time. This continuation changes the Minkowski metric, which defines the wave operator

$$\Box = -\frac{\partial^2}{\partial t^2} + \frac{\partial^2}{\partial x_1^2} + \frac{\partial^2}{\partial x_2^2} + \cdots + \frac{\partial^2}{\partial x_{d-1}^2}, \tag{6.1.1}$$

into the Euclidean metric (with the convention $x_d = ix_0$), which defines the Laplace operator

$$\Delta = \sum_{i=1}^{d} \frac{\partial^2}{\partial x_i^2}. \tag{6.1.2}$$

For this reason the imaginary time fields are called Euclidean fields. The Euclidean fields are defined by a probability measure $d\mu(\phi) = d\mu$ on the distribution space $\mathscr{D}'(R^d)$, with d the dimension of space-time. Here $d\mu$ plays the same role as does the Feynman–Kac measure in quantum

88

mechanics; see Chapter 3. Fields will be denoted ϕ. The q of Chapter 3 will be reserved for systems with a finite number of degrees of freedom.

We write

$$\phi(f) = \langle \phi, f \rangle = \int_{R^d} \phi(x) f(x)\, dx \tag{6.1.3}$$

for the canonical pairing between distributions $\phi \in \mathscr{D}'(R^d)$ and test functions $f \in \mathscr{D}(R^d) = C_0^\infty(R^d)$. The pointwise values $\phi(x)$ in the integral above have only formal significance. We introduce as generating functional

$$S\{f\} = \int e^{i\phi(f)}\, d\mu, \tag{6.1.4}$$

the inverse Fourier transform of a Borel probability measure $d\mu$ on $\mathscr{D}'(R^d)$; cf. Theorem 3.4.2.

We begin by listing axioms which characterize the probability measures $d\mu$ we are interested in. The axioms are slightly stronger than the related axioms of [Osterwalder and Schrader, 1973b, 1975]. These axioms are

OS0. Analyticity
OS1. Regularity
OS2. Euclidean invariance
OS3. Reflection (OS) positivity
OS4. Ergodicity

which we now explain in detail.

OS0 (Analyticity).   The functional $S\{f\}$ is entire analytic. Specifically, for every finite set of test functions $f \in \mathscr{D}(R^d)$, $j = 1, 2, \ldots, N$, and complex numbers $z = \{z_1, \ldots, z_N\} \in C^N$, the function

$$z \to S\left\{ \sum_{j=1}^{N} z_j\, f_j \right\}$$

is entire on C^N. In other words, $d\mu$ decays faster than any exponential.

OS1 (Regularity).   For some p, $1 \le p \le 2$, for some constant c, and for all $f \in \mathscr{D}(R^d)$,

$$|S\{f\}| \le \exp c(\|f\|_{L_1} + \|f\|_{L_p}^p). \tag{6.1.5}$$

In case $p = 2$, a further regularity axiom is assumed: The two point function (second moment of the measure $d\mu$) exists; as a function of the difference variables $x_1 - x_2$ it belongs to the space of locally integrable functions $L_1^{\mathrm{loc}}(R^d)$. In particular, it has only integrable singularities at coinciding points.[1]

---

[1] It turns out to be convenient to formulate axioms for $\phi \in \mathscr{D}'(R^d)$. With that assumption, we can establish the existence of $S\{f\}$ for $f \in C_0^\infty$, and afterwards prove the bound (6.1.5). A consequence of (6.1.5) is that $S\{f\}$ extends to a continuous functional on $\mathscr{S}(R^d)$, the Schwartz space of rapidly decreasing functions. By standard results for functional integrals, the measure $d\mu$ is therefore concentrated on $\mathscr{S}'(R^d)$ [Gelfand and Vilenkin, 1964]. In the case of free fields (i.e. Gaussian measures) with covariance operators $C : \mathscr{S} \to \mathscr{S}$, we often dispense with $\mathscr{D}'$ altogether and define $d\mu$ directly as a measure on $\mathscr{S}'$.

OS2 (Invariance). $S\{f\}$ is invariant under Euclidean symmetries E of R^d (translations, rotation, and reflections), i.e., $S\{f\} = S\{Ef\}$. Equivalently $d\mu$ is Euclidean invariant, so that $d\mu = E\,d\mu$.

Note that $E: \mathcal{D} \to \mathcal{D}$ is continuous, so E acts on $\mathcal{D}'(R^d)$ via $(E\phi)(f) = \phi(Ef)$. We use the exponential functionals

$$\mathcal{A} = \left\{ A(\phi) = \sum_{j=1}^{N} c_j \exp(\phi(f_j)), \; c_j \in C, f_j \in \mathcal{D} \right\}. \tag{6.1.6}$$

Hence an element $A \in \mathcal{A}$ maps $\phi \in \mathcal{D}'(R^d)$ into C.

By OS0, the functions in $\mathcal{A}$ are all integrable, and since $\mathcal{A}$ is an algebra, they are in L_p for all $p < \infty$. Since $d\mu$ is Euclidean invariant, Euclidean transformations define a continuous unitary group (with elements also denoted E) on $L_2(d\mu)$, and

$$(EA)(\phi) = A(E\phi), \qquad A \in \mathcal{A}. \tag{6.1.7}$$

OS3 (Reflection (OS) positivity). Let $\mathcal{A}_+ \subset \mathcal{A}$ be those $A(\phi)$ of the form (6.1.6) with $f_j \in C_0(R_+^d)$, i.e., supported in the half space $R_+^d = \{\mathbf{x}, t : t > 0\}$. Then we assume that the time reflection, $\theta : \{\mathbf{x}, t\} \to \{\mathbf{x}, -t\}$, satisfies

$$0 \le \langle \theta A, A \rangle_{L_2} = \int (\theta A)^{-} A \, d\mu. \tag{6.1.8}$$

Equivalently, in terms of $S\{f\}$, the matrix

$$M_{ij} = S\{f_i - \theta f_j\} \tag{6.1.9}$$

is positive (eigenvalues ≥ 0), for every finite sequence of functions $f_j \in \mathcal{D}_{\text{real}}(R^d)$.

OS4 (Ergodicity). The time translation subgroup $T(t)$ acts ergodically on the measure space $\mathcal{D}'(R^d, d\mu)$. In other words, for all L_1 functions $A(\phi)$,

$$\lim_{t \to \infty} \frac{1}{t} \int_0^t T(s)A(\phi)T(s)^{-1} \, ds = \int A(\phi) \, d\mu(\phi). \tag{6.1.10}$$

Note that (6.1.10) ensures the average on the left is independent of ϕ.

The axioms have physical significance, i.e., significance for the analytically continued Minkowski space quantum fields. Invariance (OS2) analytically continues to Lorentz invariance of the Minkowski quantum field when t is continued to physical values. The physical values of t are pure imaginary in the Euclidean notation used above, and so the physical region results from the substitution $\phi(t) \to \phi(-it)$ for t real.

Ergodicity is uniqueness of the vacuum. The regularity axiom is stronger than is necessary, but for scalar Boson fields seems to include the examples of interest. This axiom restricts the local singularities of the correlation functions. For a polynomial interaction of degree n, we can choose $p = n/(n-1) = n'$.

Let $\mathcal{E} = L_2(\mathscr{D}'(R^d), d\mu)$. $\mathcal{E}$ is the closure of the vectors (6.1.6) in the $L_2(d\mu)$ inner product.[2] $\mathcal{E}$ is not the quantum mechanical Hilbert space $\mathscr{H}$. Rather it is the path space for the quantum operators. Reflection positivity yields the positivity of the inner product in the Hilbert space $\mathscr{H}$ on which the Minkowski quantum field acts. We can interpret the time inversion $\mathscr{H}$ in the inner product (6.1.8) as arising from analytic continuation of the Hermitian conjugation $(e^{-itH})^* = e^{itH}$ at real time.

Formally, we have the following identifications with the language of Chapter 3:

$$\mathscr{D}'(R^d) = \text{path space},$$
$$d\mu = \text{Feynman-Kac measure on path space},$$
$$\mathscr{D}'(R^{d-1}) = \text{configuration space},$$
$$t \to \phi(\mathbf{x}, \cdot) = \text{a path with values in } \mathscr{D}'(R^{d-1}),$$
$$\mathscr{H} = L_2(\mathscr{D}'(R^{d-1}), dv) \text{ (Schrödinger representation)}.$$

Here dv is the measure defined by the ground state of the Hamiltonian operator, and dv coincides with the restriction of $d\mu$ to the time $t = 0$ subspace, i.e., to $\mathscr{H}$. The Schrödinger representation ($1 < d$) is known only for special cases.

It is convenient to bypass these formal relations and define $\mathscr{H}$ differently. Let $\mathcal{E}_+$ be the span in $\mathcal{E}$ of the vectors A in $\mathscr{A}_+$. Define the bilinear form b on $\mathcal{E}_+ \times \mathcal{E}_+$ by

$$b(A, B) = \langle \theta A, B \rangle_{L_2} = \int (\theta A)^- B \, d\mu = \langle \theta A, B \rangle_{\mathcal{E}}. \qquad (6.1.11)$$

By (6.1.8), b is positive. We call $\mathcal{E}_+$ the positive time subspace of $\mathcal{E}$. Let $\mathscr{N}$ be the set of vectors in $\mathcal{E}_+$ which are null in the inner product (6.1.11). Then $\mathscr{H}$ is the completion of the set $\mathcal{E}_+/\mathscr{N}$ of equivalence classes in the metric defined by (6.1.11).[3]

Proposition 6.1.1. *Let $d\mu$ be a probability measure on $\mathscr{D}'(R^d)$, and assume reflection positivity (OS3). Then $\mathscr{N}$ is a linear space, and (6.1.11) defines an inner product on $\mathcal{E}_+/\mathscr{N}$, which we denote $\langle \, , \, \rangle_{\mathscr{H}}$.*

PROOF. We need to show that if $A, B \in \mathcal{E}_+$ and $n \in \mathscr{N}$, then $b(A + n, B) = b(A, B)$. The Schwarz inequality

$$|b(A, B)| \equiv \left| \int \theta A^- B \, d\mu \right| \leq b(A, A)^{1/2} b(B, B)^{1/2}$$

is valid for the bilinear form b. The proposition follows from this fact.

---

[2] To see this, note that L_2 cylinder functions are dense in $L_2(\mathscr{D}', d\mu)$, so that we can restrict attention to a fixed finite dimensional subspace of $\mathscr{D}$. With $F(\phi)$ orthogonal to all exponentials, the Fourier transform $(Fd\mu)^\sim$ vanishes identically. Thus $Fd\mu$ is orthogonal to all continuous functions with compact support. The latter are dense in $L_2(dv)$ for any Radon measure dv, and so $F = 0$ as an element of L_2.

[3] Note that this definition of $\mathscr{H}$ coincides with the notion in probability theory of the conditional expectation of $\mathcal{E}_+$ onto $\mathcal{E}_- \equiv \theta\mathcal{E}_+$.

In order to distinguish vectors $A \in \mathscr{E}_+$ from the associated equivalence class $A + \mathscr{N} \in \mathscr{H}$, let $\wedge : \mathscr{E}_+ \to \mathscr{H}$ be the canonical imbedding, $A^\wedge = A + \mathscr{N}$ for $A \in \mathscr{E}_+$. Using $\wedge$, we can transfer certain operators S acting on $\mathscr{E}_+$ to become operators $S^\wedge$ acting on $\mathscr{H}$: The definition of $S^\wedge$ is

$$S^\wedge A^\wedge = (SA)^\wedge, \tag{6.1.12a}$$

as expressed by the following diagram:

$$
\begin{array}{ccc}
\mathscr{E}_+ & \xrightarrow{\ \ S\ \ } & \mathscr{E}_+ \\
\big\downarrow {\scriptstyle \wedge} & & \big\downarrow {\scriptstyle \wedge} \\
\mathscr{H} & \xrightarrow{\ \ S^\wedge\ \ } & \mathscr{H}
\end{array}
\tag{6.1.12b}
$$

Alternatively with

$$\langle A^\wedge, B^\wedge \rangle_{\mathscr{H}} = \langle \theta A, B \rangle_{\mathscr{E}}, \tag{6.1.12c}$$

define

$$\langle A^\wedge, S^\wedge B^\wedge \rangle_{\mathscr{H}} = \langle \theta A, SB \rangle_{\mathscr{E}}. \tag{6.1.12d}$$

For (6.1.12) to make sense, $S^\wedge$ must be defined on equivalence classes. In other words, we require that

$$S : \mathscr{D}(S) \cap \mathscr{E}_+ \to \mathscr{E}_+ \quad \text{and} \quad S : \mathscr{D}(S) \cap \mathscr{N} \to \mathscr{N}, \tag{6.1.13}$$

where $\mathscr{D}(S)$ is the domain of the possibly unbounded operator S.

Proposition 6.1.2. *Let $d\mu$ be a probability measure on $\mathscr{D}'$, and assume reflection positivity and reflection invariance of $d\mu$. Then $\wedge$ is a contraction. Thus $\|A^\wedge\|_{\mathscr{H}} \le \|A\|_{\mathscr{E}}$.*

PROOF.   By definition, for $A, B \in \mathscr{E}_+$,

$$\langle (A + \mathscr{N})^\wedge, (B + \mathscr{N})^\wedge \rangle_{\mathscr{H}} = \langle \theta A, B \rangle,$$

and so by the Schwarz inequality for $\mathscr{E}$,

$$\|(A + \mathscr{N})^\wedge\|_{\mathscr{H}}^2 = \langle \theta A, A \rangle_{\mathscr{E}} \le \|A\|_{\mathscr{E}} \|\theta A\|_{\mathscr{E}} = \|A\|_{\mathscr{E}}^2.$$

The next result defines the Hamiltonian, and thus the analytic continuation to Minkowski space-time. Let $T(E)$ be the unitary representation of the Euclidean group on $\mathscr{E}$ defined by the invariance axioms OS2. Also let $T(t)$ denote the time translation subgroup.

Theorem 6.1.3 (Reconstruction of quantum mechanics). *Let $d\mu$ be a probability measure on $\mathscr{D}'$. Assume reflection positivity and reflection and time translation invariance. Then for $0 \le t$, $T(t)$ satisfies (6.1.13) and*

$$T(t)^\wedge = e^{-tH}.$$

Here $0 \le H = H^*$, and for $\Omega = 1^\wedge$, $H\Omega = 0$. In other words, H is a positive self-adjoint operator with ground state Ω.

PROOF. Clearly $T(t) : \mathscr{E}_+ \to \mathscr{E}_+$. If $A \in \mathscr{N}$, then using the unitarity of T,

$$\langle \theta T(t)A,\, T(t)A \rangle_\mathscr{E} = \langle T(-t)\theta A,\, T(t)A \rangle_\mathscr{E}$$

$$= \langle \theta A,\, T(2t)A \rangle_\mathscr{E}$$

$$\le \langle \theta A,\, A \rangle_\mathscr{E}^{1/2} \langle \theta T(2t)A,\, \theta T(2t)A \rangle_\mathscr{E}^{1/2} = 0, \qquad (6.1.14)$$

by the Schwarz inequality for the form (6.1.11). Hence $T(t) : \mathscr{N} \to \mathscr{N}$, and $T(t)^\wedge$ is well defined. For convenience, write $R(t) = T(t)^\wedge$. We now verify four properties of $R(t)$:

i.   Semigroup law: $R(t)R(s) = R(t+s)$, s, $t \ge 0$.
ii. $R(t)$ is hermitian.
iii. $R(t)$ is a contraction: $\|R(t)\|_\mathscr{H} \le 1$.
iv. Strong continuity: $R(t) \to I$ as $t \to 0$.

These properties say that $R(t)$ is a strongly continuous, self-adjoint, contraction semigroup. Thus there exists a positive, self-adjoint operator H such that $R(t) = e^{-tH}$. Furthermore $T(t)1 = 1$. Thus for $\Omega \equiv 1^\wedge$,

$$e^{-tH}\Omega = (T(t)1)^\wedge = \Omega,$$

or $H\Omega = 0$.

Property i follows from the multiplication law for $T(t)$, namely by (6.1.12),

$$R(t)R(s) = (T(t)T(s))^\wedge = T(t+s)^\wedge = R(t+s).$$

For $A \in \mathscr{E}_+$,

$$\langle R(t)A^\wedge,\, A^\wedge \rangle = \langle (T(t)A)^\wedge,\, A^\wedge \rangle_\mathscr{H} = \langle \theta T(t)A,\, A \rangle_\mathscr{E} = \langle T(-t)\theta A,\, A \rangle_\mathscr{E}$$

$$= \langle \theta A,\, T(t)A \rangle_\mathscr{E} = \langle A^\wedge,\, (T(t)A)^\wedge \rangle_\mathscr{H} = \langle A^\wedge,\, R(t)A^\wedge \rangle_\mathscr{H},$$

and property ii follows.

Property iii is established as follows: By the Schwarz inequality and properties i, ii, for $A \in \mathscr{E}_+$ we have

$$\|R(t)A^\wedge\|_\mathscr{H} = \langle R(t)A^\wedge,\, R(t)A^\wedge \rangle_\mathscr{H}^{1/2} = \langle A^\wedge,\, R(2t)A^\wedge \rangle_\mathscr{H}^{1/2}$$

$$\le \|A^\wedge\|_\mathscr{H}^{1/2} \|R(2t)A^\wedge\|_\mathscr{H}^{1/2}.$$

Continuing to apply the Schwarz inequality in this fashion, we obtain after n steps

$$\|R(t)A^\wedge\|_\mathscr{H} \le \|A^\wedge\|_\mathscr{H}^{1-2^{-n}} \|R(2^n t)A^\wedge\|_\mathscr{H}^{2^{-n}}$$

$$= \|A^\wedge\|_\mathscr{H}^{1-2^{-n}} \|(T(2^n t)A)^\wedge\|_\mathscr{H}^{2^{-n}}$$

$$\le \|A^\wedge\|_\mathscr{H}^{1-2^{-n}} \|A\|_\mathscr{E}^{2^{-n}},$$

using Proposition 6.1.2 and the unitarity of $T(t)$. Letting $n \to \infty$ gives $\|R(t)A^\wedge\|_\mathscr{H} \le \|A^\wedge\|_\mathscr{H}$. Since the $A^\wedge$ are dense in $\mathscr{H}$, this proves property iii.

To establish property iv, we note that $T(t)$ is strongly continuous on $\mathscr{E}_+$ and $^\wedge$ is a contraction from $\mathscr{E}_+$ to $\mathscr{H}$. Thus $R(t)$ is strongly continuous on the dense subset of $\mathscr{H}$ of vectors $A^\wedge$, $A \in \mathscr{E}_+$. Since $\|R(t)\| \le 1$, it follows that $R(t)$ is strongly continuous on the entire Hilbert space $\mathscr{H}$, and the proof is complete.

Remark (Transfer matrix of statistical physics). In the case of lattice fields or statistical mechanics, the fields $\phi(x)$ or ξ_x are defined on a lattice, e.g., $x \in Z^d$. Hence the regularity and invariance axioms apply only in a modified sense. The reflection positivity condition is used to define the transfer matrix K which plays the role of e^{-H}.

We require invariance of $d\mu$ under the group $T(x)$ of lattice translations and lattice reflections. In particular, we are interested in reflection θ in a hyperplane Π parallel to and midway between two lattice hyperplanes. (We define Π as the $t = 0$ hyperplane.) Then the above proof shows there exists a Hilbert space of states $\mathcal{H}$ given by completion of $\mathscr{E}_+ / \mathcal{N}$ in the inner product

$$\langle A, B \rangle_{\mathcal{H}} = \int (\theta A)^- B \, d\mu.$$

Furthermore, there is a self-adjoint operator K on $\mathcal{H}$ which satisfies

$$K^t = T(t)^{\wedge}, \qquad t \in Z,$$

$$0 \le K \le I.$$

Also $\Omega = 1^{\wedge}$ is an invariant vector for K,

$$K\Omega = \Omega.$$

Likewise, space translations (in directions $\mathbf{x}$ orthogonal to the t axis) yield unitary operators

$$U(\mathbf{x}) = T(\mathbf{x})^{\wedge} = \prod_{i=1}^{d-1} U_i^{x_i},$$

with Ω as an invariant vector:

$$U(\mathbf{x})\Omega = \Omega.$$

The uniqueness of Ω (simplicity of 1 as an eigenvalue of K) is again equivalent to ergodicity of $d\mu$ under (lattice) time translations. It remains an open question for the lattice transfer matrix K when $\ln K$ exists. In other words, when is zero not an eigenvalue of K?

We now return to the continuum case and illustrate the use of the regularity axiom OS1 by constructing the Euclidean field $\phi(f)$.

Proposition 6.1.4. *Let $d\mu$ be a probability measure on $\mathcal{D}'$ which satisfies OS0. Then the measure $d\mu$ has moments of all orders. The nth moment has a density $S_n(x_1, \ldots, x_n) \in \mathcal{D}'(R^{nd})$ i.e.,*

$$\int \phi(f_1) \cdots \phi(f_n) \, d\mu = \int S_n(x_1, \ldots, x_n) \prod_{i=1}^{n} f_i(x_i) \, dx. \qquad (6.1.15)$$

Assume also OS1. Then $S_n \in L_1^{\text{loc}}(R^{nd})$.

Remark. The distributions S_n are called Schwinger functions.

PROOF. The operators $U(t) = e^{it\phi(f)}$, $f \in \mathscr{S}_{\text{real}}$, form a unitary group on $\mathscr{E}$. Their infinitesimal generator is the multiplication operator $\phi(f)$. The statement that the function 1 is in the domain of $\phi(f)^n$, for all such f, is equivalent to the statement that the moments of $d\mu$ of degree $2n$ exist. (The moments of odd degree are bounded by a Schwarz inequality by those of even degree.) To show that 1 is in the domain of $\phi(f)^n$, we use the definition $(\Delta/\Delta t)^n U(t)$ of the nth difference quotient of $U(t)$. Then

$$\left\| \left[\left(\frac{\Delta}{\Delta t_1} \right)^n - \left(\frac{\Delta}{\Delta t_2} \right)^n \right] U(t)1 \right\|^2$$

$$= \left[\left(\frac{\Delta}{\Delta t_1} \right)^n - \left(\frac{\Delta}{\Delta t_2} \right)^n \right] \left[\left(\frac{\Delta}{\Delta s_1} \right)^n - \left(\frac{\Delta}{\Delta s_2} \right)^n \right] S\{(s-t)f\} \bigg|_{s=t}.$$

Hence strong convergence of the nth difference quotient for $U(t)1$ follows from the analyticity (and hence differentiability) of S. Let f lie in a bounded, finite dimensional subset of $\mathscr{D}$. By continuity, $|S\{f\}|$ is bounded on the set. We then use the Cauchy integral formula,

$$S_n(f_1, \ldots, f_n) = \oint S\{g\} \prod_{j=1}^{n} \left(\frac{-dz_j}{2\pi z_j^2} \right),$$

where $g = \sum_{j=1}^{n} z_j f_j$ and the integrals extend over $|z_j| = 1$. This provides a bound on S_n establishing continuity on $\mathscr{D} \times \mathscr{D} \times \cdots \times \mathscr{D}$. The extension of S_n to $\mathscr{D}'(R^{nd})$ then follows from the nuclear theorem.

In case OS1 holds, S_n extends by continuity to a multilinear functional on $(L_1 \cap L_p) \times \cdots \times (L_1 \cap L_p)$. In particular, S_n is locally integrable as a function of nd variables.

(ii) Minkowski Space Axioms

By analytic continuation in the time variable, $t \to -it$, real time field operators and a representation of the Lorentz group can be constructed on $\mathscr{H}$. The Wightman and Haag–Kastler axioms can also be verified. The mathematical proof of these statements is rather technical, and so we state the results here and postpone the proof to Chapter 19.

Let $\phi = \phi_E(\mathbf{x}, t)$ denote the Euclidean field evaluated at the real Euclidean point $\mathbf{x}$, t as above. The Wightman and Haag–Kastler axioms concern the analytic continuation to real Minkowski t, $\mathbf{x}$, i.e., pure imaginary Euclidean t. To distinguish the fields at these distinct arguments, we write

$$\phi_E(\mathbf{x}, t) = \phi_M(it, \mathbf{x}),$$

and where the context is clear, ϕ may denote either ϕ_E or ϕ_M. With $x = (t, \mathbf{x})$ a vector in Minkowski space-time, the real time field operator $\phi_M(f)$ is written formally as

$$\int \phi_M(x) f(x) \, dx = \phi_M(f). \tag{6.1.16}$$

It turns out that (as an operator valued distribution)

$$\phi_M(\mathbf{x}, t) = e^{itH}\phi_M(\mathbf{x}, 0)e^{-itH}.$$

A consequence of the axioms OS1–3 is that $\phi_M(f)$ is self-adjoint for f real. (This property is not required in the Wightman axiom scheme, but it means that $\phi_M(f)$ is an observable in the sense of the postulates quantum mechanics.) It measures the strength of the field ϕ_M, as weighted over space time points x with the test function f. We now state the axioms.

W1 (Covariance). There is a continuous unitary representation of the inhomogeneous Lorentz group $g \to U(g)$ on the Hilbert space $\mathscr{H}$ of quantum theory states. The generators $(H = P_0, \mathbf{P})$ of the translation subgroup have spectrum in the forward cone $p_0^2 - \mathbf{p}^2 \geq 0$, $p_0 \geq 0$. There is a vector $\Omega \in \mathscr{H}$ (the vacuum) invariant under the operators $U(g)$.

W2 (Observables). There are field operators $\{\phi_M(f): f \in \mathscr{S}(R^d)\}$ densely defined on $\mathscr{H}$. The vector Ω is in the domain of any polynomial in the $\phi_M(f)$'s, and the subspace $\mathscr{D}$ spanned algebraically by the vectors $\{\phi_M(f_1) \cdots \phi_M(f_n)\Omega : 0 \leq n, f_i \in \mathscr{S}(R^d)\}$ is dense in $\mathscr{H}$. The field $\phi_M(f)$ is covariant under the action of the Lorentz group on $\mathscr{H}$, and depends linearly on f. In particular, $U(g)^*\phi_M(f)U(g) = \phi_M(f_g)$.

W3 (Locality). If the supports of f and h are spacelike separated, then $\phi_M(f)\phi_M(h) = \phi_M(h)\phi_M(f)$ on $\mathscr{D}$. ($(t, \mathbf{x})$ is spacelike if $|t| < |\mathbf{x}|$.)

W4. The vacuum vector Ω is the unique vector (up to scalar multiples) in $\mathscr{H}$ which is invariant under time translations.

An appeal to the Schwartz nuclear theorem shows that the field operators uniquely determine moments (Wightman n-point functions)

$$W_n = W(x_1, \ldots, x_n) = \langle \Omega, \phi_M(x_1) \cdots \phi_M(x_n)\Omega \rangle \in \mathscr{S}'(R^{nd}).$$

The axioms can equivalently be expressed as properties of the Wightman functions. The passage from Wightman functions to the field operators is the Wightman reconstruction theorem [Streater and Wightman, 1964]. Because of translation invariance, W_n can be regarded as a distribution in $\mathscr{S}'(R^{(n-1)d})$ in the difference variables $x_j - x_1$.

Theorem 6.1.5. *Let a measure $d\mu$ on $\mathscr{D}'(R^d)$ satisfy OS0–3. Then the real time field ϕ_M satisfies W1–3. Moreover OS4 is satisfied if and only if W4 is.*

PROOF. See Chapter 19.

Remark. The Schwinger functions and Wightman functions are related by analytic continuation. The precise statement, as proved in Chapter 19, is the following formula:

$$\int \phi_E(\mathbf{x}_1, t_1)\phi_E(\mathbf{x}_2, t_2) \cdots \phi_E(\mathbf{x}_n, t_n) \, d\mu$$

$$= \langle \Omega, \, \phi_M(it_1, \mathbf{x}_1)\phi_M(it_2, \mathbf{x}_2) \cdots \phi_M(it_n, \mathbf{x}_n)\Omega\rangle. \quad (6.1.17)$$

The real time Wightman functions are boundary values of analytic functions, and the Euclidean points to which they have been continued in (6.1.17) are points of analyticity whenever $x_i \neq x_j$ for all $i \neq j$.

Several standard methods of construction for quantum fields lead naturally to a measure $d\mu$ which satisfies OS0–3, but not necessarily OS4. The question of whether a specific measure $d\mu$ has a unique vacuum is usually difficult. It is tied up with the possible occurrence of a phase transition and the choice of boundary conditions in the definition of $d\mu$. There is a general theory which bypasses this difficult question and yields examples of the complete axioms OS0–4. The idea is to decompose any theory which satisfies OS0–3 into irreducible components, each with a unique vacuum, and each satisfying OS0–4, and thus W1–4. The components are labeled by a parameter ζ. They are the pure phase theories, defined on a Hilbert space $\mathcal{H}_\zeta$ such that

$$\mathcal{H} = \int \mathcal{H}_\zeta \, d\rho(\zeta).$$

Here $d\rho$ is a probability measure. Also

$$\langle \Omega, \, \phi_M(f_1) \cdots \phi_M(f_n)\Omega\rangle_{\mathcal{H}} = \int \langle \Omega_\zeta, \, \phi_{\zeta, M}(f_1) \cdots \phi_{\zeta, M}(f_n)\Omega_\zeta\rangle_{\mathcal{H}_\zeta} \, d\rho(\zeta)$$

and a similar decomposition exists for $d\mu$. Details are given in Section 19.7.

The Haag–Kastler axioms focus on the algebraic aspects of the fields, independently of their action on the specific Hilbert space $\mathcal{H}$. Thus they provide a convenient general framework for discussing $\mathcal{H}$-independent properties of the fields, such as the construction of charge superselection sectors and charge operators from observable fields and currents. The axioms are:

HK1. To each bounded open region B of space-time, there is an associated C^* algebra $\mathfrak{A}(B)$ containing the identity. The global algebra $\bigcup_B \mathfrak{A}(B)$ has a faithful, irreducible representation.

HK2. If $B_1 \subset B_2$, then $\mathfrak{A}(B_1) \subset \mathfrak{A}(B_2)$.

HK3 (Locality). If B_1 and B_2 are spacelike separated then $\mathfrak{A}(B_1)$ and $\mathfrak{A}(B_2)$ commute.

HK4 (Lorentz covariance).   Let $\{a, \Lambda\}$ be an element of the inhomogeneous Lorentz group. There is a * automorphism $\sigma_{\{a, \Lambda\}}$ of $\mathfrak{A}$ such that for a bounded region B

$$\sigma_{\{a, \Lambda\}}(\mathfrak{A}(\mathscr{B})) = \mathfrak{A}(\{a, \Lambda\}B).$$

The map $\{a, \Lambda\} \to \sigma_{\{a, \Lambda\}}$ is a representation of the Lorentz group.

We construct local algebras $\mathfrak{A}(B)$ which are also weakly closed (von Neumann algebras).

Theorem 6.1.6.   *Let $d\mu$ be a measure on $\mathscr{D}'$ satisfying OS0–3. Then for f real, $\phi_M(f)^-$ is a self-adjoint operator on $\mathscr{H}$. The von Neumann algebras $\mathfrak{A}(B)$ generated by $\exp(i\phi_M(f)^-)$ for f real, suppt $f \in B$, satisfy HK1–4.*

PROOF.   See Chapter 19.

6.2 The Free Field

The free field describes particles which do not interact. However, in spite of their trivial character, free fields play a role in the description of interacting particles. Any particle, in isolation from other particles (and external forces and fields), behaves like a free particle because there is nothing for it to interact with. Furthermore, in the $t \to \pm \infty$ asymptotic limit, the particles separate—partly due to differing velocities and partly due to the dispersion inherent in quantum mechanical time evolution. Thus as $t \to \pm \infty$, the particles behave like free particles, and are described by free fields.

The free field also serves as a starting point for calculation or construction of interacting fields, by perturbation in the coupling constant λ. In this respect, an important property of free fields is that they are explicitly solvable. Other explicitly solvable fields, such as the two dimensional Thirring field, are given as a nonlinear and nonlocal function of some free field. An alternate starting point for the construction of interacting fields is an Ising model or other statistical mechanics model at its critical point.

In the context of statistical mechanics, the Euclidean free field is known as the Gaussian model. It describes the asymptotic long distance critical behavior in the case of a canonical critical point.

We begin the analysis of free fields by applying the reconstruction theorem to a Gaussian measure $d\mu$ on $\mathscr{S}'(R^d)$. Let C be a positive continuous nondegenerate bilinear form on $\mathscr{S}(R^d) \times \mathscr{S}(R^d)$, written symbolically as

$$C(f, g) \equiv \langle f, Cg \rangle. \tag{6.2.1}$$

There is a unique Gaussian measure $d\phi_C$ on $\mathscr{S}'(R^d)$ with covariance (i.e., two point function) C, and mean zero. The generating function of $d\phi_C$ is given explicitly as

$$S\{f\} = e^{-\langle f, Cf \rangle / 2} = \int e^{i\phi(f)}\, d\phi_C. \tag{6.2.2}$$

From (6.2.2) we compute the moments of the measure $d\phi_C$, namely

$$\int \phi(f)^n\, d\phi_C = \left(-i\,\frac{d}{d\lambda}\right)^n S\{\lambda f\}\bigg|_{\lambda = 0}$$

$$= \begin{cases} 0, & n \text{ odd}, \\ (n-1)!!\, C(f, f)^{n/2}, & n \text{ even}, \end{cases} \tag{6.2.3}$$

where $n!! = n(n-2)(n-4)\cdots 1$. Observe that upon restriction to any finite number of coordinate directions in $\mathscr{S}'$, i.e., $q_j = \phi(f_j)$ for $j = 1, 2, \ldots, n$,

$$d\phi_C = (\det C)^{1/2} (2\pi)^{-n/2} e^{-(1/2)\sum_{ij} q_i C_{ij}^{-1} q_j} \prod_{i=1}^{n} dq_i, \tag{6.2.4}$$

where C is the matrix $C_{ij} = C(f_i, f_j) = \langle f_i, Cf_j \rangle$ and C^{-1} denotes the inverse matrix.

Evidently $d\phi_C$ and S satisfy time translation, reflection, or Euclidean invariance if and only if C does, and we now see that the same is true for reflection positivity.

Definition 6.2.1. A bilinear form C on $\mathscr{S}(R^d) \times \mathscr{S}(R^d)$ satisfies reflection positivity if for every $f \in \mathscr{S}(R^d)$ supported at positive times, $0 \le \langle \theta f, Cf \rangle$.

Theorem 6.2.2. *The Gaussian measure $d\phi_C$ satisfies reflection positivity if and only if C does.*

PROOF. Assume that C is reflection positive, and let

$$M_{ij} = S\{f_i - \theta f_j\} = S\{f_i\}S\{f_j\}\exp\langle \theta f_i, Cf_j \rangle,$$

with supp $t f_i \subset \{x : t > 0\}$. Then M_{ij} has positive eigenvalues (i.e. is a positive matrix) if and only if $N_{ij} = \exp\langle \theta f_i, Cf_j \rangle$ is a positive matrix. By assumption $R_{ij} = \langle \theta f_i, Cf_j \rangle$ is a positive matrix, and the positivity of $N_{ij} = e^{R_{ij}}$ follows: In fact, if A_{ij}, B_{ij} are any positive matrices, then the matrix $D_{ij} \equiv A_{ij} B_{ij}$ is also positive—from which we infer the positivity of $N_{ij} = \sum(1/r!)(R_{ij})^r$. To show that D is positive, first consider the tensor product matrix $A \otimes B$ with entries $A_{ij} B_{kl}$. Note $A \otimes B$ is positive. Hence $A \otimes B$ is also positive when restricted to the subspace of vectors α with components α_{jl} which vanish unless $j = l$. On this subspace $A \otimes B = D$.

The converse is valid for non-Gaussian measures also, and so we state it as a separate theorem.

Theorem 6.2.3. *Let $d\mu$ be a measure on $\mathscr{S}'(R^d)$ with generating functional $S\{f\}$. Assume that $S\{f\}$ is an entire analytic function of the complex variable f in the complex test function space $\mathscr{S}$. If $d\mu$ is time reflection positive, so is the two point function of $d\mu$.*

PROOF. Define M_{ij} as above and choose $f_1 = \lambda f$ to be real, $f_2 = 0$, $\alpha_1 = \lambda^{-1}$, $\alpha_2 = -\lambda^{-1}$. Then

$$0 \leq \sum_{i,j=1}^{2} \alpha_i M_{ij} \alpha_j \xrightarrow[\lambda \to 0]{} \int \phi(\theta f)\phi(f)\, d\mu$$

to complete the proof.

The Lehmann spectral formula (6.2.7), given below, characterizes two-point functions in a Lorentz-invariant field theory. Again the result is valid for non-Gaussian measures, and gives in the Gaussian case a complete classification of the measures $d\phi$ which yield a Lorentz invariant field theory. Let

$$\Theta(p_0) = \begin{cases} 1, & p_0 > 0, \\ 0, & p_0 < 0, \end{cases} \tag{6.2.5}$$

$$\Delta_m^+ = \int e^{ip \cdot x} \Theta(p_0)\, \delta(m^2 + p^2)\, dp, \tag{6.2.6}$$

where the Lorentz metric is $p \cdot x = \sum_{i=1}^{d-1} p_i x_i - p_0 x_0$.

Theorem 6.2.4. *Let the measure $d\mu$ on $\mathscr{S}'(R^d)$ satisfy OS0–3. Then the analytically continued Wightman function W_2 has a representation*

$$W_2(x, y) = \int_0^\infty d\rho(m^2)\, \Delta_m^+(x - y), \tag{6.2.7}$$

where $d\rho(m^2)$ is a positive measure of at most polynomial increase, and

$$\int_0^1 \ln m^{-2}\, d\rho(m^2) < \infty, \qquad d = 2,$$

$$\int_0^1 m^{-1}\, d\rho(m^2) < \infty, \qquad d = 1. \tag{6.2.8}$$

Furthermore, each such measure $d\rho$ arises for one and only one Gaussian measure $d\mu$ which is reflection-positive and invariant.

PROOF. By Section 6.1, $W_2 \in \mathscr{S}'(R^d)$ is a Lorentz-invariant distribution as a function of the difference variable $x - y$. The Fourier transform of W_2, as a function of the conjugate momentum dual to $x - y$ (i.e., the relative momentum), is a positive measure. This follows from the fact that

$$0 \leq \langle \phi(f)^* \phi(f) \rangle = \langle \phi(f^-)\phi(f) \rangle$$

for f in the complex test function space $\mathscr{S}(R^d)$. By the spectral condition, $0 \leq H$, and by the invariance assumption, this measure is supported in the forward cone $0 \le p_0^2 - \mathbf{p}^2$, $0 \le p_0$. Using invariance under the Lorentz group, the representation (6.2.7–8) follows. For Gaussian measures, $d\phi$ and S_2 determine one another uniquely, while in general S_2, W_2, and $d\rho$ determine one another uniquely. Thus there is at most one Gaussian measure corresponding to a given $d\rho$.

Now we let $d\rho$ be given, and we construct the corresponding Gaussian $d\phi$. First consider the case $d\rho(m^2) = \delta_{m^2}$. Then $W_2 = \Delta_m^+$. The $d\rho$ integration can be evaluated explicitly, after which analytic continuation $x_0 \to ix_0$ leads to the formula

$$S_2(x) = \int e^{ipx}(p^2 + m^2)^{-1}\, d\rho(m^2). \tag{6.2.9}$$

Thus we choose the covariance

$$C = (-\Delta + m^2)^{-1} \tag{6.2.10}$$

and construct $d\mu$ from (6.2.2).

For a general tempered measure $d\rho$, restricted by (6.2.8), we define

$$C = \int (-\Delta + m^2)^{-1}\, d\rho(m^2) \tag{6.2.11}$$

and again construct $d\mu = d\phi_C$ from (6.2.2). The convergence of (6.2.11) for small (respectively large) m^2 follows from the restrictions on $d\rho$ for such a ρ together with the known asymptotic behavior of the kernel $(-\Delta + 1)^{-1}$ for large (respectively small) x.

Definition. The quantum fields constructed from Gaussian measures $d\phi_C$ as above are called generalized free fields. In case $d\rho(m^2) = \delta_{m^2}$ as in (6.2.9–10), the field is called the free field with mass m and spin zero. It satisfies the free wave equation

$$(\partial_{x_\mu} \cdot \partial_{x_\mu} - m^2)\phi(x, t) = 0. \tag{6.2.12}$$

For the free field, the reconstruction of the physical Hilbert space of Section 6.1 can be made explicit. $\mathcal{H}$ is the L_2 space of a Gaussian measure $d\phi_0$ on $\mathscr{S}'(R^{d-1})$ defined by the time zero fields.[4] For $d = 1, 2$ we take $m^2 > 0$ in accordance with (6.2.8), although the zero mass free fields in $d = 1, 2$ dimensions can be constructed easily if a different test function space is chosen.

For $f \in \mathscr{S}(R^{d-1})$, let

$$f_t(x) = f(\mathbf{x})\,\delta_t(x_0). \tag{6.2.13}$$

and let $\mathscr{S}_0 = \{f \in \mathscr{S}(R^{d-1}): f^{\sim}(0) = 0\}$. (This last restriction allows zero mass for $d = 1, 2$.)

Proposition 6.2.5. *Let* $C = (-\Delta + m^2)^{-1}$. *Let* $\Delta = \sum_{\alpha=1}^{d-1} \partial_{x_\alpha}^2$, *and let*

$$\mu = (-\Delta + m^2)^{1/2}. \tag{6.2.14}$$

---

[4] The time zero field $\phi(h, 0)$ is $\phi(\mathbf{x}, t)$ restricted to the hyperplane $t = 0$. That $\phi(h, 0)$ is an L_p random variable follows for $d\phi_C$ introduced above. See Corollary 6.2.8, and Chapter 8 for related material.

Then for f, g ∈ $\mathscr{S}_0$ and for $0 \le s, t$,

$$\langle \theta g_s, C f_t \rangle = \tfrac{1}{2} \langle g, \mu^{-1} e^{-\mu(s+t)} f \rangle. \tag{6.2.15}$$

Moreover C is reflection-positive.

PROOF. The inner product on the left is in $L_2(R^d)$, while that on the right is in $L_2(R^{d-1})$. To establish the identity, we express the left side in Fourier transforms, evaluate the dp_0 integration by residues, and then invert the Fourier transform in the variables $p = p_1, \ldots, p_{d-1}$. For $f = g$ in $\mathscr{S}(R^d)$ and supported at positive times, the inner product in (6.2.15) is nonnegative, which implies C is reflection-positive.

Theorem 6.2.6. *With $d\mu$ Gaussian and $C = (-\Delta + m^2)^{-1}$ as above, and with $\mathscr{E}_+$ from Section 6.1, the vectors $e^{i\phi(f_0)}$, $f \in \mathscr{S}_0$, lie in $\mathscr{E}_+$. Furthermore $^\wedge$ maps the span of these vectors isometrically onto $\mathscr{H}$.*

PROOF. By direct calculation from (6.2.2), for f, g real,

$$\int |e^{i\phi(f)} - e^{i\phi(g)}|^2 \, d\phi_C = 2(1 - e^{-\langle f-g, C(f-g)\rangle/2}).$$

Choosing a sequence $g^{(n)} \in \mathscr{D}(R^d_+)$, $g^{(n)} \to f_t$, for $t > 0$ and $f \in \mathscr{S}_0$, we see that $e^{i\phi(f_t)} \in \mathscr{E}_+$ and similarly

$$e^{i\phi(f_0)} = \lim_{t \to 0} e^{i\phi(f_t)} \in \mathscr{E}_+.$$

With $f_j \in \mathscr{S}_0$, we see that $\theta\phi(f_{j,0}) = \phi(f_{j,0})$ and

$$\left\| \left(\sum c_j e^{i\phi(f_{j,0})} \right)^\wedge \right\|_{\mathscr{H}}^2 = \left\langle \sum_j (c_j e^{i\phi(f_{j,0})}), \sum_j c_j e^{i\phi(f_{j,0})} \right\rangle_{\mathscr{E}},$$

so that $^\wedge$ is isometric.

Let $\mathscr{E}_t$ be the subspace of $\mathscr{E}$ spanned by the vectors $e^{i\phi(f_s)}$, i.e., the time t subspace, and let $f^t = e^{-tH} f$ for $f \in \mathscr{S}_0$. By (6.2.2) and Proposition 6.2.5,

$$\langle A^\wedge, :\exp(i\phi(f_t)):^\wedge \rangle_{\mathscr{H}} = \langle A^\wedge, :\exp(i\phi(f_0^t)):^\wedge \rangle_{\mathscr{H}}. \tag{6.2.16}$$

This is proved first for $A = e^{i\phi(g_s)}$ and then for A in the span of $\bigcup_{t>0} \mathscr{E}_t$ by linear combinations and limits. Here $:\exp\phi(f): = \exp[\phi(f) - \tfrac{1}{2}\langle f, Cf \rangle]$. It follows that

$$(\mathscr{E}_0)^\wedge \supset \left(\bigcup_{t>0} \mathscr{E}_t \right)^\wedge.$$

The proof is now completed by showing that $\bigcup_{t>0} \mathscr{E}_t$ spans $\mathscr{E}_+$. Because the generating functional is an entire analytic function of $f \in \mathscr{S}(R^d)$ and of $f_t, f \in \mathscr{S}_0$, the moments of all orders exist. Thus the polynomials in $\{\phi(f) | f \in \mathscr{S}(R^d),$ supp$t f \subset R^d_+\}$ span $\mathscr{E}_+$. Approximating the integrals in dx_0 by Riemann sums, the same is true for polynomials in $\{\phi(f_t) | f \in \mathscr{S}_0, t > 0\}$. Hence $\bigcup_{t>0} \mathscr{E}_t$ spans $\mathscr{E}_+$.

Corollary 6.2.7. $e^{-tH} :\exp(i\phi(f_0)):^\wedge = :\exp(i\phi(f_0^t)):^\wedge$ *for* $f \in \mathscr{S}_0$.

PROOF. Combine (6.2.16) with the formula $T(t)e^{i\phi(f_0)} = e^{i\phi(f_t)}$ and the definition $e^{-tH} = T(t)^\wedge$ of H.

Corollary 6.2.8. *Identifying $\mathscr{H}$ with $\mathscr{E}_0$, the inner product in $\mathscr{H}$ is defined by a Gaussian measure on $\mathscr{S}'(R^{d-1})$ whose generating functional is*

$$S\{f\} = e^{-\langle f, (2\mu)^{-1}f\rangle/2}. \tag{6.2.17}$$

This result can be summarized in the formula

$$\mathscr{H} = L_2(\mathscr{S}'(R^{d-1}), d\phi_{1/(2\mu)}). \tag{6.2.18}$$

In this representation for $\mathscr{H}$, the time zero field operators act as multiplication operators, and hence are diagonalized. These field operators are the linear coordinate functions on $\mathscr{S}'(R^{d-1})$, and $\mathscr{S}'(R^{d-1})$ is the space of configurations of the classical field. Thus (6.2.18) provides a quantization of the free field which is the immediate generalization of the quantum mechanics with a finite number of degrees of freedom discussed in Chapter 1. See Section 6.4.

The mechanics defined by Corollaries 6.2.7–8 is the mechanics of a system of uncoupled oscillators. In fact any Gaussian measure can be regarded as a tensor product of one dimensional Gaussian measures, in the same sense that any self-adjoint operator is a direct sum or direct integral of one dimensional operators by the spectral theorem. Specifically, in the case of (6.2.18), μ^{-1} is diagonalized by the Fourier transform in the space variables $\mathbf{p} \in R^{d-1}$. The decomposition into Fourier (normal) modes is technically easier in the case of a discrete direct sum and integral, and so we consider only that case.

Let Δ_D denote the Laplace operator with zero Dirichlet data on the boundary and exterior of the strip

$$-\infty \leq t \leq \infty, \qquad 0 \leq x_i \leq L.$$

Also let $\mathbf{\Delta}_D = \Delta_D - d^2/dt^2$ be the corresponding spatial Laplacian, and let $\mu_D = (-\mathbf{\Delta}_D + m^2)^{1/2}$. The functions

$$e_{\mathbf{k}}(\mathbf{x}) = \left(\frac{2}{L}\right)^{(d-1)/2} \prod_{\alpha=1}^{d-1} \sin(k_\alpha x_\alpha), \qquad k_\alpha \in \frac{\pi}{L}Z_+ ,$$

are eigenfunctions of $-\mathbf{\Delta}_D$, and a complete set of eigenfunctions of $\mathbf{\Delta}_D$ and μ_D, when the latter act on $L_2([0, L]^{d-1})$. The eigenvalue of $\mathbf{\Delta}_D$ corresponding to $e_{\mathbf{k}}$ is

$$\lambda_{\mathbf{k}} = \mathbf{k} \cdot \mathbf{k}.$$

Starting with the Gaussian free field defined by the covariance $-\Delta_D^{-1}$, we construct the Euclidean and quantum mechanical Hilbert space $\mathscr{E}_D$ and $\mathscr{H}_D$, and as above show that

$$\mathscr{H}_D = L_2(\mathscr{S}'(R^{d-1}, d\phi_{(2\mu_D)^{-1}})).$$

Diagonalizing μ_D as above, we write

$$\mathscr{H}_D = \underset{\mathbf{k} \in ((\pi/L)Z_+)^{d-1}}{\bigtimes} \mathscr{H}_{\mathbf{k}}$$

and

$$\mathscr{H}_{\mathbf{k}} = L_2(R, d\phi_{2^{-1}(\mathbf{k}^2 + m^2)^{-1/2}}).$$

In other words, $\mathscr{H}_{\mathbf{k}}$ is the L_2 space of the one dimensional Gaussian measure with covariance $\frac{1}{2}(\mathbf{k}^2 + m^2)^{-1/2}$. We also decompose $\mathscr{E}_D$ as a tensor product

$$\mathscr{E}_D = \underset{\mathbf{k} \in ((\pi/L)Z_+)^{d-1}}{\text{X}} \mathscr{E}_{\mathbf{k}},$$

and inverting the argument leading from (6.2.10) to (6.2.14), we see that

$$\mathscr{E}_{\mathbf{k}} = L_2(\mathscr{S}'(R), d\phi_{C(\mathbf{k})}),$$

where

$$C(\mathbf{k}) = \left(-\frac{d^2}{dt^2} + \mathbf{k}^2 + m^2\right)^{-1}.$$

In other words, $\mathscr{E}_{\mathbf{k}}$ and $\mathscr{H}_{\mathbf{k}}$ are respectively the Euclidean and quantum mechanical Hilbert spaces for a single harmonic oscillator.

6.3 Fock Space and Wick Ordering

To continue the analysis of the free field, we want to find the particles in

$$\mathscr{H} = L_2(\mathscr{S}'(R^{d-1}), d\phi_{(2\mu)^{-1}}). \tag{6.3.1}$$

As in the case of one degree of freedom (Section 1.5), it is equivalent to expand $\mathscr{H}$ in terms of Hermite polynomials, since the n-particle states are exactly the span of the degree n Hermite polynomials. This expansion leads to the familiar construction of Fock space. This Hermite–Fock representation applies to any Gaussian measure, and so for increased generality we consider the Gaussian measure $d\phi_C$ of (6.2.2). As technical preparation, we need an integration by parts formula. Let

$$\left\langle f, C\frac{\delta}{\delta\phi} \right\rangle = \int f(x)C(x, y)\frac{\delta}{\delta\phi(y)} dx\, dy \tag{6.3.2}$$

Here $C(x, y)$ is the integral kernel of the covariance operator C. Also $\delta/\delta\phi(y)$ is the derivative with respect to ϕ; see Section 9.1.

Theorem 6.3.1. *Let $A(\phi)$ be a polynomial function defined on $\mathscr{S}'(R^d)$, and let C be a continuous bilinear form on $\mathscr{S}(R^d) \times \mathscr{S}(R^d)$. Then*

$$\int \phi(f)A(\phi)\, d\phi_C = \int \left\langle f, C\frac{\delta}{\delta\phi} \right\rangle A(\phi)\, d\phi_C. \tag{6.3.3}$$

PROOF. We first prove the formula for $A = e^{i\phi(g)}$. The case of polynomials follows by sums and limits. Let

$$F(\lambda) = \int e^{i\phi(g + \lambda f)}\, d\phi_C = e^{-\langle g + \lambda f, C(g + \lambda f)\rangle/2}.$$

Then

$$i \int \phi(f) e^{i\phi(g)} \, d\phi_C = F'(0) = -\langle f, Cg \rangle e^{-\langle g, Cg \rangle/2}$$

$$= -\langle f, Cg \rangle \int e^{i\phi(g)} \, d\phi_C$$

$$= i \int \left\langle f, C \frac{\delta}{\delta\phi} \right\rangle e^{i\phi(g)} \, d\phi_C \, .$$

Remark. Integration by parts extends by continuity to a much larger class of functions $A(\phi)$, and is very useful for a variety of purposes. See Section 9.1.

Definition 6.3.2. Let $\mathscr{S}$ be a real pre-Hilbert space. A representation of the canonical commutation relations over $\mathscr{S}$ is a pair of linear maps

$$f \to a(f), \qquad g \to a^*(g)$$

from $\mathscr{S}$ to operators $a(f)$ and $a^*(g)$ defined on a dense domain $\mathscr{D}$ in a complex Hilbert space $\mathscr{H}$ such that

$$a(f)\mathscr{D} \subset \mathscr{D}, \qquad a^*(g)\mathscr{D} \subset \mathscr{D}, \tag{6.3.4a}$$

$$\langle A_1, a(f)A_2 \rangle = \langle a^*(f)A_1, A_2 \rangle, \tag{6.3.4b}$$

$$[a(f), a(g)] = [a^*(f), a^*(g)] = 0, \tag{6.3.4c}$$

$$[a(f), a^*(g)]A = \langle f, g \rangle A \tag{6.3.4d}$$

for all A, A_1, A_2 in $\mathscr{D}$ and all f, g in $\mathscr{S}$. The representation is called a Fock representation if there is a unit vector $\Omega \in \mathscr{H}$ such that

$$a(f)\Omega = 0$$

for all $f \in \mathscr{S}$ and such that $\mathscr{D}$ is the algebraic span of the vectors $a^*(f_1) \cdots a^*(f_n)\Omega$, $n = 0, 1, \ldots$.

EXAMPLE. Let $\mathscr{F}_n$ be the space of symmetric L_2 functions on R^{nd}; $\mathscr{F}_0$ is the complex numbers. Let

$$\mathscr{F} = \sum_{n=0}^{\infty} \mathscr{F}_n, \qquad \Omega = 1 \in \mathscr{F}_0. \tag{6.3.5a}$$

Let S_n be the symmetrizing projection of $L_2(R^{nd})$ onto $\mathscr{F}_n$, and let $\mathscr{D}$ be the algebraic span in $\mathscr{F}$ of Ω and vectors of the form

$$f(x_1, \ldots, x_n) = S_n \, f_1(x_1) \cdots f_n(x_n),$$

where $f_j \in \mathscr{S}(R^d)$ and $n = 1, 2, \ldots$. Then with f given as above, we define a, a^* on $\mathscr{D}$ by the formulas

$$(a^*(g)f)(x_1, \ldots, x_{n+1}) = (n+1)^{1/2} S_{n+1} g(x_{n+1})f(x_1, \ldots, x_n),$$

$$(a(g)f)(x_1, \ldots, x_{n-1}) = n^{1/2} \int g(x_n)f(x_1, \ldots, x_n) \, dx_n. \tag{6.3.5b}$$

A direct calculation shows that (6.3.5) defines a Fock representation of the canonical commutation relations.

For the second example of a Fock representation, take $\mathscr{H} = L_2(\mathscr{S}'(R^d), d\phi_C)$, and let $\Omega = 1$, while $\mathscr{D}$ is the space of polynomial functions on $\mathscr{S}'(R^d)$. Let $\phi(f)$ (which is a linear coordinate function on $\mathscr{S}'(R^d)$) act as a multiplication operator on $\mathscr{H}$, with the domain $\mathscr{D}$. Then

$$\phi(f)\mathscr{D} \subset \mathscr{D} \subset \mathscr{H}.$$

As in (6.3.2), let

$$\pi(f) = -i\left\langle f, \frac{\delta}{\delta\phi} \right\rangle + 2^{-1}i\phi(C^{-1}f).$$

If C^{-1} is an operator from $\mathscr{S}(R^d)$ into $\mathscr{S}(R^d)$, then

$$\pi(f)\mathscr{D} \subset \mathscr{D}$$

also, and one can check that as operators on $\mathscr{D}$, ϕ and π are symmetric and satisfy

$$[\phi(f), \pi(g)] = i\langle f, g \rangle I, \tag{6.3.6a}$$

$$[\phi(f), \phi(g)] = 0 = [\pi(f), \pi(g)]. \tag{6.3.6b}$$

Theorem 6.3.3. *Suppose that C has a square root $C^{1/2}$ and that $C^{\pm 1/2}$ are operators from $\mathscr{S}(R^d)$ onto $\mathscr{S}(R^d)$, and define*

$$a(f) = 2^{-1}\phi(C^{-1/2}f) + i\pi(C^{1/2}f), \tag{6.3.7a}$$

$$a^*(f) = 2^{-1}\phi(C^{-1/2}f) - i\pi(C^{1/2}f). \tag{6.3.7b}$$

Then a, a^* define a Fock representation of the canonical commutation relations.

PROOF. The commutation relations for a, a^* follow from those for ϕ, π. Since $\Omega \equiv 1$ and

$$a(f) = \left\langle C^{-1/2}f, \frac{\delta}{\delta\phi} \right\rangle$$

we have $a(f)\Omega = 0$. The adjoint relation between a and a^* is a consequence of the facts that ϕ is symmetric (as a real multiplication operator) and that π is symmetric (by Theorem 6.3.1).

Finally we obtain the Hermite expansion from the identifications

$$\mathscr{H} = L_2(\mathscr{S}'(R^d), d\phi_C) = \mathscr{F}$$

allowed by the uniqueness of the Fock representation proved below.

Theorem 6.3.4. *If $\{a_i, a_i^*\}$, $i = 1, 2$, are two Fock representations of the canonical commutation relations over $\mathscr{S}$, with vacuums Ω_i, then they are unitarily equivalent, and moreover the equivalence operator U is uniquely determined if we also require $U\Omega_1 = \Omega_2$.*

PROOF. Let

$$A_i = a_i^*(f_1) \cdots a_i^*(f_n)\Omega_i ,$$

$$B_i = a_i^*(g_1) \cdots a_i^*(g_m)\Omega_i .$$

We compute $\langle A_1, B_1 \rangle = \langle A_2, B_2 \rangle$ so that U defined by $U A_1 = A_2$ extends by linearity to a unitary operator from $\mathscr{H}_1$ to $\mathscr{H}_2$ giving the required equivalence. If U' is any other unitary equivalence operator with $U'\Omega_1 = \Omega_2$, then necessarily $U'A_1 = A_2$, and so $U = U'$ and U is unique.

From the commutation relations and the fact that $a_i \Omega_i = 0$,

$$\langle A_i , B_i \rangle = \sum_{j=1}^{m} \langle f_1, g_j \rangle$$

$$\times \langle a_i^*(g_1) \cdots a_i^*(g_{j-1})a_i^*(g_{j+1}) \cdots a_i^*(g_m)\Omega_i \rangle \, a_i^*(f_2) \cdots a_i^*(f_n)\Omega_i \rangle.$$

Thus by induction on m, $\langle A_1, B_1 \rangle = \langle A_2, B_2 \rangle$.

Definition 6.3.5. $\mathscr{F}_n$ is the n particle subspace of $\mathscr{F}$, and if $f = \{f_n\}_{n=0}^{\infty} \in \mathscr{F}$, with $f_n \in \mathscr{F}_n$, then the operator

$$Nf = \{nf_n\}_{n=0}^{\infty}$$

with domain

$$\mathscr{D}(N) = \{f : f \in \mathscr{F}, \sum n^2 \|f_n\|^2 < \infty\}$$

is called the number operator.

The Fock representation facilitates the introduction of orthogonal (Hermite) polynomials. Let E_n be the orthogonal projection of $\mathscr{F}$ onto $\mathscr{F}_n$. Because $\phi = c^{1/2}(a^* + a)$, degree n polynomials span $\sum_{j=0}^{n} \mathscr{F}_j$. Thus the orthogonalization of the degree n monomials can be defined by the formula

$$:\phi(f_1) \cdots \phi(f_n): = E_n \phi(f_1) \cdots \phi(f_n), \tag{6.3.8}$$

and in particular

$$:\phi(f)^n: = c^{n/2} P_n(c^{-1/2}\phi(f)), \tag{6.3.9}$$

where $P_n(x)$ is the Hermite polynomial in a single variable, given by (1.5.12), (1.5.16), and where the normalization

$$c = \langle f, Cf \rangle = \int \phi(f)^2 \, d\phi_C$$

replaces the normalization $\langle Q^2 \rangle = \frac{1}{2}$ of (1.5.18). We extend the Wick dots $: :$ by linearity to polynomials and convergent power series, so that by (1.5.12)

$$:e^{\phi(f)}: = \sum \frac{:\phi(f)^n:}{n!} = e^{-c/2}e^{\phi(f)}. \tag{6.3.10}$$

Much of the combinatorics of Wick monomials is summarized by

$$\int :e^{\phi(f)}: :e^{\phi(g)}: \, d\phi_C = e^{-(\langle f, Cf \rangle + \langle g, Cg \rangle)/2} \int e^{\phi(f+g)} \, d\phi_C. \tag{6.3.11}$$

In particular, with $f = \sum \alpha_i f_i$, $g = \sum \beta_j g_j$, we apply

$$\prod_{i=1}^{n} (d/d\alpha_i) \prod_{j=1}^{m} (d/d\beta_j)$$

to both sides of (6.3.11) and then evaluate at $\alpha = \beta = 0$. This procedure yields the identity

$$\int : \prod_{i=1}^{n} \phi(f_i) : : \prod_{j=1}^{m} \phi(g_j) : d\phi_C$$

$$= \delta_{nm} \sum_{\pi \in \mathfrak{S}_n} \langle f_1, C g_{\pi(1)} \rangle \cdots \langle f_n, C g_{\pi(n)} \rangle, \quad (6.3.12)$$

where $\mathfrak{S}_n$ is the group of $n!$ permutations π on n elements. The comparison with the Fock representation is given by the formula

$$: \prod_{i=1}^{n} \phi(f_i) : \Omega = \prod_{i=1}^{n} (a^*(C^{1/2} f)) \Omega. \quad (6.3.13)$$

We return to the free field covariance $C = (-\Delta + m^2)^{-1}$ and (6.2.18) to write

$$\mathcal{H} = \mathcal{E}_0 = \mathcal{F} = L_2(\mathcal{S}'(R^{d-1}), d\phi_{(2\mu)^{-1}}), \quad (6.3.14)$$

where $\mathcal{F} = \sum \mathcal{F}_n$ is the Fock representation of the time zero fields, defined by the Gaussian measure on $\mathcal{S}'(R^{d-1})$ with covariance $(2\mu)^{-1}$, $\mu = (-\Delta + m^2)^{1/2}$. Let μ_j denote μ acting on the jth variable of $f_n \in \mathcal{F}_n$. For $f \in \mathcal{D}(H)$, where

$$\mathcal{D}(H) = \left\{ f \in \mathcal{F} : f_n \in \mathcal{D}(\mu_j), \sum_{n=1}^{\infty} \left\| \sum_{j=1}^{n} \mu_j f_n \right\|^2 < \infty \right\}, \quad (6.3.15a)$$

let

$$Hf = \{Hf_n\}_{n=0}^{\infty} = \left| \sum_{j=1}^{n} \mu_j f_n \right|_{n=0}^{\infty}. \quad (6.3.15b)$$

Theorem 6.3.6. *The free field Hamiltonian, in the Fock representation (6.3.14), is given by (6.3.15), and moreover*

$$e^{-itH} f_n = \left(\prod_{j=1}^{n} e^{-it\mu_j} \right) f_n. \quad (6.3.16)$$

Remark. Equation (6.3.16) expresses the fact that each of the particles in f_n moves independently of the other $n - 1$ particles under the free field dynamics, and furthermore its motion coincides with the single particle dynamics $e^{-it\mu}$ in $\mathcal{F}_1$. Symbolically $H = \int \mu(k) a^*(k) a(k) \, dk$.

PROOF. With all functions taken at time zero,

$$e^{-itH} e^{i\phi(f)} = \exp(i\phi(e^{-it\mu} f))$$

by analytic continuation and Corollary 6.2.7. Thus by (6.3.10),

$$e^{-itH} : e^{i\phi(f)} := \exp[(\langle e^{-it\mu}f, (2\mu)^{-1}e^{-it\mu}f \rangle - \langle f, (2\mu)^{-1}f \rangle)/2] : \exp(i\phi(e^{-it\mu}f)):$$

Because $e^{-it\mu}$ is unitary, the first exponential on the right does not contribute. Expanding according to (6.3.10) gives

$$e^{-itH} : \phi(f)^n := : \phi(e^{-it\mu}f)^n:,$$

and using (6.3.14) to identify $\mathscr{F}$ and $L_2(\mathscr{S}'(R^{d-1}))$, we conclude (6.3.16).

Theorem 6.3.7. *On $\mathscr{H} = \mathscr{F}$, the free field Hamiltonian can be written*

$$H = \int_{t=0} H(\mathbf{x}) \, d\mathbf{x} = \int_{t=0} a^*(\mathbf{x})\mu a(\mathbf{x}) \, d\mathbf{x},$$

where the energy density is

$$H(\mathbf{x}) = \tfrac{1}{2} : \pi^2(x): + \tfrac{1}{2} : (\nabla\phi)^2(x): + \tfrac{1}{2}m^2 : \phi^2(x):.$$

PROOF. Expanding $a(\mathbf{x})^*a(\mathbf{x})$ using (6.3.7) yields $H(\mathbf{x})$, using $C = (2\mu)^{-1}$.

6.4 Canonical Quantization

We begin by summarizing Sections 6.2–3 for the case $d = 1$, and comparing the construction of $\mathscr{H}$, $\mathscr{F}$, H, π, etc. with Chapters 1, 3. In fact, the abstract construction of Fock space reproduces the quantum mechanics of a harmonic oscillator. In general, different $d = 1$ measures would yield the quantum mechanics of other systems with one degree of freedom. With a vector valued $\phi(t)$, we could obtain all the Schrödinger Hamiltonians of Chapter 1.

In particular, we begin with the Gaussian measure $d\phi_C$ on $\mathscr{S}'(R)$ with mean zero and covariance $C = (-d^2/dt^2 + m^2)^{-1}$. By Corollary 6.2.8, we obtain for $\mathscr{H}$ the Hilbert space

$$\mathscr{H} = L_2(R, dv), \tag{6.4.1}$$

where dv is the Gaussian measure on R with mean zero and covariance $(2\mu)^{-1} = 1/2m = \text{const.}$ In the standard representation,

$$dv(q) = (m/\pi)^{1/2}e^{-mq^2} \, dq \tag{6.4.2}$$

where dq is Lebesgue measure, and $\Omega = 1$. We now write down the corresponding representation of the canonical commutation relations, defined in Theorem 6.3.3 and shown unique in Theorem 6.3.4. To use these theorems, we substitute $(2\mu)^{-1} = (2m)^{-1}$ for C in (6.3.7), q for ϕ, p for π, etc. Then

$$a = \frac{1}{\sqrt{2}}(m^{1/2}q + im^{-1/2}p),$$
$$\tag{6.4.3}$$
$$a^* = \frac{1}{\sqrt{2}}(m^{1/2}q - im^{-1/2}p),$$

and inverting,

$$q = (2m)^{-1/2}(a^* + a),$$
$$p = (m/2)^{1/2}i(a^* - a). \tag{6.4.4}$$

Also

$$p = p^* = -i\frac{d}{dq} + imq, \qquad [p, q] = -i,$$

$$a = (2m)^{-1/2}\frac{d}{dq}, \qquad a^* = (2m)^{1/2}q - (2m)^{-1/2}\frac{d}{dq}, \tag{6.4.5}$$

$$[a, a^*] = 1.$$

These are the 'canonical' commutation relations (CCR). By (6.3.9) and $c = (2m)^{-1}$,

$$a^{*n}\Omega = P_n(q\sqrt{2m}) \tag{6.4.6}$$

is an eigenstate of H with eigenvalue nm. Comparing our formulas with Section 1.5, we find they agree completely for $m = 1$ in the representation where $\Omega_0 \equiv 1$. Thus the Gaussian measure for $d = 1$ gives the harmonic oscillator quantum mechanics. Written in terms of $(q, d/dq)$, the Hamiltonian ma^*a is

$$H_{\text{osc}} = -\frac{1}{2}\frac{d^2}{dq^2} + mq\frac{d}{dq}. \tag{6.4.7}$$

To relate H to its more familiar form on $L_2(R, dq)$, use the similarity transformation

$$H_{\text{Leb}} = \exp\left(-\frac{mq^2}{2}\right)H\exp\left(\frac{mq^2}{2}\right) = \tfrac{1}{2}m\left(-\frac{d^2}{dq^2} + q^2 - 1\right). \tag{6.4.8}$$

This concludes our discussion of the Gaussian measure and the canonical quantization of a harmonic oscillator.

Let us consider now a perturbation of H, and establish the Feynman–Kac formula.

Theorem 6.4.1. *Let $d\mu$ be the Gaussian, $d = 1$ measure above, let $V(q)$ be real, continuous, and bounded below, and let $H + V$ be self-adjoint. Then*

$$e^{-t(H+V)} = \left(\exp\left[-\int_0^t V(\phi(s))\,ds\right]T(t)\right)^{\wedge}$$

$$= \left(\exp\left[-\int_0^t V(\phi(s))\,ds\right]\right)^{\wedge}e^{-tH}. \tag{6.4.9}$$

PROOF. Write

$$\int_0^t V(\phi(s))\,ds = \lim_{n\to\infty}\frac{t}{n}\sum_{j=0}^{n-1}V\left(\phi\left(\frac{jt}{n}\right)\right),$$

so

$$\exp\left[-\int_0^t V(\phi(s))\, ds\right] T(t) = \lim_{n\to\infty}\left(e^{-(t/n)V(\phi(0))} T\left(\frac{t}{n}\right)\right)^n. \tag{6.4.10}$$

But

$$\left[\left(e^{-(t/n)V(\phi(0))} T\left(\frac{t}{n}\right)\right)^n\right]^{\wedge} = (e^{-(t/n)V(q)} e^{-(t/n)H})^n.$$

The right side converges to $e^{-t(H+V)}$, using (3.2.4) as $n \to \infty$. The left side converges pointwise on $\mathscr{E}$ to (6.4.10). Thus (6.4.9) holds.

Furthermore, let $\Omega(q)$ be the ground state vector of $H + V$ in the representation for which $H1 = 0$. Then $(H + V)\Omega = E\Omega$. With

$$p = -i\frac{d}{dq} + imq - i\frac{d}{dq}\ln \Omega(q), \tag{6.4.11}$$

the operators p, q yield a representation of the canonical commutation relations on $L_2(R, \Omega^2\, dv)$ and are self-adjoint. The measure

$$d\mu = \lim_{t\to\infty} Z(t)^{-1} \exp\left[-\int_{-t}^t V(q(s))\, ds\right] d\phi_C \tag{6.4.12}$$

also determines this representation of the canonical commutation relations. The Hamiltonian defined by this measure is $H + V - E$.

Given the canonical commutation relations, we notice that for the case of one degree of freedom,

$$\phi(t)^{\wedge} = e^{-tH}qe^{tH} \tag{6.4.13}$$

and

$$q(t) \equiv e^{itH}qe^{-itH} \tag{6.4.14}$$

are related by the analytic continuation $t \to -it$. Note that (6.4.14) satisfies

$$\dot{q}(t) = p(t)/m = e^{itH}[iH, q]e^{-itH}$$

(in units with $\hbar = 1$), while the second derivative gives Newton's equation

$$m\ddot{q}(t) = F(q(t)),$$

with $F(q) = -[iV, p] = -dV/dq$ and where V is the total potential energy (assumed a function of q).

We now generalize these ideas to fields, using the analysis of free fields in Sections 6.2 and 6.3. The construction of the Hilbert space $\mathscr{H} = \mathscr{E}_+/\mathscr{N}$ of quantum mechanical states in Section 6.1 can be simplified, at least on a formal level. For the free field, $\mathscr{H}$ is an L_2 space formed out of functions of the time $t = 0$ fields (Section 6.2). We expect the same statement is true for $P(\phi)_2$ and ϕ_d^4 fields, $d \leq 4$, and accordingly assume

$$\mathscr{H} = L_2(\mathscr{S}'(R^{d-1}), dv) \tag{6.4.15}$$

where dv is the measure on $\mathscr{S}'(R^{d-1})$ defined by restriction from the Euclidean vacuum state Ω and measure $d\mu$. Thus

$$\int F(\phi)\, dv = Z^{-1} \int F(\phi) e^{-\mathscr{A}_I(\phi)}\, d\phi_C = \int F(\phi)\, d\mu, \qquad (6.4.16)$$

where Z is the normalizing factor and

$$\mathscr{A}_I(\phi) = \lambda \int : \phi^4(x) : d^d x$$

(with appropriate counterterms implied; see Sections 9.4, 14.3). Formally, we go one step further and write the Gaussian integral $d\phi_C$ as a density times Lebesgue measure on $\mathscr{S}'(R^d)$, so that with a new Z

$$d\mu = Z^{-1} e^{-\mathscr{A}(\phi)} \prod_{x \in R^d} d\phi(x), \qquad (6.4.17)$$

where

$$\mathscr{A}(\phi) = \int : (\tfrac{1}{2} \nabla\phi^2(x) + \tfrac{1}{2} m^2\phi^2(x) + \lambda\phi^4(x)) : d^d x. \qquad (6.4.18)$$

The classical ϕ^4 field (in Minkowski space) is a solution of the nonlinear hyperbolic wave equation

$$-\Box\phi + m^2\phi + 4\lambda\phi^3 = 0. \qquad (6.4.19)$$

The field ϕ is uniquely determined by its Cauchy data:

$$\phi|_{t=0}, \qquad \partial_t\phi = \phi|_{t=0}. \qquad (6.4.20)$$

This set of pairs of functions is thus the state space of a classical nonlinear system defined by (6.4.19). The configuration space is the set functions $\phi|_{t=0}$ (time zero field configurations). The precise class of functions $\phi|_{t=0}$ is not determined by this reasoning, but from the point of view of integration theory, as in (6.4.15), the only essential requirement is to choose a big enough space. $\mathscr{S}'(R^{d-1})$ is very big, and as it turns out, big enough.

The quantum mechanical ϕ is the multiplication operator acting on functions of classical field configurations, and the conjugate momentum π is the generator of a unitary group, which is a similarity transform of the translation group $\phi(x) \to \phi(x) + g(x)$.

Since ϕ and the conjugate momentum $\pi = -i\,\delta/\delta\phi + f(\phi)$ are a complete set of observables, it is possible to express the Hamiltonian H and the measure $d\mu | \mathscr{S}'(R^{d-1})$ in terms of them. We have solved this problem already for the free field measure $d\phi_C$ in Section 6.2, and specifically in Theorem 6.3.7 for the free field Hamiltonian H_0. Then (6.4.16) is the Feynman–Kac formula for a perturbation of the Hamiltonian H_0. To make this relation clear, we write $\phi = \phi(t)$ (suppressing the dependence on $\mathbf{x} \in R^{d-1}$), $\mathscr{A}_I(\phi(t)) = \lambda \int : \phi^4(x) : d^{d-1}\mathbf{x}$, and

$$\mathscr{A}_I(\phi) = \int_0^\infty \mathscr{A}_I(\phi(t))\, dt + \int_{-\infty}^0 \mathscr{A}_I(\phi(t))\, dt.$$

Let Ω_0 be the ground state for H_0, and Ω the ground state for $H = H_0 + \mathscr{A}_I(\phi(t = 0))$. Then (6.4.16), combined with the Feynman–Kac formula (Sections 3.2–4), states that

$$\int F(\phi)\, dv = \int F(\phi)\, d\mu$$

$$= Z^{-1} \exp\left(-\int_{-\infty}^{0} \mathscr{A}_I(\phi(t))\, dt\right) F(\phi) \exp\left(-\int_{0}^{\infty} \mathscr{A}_I(\phi(t))\, dt\right) d\phi_C$$

$$= \lim_{t \to \infty} Z^{-1} \langle \exp[-t(H_0 + \mathscr{A}_I(\phi(t = 0)))]\Omega_0,$$

$$F(\phi) \exp[-t(H_0 + \mathscr{A}_I(\phi(t = 0)))]\Omega_0 \rangle = \langle \Omega, F(\phi)\Omega \rangle.$$

Thus the abstract vacuum state defined by the reconstruction of Section 6.1 coincides with ground state Ω of the Hamiltonian H defined by

$$H = \int \mathscr{H}(\mathbf{x})\, d^{d-1}\mathbf{x},$$

where

$$\mathscr{H}(\mathbf{x}) = \,:\!(\tfrac{1}{2}\pi(x)^2 + \tfrac{1}{2}\nabla\phi(\mathbf{x}) + \tfrac{1}{2}m^2\phi(\mathbf{x})^2 + \lambda\phi(\mathbf{x})^4)\!:\,.$$

Similarly the abstract dynamics of Section 6.1 is given by H, in analogy with Chapter 3.

6.5 Fermions

Particles are divided into two groups by spin and statistics. Electrons and quarks have half-integer spin and Fermi–Dirac statistics; they are called fermions. The same applies to bound states made out of an odd number of fermions, such as the proton, neutron, or H^3 nucleus. The photon has integer spin and Bose–Einstein statistics, as do bound states composed of an even number of fermions, such as mesons (in a quark–gluon theory), the H^2 nucleus, or Cooper pairs of electrons in a superconductor. In Sections 6.2, 6.3 we described the free Klein–Gordon field, defined by a spin zero representation of the Lorentz group and by boson statistics. Here we describe the quantization of the free Dirac field, defined by a spin $\frac{1}{2}$ (double valued) representation of the Lorentz group and by fermion statistics.

There are two main ideas in this section. The first is the construction of the fermion Fock space and representation of the canonical anticommutation relations, and the second is the use of antiparticles to eliminate negative energy states in the quantization of the Dirac equation. See Section 15.3 for the Dirac equation itself.

Given a Hilbert space $\mathscr{F}_1$, let $\mathscr{F}_n$ be the antisymmetric tensor product of n copies of $\mathscr{F}_1$,

$$\mathscr{F}_n = A_n \otimes^n \mathscr{F}_1, \tag{6.5.1}$$

where A_n is the antisymmetrizing projection operator. Then the fermion Fock space is

$$\mathscr{F} = \sum_{n=0}^{\infty} \mathscr{F}_n. \tag{6.5.2}$$

The subspace $\mathscr{F}_n \subset \mathscr{F}$ defines states with n particles. Normally $\mathscr{F}_1$ is an L_2 space of vector valued functions from R^d or R^{d-1} to C^v for some v, and then $\mathscr{F}_n$ is the subspace of vector (C^{nv}) valued $L_2(R^{nd})$ or $L_2(R^{n(d-1)})$ functions which are antisymmetric under permutation of variables. The annihilation and creation operators are defined as in (6.3.5b), with S_{n+1} replaced by A_{n+1}. For a Lorentz-invariant free field theory, the Lorentz or Euclidean groups act on $\mathscr{F}$ and on each $\mathscr{F}_n$ by a tensor product of the transformation law defined on the single particle space $\mathscr{F}_1$. In other words, the transformation law for n-particle states is specified as transforming each of the n particles in this state independently, according to the $\mathscr{F}_1$ transformation law. The finite dimensional space C^v is called spin space. It carries a representation of SU(2) (the covering group of the space rotation group O(3), contained in the Lorentz group). This representation is labeled by the dimension v, and the spin is $(v-1)/2$. The scalar boson fields of Sections 6.2, 6.3 had $v=1$ and spin zero. The photon has spin one ($v=3$).

The free field is completely specified by its single particle space $\mathscr{F}_1$, its statistics, and the group invariance, acting on $\mathscr{F}_1$, as we have seen above. We now give this construction explicitly. The annihilation and creation operators a and a^* are defined by the formulas

$$a^*(g)f(x_1, \ldots, x_{n+1}) = (n+1)^{1/2} A_{n+1} g(x_{n+1}) f(x_1, \ldots, x_n),$$

$$a(g)f(x_1, \ldots, x_{n-1}) = n^{1/2} \int g(x_n)^- f(x_1, \ldots, x_n)\, dx_n. \tag{6.5.3}$$

Here $f \in \mathscr{F}_n$ and $g \in \mathscr{F}_1$. In particular f and g have a spin index (implicit in (6.5.3)) for each variable x_j, and the $\int \cdots dx_n$ inner product includes a summation over these indices. A direct calculation establishes the anticommutation relations

$$\{a(f), a(g)\} = \{a^*(f), a^*(g)\} = 0, \tag{6.5.4a}$$

$$\{a(f), a^*(g)\} = \langle f, g \rangle, \tag{6.5.4b}$$

where the inner product is in $\mathscr{F}_1$, and

$$\langle A_1, a(f)A_2 \rangle = \langle a^*(f)A_1, A_2 \rangle \tag{6.5.4c}$$

with the inner product in $\mathscr{F}$, and $A_1, A_2 \in \mathscr{F}$. Note that $\mathscr{F}_0$ is one dimensional, consisting of all scalar multiples of the vacuum state Ω, and

$$a(f)\Omega = 0, \qquad a^*(g)\Omega = g \in \mathscr{F}_1. \tag{6.5.4d}$$

A simple algebraic consequence of (6.5.4b) is the inequality

$$0 \le a(f)a^*(f) \le \{a(f), a^*(f)\} = \|f\|^2,$$

so that $a(f)$ and $a^*(f)$ are bounded operators on $\mathscr{F}$.

The fact that a and a^* are bounded operators is in distinction to the boson case, and is easy to understand from the point of view of physics. The boson operators a and a^* become unbounded due to the presence of an unbounded number of boson particles which may occupy the same state $f \in \mathscr{F}_1$. For bosons, $a(f)$ and $a^*(f)$ have the approximate size $n^{1/2}$ for quantum field states which contain n particles, each in the state $f \in \mathscr{F}_1$. For fermions, the maximum value of n is 1. The antisymmetric Fermi–Dirac statistics gives rise to the Pauli exclusion principle, whereby each state θ in the Fock space $\mathscr{F}$ contains at most one particle in a given state $f \in \mathscr{F}_1$. It is easy to see this from (6.5.4). In fact,

$$a(f)^2\theta = \tfrac{1}{2}\{a(f), a(f)\}\theta = 0$$

for any $\theta \in \mathscr{F}$, so that $a(f)\theta$ has no particles in the state $f \in \mathscr{F}_1$ and θ has at most one such particle.

Linear transformations of the a's and a^*'s among themselves are called Bogoliubov transformations. These transformations are useful for a variety of purposes. They allow linear quantum field problems (e.g., a field interacting with an external source or potential) to be expressed in terms of a free field and the solution of a (classical) partial differential equation (in $\mathscr{F}_1$). We discuss here a specific Bogoliubov transformation which occurs in the quantization of the Dirac field. The transformation is

$$\begin{aligned} a^*(f) &\to a(f), \\ a(f) &\to a^*(f). \end{aligned} \tag{6.5.5}$$

With the transformation acting on states, rather than the operators a and a^*, we obtain the transformed version of (6.5.4d):

$$a^*(f)\Omega' = 0, \qquad a(g)\Omega' = g' \in \mathscr{F}'_1, \tag{6.5.4d'}$$

where Ω', g', and $\mathscr{F}'_1$ are transformed states. Just as (6.5.4d) identifies Ω as the no particle state, (6.5.4d') identifies Ω' as the *all* particle state. When $\mathscr{F}_1$ is infinite dimensional, as is normally the case, the transformed Fock space $\mathscr{F}'$ is disjoint from $\mathscr{F}$, in the sense that the actions of a and a^* on $\mathscr{F}$ and $\mathscr{F}'$ are unitarily inequivalent. $\mathscr{F}'_1$ is thought of as a space of holes, and Ω' a state of the field in which all particle states are occupied. $\mathscr{F}'$ is the Fock space generated by states with a finite number of holes (created by the hole creation operator a), just as $\mathscr{F}$ is the Fock space generated by states with a finite number of particles.

The Dirac field quantization is based on neither $\mathscr{F}$ nor on $\mathscr{F}'$, but on a mixture of these two spaces. If we write

$$\mathscr{F}_1 = \mathscr{F}_1^{(\mathrm{pos})} \oplus \mathscr{F}_1^{(\mathrm{neg})} \tag{6.5.6}$$

to show the decomposition of $\mathscr{F}_1$ under the single particle Dirac energy operator into positive and negative energy states, then the Dirac field Fock space $\mathscr{F}''$ is

$$\mathscr{F}'' = \sum_{m,\,n} \mathscr{F}_m^{(\mathrm{pos})} \otimes \mathscr{F}_n^{(\mathrm{neg})'}. \tag{6.5.7}$$

In other words, $\mathscr{F}''$ is spanned by particles from positive energy states in $\mathscr{F}_1^{\text{pos}}$ and holes (also called antiparticles) from negative energy states in $\mathscr{F}_1^{(\text{neg})}$. This leads to the equations

$$a(f)\Omega'' = 0, \qquad a^*(f)\Omega'' = f'' \in \mathscr{F}_1^{(\text{pos})''},$$

$$a^*(g)\Omega'' = 0, \qquad a(g)\Omega'' = g'' \in \mathscr{F}_1^{(\text{neg})''}, \qquad (6.5.4d'')$$

and the interpretation that Ω'' is the Dirac sea: a state in which all negative energy particles are present, but no positive energy particles are.

In order to make the decomposition (6.5.6) explicit, we must introduce the Dirac energy operator (see Section 15.3). However, for the present we merely observe that the interpretation of classical negative energy states as antiparticles with *positive* energy gives a quantum energy operator which is positive. In fact, if $E(f)$ is the classical energy of a state f, then

$$E(f)a^*(f)a(f) = -E(f)a(f)a^*(f) + E\|f\|^2$$

by the anticommutation relations. The additive constant $E\|f\|^2$ can be subtracted as a normalization, and then for $E < 0$, $-Eaa^*$ is a positive operator.

6.6 Interacting Fields

The main interest lies in interacting fields. The program of constructing quantum fields increases in difficulty with the dimension d of space-time. Here we present a program which is intended to be adequate for the physical case ($d = 4$). It is not an introduction to Part II, where the construction is carried out for $d = 2$. In fact the restriction to $d = 2$ introduces simplifications which are utilized in Part II.

The interacting fields are constructed from lattice fields (Chapters 2 and 4) with lattice spacing ε, in the limit $\varepsilon \to 0$. There is a trivial rescaling which multiplies all lengths (lattice spacing, correlation length, etc.) by s. This rescaling is a unitary transformation and maps theories (with lengths) into distinct but equivalent theories. Taking advantage of this point of view, we see that it is immaterial whether we consider the limit

$$\text{lattice spacing} \to 0, \qquad \text{correlation length} = \text{const} \qquad (6.6.1)$$

or the limit

$$\text{lattice spacing} = \text{const}, \qquad \text{correlation length} \to \infty. \qquad (6.6.2)$$

The latter of these equivalent points of view defines the critical point of a lattice theory. The trivial rescaling, which maps it onto the former, defines a scaling limit of the critical point. Thus we see that a field theory can be constructed as a scaling limit of a lattice field theory, and that this construction is equivalent to the conventional construction of taking the lattice spacing to zero. The resulting field theory may still have several length parameters, as we know from $P(\phi)_2$ theories, and when all of these except the correlation length are set equal to zero, we obtain what is known as *the scaling limit*.

We illustrate these ideas with the familiar one dimensional examples. The lattice theory describes a random walk, while the continuum limit, a one dimensional field theory, is a continuous time stochastic process with a harmonic or anharmonic oscillator as its generator. One length is the correlation length, and a second length scale is defined by the maximum distance on which the process is essentially harmonic (Gaussian). When this second length scale is taken to zero, the scaling limit is obtained [Isaacson, 1977]. It is not harmonic on any distance scale, and is very much simpler than the ϕ_1^4 anharmonic oscillator process. In fact it is the Poisson process. Its generator is a 2×2 matrix.

In dimensions $2 \leq d$, the ϕ_d^4 scaling limit should be regarded as a generalization of the Poisson process. Just as the Poisson process describes random events (points, or intervals between points) on the line, the ϕ_d^4 scale limits should describe random events (connected sets) in R^d. It is presumed that this field theory coincides with the scaling limit of the Ising model critical point. For $d = 2$, the latter has been exhibited in closed form in a remarkable asymptotic calculation [McCoy, Tracy, and Wu, 1977].

In the limits (6.6.1)–(6.6.2), the critical point is Gaussian (also called canonical, or trivial) if and only if the corresponding field theory is Gaussian (i.e., free, see Sections 6.2–3). Thus the existence problem for nontrivial quantum fields may be reformulated as: construct a scaling limit for a nontrivial critical point. We see that there is an underlying identity of mathematical structure in these two problems—quantum fields and the critical point—which arise in very distinct parts of physics.

The existence of quantum fields has been established for a variety of two dimensional problems, and for the ϕ_3^4 ($d = 3$) interaction. In addition, a considerable amount of detailed structure has been extracted, including, in favorable cases, verification of the Wightman axioms, an S matrix not equal to I, asymptotic validity and Borel summability of perturbation theory, phase transitions, and critical exponent inequalities.

For the ϕ_4^4 interaction, a number of partial results related to existence have been obtained (see [Glimm and Jaffe, 1974d, 1975c]). In fact, a bound on the renormalized two point function, uniform in ε, would complete the proof; this bound is a consequence of a conjectured correlation inequality. The ϕ_4^4 interaction should require a coupling constant renormalization. This means that $\lambda\phi^4$ occurs in the action, with the parameter $\lambda = \lambda(\varepsilon)$ dependent on ε. As ε changes, the nature of the lattice critical point may change, and in particular the length scale on which there is nontrivial critical behavior may change. At $\lambda = 0$, the critical point (as well as the entire theory) is Gaussian, while for ε fixed at $\lambda = \infty$, the lattice theory coincides with the Ising model. Let $\lambda_{\text{phys}}(\lambda, \varepsilon)$ be the physical coupling constant, defined for example as in Chapter 14. If

$$\limsup_{\varepsilon \to 0} \sup_{\lambda \geq 0} \lambda_{\text{phys}}(\lambda, \varepsilon) = 0.$$

then the resulting ϕ_4^4 field theory will be trivial; otherwise construction of a nontrivial theory would be expected. At present the arguments favoring triviality seem to be stronger, but a definitive answer seems to be out of reach of available methods. These arguments apply equally to the four-dimensional Yukawa and electrodynamic (QED) interactions. If these interactions are all trivial, it would mean that a short distance cutoff resulting from the quark interactions is essential to a theory of protons, photons, mesons, and electrons as elementary particles. Since it is known experimentally that protons and mesons are not elementary particles, but composites formed from quarks and gauge fields ("gluons"), such a short distance cutoff, set at the proton radius, for example, would not violate experimental facts of physics.

Most current work on the $d = 4$ existence problem is devoted to gauge fields. Gauge fields are believed to have two advantages, both from the point of view of mathematics and from the point of view of physics. The first advantage is called asymptotic freedom, and means that the short distance behavior is approximately Gaussian, as with the ϕ_1^4 anharmonic oscillator, and in contrast to the Poisson process. The second advantage is called quark confinement, and is reflected in the idea that the relevant critical point occurs at zero temperature, as with many one dimensional processes.

By definition, a classical gauge field is a connection form on a principle bundle, in the sense of differential geometry. In other words, a gauge field is (locally) a vector valued function, with values in a Lie algebra, and a covariant vector transformation law.

Such a field A gives rise to a curvature two form

$$F = DA \equiv dA + A \wedge A, \tag{6.6.3}$$

where D is the covariant derivative defined by A. The action is then

$$\mathscr{A} = \int \mathrm{Tr}\, F^2(x)\, dx, \tag{6.6.4}$$

and the quantization problem is to define a suitable measure on the space of connections, formally proportional to

$$d\mu = e^{-\mathscr{A}} \prod_x dA(x). \tag{6.6.5}$$

We note that (6.6.5) is not Gaussian because (6.6.4) is not quadratic in its dependence on A. Classical gauge fields are solutions of the Yang–Mills equation

$$D*F = 0. \tag{6.6.6}$$

This equation is the Euler equation, and it results from setting the first variation $\delta\mathscr{A}$ to zero in (6.6.4). We resume this discussion in Section 20.9.

In conclusion, it is clear that $d = 4$ quantum fields remain an open problem, from the point of view of mathematics as well as of physics. If extrapolations from the past are used to predict the future, we may expect that further efforts to solve this problem may lead to interesting mathematical structures, as well as a deeper understanding of the mathematics and physics of systems with an infinite number of degrees of freedom.

REFERENCES

[Bjorken and Drell, 1964–5], [Kastler, 1961], [Schweber, 1961], [Bogoliubov and Shirkov, 1959], [Thirring, 1958], [Itzykson and Zuber, 1980].

PART II

FUNCTION SPACE INTEGRALS

Part II provides a self-contained construction of certain non-Gaussian measures on function space. The examples satisfy the axioms of Chapter 6 and yield nonlinear quantum fields. The material is developed in logical order. Conceptual topics of broad interest are interspersed with technical material special to this construction. The material of broader interest concerns Feynman diagrams, perturbation theory, and calculus on path space. It is located at the beginnings of Chapters 8–10 and can be read independently of the rest of Part II. Some aspects of the construction given here are technically new, especially the use of nonsymmetric multiple reflections in Chapters 10 and 12.

The technical estimates of Part II are developed for $P(\phi)_{d=2}$ boson interactions. The vacuum energy estimates (Chapter 8) have been extended with renormalization to $\phi_{d=3}^4$ and $Yukawa_{d=2,3}$ models. Formally they hold for all super-renormalizable interactions. The remaining methods presented in Part II—multiple reflections and monotonicity—are dimension independent. The multiple reflection methods, as well as estimates uniform in the volume of interaction, also extend to particles with spin. But monotonicity properties and the proof of convergence of the infinite volume limit (e.g., properties based on correlation inequalities) are spin dependent. In Chapter 18 we establish the convergence of the infinite volume limit for certain values of the coupling constants by an alternative method which is independent of spin.

CHAPTER 7

Covariance Operator
= Green's Function
= Resolvent Kernel
= Euclidean Propagator
= Fundamental Solution

7.1 Introduction

The covariance operators C for the free field Gaussian measures of Chapter 6 are basic to a number of subjects. Their kernels $C(x, y)$ are characterized as solutions of the Laplace equation

$$(-\Delta + m^2)C(x, y) = \delta(x - y). \tag{7.1.1}$$

The essential properties of C coincide with the axioms of Chapter 6.1, namely Euclidean invariance, OS positivity, and regularity. Since $C(x, y)$ is C^∞ except at $x = y$, regularity is expressed in terms of asymptotic power laws for large and small values of $x - y$. For $m|x - y|$ small

$$C(x, y) \sim \begin{cases} \alpha|x - y|^{-d+2}, & d \geq 3, \\ -\dfrac{1}{2\pi} \ln(m|x - y|), & d = 2. \end{cases} \tag{7.1.2}$$

where the constant $\alpha = \alpha(d)$ is given by

$$\alpha = [(d - 2)|S^{d-1}|^{-1}] = 4^{-1}\pi^{-d/2}\Gamma\left(\frac{d - 2}{2}\right).$$

Here $|S^n|$ is the volume of the n-sphere, and the gamma function is $\Gamma(s) = \int_0^\infty e^{-t}t^{s-1}\, dt$. Thus $\alpha(3) = (4\pi)^{-1}$, $\alpha(4) = (4\pi^2)^{-1}$, etc.

On the other hand, for $m|x - y|$ large,

$$C(x, y) \sim \left(\frac{\pi}{2}\right)^{1/2} (2\pi)^{-d/2}m^{(d-3)/2}|x - y|^{-(d-1)/2} \exp(-m|x - y|). \tag{7.1.3}$$

The same properties—invariance, positivity, and regularity—apply (or are presumed to apply) to the two point function of the physical measure, except that either long or short distance exponents may differ from their canonical values $-(d-1)/2$ and $-d+2$ given in (7.1.2–3). In that case, the asymptotic regions are said to be governed by a nontrivial critical point (see Chapter 17).

The positivity properties of C are spin-dependent. Special to the spin zero Bosons considered here are the pointwise and operator inequalities

$$0 \le C(x, y) \tag{P1}$$

and

$$0 \le C \le m^{-2}I. \tag{P2}$$

The operator positivity (P2) is equivalent to the existence of a Gaussian measure, and thus is fundamental to the methods developed here. Integration is also defined for Fermi fields, but has a different character; see [Berezin, 1966] and [Osterwalder and Schrader, 1973a]. The OS or reflection positivity is positivity of the inner product in the Hilbert space of physical states; see Theorem 6.2.3. Let Π be a hyperplane in R^d, and let θ_Π be reflection about Π. Then reflection positivity for the hyperplane Π is

$$0 \le \langle \theta_\Pi \, f, \, Cf \rangle_{L_2} = \int \overline{(\theta_\Pi \, f)(y)} \, C(x, y) f(x) \, dx \, dy \tag{P3}$$

for all f supported on one side of Π.

We study C with various classical boundary conditions imposed, i.e. free, Dirichlet, Neumann, and periodic. The boundary conditions violate some of the axioms, but do not change the local singularity (7.1.2), nor the positivity (P1), (P2). The condition (P3) requires a reflection symmetry of the boundary conditions.

In *two dimensions*, C has only very weak (i.e. logarithmic) singularities. For this reason it is efficient to axiomatize its local regularity properties in terms of L_p spaces. Let ζ denote the operator of multiplication by the function $\zeta \in C_0^\infty(R^2)$. The first local regularity axiom is

$$\sup_x \|(C\zeta)(x, \cdot)\|_{L_q} < \infty \quad \text{for all } q < \infty. \tag{LR1}$$

Next we define an approximate Dirac delta function $\delta_\kappa(x)$. First let $h \in C_0^\infty$, and let h satisfy

$$0 \le h, \quad 0 < h(0), \quad \text{and} \quad \int h(x) \, dx = 1. \tag{7.1.4}$$

Then define

$$\delta_\kappa(x) = \kappa^2 h(\kappa x), \tag{7.1.5}$$

which we regard as a convolution operator acting on L_2, $(\delta_\kappa f)(x) = \int \delta_\kappa(x - y) f(y) \, dy$. In terms of δ_κ, we have two more axioms: For any $q < \infty$ there exists $\varepsilon = \varepsilon(q) > 0$ such that

$$\|\zeta \, \delta_\kappa \, C \, \delta_\kappa \, \zeta - \zeta C \zeta\|_{L_q(R^2 \times R^2)} \le O(\kappa^{-\varepsilon}), \quad \kappa \to \infty. \tag{LR2}$$

In other words, $\delta_\kappa C \delta_\kappa \to C$ in L_q^{loc} with a rate $O(\kappa^{-\varepsilon})$. Also the singularity of the Green's function on the diagonal is characterized by

$$\sup_x (\delta_\kappa C \delta_\kappa)(x, x) \le O(\ln \kappa). \tag{LR3}$$

The constants in the LR axioms depend on ζ, since the free covariance $C(x, y)$ is a function of x-y and hence not in any $L_q(R^2 \times R^2)$ space.

Cover R^d by a grid of unit cubes Δ, and impose boundary conditions on segments of the boundary $\partial\Delta$ of Δ. For example, if $d = 2$, then $\{\Delta\}$ is a cover of R^2 by unit squares, and we impose boundary conditions on certain line segments bounding the Δ's. For $d = 3$, the grid is composed of unit 3-cubes, the boundary segments are unit squares, etc. Let Γ denote a collection of hypersurface boundary segments in the lattice. We impose the classical boundary (free, periodic, Dirichlet, or Neumann) conditions "B" for the Laplace operator $-\Delta$ on Γ, yielding a self-adjoint, positive operator $-\Delta_{B(\Gamma)}$ on L_2. We study the covariance operator $C_{B(\Gamma)} = (-\Delta_{B(\Gamma)} + m^2)^{-1}$ by two methods: the method of images and the method of Wiener integrals. In each case, we relate $C_{B(\Gamma)}$ to the free covariance $C = C_\varnothing$ estimated in Proposition 7.2.1.

Let $\mathscr{C}_m$ denote the convex class of covariance operators generated by the particular C_B studied in the following sections, with a mass of m or bigger. (A precise definition can be found in Section 7.9.) The main results of this chapter are summarized in the following theorems.

Theorem 7.1.1. *Let $C \in \mathscr{C}_m$. Then C satisfies local regularity (7.1.2) and positivity (P1), (P2). If $d = 2$, then C satisfies (LR1–3) with bounds independent of $C \in \mathscr{C}_m$.*

Theorem 7.1.2. *Suppose the periodic covariance does not contribute to the convex sum $C \in \mathscr{C}_m$. Then C satisfies the decay estimate (7.1.3). If $C = C_B$ (no convex sum) and if C commutes with θ_Π, then C satisfies reflection positivity (P3). In the periodic case $C = C_p$, a modified version of reflection positivity is satisfied; cf. Section 7.10.*

Proofs of these theorems will be developed throughout this chapter. In particular, Section 7.9 is devoted to regularity and Section 7.10 to reflection positivity.

7.2 The Free Covariance

The simplest covariance operator is the bare Euclidean propagator or free covariance $C = C_\varnothing$. It is the Green's function defined by (7.1.1), and alternatively by the Fourier transform

$$C(x, y) = C(x - y) = (2\pi)^{-d} \int e^{-ip(x-y)}(p^2 + m^2)^{-1} \, dp. \tag{7.2.1}$$

The m dependence of C is given by the relation

$$C(m; x - y) = m^{d-2}C(1; m(x - y))$$

$$= (2\pi)^{-d/2}\left(\frac{m}{|x - y|}\right)^{(d-2)/2} K_{(d-2)/2}(m|x - y|). \quad (7.2.2)$$

Here $K_v(x) > 0$ is the modified Bessel function; up to a factor, K_v is a Hankel function of imaginary argument (cf. [Erdélyi et al., 1953]). For $d = 3$,

$$C(x - y) = \frac{1}{4\pi|x - y|} e^{-m|x - y|}.$$

Many properties of the covariance operators can be deduced from (7.2.2) and the known properties of Bessel functions: the singularity as $|x - y| \to 0$, the exponential decay as $m|x - y| \to \infty$ (cf. (7.1.2–3)), etc. However, we present direct proofs, whose methods have wider applications.

Proposition 7.2.1. *The free covariance $C(m; x - y) = C(x, y) = C(x - y)$ has the following properties:*

(a) *$C(x, y)$ is the kernel of a positive operator on L_2.*
(b) *$0 < C(x, y)$.*
(c) *For $m|x - y|$ bounded away from zero,*

$$C(x, y) \le O(1)m^{(d-3)/2}|x - y|^{-(d-1)/2} \exp(-m|x - y|). \quad (7.2.3)$$

(d) *For $d \ge 3$ and $m|x - y|$ in a neighborhood of zero,*

$$C(x, y) \sim |x - y|^{-d+2} \quad (7.2.4)$$

(e) *For $d = 2$ and $m|x - y|$ in a neighborhood of zero,*

$$C(x, y) \sim -\ln(m|x - y|). \quad (7.2.5)$$

PROOF. By (7.2.2), we may suppose $m = 1$. Positivity (a) follows by Fourier transformation: $\langle f, Cf \rangle_{L_2} = \int_{R^d} (p^2 + 1)^{-1}|f^{\sim}(p)|^2 \, dp$. Define $g(t)$ by the formula

$$0 < g(t) \equiv \int_{R^{d-1}} e^{-t\mu}\mu^{-1} \, d\mathbf{p} = \int d\omega \int_0^\infty e^{-t\mu}\mu^{-1}k^{d-2} \, dk. \quad (7.2.6)$$

Here $\mathbf{p} \in R^{d-1}$, $|\mathbf{p}| = k \in R$, $\mu = \mu(k) = (k^2 + 1)^{1/2}$, and $d\omega$ indicates (angular) integration over $\mathbf{p}/k$ in the $(d - 2)$-sphere of unit radius. The identity $C(x, y) = (2\pi)^{-d+1}2^{-1}g(|x - y|)$, which follows from the Cauchy integral formula to evaluate the partial inverse Fourier transform in (7.2.1) in the direction $x - y$, establishes positivity (b).

The detailed estimates (c)–(e) can be proved as follows. For sufficiently small $\varepsilon > 0$,

$$\mu(k) \ge \begin{cases} |1 + \varepsilon|k|^2 & \text{for } |k| \le 1, \\ |1 + \varepsilon|k| & \text{for } 1 \le |k|. \end{cases} \quad (7.2.7)$$

Thus

$$g(t) \leq \text{const } e^{-t}\left(\int_0^1 e^{-tek^2}k^{d-2}\,dk + \int_1^\infty e^{-tek}k^{d-2}\,dk\right)$$

$$\leq \text{const } e^{-t}(t^{-(d-1)/2} + t^{-(d-1)}). \tag{7.2.8}$$

For t bounded away from zero, we infer

$$g(t) \leq \text{const } e^{-t}t^{-(d-1)/2}, \tag{7.2.9}$$

and (c) follows.

To study the local singularity of g, write $k^2 = s^2 t^{-2}$, so that

$$t^{d-2}g(t) = |S^{d-2}|\int_0^\infty \exp[-s(1 + t^2 s^{-2})^{1/2}]s^{d-3}(1 + t^2 s^{-2})^{-1/2}\,ds. \tag{7.2.10}$$

In case $d \geq 3$, the integral (7.2.10) is bounded by its value at $t = 0$, and $t^{d-2}g(t)$ converges monotonically to this nonzero limit. Thus

$$g(t) \sim t^{-d+2}, \qquad t \to 0,$$

and (d) follows. In case $d = 2$, write $s = t\mu$, so that

$$g(t) = \int_t^\infty e^{-s}(s^2 - t^2)^{-1/2}\,ds \sim -\ln t \tag{7.2.11}$$

as $t \to 0$, to complete the proof. (In fact, use of the relation $|S^{n-1}| = 2\pi^{n/2}/\Gamma(n/2)$ yields the constant α in (7.1.2), while the $d = 2$ constant $1/2\pi$ follows by inspection of (7.2.11). The constant in (7.1.3) similarly comes from an analysis of the constants in (7.2.8–9).)

7.3 Periodic Boundary Conditions

We choose periodic boundary conditions with periods $L \equiv (L_1, \ldots, L_d)$, $L_j \in Z_+$. Then functions f in the domain of the Laplace operator Δ_p have the form $f(x) = f(x + n_L)$, where $n_L = \{n_j L_j : j = 1, 2, \ldots, d\} \in Z^d$. For example, with $d = 2$, we impose periodic boundary conditions on the lattice lines Γ bounding translates of $L_1 \times L_2$ rectangles which cover the plane.

Proposition 7.3.1. *For $0 < m$, the periodic covariance is given by*

$$C_p(x, y) = \sum_{n_L \in Z^d} C(x - y + n_L). \tag{7.3.1}$$

An infinite period, $L_j = \infty$, denotes the lack of boundary conditions in the jth coordinate direction. The series (7.3.1) is not convergent for $m = 0$ and estimates based on (7.3.1) are not uniform as $mL_j \to 0$. For this reason we suppose that mL_j is bounded away zero. The same restriction is imposed on the Dirichlet and Neumann covariance operators.

PROOF. The series converges exponentially by (7.2.3). The sum C_p is periodic by inspection. Each lattice hypercube contains one n_L. Thus $(-\Delta + m^2)C_p =$

$\sum_{n_L} \delta(x - y + n_L)$, and C_p is the periodic covariance by uniqueness of solutions to the linear boundary value problem.

Corollary 7.3.2.   *The periodic covariance C_p satisfies:*

(a) C_p *is a positive operator on L_2.*
(b) $0 < C(x, y) < C_p(x, y) = C_p(y, x)$.
(c) *For $m|x - y|$ in a neighborhood of zero,*

$$C_p(x, y) \sim \begin{cases} |\,|x - y|^{-d+2} & \text{if } d \geq 3, \\ -\ln(m|x - y|) & \text{if } d = 2. \end{cases}$$

PROOF.   The positivity of C_p on L_2 follows from the fact that in Fourier transform space, C_p is the operator of multiplication by $(p^2 + m^2)^{-1} \geq 0$. The positivity (b) of the kernel and the bound (c) follow by (7.3.1) and Proposition 7.2.1.

7.4. Neumann Boundary Conditions

For Neumann boundary conditions, we take the boundary Γ to be a union of periodically spaced lattice hyperplanes (as in the periodic case above). The Neumann boundary conditions on a function f in the domain of $-\Delta_N$ require the vanishing on Γ of the derivative of $f(x)$ in the direction normal to Γ.

We now obtain a simple formula for the Neumann covariance operator $C_N(x, y)$ by the method of images. Let $y \equiv y_0$ be given. Define a set of image points $\{y_j\}, j = 0, 1, 2, \ldots$, by the following two requirements:

1. $y_0 \in \{y_j\}$.
2. The set $\{y_j\}$ is invariant under reflection in any lattice hyperplane belonging to Γ.

Let Λ denote a connected component of $R^d \backslash \Gamma$. With the above definition, each $\Lambda = \Lambda_j$ contains exactly one y_j.

Proposition 7.4.1.   *For $0 < m$, the Neumann covariance is given by*

$$C_N(x, y) = \begin{cases} \sum_{j=0}^{\infty} C(x - y_j) & \text{if } x, y \in \Lambda, \\ 0 & \text{if } x \in \Lambda, y \in \Lambda' \neq \Lambda. \end{cases} \tag{7.4.1}$$

PROOF.   Since each Λ contains exactly one y_j, the right side of (7.4.1), acted on by $(-\Delta_N + m^2)$, yields $\delta(x - y)$. Furthermore, the invariance of $\{y_j\}$ under reflections in hyperplanes Γ ensures the Neumann boundary conditions. Hence (7.4.1) is C_N by uniqueness of solutions to the linear boundary value problem.

Uniqueness can also be used to establish the symmetry $C_N(x, y) = C_N(y, x)$ in the corollary below.

Corollary 7.4.2. *The Neumann covariance satisfies:*

(a) C_N *is a positive operator on* L_2.
(b) $0 < C(x, y) < C_N(x, y) = C_N(y, x)$, $x, y \in \Lambda$.
(c) *For* $m|x - y|$ *in a neighborhood of zero,*

$$C_N(x, y) \sim \begin{cases} |x - y|^{-d+2} & \text{if } d \geq 3, \\ -\ln(m|x - y|) & \text{if } d = 2. \end{cases}$$

7.5 Dirichlet Boundary Conditions

We define the Laplace operator Δ_Γ with Dirichlet boundary conditions on Γ by the requirement that functions $f(x)$ in the domain of Δ_Γ vanish for $x \in \Gamma$. Also $-\Delta_\Gamma$ is the Friedrichs extension [Kato, 1966] of $-\Delta$ restricted as a bilinear form to C^∞ functions with support contained in $R^d \backslash \Gamma$. Let $C_\Gamma = (-\Delta_\Gamma + m^2)^{-1}$. The case of general Γ is analyzed in Sections 7.7–8. Here we restrict Γ to be the union of periodic lattice hyperplanes, as in the periodic and Neumann cases above. Let C_D denote C_Γ in this special case.

Using the method of images, we obtain an explicit formula for C_D. Let $\{y_j\}$ be the collection of image points used in the study of C_N above. Let y_j be obtained from $y = y_0$ by ε_j reflections in lattice hyperplanes. Then $(-1)^{\varepsilon_j}$ is independent of the choice of reflections.

Proposition 7.5.1.

$$C_D(x, y) = \begin{cases} \displaystyle\sum_{j=0}^{\infty} (-1)^{\varepsilon_j} C(x - y_j) & \text{if } x, y \in \Lambda, \\ 0 & \text{if } x \in \Lambda, y \in \Lambda' \neq \Lambda. \end{cases} \tag{7.5.1}$$

PROOF. The factors $(-1)^{\varepsilon_j}$ are chosen so that the sum (7.5.1) vanishes for $x \in \partial\Lambda$. Since each Λ_j contains only one y_j, $-\Delta_D + m^2$ applied to the sum yields $\delta(x - y)$ for x in the interior. Thus by uniqueness, (7.5.1) equals C_D.

Corollary 7.5.2. *The Dirichlet covariance satisfies:*

(a) C_D *is a positive operator on* L_2.
(b) $0 \leq C_D(x, y) = C_D(y, x) < C(x, y)$.
(c) *For* $m|x - y|$ *in a neighborhood of zero,*

$$C_D(x, y) \sim \begin{cases} |x - y|^{-d+2} & \text{if } d \geq 3, \\ -\ln(m|x - y|) & \text{if } d = 2. \end{cases}$$

PROOF. The positivity (a) follows from the fact that C_D is the inverse of a positive operator $-\Delta_D + m^2$. Local regularity (c) follows from (7.5.1). To prove (b), we use the maximum principle: A function $f(x)$, continuous on a closed bounded region Ω and satisfying the equation $(-\Delta + m^2)f(x) = 0$ in the interior of Ω, achieves its positive maximum and negative minimum on the boundary $\partial\Omega$. Let $\Omega = \Lambda$ be a connected subset of $R^d \backslash \Gamma$, and let $f(x) = C(x, y) - C_D(x, y)$ for

$y \in \Lambda$. Since $f(x) = C(x, y) > 0$ on $\partial\Lambda$, f cannot have a negative or zero minimum, i.e. $f(x) > 0$. This proves the upper bound in (b).

The lower bound is trivial for $y \in \partial\Lambda$, because then $C_D(x, y) = 0$. Let $y \in \text{Int } \Lambda$, and let Λ_ε denote Λ minus an ε neighborhood of y. By (c), $0 \le C_D(x, y)$ for $x \in \partial\Lambda_\varepsilon$, for ε sufficiently small. On Λ_ε, $(-\Delta + m^2)C_D = 0$. Since C_D has no singularities on Λ_ε, the maximum principle can be applied again. Thus C_D does not have a negative interior minimum, and hence $0 \le C_D$. This completes the proof.

7.6 Change of Boundary Conditions

Even though the kernel $C_B(x, y)$, with boundary conditions $B = \varnothing, N, D, p$, etc., on Γ is singular on the diagonal $x = y$, the difference between two such kernels is singular only on the boundary Γ. We meet such differences in discussing the change of Wick ordering, so estimates on these differences are of interest. Let

$$\delta c_B(x) = \lim_{y \to x} [C(x, y) - C_B(x, y)] \qquad (7.6.1)$$

for $B = N, D$, or p. ($c(x) \equiv C(x, x)$ is identically infinite.) Again let Λ denote a connected component of $R^d\backslash\Gamma$.

Proposition 7.6.1. *For* $x \in \Lambda$,

$$0 \le \delta c_D(x) \le -\delta c_N(x) \le \begin{cases} \text{const dist}(x, \partial\Lambda)^{-d+2} & \text{if } d \ge 3 \\ \text{const } (1 + |\ln \text{dist}(x, \partial\Lambda)|) & \text{if } d = 2. \end{cases} \qquad (7.6.2)$$

In addition $-\delta c_p(x)$ *is positive and bounded above as in* (7.6.2).

PROOF. The positivity of $\delta c_D(x)$ follows from Proposition 7.5.1 and the definition of δc_D. The second inequality follows by (7.6.1) and the positivity of $C(x, y)$. Namely,

$$\delta c_D(x) = -\sum_{j=1}^{\infty} (-1)^{\varepsilon_j} C(x - x_j) \le \sum_{j=1}^{\infty} C(x - x_j) = -\delta c_N(x). \qquad (7.6.3)$$

We note that as $x \to \partial\Lambda$, at least one x_j and at most $2^d - 1$ of the x_j's (at a corner of Λ) approach x. The upper bound of (7.6.2) then follows by Proposition 7.2.1, which gives the local singularity of $C(x - y)$. The proof for c_p is similar.

7.7 Covariance Operator Inequalities

Above we have compared the kernels of covariance operators with different boundary conditions. Here we compare the operators themselves, as bilinear forms on $L_2 \times L_2$. We sketch the main ideas, but omit mathematical proofs. Note that $A \leq B$ means $\mathcal{D}_B \subseteq \mathcal{D}_A$ (form domains) and $\langle x, Ax \rangle \leq \langle x, Bx \rangle$ for all $x \in \mathcal{D}_B$. The form domains for Δ_B are as follows:

$$\mathcal{D}_\Delta = \{ f \in L_2 : \nabla f \in L_2 \}.$$

Here ∇f is the distribution derivative, so that $\nabla f \in \mathcal{D}'$ is defined for all $f \in L_2$. With $\nabla f \in L_2$, one can show that $f \mid \Gamma$ is defined and is in L_2 of the hypersurface. As an element of L_2 in the tangent variables, such an f is continuous in the normal variable. Moreover the same analysis applies to one sided derivatives; let $\nabla_{+/-} f$ denote a gradient which is two sided in the interior $R^d \backslash \Gamma$ and one sided in the normal direction on Γ. Then

$$\mathcal{D}_{\Delta_N} = \{ f \in L_2 : \nabla_{+/-} f \in L_2 \}. \tag{7.7.1}$$

Note that $f \in \mathcal{D}_{\Delta_N}$ is one sided continuous in the normal variable (as an element of L_2 of tangential variables), but may have a jump discontinuity across Γ. Also

$$\mathcal{D}_{\Delta_\Gamma} = \{ f \in L_2 : \nabla f \in L_2, f \mid \Gamma = 0 \}. \tag{7.7.2}$$

Since

$$\mathcal{D}_{\Delta_\Gamma} \subset \mathcal{D}_\Delta \subset \mathcal{D}_{\Delta_N},$$

we see that

$$-\Delta_N \leq -\Delta \leq -\Delta_{\Gamma_1} \leq -\Delta_{\Gamma_2} \tag{7.7.3}$$

with $\Gamma_1 \subset \Gamma_2$. Taking inverses establishes

$$0 \leq C_{\Gamma_2} \leq C_{\Gamma_1} \leq C \leq C_N. \tag{7.7.4}$$

The operators are defined on domains

$$\mathcal{D}(\text{operator}) = \{ f \in \mathcal{D}_{\text{form}} : |\langle \nabla f, \nabla g \rangle_{L_2}| \leq \text{const} \|g\|_{L_2} \text{ for all } g \in \mathcal{D}_{\text{form}} \}.$$

Integration by parts shows that this definition coincides with conventional definitions. For example, with f in the operator domain of Δ_N and $g \in C^\infty$ (except for jumps across Γ), $g \in \mathcal{D}_{\Delta_N}$, we have

$$\int \nabla f(x) \nabla g(x)\, dx = \int (-\Delta f(x)) g(x)\, dx - \int_{\Gamma_+} (\mathbf{n} \cdot \nabla f) g\, dx + \int_{\Gamma_-} (\mathbf{n} \cdot \nabla f) g\, dx.$$

Here Γ_+, Γ_- denote the two sides of Γ, so that g, with a jump discontinuity across Γ, is single valued on Γ_+ and on Γ_-. Since each term above is independent, each defines a continuous functional on $g \in L_2$. For example, choosing g continuous across Γ causes the sum of the second and third terms to vanish. Thus $\Delta f \in L_2$ (one sided second derivatives on Γ), ∇f is one sided continuous in the normal variable, and in the second and third terms, $\nabla f \mid \Gamma_{+/-} \in L_2(\Gamma)$. However, $g \to g \mid \Gamma_{+/-}$ is not continuous in L_2, and since the second and third terms are each continuous in L_2, they must vanish. Thus $\nabla f \mid \Gamma = 0$.

7.8 More General Dirichlet Data

For application to the cluster expansion in Chapter 18, we consider Dirichlet data on Γ, where Γ is a union of elementary lattice hypersurfaces. For application to the infinite volume limit in Chapters 11, 18, we take Γ to consist of the surface of two cubes, one inside the other and bounded away from each other, but not aligned. It is convenient in studying these covariance operators to have the representations for C_Γ which we present here. In general, an elementary series for C_Γ (such as (7.5.1)) does not exist, but the results of Corollary 7.5.2 remain valid.

The formula for C_Γ uses a Wiener integral representation. This method is conceptually simple but technically complicated, so we include an elementary proof. Let dW_{xy} denote the conditional Wiener measure for continuous trajectories $\omega(\tau)$ from x to y such that $\omega(0) = x$, $\omega(t) = y$. Let χ_Γ denote the characteristic function on path space for the set of paths which do not cross Γ. In other words

$$\chi_\Gamma(\omega) = \begin{cases} 0 & \text{if } \omega(\tau) \in \Gamma \text{ for some } \tau, 0 \le \tau \le t, \\ 1 & \text{otherwise} \end{cases} \tag{7.8.1}$$

We show that the kernel of $e^{t\Delta_\Gamma}$ is obtained by considering trajectories which do not cross Γ, namely

$$e^{t\Delta_\Gamma}(x, y) = \int \chi_\Gamma(\omega) \, dW_{xy}^t(\omega). \tag{7.8.2}$$

In particular, if $\Gamma = \phi$, then (7.8.2) reduces to the standard formula for free boundary conditions, namely (3.1.14). Taking the Laplace transform of (7.8.2), we obtain C_Γ with kernel

$$C_\Gamma(x, y) = \int_0^\infty dt \, e^{-t(m^2 - \Delta_\Gamma)}(x, y) \tag{7.8.3}$$

$$= \int_0^\infty dt \, e^{-tm^2} \int \chi_\Gamma(\omega) \, dW_{xy}^t(\omega).$$

For a detailed discussion of Wiener integral representations with Dirichlet (and Neumann) boundary conditions, see [Ginibre, 1971].

Our proof of (7.8.3) is based on a second definition of Dirichlet boundary conditions. We introduce Dirichlet boundary conditions on the boundary $\Gamma = \partial\Lambda$ of a region Λ, by considering a constant potential on $R^d \backslash \Lambda$, and letting the height of the potential tend to ∞. We then obtain Δ_Λ, the Laplace operator on $L_2(\Lambda)$ with Dirichlet boundary conditions on $\partial\Lambda$. Similarly we obtain $\Delta_{R^d \backslash \Lambda}$; then Δ_Γ is the sum of the two commuting operators Δ_Λ and $\Delta_{R^d \backslash \Lambda}$ on $L_2(R^d)$. In case Γ does not enclose a region Λ, we obtain Δ_Γ by "thickening" Γ to enclose a region, and then performing a limiting operation as the "thickness" tends to zero.

Proposition 7.8.1. *Let Λ be a domain in R^d with boundary $\partial\Lambda$, and let $\chi_\Lambda(x)$ be the characteristic function of Λ. Then*

$$\chi_\Lambda(-\Delta_{\partial\Lambda} + m^2)^{-1}\chi_\Lambda = \lim_{\lambda\to\infty}(-\Delta + m^2 + \lambda\chi_{\sim\Lambda})^{-1}. \qquad (7.8.4)$$

Corollary 7.8.2. *For Λ a domain with $\Gamma = \partial\Lambda$,*

$$\begin{aligned} C_\Gamma &= \lim_{\lambda\to\infty}(-\Delta + m^2 + \lambda\chi_\Lambda)^{-1} + \lim_{\lambda\to\infty}(-\Delta + m^2 + \lambda\chi_{\sim\Lambda})^{-1} \\ &= (-\Delta_\Gamma + m^2)^{-1}. \end{aligned} \qquad (7.8.5)$$

Lemma 7.8.3. *The operator $C(\lambda) \equiv (-\Delta + m^2 + \lambda\chi_{\sim\Lambda})^{-1}$ decreases monotonically to a strong limit C as $\lambda \uparrow \infty$. The kernel $C(\lambda, x, y)$ decreases to $C(x, y)$ for $x \neq y$. Furthermore C is self-adjoint.*

Remark. Below, C is identified with the left side of (7.8.4).

PROOF. Monotonicity is a consequence of the identity

$$\frac{dC(\lambda)}{d\lambda} = -C(\lambda)\chi_{\sim\Lambda}C(\lambda) \leq 0. \qquad (7.8.6)$$

Since C is a positive operator, $\lim\langle f, C(\lambda)f\rangle$ exists for each $f \in L_2$, and $C = \text{weak lim } C(\lambda)$ is a bounded, self-adjoint operator. Let $B(\lambda) = C(\lambda) - C \geq 0$. Then for all f, $\|B(\lambda)^{1/2}f\| \to 0$, so st lim $B(\lambda)^{1/2} = 0$ and st lim $B(\lambda) = 0$. Thus $C(\lambda)$ converges strongly to C, as claimed.

As in Section 7.5, $C(\lambda)$ has a strictly positive kernel, and (7.8.6) ensures pointwise monotonic decrease of $C(\lambda, x, y)$. Hence $\lim C(\lambda, x, y) \equiv C(x, y) \geq 0$ exists for all $x \neq y$.

The following result is a special case of graph convergence. For a general discussion see [Glimm and Jaffe, 1969, 71b].

Lemma 7.8.4. *The operator $C = \text{st lim } C(\lambda)$ maps $L_2(\Lambda)$ into itself. Furthermore $C = \chi_\Lambda C\chi_\Lambda$ has a self-adjoint inverse on $L_2(\Lambda)$.*

PROOF. We wish to define an inverse of C, and we now show that Null $C = L_2(R^d\backslash\Lambda) = (\text{Range } C)^\perp$. It follows from these two equalities that C has a densely defined inverse C^{-1} acting on $L_2(\Lambda)$. Since C is self-adjoint (and bounded), Range $C^{-1} = \text{Domain } C = L_2(\Lambda)$ and C^{-1} is self-adjoint.

We first show $L_2(R^d\backslash\Lambda) \subset \text{Null } C$. Since C is a bounded operator, Null C is closed. Hence we need only verify that a dense subset of $L_2(R^d\backslash\Lambda)$ is contained in Null C. For this subset we choose functions f which are C^2 and compactly supported in $R^d\backslash\Lambda$. For such an $f \in C_0^2(R^d\backslash\Lambda)$ define

$$f(\lambda) \equiv \lambda^{-1}C(\lambda)^{-1}f = \lambda^{-1}(-\Delta + m^2)f + f. \qquad (7.8.7)$$

Thus $\|f(\lambda) - f\| = \lambda^{-1}\|(-\Delta + m^2)f\| = O(\lambda^{-1})$ as $\lambda \to \infty$. It follows from $C(\lambda)f(\lambda) = \lambda^{-1}f$ that

$$
\begin{aligned}
\|C(\lambda)f\| &= \|C(\lambda)\{f - f(\lambda)\} + C(\lambda)f(\lambda)\| \\
&\leq \|C(\lambda)\|\,\|f - f(\lambda)\| + \lambda^{-1}\|f\| \\
&\leq O(\lambda^{-1}) \quad \text{as } \lambda \to \infty.
\end{aligned}
\tag{7.8.8}
$$

Since $C(\lambda) \to C$ strongly, $Cf = 0$ as claimed.

Conversely, we prove Null $C \subset L_2(R^d\backslash\Lambda)$. Let $f \in$ Null C and $g \in C_0^2(\Lambda)$. Then

$$
\begin{aligned}
0 = \langle(-\Delta + m^2)g,\, Cf\rangle &= \langle(-\Delta + m^2 + \lambda\chi_{\sim\Lambda})g,\, Cf\rangle \\
&= \langle C(\lambda)^{-1}g,\, Cf\rangle = \lim_{\lambda\to\infty}\langle C(\lambda)^{-1}g,\, C(\lambda)f\rangle \\
&= \langle g, f\rangle.
\end{aligned}
\tag{7.8.9}
$$

Since the vectors g are dense in $L_2(\Lambda)$, it follows that $f \in L_2(\Lambda)^\perp = L_2(R^d\backslash\Lambda)$, and Null $C = L_2(R^d\backslash\Lambda)$.

Because C is self-adjoint, Null $C = (\text{Range } C)^\perp$. In fact, for $g \in$ Null C and $f \in L_2(R^d)$,

$$
\begin{aligned}
\langle g,\, Cf\rangle &= \lim_{\lambda\to\infty}\langle g,\, C(\lambda)f\rangle = \lim_{\lambda\to\infty}\langle C(\lambda)g, f\rangle \\
&= \langle Cg, f\rangle = 0,
\end{aligned}
\tag{7.8.10}
$$

so $g \in (\text{Range } C)^\perp$ and Null $C \subset (\text{Range } C)^\perp$. The reverse inclusion is similar, and the lemma is proved.

PROOF OF PROPOSITION 7.8.1. It remains to identify C^{-1} with $-\Delta_\Lambda + m^2$. For this purpose, we characterize the Dirichlet Laplacian as the Friedrichs extension of Δ, considered as a bilinear form on $C_0^{(2)}(\Lambda) \times C_0^{(2)}(\Lambda)$. For $g \in C_0^{(2)}(\Lambda)$ and $f \in L_2(R^d)$,

$$
(-\Delta + m^2)g = (-\Delta + m^2 + \lambda\chi_{\sim\Lambda})g = C(\lambda)^{-1}g
\tag{7.8.11}
$$

and

$$
\begin{aligned}
\langle(-\Delta + m^2)g,\, Cf\rangle &= \lim_{\lambda\to\infty}\langle C(\lambda)^{-1}g,\, C(\lambda)f\rangle = \langle g, f\rangle \\
&= \lim_{\lambda\to\infty}\langle C(\lambda)(-\Delta + m^2)g, f\rangle \\
&= \langle C(-\Delta + m^2)g, f\rangle.
\end{aligned}
\tag{7.8.12}
$$

Hence

$$
g = C(-\Delta + m^2)g \in \text{Range } C = \mathscr{D}(C^{-1})
$$

and

$$
C^{-1} \supset -\Delta + m^2\Big|_{C_0^{(2)}(\Lambda)}
$$

Since C^{-1} is self-adjoint and positive, it is closed both as an operator and as a bilinear form. Thus C^{-1} is an extension of the bilinear form $-\Delta_\Lambda + m^2$, when the latter is restricted to the domain $C_0^{(2)}(\Lambda)$ and then closed.

The bilinear form domain of a positive self-adjoint operator is equal to the operator domain of its square root (see [Kato, 1966]). Hence we complete the proof by showing the operator domain inclusion

$$\mathscr{D}(C^{-1/2}) \subset \mathscr{D}((-\Delta_\Lambda + m^2)^{1/2}). \qquad (7.8.13a)$$

Because $-\Delta \le C(\lambda)^{-1}$, we have $\|\nabla C(\lambda)^{1/2}f\|_{L_2} \le \|f\|_{L_2}$, and so the sequence $\{C(\lambda)^{1/2}f\}$ is bounded in λ in the gradient norm. For $h \in \mathscr{D}(\Delta)$, $\langle \nabla h, \nabla C(\lambda)^{1/2}f \rangle \to \langle -\Delta h, C^{1/2}f \rangle$. Thus $C(\lambda)^{1/2}f$ is weakly convergent on a dense domain in the gradient norm and is uniformly bounded. Hence $C(\lambda)^{1/2}f$ is weakly convergent on the full gradient Hilbert space, with limit $C^{1/2}f$, and

$$\|\nabla C^{1/2}f\|_{L_2} \le \limsup_\lambda \|\nabla C(\lambda)^{1/2}f\|_{L_2} \le \|f\|_{L_2}.$$

Thus

$$\mathscr{D}(C^{-1/2}) = \text{Range } C^{1/2} \subset \mathscr{D}(-\Delta^{1/2}).$$

In other words, $C^{1/2}f$ has a finite Dirichlet norm, with the gradient taken in $L_2(R^d)$. From Lemma 7.8.4, $C^{1/2}f$ vanishes on $\sim \Lambda$. Next we estimate the $L_2(R^{d-1})$ norm $\| \cdot \|_\varepsilon$ of $C^{1/2}f$ in the direction tangential to the boundary of the domain Λ and at a distance ε from the boundary. Let n denote the coordinate normal to $\partial\Lambda$ and let p denote the coordinate parallel to $\partial\Lambda$. In this notation $\|f\|_n^2 = \int |f(n, p)|^2 \, dp$ and

$$\|C^{1/2}f\|_\varepsilon = \int_0^\varepsilon \left(\frac{d}{dn} \|C^{1/2}f\|_n \right) dn.$$

Using the triangle inequality on $L_2(R^{d-1})$,

$$\left| \frac{d}{dn} \|C^{1/2}f\|_n \right| \le \left\| \left| \frac{d}{dn} C^{1/2}f \right| \right\|_n.$$

Substitute this bound, and use the Schwarz inequality in the n-integration to obtain

$$\|C^{1/2}f\|_\varepsilon \le \varepsilon^{1/2} \|\nabla C^{1/2}f\|_{L_2} \le \varepsilon^{1/2} \|f\|_{L_2}. \qquad (7.8.13b)$$

Using the convolution operators in a standard way, $C^{1/2}f$ is approximated by smooth functions in the gradient norm; if these convolution operators are slightly displaced near the boundary, we approximate $C^{1/2}f$ by $C_0^{(2)}(\Lambda)$ functions in the gradient norm. This shows that

$$\mathscr{D}(-\Delta^{1/2}) \cap L_2(\Lambda) \subset \mathscr{D}((-\Delta_\Lambda + m^2)^{1/2})$$

and proves (7.8.13a).

Remark 1. The inequality (7.8.13b) shows that functions in $\mathscr{D}(-\Delta^{1/2}) \cap L_2(\Lambda)$ are Hölder continuous in a variable normal to $\partial\Lambda$, when considered as L_2 functions of the variables tangential to Λ. Thus we obtain vanishing Dirichlet boundary data in the naive sense. See [Agmon, 1965] for related material.

Remark 2. Let $\Gamma = \partial\Lambda$. By the Feynman–Kac formula of Chapter 3,

$$e^{-tC(\lambda)^{-1}}(x, y) = e^{-t(-\Delta + m^2 + \lambda\chi_{\sim\Lambda})}(x, y)$$

$$= e^{-tm^2} \int \exp\left[-\lambda \int_0^t \chi_{\sim\Lambda}(\omega(\tau))\, d\tau\right] dW_{xy}^t(\omega). \quad (7.8.14)$$

As $\lambda \to \infty$, the right side tends to

$$e^{-tm^2} \int \chi_\Lambda(\omega)\, dW_{xy}^t(\omega)$$

by the Lebesgue bounded convergence theorem. The left side, by the proposition, tends to the kernel of $\chi_\Lambda \exp[t\Delta_{\partial\Lambda} - m^2)]\chi_\Lambda$. Adding this expression and the corresponding formula for the region $R^2\backslash\Lambda$, we obtain (7.8.3) as an identity of L_2 operator kernels.

Remark 3. We may introduce Dirichlet data successively on contours Γ, Γ' enclosing regions Λ, Λ' by the formula

$$(-\Delta_{\Gamma \cup \Gamma'} + m^2)^{-1}$$

$$= \lim_{\lambda \to \infty} [(-\Delta_\Gamma + m^2 + \lambda\chi_{\sim\Lambda'})^{-1} + (-\Delta_\Gamma + m^2 + \lambda\chi_{\Lambda'})^{-1}]. \quad (7.8.15)$$

The proof follows the proof of the proposition above.

Remark 4. We may introduce Dirichlet data on a "bond" b not bounding a region as follows. Let b_ε be an ε neighborhood of b, and use the above method to introduce Dirichlet data on ∂b_ε. Monotonicity in ε follows from (7.8.6) as in Remark 3. Then

$$(-\Delta_{\Gamma \cup b} + m^2)^{-1}(x, y) = \sup_\varepsilon (-\Delta_{\Gamma \cup \partial b_\varepsilon} + m^2)^{-1}(x, y). \quad (7.8.16)$$

As above, we show that vectors $((-\Delta_{\Gamma \cup b} + m^2)^{-1}f)(x)$ in the range of (7.8.16) vanish continuously in a normal variable as $x \to b$.

Remark 5. From the above representations, we conclude monotonicity of the Dirichlet covariance operators C_Γ in Γ, generalizing Proposition 7.6.1. Let

$$\delta c_\Gamma(x) = \lim_{y \to x} [C_\varnothing(x, y) - C_\Gamma(x, y)]. \quad (7.8.17)$$

Proposition 7.8.5. *For* $\Gamma_1 \subset \Gamma$,

$$0 \le C_\Gamma(x, y) \le C_{\Gamma_1}(x, y) \le C(x, y), \quad (7.8.18)$$

$$0 \le \delta c_{\Gamma_1}(x) \le \delta c_\Gamma(x) \le \begin{cases} \text{const}\,(1 + \text{dist}(x, \Gamma)^{-d+2}) & \text{if } d \ge 3, \\ \text{const}\,(1 + |\ln \text{dist}(x, \Gamma)|) & \text{if } d = 2. \end{cases} \quad (7.8.19)$$

PROOF. We need only establish the upper bound in (7.8.19). Let x belong to the lattice square Δ, and let $b \subset \partial\Delta$ be the hypersurface (bond) closest to x. If $b \subset \Gamma$, then since $\delta c_\Gamma \leq \delta c_D$, the upper bound follows by (7.6.2). Now suppose b is not contained in Γ, and let $\Lambda = \Delta \cup \Delta'$ be the rectangle formed from two adjacent lattice squares, so that $b \subset \partial \Delta \cup \partial \Delta'$. Translates of Λ cover R^d, and yield a Dirichlet covariance $C_{\partial \Lambda} = C_D$. (For a suitable choice of lattice, this C_D coincides with the C_D of Section 7.5.) Hence for $x, y \in \Lambda$, $C_{\partial \Lambda}(x, y) \leq C_\Gamma(x, y)$ by (7.8.18), and $\delta c_\Gamma(x) \leq \delta c_{\partial \Lambda}(x)$ for $x \in \Lambda$.

Let b' be the hypersurface of Λ closest to x. If $b' \cap \Gamma \neq \varnothing$, then the upper bound follows as before from (7.6.2). If not, we form as above a new rectangle $\Lambda \cup \Lambda'$ and repeat the above argument. The argument terminates after d steps, or else x is bounded away from Γ. The latter case is also reduced to (7.6.2) by monotonicity.

We complement the short distance bound (7.8.19) with a large distance decay estimate.

Proposition 7.8.6. *Let δc_Γ be as in (7.8.17). Let $\mathrm{dist}(x, \Gamma)$ be bounded away from zero. Then*
$$0 \leq \delta c_\Gamma(x) \leq O(1)e^{-2m\,\mathrm{dist}(x,\,\Gamma)/\sqrt{d}} \tag{7.8.20}$$

PROOF. With $\mathrm{dist}(x, \Gamma) = r$, we can center a d-cube B of side length $l = 2r/\sqrt{d}$ at x, such that $B \cap \Gamma = \varnothing$. Let ∂B and its translates generate a lattice Γ_1. By Proposition 7.8.5, for $x \in B$,
$$0 \leq \delta c_\Gamma(x) \leq \delta c_{\Gamma_1}(x).$$

But $\delta c_{\Gamma_1}(x)$ is given explicitly by (7.6.3), with $x - x_j \geq l$. Hence $\delta c_\Gamma \leq O(1)e^{-2m\,\mathrm{dist}(x,\,\Gamma)/\sqrt{d}}$.

Proposition 7.8.7. *Let $d = 2$. For $1 \leq q \leq \infty$, there is a constant $K = K(q)$, independent of $m \geq 1$ and of $C \in \mathscr{C}_m$ defined in Section 7.9 below, such that for any lattice square Δ,*
$$\|\delta c_\Gamma(x)\|_{L_q(\Delta)} \leq Km^{-1/q}. \tag{7.8.21}$$

PROOF. By scaling, we take $m = 1$. The result then follows from Propositions 7.8.5 and 7.8.6 and from (7.2.2).

7.9 Regularity of C_B

Definition. Let $\mathscr{C}_m$ be the set of convex combinations of covariance operators considered in Sections 7.1–8 with a mass m or larger. Explicitly, $\mathscr{C}_m$ includes $C_B = (-\Delta_B + m_1^2)^{-1}$, $m_1 \geq m$ and $B = \varnothing$, p, D, N, and Γ. Furthermore let $\mathscr{C}_m^M \subset \mathscr{C}_m$ denote the subset of convex combinations with $m \leq m_1 \leq M$.

We drop the subscript m and superscript M when not necessary. In this section we complete the proof of Theorem 7.1.1. As a summary of previous sections of this chapter, we note that the local regularity condition (7.1.2) and the pointwise positivity condition (P1) of Section 7.1 follow from (7.2.4–5), Corollaries 7.3.2, 7.4.2, and 7.5.2, and (7.8.18). The operator positivity condition (P2) of Section 7.1 follows from the definition of C_B.

It remains to prove the local regularity properties (LR1–3) of Section 7.1. In case the periodic covariance C_p does not contribute to the convex sum C, we also prove long distance decay. Let $\{\Delta_i\}$ be a cover of R^2 by unit squares with lower left corner at $i \in Z^2$. Recall that h, δ_κ, and ζ were introduced in Section 7.1.

Proposition 7.9.1. *Let $C \in \mathscr{C}_m$, let m be bounded away from zero, let $p < \infty$, and let $d = 2$. Then*

$$\sup_x \|(C\zeta)(x, \cdot)\|_{L_p} \leq \text{const}, \tag{7.9.1}$$

$$\|(\delta_\kappa C \delta_\kappa)(x, x)\|_{L_\infty} \leq \text{const } \ln \kappa. \tag{7.9.2}$$

If C_p does not contribute to C, then $(7.1.3)$ is valid and also

$$\sup_x \|C(x, \cdot)\|_{L_p} \leq \text{const}, \tag{7.9.3}$$

$$\|C\|_{L_p(\Delta_i \times \Delta_j)} \leq \text{const } m^{-2/p} e^{-m \,\text{dist}(\Delta_i, \Delta_j)}. \tag{7.9.4}$$

All constants above are independent of C and κ.

PROOF. All of these bounds follow from the bound $(7.2.3)$ and $(7.2.5)$ on $C_\varnothing$; the method of images formulas $(7.3.1)$, $(7.4.1)$, and $(7.5.1)$ to get similar bounds on C_B, $B = p$, N, or D; and the pointwise upper bound $0 \leq C_\Gamma \leq C_\varnothing$ of $(7.8.18)$. To prove $(7.9.4)$ we also use the scaling relation $(7.2.2)$, so that for $C = C_\varnothing$,

$$\left(\int_{\Delta_i \times \Delta_j} C(m; x - y)^p \, dx \, dy \right)^{1/p} = \left(\int_{\Delta_i \times \Delta_j} C(1; m(x - y))^p \, dx \, dy \right)^{1/p}$$

$$= \left(\int_{m\Delta_i \times m\Delta_j} m^{-4} C(1; x - y)^p \, dx \, dy \right)^{1/p}, \tag{7.9.5}$$

where $m\Delta_i = \{mx : x \in \Delta_i\}$. By $(7.2.3)$ and $(7.2.5)$, $(7.9.5)$ is bounded by

$$\left(\int_{m\Delta_i} m^{-4} \, dx \right)^{1/p} e^{-m \,\text{dist}(\Delta_i, \Delta_j)} = \text{const } m^{-2/p} e^{-m \,\text{dist}(\Delta_i, \Delta_j)}.$$

This bound applies to C_N and C_D by explicit formulas as above and to C_Γ by a pointwise upper bound.

The proof of Theorem 7.1.1 is now reduced to the property (LR2), which in effect depends on fractional derivatives of C belonging to L_q^{loc}. This property is more delicate, because fractional derivatives cannot be inferred from a pointwise upper bound.

Theorem 7.9.2. *Let $C \in \mathscr{C}_m$, let m be bounded away from zero, let $p < \infty$, and let $d = 2$. Then for some $\varepsilon > 0$ and for $\kappa_1 \leq \kappa_2$,*

$$\|\zeta_1 C(\delta_{\kappa_1} - \delta_{\kappa_2})\zeta_2\|_{L_p(R^2 \times R^2)} \leq \text{const } \kappa_1^{-\varepsilon}$$

as $\kappa_1 \to \infty$. The constant is independent of C, but depends on p, ε, ζ_1, and ζ_2.

Remark. Given $\zeta \in C_0^\infty$, we choose $\zeta_1 \in C_0^\infty$ to be equal to 1 on a neighborhood of the support of ζ. Then $\zeta \delta_\kappa C = \zeta \delta_\kappa \zeta_1 C$ for κ sufficiently large. In proving (LR2), we change the κ's on the two sides of C independently. Thus

$$\|\zeta \delta_\kappa C \delta_{\kappa_1} \zeta - \zeta \delta_\kappa C \delta_{\kappa_2} \zeta \|_{L_p} = \|\zeta \delta_\kappa \zeta_1 C(\delta_{\kappa_1} - \delta_{\kappa_2}) \zeta\|_{L_p}$$
$$\leq \text{const} \|\zeta_1 C(\delta_{\kappa_1} - \delta_{\kappa_2}) \zeta\|_{L_p},$$

so that Theorem 7.9.2 implies (LR2).

PROOF FOR $p = 2$, $C = C_\varnothing$. In this case, δ_κ and C commute, so the theorem is contained in the following two lemmas. Recall that the Hilbert–Schmidt norm $\|A\|_{HS}$ of an operator A is the L_2 norm of the kernel of A, and that

$$\|AB\|_{HS} \leq \|A\| \, \|B\|_{HS}.$$

Lemma 7.9.3. Let $C = C_\varnothing$. Then for $0 \leq a \leq \frac{1}{2}$,

$$\|C^a(\delta_{\kappa_1} - \delta_{\kappa_2})\| \leq O(\kappa_1^{-2a}).$$

PROOF. The operator norm above is the L_∞ norm in Fourier transform space. Using the definitions (7.1.4–5), the norm is evaluated as

$$\sup_p (p^2 + 1)^{-a} |\tilde{h}(p/\kappa_1) - \tilde{h}(p/\kappa_2)|.$$

For $|p| \leq \kappa_1$, substitute

$$|\tilde{h}(p/\kappa_1) - \tilde{h}(p/\kappa_2)| \leq O(1)|p/\kappa_1|^{2a}, \tag{7.9.6}$$

and for $|p| \geq \kappa_1$, substitute

$$|\tilde{h}(p/\kappa_1) - \tilde{h}(p/\kappa_2)| \leq O(1). \tag{7.9.7}$$

The constant $O(1)$ depends on h, but h is fixed.

Lemma 7.9.4. Let $C = C_\varnothing$. Then for $a > \frac{1}{4}$ and $A = C^a \zeta$, A^*A is Hilbert–Schmidt. Also for $a > \frac{1}{2}$, A is Hilbert–Schmidt.

PROOF.

$$\|A^*A\|_{HS} = \|\zeta C^{2a} \zeta\|_{HS} \leq \|\zeta\| \, \|C^{2a} \zeta\|_{HS}$$

The operator $C^{2a} \zeta$ has a Fourier transformed kernel

$$(p^2 + 1)^{-2a} \tilde{\zeta}(p - q)$$

which is in $L_2(R^4)$ for $a > \frac{1}{4}$.

PROOF OF THEOREM 7.9.2 FOR $p = 2$, $C = C_p$, C_D, OR C_N. Because ζ_i has compact support, we may consider $\chi_\Lambda C \chi_\Lambda$ in place of C. Let

$$F_\gamma(p) = (p_1^2 + 1)^{-(1/2)+\gamma}(p_2^2 + 1)^{-(1/2)+\gamma}$$

and let $C^\sim(p, q)$ be the Fourier transformed kernel of $\chi_\Lambda C \chi_\Lambda$. The proof is contained in the following two lemmas.

Lemma 7.9.5. *For any $\gamma > 0$,*

$$|C^{\sim}(p, q)| \leq \text{const } F_\gamma(p)F_\gamma(q).$$

PROOF. C is a sum of reflected translates of $C_\varnothing$. Let C_j be a single term. Translation in x space does not affect $|C_j^{\sim}|$, while reflection produces the substitution $p \to -p$. Multiplication by χ_Λ gives convolution by $\tilde{\chi_\Lambda}$, and $|\tilde{\chi_\Lambda}| \leq$ const F_γ. Thus

$$|C_j^{\sim}| \leq \text{const} \int F_\gamma(p - r)(r^2 + 1)^{-2}F_\gamma(r - q)\, dr \qquad (7.9.8)$$

$$\leq \text{const } F_{2\gamma}(p)F_{2\gamma}(q).$$

To obtain convergence of the sum over j, we argue differently. By the representation (7.2.6) for g and by the identity $C_\varnothing(x, y) = \text{const } g(|x - y|)$, we deduce exponential decay for all derivatives of $C_\varnothing$, for $|x - y|$ large. Thus $C_j(x, y) = \chi_\Lambda(x)v_j(x, y)\chi_\Lambda(y)$ where

$$|\tilde{v}_j(r, s)| \leq \text{const}(1 + r^2 + s^2)^{-n}e^{-0(\text{dist}(\Lambda, y_j))}$$

for any finite n. Then (7.9.8) can be modified so that the final constant is exponentially small in the translation defining C_j. Thus we can sum over j to complete the proof.

Lemma 7.9.6. *For any $\gamma > 0$*

$$\|\chi_\Lambda C\chi_\Lambda(\delta_{\kappa_1} - \delta_{\kappa_2})\zeta\|_{HS} \leq \text{const } \kappa_1^{-1/2 + \gamma}.$$

PROOF. Let c be the Fourier transformed kernel of $\chi_\Lambda C\chi_\Lambda(\delta_{\kappa_1} - \delta_{\kappa_2})\zeta$. By (7.9.6), (7.9.7), and Lemma 7.9.5,

$$|c| \leq \text{const } F_\gamma(p) \int F_\gamma(r)|\tilde{h}(r/\kappa_1) - \tilde{h}(r/\kappa_2)|\zeta^{\sim}(r - q)\, dr$$

$$\leq \text{const } F_\gamma(p) \int F_\gamma(r)^{1-a}\kappa_1^{-a(1-\gamma)}(|r - q|^2 + 1)^{-2}\, dr$$

$$\leq \text{const } F_\gamma(p)F_\gamma(q)^{1-a}\kappa_1^{-a(1-\gamma)}.$$

We choose $a < \frac{1}{2}$, $(1 - a)(1 - \gamma) > \frac{1}{2}$ and then $\|c\|_{L_2} = O(\kappa_1^{-a(1-\varepsilon)})$ to prove the lemma.

PROOF OF THEOREM 7.9.2 FOR $p = 2$, $C = C_\Gamma$. In this case $0 \leq C \leq C_\varnothing$, so

$$0 \leq C_\varnothing^{-1/2}CC_\varnothing^{-1/2} \leq 1.$$

Let $A = \zeta_1 C_\varnothing^{1/2}C_\varnothing^{-1/2}CC_\varnothing^{-1/2}$, $B = C_\varnothing^{1/2}(\delta_{\kappa_1} - \delta_{\kappa_2})\zeta$. Since

$$\|AB\|_{HS}^2 = \text{Tr}(AB)^*AB = \text{Tr } B^*A^*AB = \text{Tr } A^*ABB^*$$

$$\leq (\text{Tr}(A^*A)^2)^{1/2}(\text{Tr}(B^*B)^2)^{1/2} = \|A^*A\|_{HS}\|B^*B\|_{HS},$$

$$\|\zeta_1 C(\delta_{\kappa_1} - \delta_{\kappa_2})\zeta\|_{HS} = \|AB\|_{HS} \leq \|A^*A\|_{HS}^{1/2}\|B^*B\|_{HS}^{1/2}$$

By Lemmas 7.9.3 and 7.9.4, $\|B^*B\|_{HS} \leq$ const $\kappa_1^{-1/2 + \varepsilon}$ for $\varepsilon > 0$. Also,

$$\|A^*A\|_{HS} \leq \|C_\varnothing^{-1/2}CC_\varnothing^{-1/2}\|^2 \|C_\varnothing^{1/2}\zeta_1^2 C_\varnothing^{1/2}\|_{HS} \leq \|\zeta_1 C_\varnothing \zeta_1\|_{HS} \leq \text{const}$$

by Lemma 7.9.4.

PROOF OF THEOREM 7.9.2, GENERAL CASE. For $p = 2$, we take convex sums in the bound of the theorem. Thus the previous cases are sufficient to prove the theorem for $p = 2$. Let c_κ denote the kernel of $\zeta_1 C \delta_\kappa \zeta_2$. Then $c_\kappa \in L_p(R^2 \times R^2)$ with a norm which is bounded uniformly in κ, by Proposition 7.9.1. Thus with $\delta c = c_{\kappa_1} - c_{\kappa_2}$,

$$\|\delta c\|_{L_{2n}}^{2n} = \int \delta c^{2n} \, dx \leq \left(\int \delta c^2 \, dx \right)^{1/2} \left(\int \delta c^{4n-2} \, dx \right)^{1/2}$$

$$\leq \text{const } \kappa_1^{-\varepsilon} (\|c_{\kappa_1}\|_{L_{4n-2}} + \|c_{\kappa_2}\|_{L_{4n-2}})^{2n-1}$$

$$\leq \text{const } \kappa_1^{-\varepsilon}$$

by the previous cases, $p = 2$.

7.10 Reflection Positivity

Positivity of the inner product in the Hilbert space $\mathcal{H}$ of quantum mechanical states was derived from the reflection positivity property of the covariance operator C for the Gaussian measure $d\phi_C$; see Theorem 6.2.2. In Chapter 6, we established this positivity for the free covariance $(-\Delta + I)^{-1}$ by a direct computation in Proposition 6.2.5. The corresponding positivity for non-Gaussian quantum fields relies on positivity in the Gaussian case, as we will see in Chapter 10.4.

In order to construct nonlinear quantum fields, it is convenient to establish reflection positivity for other measures which occur as intermediate steps in the construction. Thus we present here an analysis of the corresponding covariance operators $C = (-\Delta_B + I)^{-1}$, where B denotes Dirichlet and/or Neumann data on a union Γ of piecewise smooth hypersurfaces. We assume that the boundary conditions are reflection invariant, and we give a general proof of reflection positivity for C. In fact we show that reflection positivity is equivalent to monotonicity of C with respect to the introduction of Dirichlet and Neumann boundary data on the reflection hyperplane. The method we present also establishes reflection positivity for periodic boundary conditions and for lattice approximations to C.

We formulate reflection positivity for $C = (-\Delta_B + I)^{-1}$ as follows: Let Π denote a hyperplane in R^d, and let $\theta = \theta_\Pi$ denote reflection of R^d about Π. Let $R_\pm^d$ denote the two connected components of $R^d \backslash \Pi$. Note that on $L_2(R^d)$, θ is self-adjoint, and $\theta^2 = I$. Let $\Pi_\pm$ denote the orthogonal projection onto $L_2(R_\pm^d)$.

Definition 7.10.1. An operator C is θ-invariant (i.e. reflection-invariant with respect to Π) if $[\theta, C] = 0$.

Definition 7.10.2. A θ-invariant operator C is reflection-positive with respect to θ if

$$0 \leq \Pi_+ \theta C \Pi_+ . \tag{7.10.1}$$

Remark 1. Since $[\theta, C] = 0$, (7.10.1) is equivalent to $0 \leq \Pi_- \theta C \Pi_-$. The definition is a reformulation of Definition 6.2.1.

Remark 2. By Theorem 6.2.2, a reflection positive C is the covariance of a reflection-positive Gaussian measure $d\phi_C$.

Theorem 7.10.1. *Let $C = (-\Delta_B + I)^{-1}$ be θ-invariant, and let B denote Dirichlet and/or Neumann data on Γ. Then C is reflection-positive.*

PROOF. Let $P_\pm = \frac{1}{2}(I \pm \theta)$ be the projections onto the even and odd eigenspaces of θ. Note that $(I - \theta)C = 2P_- C = 2CP_-$. Thus the positivity condition (7.10.1) can be written

$$0 \leq \Pi_+[C - 2P_- C]\Pi_+ . \tag{7.10.2}$$

We assert first that $2P_- C$, restricted to $\Pi_\pm$, equals C_D restricted to $\Pi_\pm$. Here $C_D = (-\Delta_{B'} + I)^{-1}$ where $\Delta_{B'}$ is the Laplace operator with Dirichlet data on Π and B data on $\Gamma \backslash \Pi$. Secondly we assert $C_D \leq C$. Then positivity (7.10.2) is a consequence of these two assertions.

The first assertion follows from the observation that $(I - \theta)C(x, y)$ vanishes on Π, and that $(I - \theta)C$ satisfies $(-\Delta + I)(I - \theta)C(x, y) = \delta(x - y)$. The second assertion follows from the characterization of the bilinear form $\Pi_\pm C_D^{-1}\Pi_\pm$ as a restriction of the bilinear form C^{-1} to functions vanishing on Π and supported in $\Pi_\pm$. See Section 7.7 for a more extensive discussion.

Remark 1. The free covariance $\Gamma = \varnothing$ is a special case. Then using (7.5.1), with $\Gamma = \Pi$, we find that C_D is the ordinary covariance $(-\Delta_\Pi + I)^{-1}$ with Dirichlet data on Π only.

Remark 2. Define

$$C_N = (I + \theta)C = 2P_+ C = 2CP_+ . \tag{7.10.3}$$

In place of (7.10.2), one can write reflection positivity (7.10.1) as

$$0 \leq \Pi_+ \theta C \Pi_+ = \Pi_+(C_N - C)\Pi_+ , \tag{7.10.4}$$

or equivalently

$$0 \leq \Pi_-(C_N - C)\Pi_- .$$

We state, without proof, that C_N is the covariance with Neumann data on Π and B data on $\Gamma \backslash \Pi$. The free covariance is a special case where the identification of C_N is straightforward, and was given in (7.4.1).

Remark 3. From (7.10.2-4) we find that reflection positivity is equivalent to

$$0 \leq \Pi_+ \theta C \Pi_+ = \frac{1}{2}\Pi_+(C_N - C_D)\Pi_+$$

or to

$$0 \leq \Pi_-(C_N - C_D)\Pi_- .$$

Since $\Pi_+(C_N - C_D)\Pi_- = 0$, we infer

$$C_D \leq C_N. \tag{7.10.5}$$

We stated in the previous remark that C_N is the covariance C modified by Neumann data on Π. Assuming this identification, we have proved a generalization of the monotonicity statements established in Sections 7.7–8, namely: If $C = (-\Delta_B + I)^{-1}$ is θ-invariant, then $C_D \leq C_N$; here D and N refer to Dirichlet or Neumann data on Π and B data on $\Gamma\backslash\Pi$. Furthermore this monotonicity inequality (7.10.5) is equivalent to reflection positivity with respect to the hyperplane Π.

We now allow periodic boundary conditions. Consider R^d replaced by the torus T^d, and replace Π by a cross section dividing T^d in two equal parts. This is easily visualized by imbedding T^d in R^{d+1} and extending Π to a hyperplane in R^{d+1}, as in Figure 7.1, drawn for $d = 1$. Now $\Pi_\pm$ are

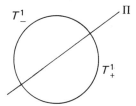

Figure 7.1. Hyperplane reflection on T^1. $T^1_\pm$ are the two components of $T^1\backslash\Pi$.

the projections onto L_2 functions supported on either side of Π. We allow Dirichlet and/or Neumann data B as well. With these replacements, Definitions 7.10.1–2 are made. As above, we show

Theorem 7.10.2. *Let $C = (-\Delta_B + I)^{-1}$ be θ-invariant on $L_2(T^d)$. Then C is reflection-positive.*

Finally we remark that R^d or T^d can be replaced by the lattice Z^d or the finite periodic lattice Z^d_n with spacing δ. In this case we obtain OS positivity for classical θ-invariant boundary conditions on a lattice. Again we use the inequality $C_D \leq C$, which can also be established by introducing Dirichlet data with a mass perturbation of C^{-1} on the lattice; cf. Sections 7.8 and 9.5.

Theorem 7.10.3. *The lattice analogs of Theorems 7.10.1 and 7.10.2 hold.*

CHAPTER 8

Quantization = Integration over Function Space

8.1 Introduction

The construction of $P(\phi)_2$ fields begins with this chapter; Chapters 11–12 conclude the construction. We develop efficient methods for computing and estimating integrals

$$\int A \, d\phi = \sum_{G=\text{graph}} I(G) \tag{8.1.1}$$

of polynomials $A = A(\phi)$ with respect to a Gaussian measure $d\phi$. There are a variety of equivalent methods for computing Gaussian integrals of polynomials, such as integration by parts, expansion in Hermite polynomials, or the use of annihilation and creation (raising and lowering) operators. These methods are purely algebraic. The Feynman graph G is a mnemonic device to record and label the terms generated by this method. Each term $I(G)$ is an integral over a finite dimensional space. The integral depends on the covariance C of $d\phi$; thus estimates on $I(G)$ follow from Hölder's inequality together with estimates on C. The required estimates, e.g.

$$C(x - y) \equiv \text{kernel}(-\Delta + m^2)^{-1} \in L_p(x - y),$$

depend strongly on the dimension d of space time. The estimates of this chapter are appropriate for $d = 2$ (and $d = 1$ is included as an easy special case). For $d = 2$, the renormalizations of quantum field theory are simple. In particular, for $P(\phi)_2$ fields, only Wick ordering subtractions are required. We show that this infinite subtraction is well defined. Thus for a bounded region Λ,

$$V = \int_\Lambda :P(\phi(x)): dx \in L_2(d\phi)$$

is meaningful, while e.g. $\int_\Lambda \phi(x)^2 \, dx$ is infinite for a.e. $\phi \in \mathcal{D}'$. Now restrict P to be bounded from below. Although $\int :P(\phi)\,(x): dx$ is no longer bounded from below (because of the Wick ordering subtractions), it is semibounded except on sets of small measure. This fact allows us to show that

$$e^{-V} \in L_p(d\phi), \quad \text{all } p < \infty.$$

8.2 Feynman Graphs

Feynman graphs (or diagrams) are mnemonics for reducing Gaussian functional integrals to finite dimensional integrals. They result from repeated application of the integration by parts formula

$$\int \phi(f)A(\phi) \, d\phi_C = \int \left\langle Cf, \frac{\delta A}{\delta \phi} \right\rangle d\phi_C \tag{8.2.1}$$

of Section 6.3. After multiple integrations by parts, integrals $\int A \, d\phi_C$, for polynomial functions $A(\phi)$, are reduced to a sum of finite dimensional integrals,

$$\int A \, d\phi_C = \sum_{\{G\}} I(G), \tag{8.2.2}$$

where G ranges over a set of graphs and $I(G)$ is a number (defined by a finite dimensional definite integral) assigned to G. Here C denotes a general covariance of a Gaussian measure. (While some authors define diagrams as equivalence classes of graphs, we use these two words interchangeably.)

The simplest example, and the one from which the examples below follow, is the integral of a product $A(\phi) = \phi(f_1)\phi(f_2) \cdots \phi(f_r)$. Such an integral $\int A \, d\phi_C$ is a moment of $d\phi_C$. In particular, define

$$(f, g)_C = \int f(x)C(x, y)g(y) \, dx \, dy. \tag{8.2.3}$$

For r odd, $\int \phi(f_1) \cdots \phi(f_r) \, d\phi_C = 0$; for r even,

$$\int A(\phi) \, d\phi_C = \int \phi(f_1) \cdots \phi(f_r) \, d\phi_C$$

$$= \sum_{\text{pairings}} (f_{i_1}, f_{i_2})_C \cdots (f_{i_{r-1}}, f_{i_r})_C . \tag{8.2.4}$$

The sum in (8.2.4) extends over the $(r-1)(r-3)(r-5) \cdots 1 = (r-1)!!$ distinct ways of choosing the pairs $\{f_{i_j}, f_{i_{j+1}}\}$.

We assign a Feynman graph G to each term in the sum (8.2.4), and we define $I(G)$ to be equal to the term. In general, we have a set of rules of associating integrals to graphs, and a second set of rules giving the graphs corresponding to an integral $\int A(\phi) \, d\phi_C$.

In particular, a graph is a collection of vertices (represented as points in R^d), lines (represented as line segments joining vertices), and legs (represented as line segments which attach at one endpoint only to a vertex). Each leg represents a factor ϕ in the integrand; the vertex it joins represents the argument x of that $\phi(x)$. The integration by parts formula (8.2.1) can be interpreted graphically as the pairing (joining) of two legs to form a line. Finally each line gives rise to a propagator $C(x, y)$ in the integral which defines $I(G)$.

An advantage of the graphical representation is that it allows one to see by inspection whether singular (infinite) terms occur as an $I(G)$ contributing to (8.2.2). For $d = 2$, Wick ordering eliminates singular terms, as we will see in Section 8.3.

Let us consider as an example the evaluation of

$$\int \phi(x_1) \cdots \phi(x_6) \, d\phi_C . \tag{8.2.5}$$

Here we have taken the limit $f_j = \delta_{x_j}$. We assign a graph to the integrand with a vertex for each x_j and a leg at each vertex denoting $\delta(x_j)$ (see Figure 8.1). The integral $\int A \, d\phi_C$ is given by a set of graphs with no free legs (i.e. a constant function on path space). These 15 graphs (in the case of (8.2.5)) are obtained by pairing the legs in Figure 8.1 in all possible manners. In

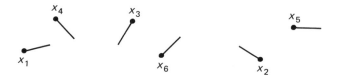

Figure 8.1. The graph for $\phi(x_1) \cdots \phi(x_6)$.

particular, if we use (8.2.1) to integrate by parts the factor $\phi(x_4)$, the right side of (8.2.4) is given by the sum of graphs with the x_4 leg paired to one of the other five legs (Figure 8.2). The particular graph in Figure 8.2 corresponds to the integrand $C(x_1, x_4)\phi(x_2)\phi(x_3)\phi(x_5)\phi(x_6)$. After three integrations by parts, the integral of $\phi(x_1) \cdots \phi(x_6)$ is reduced to a sum of

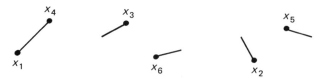

Figure 8.2. The integrand after integrating $\phi(x_4)$ by parts: one of five possible graphs.

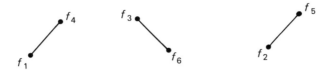

Figure 8.3. A Feynman graph contributing to $\int \phi(f_1) \cdots \phi(f_6)\, d\phi_C$.

constants, each represented by a graph (Figure 8.3). We call graphs with no legs "fully contracted", or "paired". The factors $f(x_1) \ldots f(x_6)$ can be reintroduced, and $\int \phi(f_1) \cdots \phi(f_6)\, d\phi_C$ is then expressed as a sum

Figure 8.4. The elementary graph for $\int \phi(f)\phi(g)\, d\phi_C$.

over $I(G)$. The graph of Figure 8.4 is assigned $I(G) = (f, g)_C = \int f(x)C(x, y)g(y)\, dx\, dy$. A fully contracted graph G, which is the union of connected components, $G = G_1 \cup G_2 \cup \cdots \cup G_k$, is assigned the value

$$I(G) = I(G_1)I(G_2)\cdots I(G_k) = \prod_{\substack{\text{connected} \\ \text{components}}} I(G_j). \qquad (8.2.6)$$

Let G be a graph with lines l, where l denotes a line connecting space-time points x_{l_1}, x_{l_2}. We also let l_1 denote the vertex corresponding to the space-time point x_{l_1}. Thus we can write

$$I(G) = \int \left(\prod_{\substack{\text{lines} \\ l \text{ in } G}} C(x_{l_1}, x_{l_2}) \right) \prod_{\substack{\text{vertices} \\ l_j \text{ in } G}} (f_j(x_{l_j})\, dx_{l_j}). \qquad (8.2.7)$$

8.3 Wick Products

It is important to integrate more general functions, and in particular Wick monomials and their products. Let

$$:\phi^n(f):_C = \int :\phi(x_1) \cdots \phi(x_n):_C f(x_1, \ldots, x_n)\, dx_1 \cdots dx_n,$$

and

$$A(\phi) = \prod_{i=1}^{r} :\phi^{n_i}(f_i):_{C_i}. \qquad (8.3.1)$$

Here $:\ :_C$ denotes Wick ordering, as defined in Section 6.3, using the covariance operator C. We also define

$$\delta C_i(x, y) = C(x, y) - C_i(x, y), \qquad \delta c_i(x) = \delta C_i(x, x). \qquad (8.3.2)$$

In deriving rules for $\int A(\phi)\, d\phi_C$, it is convenient to use diagrams in which $:\phi^n(f):_C$ is represented by a single vertex (labeled by f, or $(x_1, \ldots, x_n)$) with

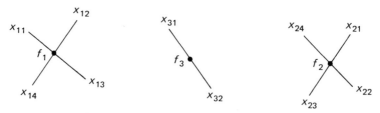

Figure 8.5. The graph for $A = : \phi^4(f_1) :_{C_1} : \phi^4(f_2) :_{C_2} : \phi^2(f_3) :_{C_3}$.

n free legs, labeled by $x_1, \ldots, x_n$. Consider the example of $A(\phi)$, in the case $r = 3, n_1 = n_2 = 4, n_3 = 2$; then A is represented by the graph of Figure 8.5. The integral $\int A \, d\phi_C$ is represented by a sum of graphs, obtained by pairing the legs in the graph for A in all possible manners. (In this example, there are $9!!$ such graphs.)

We must now distinguish two types of lines in the graphs contributing to $\int A \, d\phi_C$, namely "self-lines" obtained by pairing two legs at a given vertex, and "interaction lines" obtained by pairing two legs from different vertices. Let $\mathscr{L}_s, \mathscr{L}_i$ denote the set of self-lines and interaction lines in a graph G. For example, three graphs contributing to the integral of A in Figure 8.5 are shown in Figure 8.6(a–c). The general formula for $\int A \, d\phi_C$ is the following:

Proposition 8.3.1. *For $A(\phi)$ of the form* (8.3.1),

$$\int A(\phi) \, d\phi_C = \sum_{\{G\}} I(G), \tag{8.3.3}$$

where $\{G\}$ ranges over the $\left(\sum_{i=1}^{r} n_i - 1\right)!!$ fully contracted graphs with r vertices and n_i legs at the ith vertex. Furthermore,

$$I(G) = \int \left(\prod_{l \in \mathscr{L}_i} C(x_{l_1}, x_{l_2}) \right) \left(\prod_{l \in \mathscr{L}_s} \delta C_{l_1}(x_{l_1}, x_{l_2}) \right)$$

$$\times \prod_{i=1}^{r} f_i(x_{i1}, \ldots, x_{in_i}) \prod_{j=1}^{n_i} dx_{ij} . \tag{8.3.4}$$

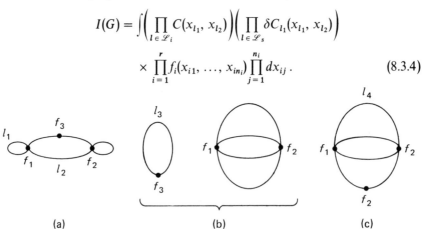

(a) (b) (c)

Figure 8.6. Three graphs, (a), (b), and (c), contributing to the integral of Figure 8.5. The graph (b) is disconnected. Lines l_1, l_3 are self-lines, while lines l_2, l_4 are interaction lines. The graph (c) has no self-lines.

Remark 1. In (8.3.3), we have regarded the legs of the graphs $G \in \mathscr{G}$ as labeled. This is consistent with there being no symmetry assumption on the functions $f_i(x_{ij})$. In case certain f_i are symmetric under permutation of their n_i arguments, we can rewrite (8.3.3) as

$$\int A(\phi) \, d\phi_C = \sum_{\{G\}'} I(G)c(G), \tag{8.3.3'}$$

where $\{G\}'$ is an equivalence class of graphs, each giving the same $I(G)$, because they differ only by a relabeling of the legs at each vertex, and $c(G)$ is the number of (labeled) graphs in the equivalence class.

PROOF. We state the Hermite recursion relation:

$$:\phi(x_1) \cdots \phi(x_n):_{C_i} = \phi(x_1):\phi(x_2) \cdots \phi(x_n):_{C_i}$$

$$- \sum_{j=2}^{n} C_i(x_1, x_j):\phi(x_2) \cdots \hat{\phi}(x_j) \cdots \phi(x_n):_{C_i}, \tag{8.3.5}$$

where $\hat{\phi}(x_j)$ denotes the omission of the factor $\phi(x_j)$; see (9.1.7–8).
    We use a second identity (cf. (9.1.3))

$$\frac{\delta}{\delta\phi(y)}:\phi(x_2) \cdots \phi(x_n):_{C_i} = \sum_{j=2}^{n} \delta(y - x_j):\phi(x_2) \cdots \hat{\phi}(x_j) \cdots \phi(x_n):_{C_i}. \tag{8.3.6}$$

Using (8.3.5–6) and the integration formula (8.2.1), we can integrate by parts the factor $\phi(x_{11})$ in the integral

$$\int :\phi(x_{11}) \cdots \phi(x_{1n_1}):_{C_1} B(\phi) \, d\phi_C.$$

The $\delta/\delta\phi$ derivative will act on the product $:\phi(x_{12}) \cdots \phi(x_{1n_1}):_{C_1} B(\phi)$. In the derivative $\delta B/\delta\phi$, the integration by parts formula introduces the covariance C of the Gaussian measure and an interaction line in G. The derivative of $:\phi(x_{12}) \cdots \phi(x_{1n_1}):_{C_1}$, however, combines with the second term on the right of (8.3.5) to yield a covariance δC_1 and a self-line in G. Iterating this procedure until all ϕ's have been eliminated by integration by parts yields (8.3.3–5).

Remark 2. If $C_1 = \cdots = C_r = C$, then all the $\delta C_i = 0$. This eliminates all self-lines and gives

Corollary 8.3.2. *For*

$$A(\phi) = \prod_{i=1}^{r} :\phi^{n_i}(f_i):_C, \tag{8.3.7}$$

then

$$\int A \, d\phi_C = \sum_{\{G\}} I(G), \tag{8.3.8}$$

where G runs over graphs without self-lines, and

$$I(G) = \int \left(\prod_{l \in \mathscr{L}_i = \mathscr{L}} C(x_{l_1}, x_{l_2}) \right) \prod_{i=1}^{r} f_i(x_{i1}, \ldots, x_{in_i}) \prod_{j=1}^{n_i} dx_{ij}. \tag{8.3.9}$$

Remark 3.   In the limit (justified in Section 8.5)

$$f_i(x_{i1}, x_{i2}, \ldots, x_{ir}) = \int f_i(x)\delta(x - x_{i1}) \cdots \delta(x - x_{ir}) \, dx, \quad (8.3.10)$$

we consider Wick monomials

$$:\phi^{n_i}(f_i):_{C_i} = \int :\phi^{n_i}(x):_{C_i} f_i(x) \, dx \quad (8.3.11)$$

and their products. In this case, we obtain

Corollary 8.3.3.   *Let*

$$A(\phi) = \prod_{i=1}^{r} \int :\phi^{n_i}(x):_{C_i} f_i(x) \, dx. \quad (8.3.12)$$

Then

$$\int A(\phi) \, d\phi_C = \sum_{\{G\}} I(G), \quad (8.3.13)$$

where G ranges over all graphs as in the proposition, and

$$I(G) = \int \prod_{l \in \mathscr{L}_i} C(x_{l_1}, x_{l_2}) \prod_{l \in \mathscr{L}_s} \delta c_{l_1}(x_{l_1}) \prod_{i=1}^{r} f_i(x_i) \, dx_i. \quad (8.3.14)$$

8.4 Formal Perturbation Theory

We are interested not only in Gaussian measures, but also in integrals of Wick products $A(\phi)$ in a measure such as

$$d\mu = d\mu_C = \frac{1}{Z} e^{-V} \, d\phi_C, \quad (8.4.1)$$

where Z is chosen to normalize $d\mu$, $\int d\mu = 1$. In this case, integration by parts of functions $A(\phi)$ does not reduce A to a sum of finite dimensional integrals. Rather, the integration by parts formula

$$\int \phi(f)A(\phi) \, d\mu = \int \left\langle Cf, \left(\frac{\delta A}{\delta \phi} - A \frac{\delta V}{\delta \phi} \right) \right\rangle d\mu \quad (8.4.2)$$

has an extra term arising from differentiation of the exponent e^{-V}. Multiple applications of (8.4.2) give a series in powers of V, integrated in $d\mu$. As in Section 8.2, we represent each such integrand by a sum of Feynman graphs. The graph for a monomial $:\phi^n:_C$, as before, is a vertex with n free legs. The legs are paired to form lines by the first term in (8.4.2). The second term in (8.4.2) adds a new vertex (with one leg paired to the leg of ϕ).
    Iteration of (8.4.2) can be interpreted as follows: The existing legs of an

integrand either contract with each other, or contract to the "exponent", producing a new factor $\delta V/\delta\phi$.

Now replace V by λV and write

$$\int A(\phi)\, d\mu \sim \sum_l a_l \lambda^l. \tag{8.4.3}$$

Because $d\mu$ is in general a singular perturbation of $d\phi_C$, $\int A(\phi)\, d\mu$ is not in general an analytic function of λ. In such cases (8.4.3) is interpreted as asserting that $\int A(\phi)\, d\mu$ is C^∞ in λ at $\lambda = 0$, with lth derivatives at $\lambda = 0$ given by $a_l l!$.

We now show that each coefficient a_l is represented as a sum of Feynman diagrams—explicitly,

$$a_l = \sum_{G \in \mathscr{G}_l} I(G), \tag{8.4.4}$$

where $\mathscr{G}_l$ is a set of fully contracted labeled graphs (having lines and vertices but no unpaired legs). By way of definition, each graph $G \in \mathscr{G}_l$ has l V-vertices as well as the vertices defined by $A(\phi)$. Also each $G \in \mathscr{G}_l$ is connected, in the sense that each connected component of the graph contains at least one $A(\phi)$ vertex; $\mathscr{G}_l$ is the set of all such graphs. $I(G)$ is given by (8.3.4).

To establish (8.4.4) up to order l, we integrate by parts, if possible, until $l+1$ factors $\lambda\, \delta V/\delta\phi$ have been produced by $l+1$ differentiations of the exponent. All such terms are $O(\lambda^{l+1})$, and can be neglected. The remaining terms are those for which repeated integration by parts is not possible. These are the terms in which the polynomial integrand has reduced to a constant. They are evaluated as

$$\frac{1}{Z}\int \text{const }d\mu = \frac{Z}{Z} \times \text{const} = \text{const}, \tag{8.4.5}$$

where the constant is $I(G)$ for some G. The graphs G produced in this manner are exactly the G in $\mathscr{G}_l$ as defined above.

Conventional discussions of the formula (8.4.4) include an expansion of $1 = Z/Z$ in (8.4.5) in powers of λ, which, after cancellation of disconnected vacuum diagrams, gives the same result. Normally this formula is written in terms of unlabeled diagrams, and then includes combinatorial factors $c(G)$ as in (8.3.4). Note that the zero order term in the series is

$$a_0 = \int A(\phi)\, d\phi_C = (8.2.2),$$

the sum of all graphs with only $A(\phi)$ vertices.

There are a variety of connectivity properties, useful in characterizing the graphs which contribute to various truncated expectations. A simple example is (8.4.4–5), and another occurs in the truncated expectation

$$\int AB\, d\mu - \int A\, d\mu \int B\, d\mu \equiv \langle AB \rangle^T. \tag{8.4.6}$$

It follows from (8.4.3–4) that

$$\langle AB \rangle^T = \sum b_l \lambda^l, \tag{8.4.7}$$

where

$$b_l = \sum_{G \in G_l{}^T} I(G). \tag{8.4.8}$$

Here we define $\mathscr{G}_1 = \mathscr{G}_1(AB)$ to be the family of graphs which contribute to $\int AB \, d\mu$ as in (8.4.3–4). Then $\mathscr{G}_l^T \subset \mathscr{G}_l(AB)$ consists of the graphs connected as above and connected also in the stronger sense that at least one connected component must contain vertices from both A and B. Higher connectivity properties define various Bethe–Salpeter kernels (see Chapter 14) and Legendre transforms.

8.5 Estimates on Gaussian Integrals

In two dimensions, the Wick polynomials and exponentials are in $L_p(d\phi_C)$ for all $p < \infty$. In dimension two (and higher), the Gaussian measure $d\phi_C$ is supported in a space of distributions, so pointwise operations such as $\phi(x) \rightarrow \phi(x)^2$ are suspect. In fact,

$$\phi(x)^2 = +\infty, \quad \text{a.e. with respect to } d\phi_C, \tag{8.5.1}$$

because

$$\int_\Lambda \, :\phi(x)^2:_C \, dx \in L_2(d\phi_C). \tag{8.5.2}$$

In (8.5.2), the Wick ordering constant c is logarithmically infinite for $d = 2$, by (7.2.5), since $c = C(x, x)$. But formally

$$\int_\Lambda \phi(x)^2 \, dx = \int_\Lambda \, :\phi(x)^2: \, dx + c|\Lambda|.$$

In fact, the functions which are locally integrable in $\mathscr{D}'$ form a measure zero subset of the support of $d\phi_C$. It follows that functions do not influence properties of measure $d\phi_C$. To make the foregoing mathematically precise, we introduce a momentum cutoff or mollifier. As in Section 7.1, choose $h \in C_0^\infty(R^2)$, with

$$0 \le h(y) \quad \text{and} \quad \int h(y) \, dy = 1,$$

and define the approximate δ function centered at $x \in R^2$ by

$$\delta_{\kappa, x}(y) = \kappa^2 h(\kappa(x - y)). \tag{8.5.3}$$

Then we define the momentum cutoff field ϕ_κ as

$$\phi_\kappa(x) = \phi(\delta_{\kappa, x}) = \int \phi(y)\delta_{\kappa, x}(y) \, dy. \tag{8.5.4}$$

The momentum cutoff Wick monomials are defined as the orthogonal polynomials with respect to the Gaussian measure $d\phi_C$. In particular, the nth Wick power of $\phi_\kappa(x)$ is defined as the nth Hermite polynomial

$$: \phi_\kappa(x)^n :_C = \sum_{j=0}^{[n/2]} \frac{(-1)^j n!}{(n-2j)! j! 2^j} \, c_\kappa(x)^j \phi_\kappa(x)^{n-2j}, \qquad (8.5.5)$$

where $c_\kappa(x) = \langle \delta_{\kappa,x}, C\delta_{\kappa,x} \rangle$ and

$$: \phi_\kappa^n(f) :_C = \int : \phi_\kappa(x)^n :_C f(x) \, dx. \qquad (8.5.6)$$

Note that (8.5.5) is identical with the definition (6.3.9) using (8.5.4) and with $c = c_\kappa$. This can be seen from the discussion of the single harmonic oscillator; cf. (1.5.12–16). Here C is any covariance operator in $\mathscr{C}_m$, namely convex combinations of the Euclidean propagators $(-\Delta_B + m_i^2)^{-1}$ with classical boundary conditions and $m \leq m_1$; see Section 7.9.

Proposition 8.5.1. *Let $C \in \mathscr{C}_m$ and $f \in L_p(R^2)$ for some $p > 1$, and let f have compact support. Then $: \phi_\kappa^n(f) :_C \in L_2(d\phi_C)$, and converges in L_2 as $\kappa \to \infty$. Let $: \phi^n(f) :_C$ denote this limit. Then for some $\delta > 0$*

$$\| : \phi_\kappa^n(f) :_C - : \phi^n(f) :_C \|_{L_2(d\phi_C)} \leq O(\kappa^{-\delta}) \qquad (8.5.7)$$

as $\kappa \to \infty$.

PROOF. As in Section 8.4, we represent $\int : \phi_\kappa^n(f) :_C^2 \, d\phi_C$ as a sum of $n!$ (identical) Feynman diagrams, each shown in Figure 8.7.

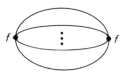

Figure 8.7. The Feynman graph for $\int : \phi_\kappa^n(f) :_C^2 \, d\phi_C$.

For this graph

$$I(G) = \int f(x) C_\kappa(x, y)^n f(y) \, dx \, dy, \qquad (8.5.8)$$

where

$$C_\kappa(x, y) = \delta_\kappa C \delta_\kappa \equiv \int \delta_\kappa(x - x') C(x', y') \delta_\kappa(y' - y) \, dx' \, dy'. \qquad (8.5.9)$$

Let ζ be a continuous function with compact support, and let $\zeta \equiv 1$ on supp f. Let $I(G, \kappa)$ denote the integral (8.5.8). Let $\delta C = C_{\kappa_1} - C_{\kappa_2}$. It follows by Hölder's inequality and property LR2 of Theorem 7.1.1 that for $\kappa = \min(\kappa_1, \kappa_2)$

$$|I(G, \kappa_1) - I(G, \kappa_2)| \leq \text{const} \| f \|_{L_p}^2 \| \zeta \, \delta C \, \zeta \|_{L_{p'}} \leq O(\kappa^{-\delta}) \qquad (8.5.10)$$

for some $\delta > 0$. Hence $: \phi^n(f) :_C$ converges as $O(\kappa^{-\delta})$ in L_2, and the proof is complete.

We now systematize the use of Hölder's inequality for application to more complicated graphs. We consider a function of the form

$$F(x_1, \ldots, x_n) = \prod_{i=1}^{n} f_i(x_i) \prod_l F_l(x_{i_1(l)}, x_{i_2(l)}), \qquad (8.5.11)$$

where for each index l,

$$1 \leq i_1(l) < i_2(l) \leq n.$$

Such an F arises in a graph with n vertices labeled by i, and lines l which join vertices $i_1(l)$ to $i_2(l)$. The f_i represent either self-lines or test functions for the vertex., while F_l is the covariance operator for the line l restricted to the support of $f_{i_1(l)} \times f_{i_2(l)}$. We can regard the vertices as ordered (by i) and the lines as directed (from $i_2(l)$ to $i_1(l)$). Then

$$\mathcal{L}_i^{\text{in}} = \{l : i_1(l) = i < i_2(l)\}$$

and

$$\mathcal{L}_i^{\text{out}} = \{l : i_1(l) < i = i_2(l)\}$$

are the incoming and outgoing lines respectively at the ith vertex; cf. Figure 8.8.

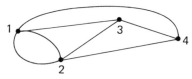

Figure 8.8. Ordering vertices from left to right, "in" lines leave vertices on right, while "out" lines leave vertices on left. Vertex 1 has only "in" lines, vertex 4 only "out" lines.

Lemma 8.5.2. *In the above notation,*

$$\|F\|_{L_q} \leq \prod_i \|f_i\|_{L_{q_i}} \prod_l \|F_l\|_{L_{q_l}},$$

provided

$$q_i^{-1} + \sum_{l \in \mathcal{L}_i^{\text{in}} \cup \mathcal{L}_i^{\text{out}}} q_l^{-1} = q^{-1}$$

for each i.

PROOF. With the substitution $F \to F^q$, we are reduced to the case $q = 1$. We apply Hölder's inequality successively in each of the n variables of F, starting with $i = n$, and proceed by a finite induction to $i = 1$. Thus the ith inequality is

$$\int_{R^2} |f_i(x_i)| \prod_{l \in \mathcal{L}_i^{\text{out}}} |F_l(x_{i_1(l)}, x_i)| \prod_{l \in \mathcal{L}_i^{\text{in}}} \|F_l(x_i, \cdot)\|_{L_{q_l}(R^2)} \, dx_i$$

$$\leq \|f_i\|_{L_{q_i}(R^2)} \prod_{l \in \mathcal{L}_i^{\text{out}}} \|F_l(x_{i_1(l)}, \cdot)\|_{L_{q_l}(R^2)} \prod_{l \in \mathcal{L}_i^{\text{in}}} \|F_l\|_{L_{q_l}(R^2 \times R^2)},$$

and the proof is complete.

Next we consider

$$R = \int w(x) \prod_{i=1}^{r} : \phi(x_i)^{n_i} :_C dx \qquad (8.5.12)$$

and the momentum cutoff approximation

$$R_\kappa = \int w(x) \prod_{i=1}^{r} : \phi_\kappa(x_i)^{n_i} :_C dx. \qquad (8.5.13)$$

Theorem 8.5.3. *Let* $w \in L_p$ *for some* $p > 1$ *and let* w *have compact support. Then* R_κ, $R \in L_q(d\phi_C)$ *for all* $q < \infty$, *with bounds uniform in* κ. *Moreover for* j *an integer and for some* $\varepsilon > 0$,

$$\left| \int (R - R_\kappa)^j \, d\phi \right| \leq j!^{\Sigma n_i/2} (\text{const} \, \|w\|_{L_p} \kappa^{-\varepsilon})^j. \qquad (8.5.14)$$

Here ε *and the constant are independent of* j *and of* $C \in \mathscr{C}_m$, *but the constant is proportional to* $(\sum n_i/2)!$ *and depends on* suppt w.

PROOF. Expand $\int R^j$ and $\int R_\kappa^j$ into a sum of Feynman graphs, according to Proposition 8.3.1. In this expansion, we have at most $[(j \sum n_i) - 1]!! \leq j!^{\Sigma n_i/2} (\text{const})^j$ such graphs. Here the constant is proportional to $(\sum n_i/2)!$. We bound each graph $I(G)$ as follows: In general

$$I(G) = \int \left(\prod_{k=1}^{j} w(x^k) \right) F(x^1, \ldots, x^j) \, dx^1 \cdots dx^j,$$

where each $x^k \in R^{2r}$ denotes $2r$ variables. By Hölder's inequality,

$$|I(G)| \leq (\|w\|_{L_p})^j \|F\|_{L_{p'}},$$

and $\|F\|_{L_{p'}}$ is dominated using Lemma 8.5.2. The f_i are characteristic functions of the support of w in the variable x_i. The F_l are propagators or differences of propagators (restricted to the support of w). The L_{q_l} norm of F_l is bounded by Proposition 7.9.1. These bounds are uniform in κ for R_κ.

Similarly, we bound $(R - R_\kappa)^j$. In this case each $R - R_\kappa$ factor can be written

$$R - R_\kappa = \sum_{j=1}^{r} \int \left(\prod_{i=1}^{j-1} : \phi(x_i)^{n_i} :_C \right) \left(: \phi(x_j)^{n_j} :_C - : \phi_\kappa(x_j)^{n_j} :_C \right)$$

$$\left(\prod_{i=j+1}^{r} : \phi_\kappa(x_i)^{n_i} :_C \right) w(x) \, dx,$$

where

$$: \phi(x_j)^{n_j} :_C - : \phi_\kappa(x_j)^{n_j} :_C = \sum_{j=1}^{n_j} : \phi(x_j)^{j-1} \{ \phi(x_j) - \phi_\kappa(x_j) \} \phi_\kappa(x_j)^{n_j - j} :.$$

Thus each $R - R_\kappa$ factor introduces a $\delta - \delta_\kappa$ into one F_l. Since each F_l has at most two such factors, these $\delta - \delta_\kappa$ occur in at least $j/2$ lines. By Theorem 7.1.1,

$$\|F_l\|_{L_{q_l}} \leq O(\kappa^{-\varepsilon})$$

in case F_l has a δC factor and compact support. Thus we obtain $\kappa^{-\varepsilon j}$ in (8.5.14) and the proof is complete.

Corollary 8.5.4. *Let* $R^{(l)}$, $l = 1, 2, \ldots, j$ *be a sequence of R's satisfying the hypotheses of the theorem; let* p_l *be bounded away from 1 and* $\sum_i n_i^{(l)} \leq n$ *for all l. Then for* κ *sufficiently large*

$$\left| \int \prod_{l=1}^{j} (R^{(l)} - R_\kappa^{(l)}) \, d\phi_C \right| \leq j!^{n/2} \kappa^{-\varepsilon j} \prod_{l=1}^{j} \|w^{(l)}\|_{L_p}.$$

PROOF.   Apply Hölder's inequality to yield

$$\int \prod_{l=1}^{j} (R^{(l)} - R_\kappa^{(l)}) \, d\phi_C \leq \prod_{l=1}^{j} \|R^{(l)} - R_\kappa^{(l)}\|_{L_j},$$

and apply the theorem to each factor. The constant in the theorem is bounded uniformly for p bounded away from 1. The factor $(\text{const})^j$ can be replaced by 1 by choosing ε slightly smaller and κ large.

In these estimates, we have so far not taken advantage of the fact that the individual graphs have space-time-localized kernels. With R above there are $O(\sum n_i/2!)$ graphs contributing to $\int R \, d\phi_C$. But most of these graphs are quite small, because of the exponential decay in $C(x, y)$ as $|x - y| \to \infty$. To take advantage of this fact, we let Δ_i denote a cover of R^2 by unit squares, indexed by $i \in Z^2$. Suppose that $w(x)$ in (8.5.12) is supported in $\Delta_{i_1} \times \cdots \times \Delta_{i_r}$. Let

$$N(\Delta) = \sum_j \{n : \Delta_{i_j} = \Delta\} \tag{8.5.15}$$

denote the number of legs (linear factors of $\phi(x)$) of R localized in Δ.

Theorem 8.5.5. *Let R be given by (8.5.12) with* suppt $w \subset \Delta_{i_1} \times \cdots \times \Delta_{i_r}$ *and $n_i \leq n$. Let $C \in \mathscr{C}_m$, and let C have no contribution from the periodic covariance. Then*

$$\left| \int R \, d\phi_C \right| \leq \|w\|_{L_p} \prod_\Delta N(\Delta)! \, (\text{const } m^{-1/q})^{N(\Delta)},$$

where $q = p'n$ and p' is the conjugate Lebesgue index to p.

PROOF.   We repeat the estimate of Theorem 8.5.3, but write

$$\left| \int R \, d\phi_C \right| \leq \sum |I(G)|.$$

By Proposition 7.9.1,

$$\|C\|_{L_q(\Delta_i \times \Delta_j)} \leq O(m^{-2/q}) e^{-m \, \text{dist}(\Delta_i, \Delta_j)}$$

Hence, since each line has two legs,

$$|I(G)| \leq O\left(\prod_\Delta (m^{-1/q})^{N(\Delta)} \right) e^{-m \, \text{dist}(G)} \|w\|_{L_p},$$

where $\text{dist}(G) = \sum_{l \in G} \text{dist}(\Delta_{i_1(l)}, \Delta_{i_2(l)})$. We need to show that

$$\sum_G e^{-m \, \text{dist}(G)} \leq \prod_\Delta (\text{const}^{N(\Delta)} N(\Delta)!).$$

First we note that within a given Δ, permutations of the $N(\Delta)$ legs may give rise to distinct graphs. Furthermore, taking these $\prod_\Delta N(\Delta)!$ possibilities into account includes all graphs with the specified localization of lines connecting different Δ. Thus we need only sum over these localizations and show

$$\sum_{i_2(l)} e^{-m\,\text{dist}(G)} = \sum_{i_2(l)} \prod_l \exp[-m\,\text{dist}(\Delta_{i_1(l)}, \Delta_{i_2(l)})] \leq \text{const}^{N(\Delta)}.$$

But we can increase this sum by summing over all Δ_{i_2} for each line. Hence our sum satisfies

$$\sum_{i_2(l)} e^{-m\,\text{dist}(G)} \leq \left[\sum_{\Delta'} e^{-m\,\text{dist}(\Delta,\,\Delta')} \right]^{\text{number of lines}}$$

$$\leq (\text{const})^{\text{number of lines}}$$

$$= \prod_\Delta \text{const}^{N(\Delta)}.$$

This completes the proof.

8.6 Non-Gaussian Integrals, $d = 2$

In this section we present the main result of this chapter, namely the proof that in two dimensions and for P a semibounded polynomial, $e^{-:P:}$ is integrable. This is a key step in the construction of the $P(\phi)_2$ model, whose function space measure has the form $e^{-:P:}\,d\phi$, up to a normalizing constant. Our results here concern $:P: = \int_\Lambda :P(\phi(x)):_C\,dx$ and $Z = \int e^{-:P:}\,d\phi$, namely a finite volume interaction P. In Chapter 11 (and by another method in Chapter 18) we pass to the infinite volume limit.

For later use, we consider a change of covariance. For $C_1, C_2 \in \mathscr{C}_m$, let $\delta c(x) = \lim_{y \to x} [C_2(x, y) - C_1(x, y)]$. Then Wick reordering is the formula

$$:\phi(x)^n:_{C_1} = \sum_{j=0}^{[n/2]} \frac{n!}{(n-2j)!\,j!\,2^j} \delta c(x)^j : \phi(x)^{n-2j}:_{C_2}. \qquad (8.6.1)$$

Now let $f = \{f_0, \ldots, f_n\}$ be a sequence of coefficient functions, and define $P(\xi, f) = \sum_{j=0}^n f_j(x)\xi^j$ and

$$:P(\phi, f):_C \equiv \sum_{j=0}^n :\phi^j(f_j):_C. \qquad (8.6.2)$$

The Wick reordering can be regarded as a transformation T on the coefficients f defined by the identity

$$:P(\phi, f):_{C_1} = :P(\phi, Tf):_{C_2}. \qquad (8.6.3)$$

Since we are interested mainly in finite volume approximations to translation invariant interactions, the coefficient functions $f_j = \text{const }\chi_\Lambda$, where χ_Λ is the characteristic function of a bounded region Λ, would be a natural choice. However, the changes introduced by Wick reordering, (8.6.1–3), show that this class is too narrow. The same conclusion derives from the definition of the characteristic functional $S\{f\}$ in Section 6.1. In fact $S\{g\}$ is defined by a perturbation $f_1 \to f_1 + ig = f'_1$. Then $S\{g\} = Z\{f'\}/Z\{f\}$.

To systematize our estimates for the f_j, we require

$$\operatorname{supp} f_j \subset \operatorname{supp} f_n = \Lambda, \quad \Lambda \text{ bounded},$$

$$f_j/f_n \in L_{n/(n-j)},$$ (8.6.4)

$$0 \le f_n \in L_\infty, \quad 0 \le 1/f_n \in L_\infty(\Lambda), \qquad n = \deg P \text{ even}.$$

We now introduce two measures on the size of f. Let

$$N(f) = \sum_{j=0}^{n-1} \left\| \frac{f_j}{f_n} \right\|_{L_{n/(n-j)}}^{n/(n-j)},$$ (8.6.5)

$$M(f) = \sum_{j=1}^{n} \| f_j \|_{L_{n/(n-j)}}.$$ (8.6.6)

Proposition 8.6.1. *The re-Wick-ordering transformation $f \to Tf$ of (8.6.3) with C_1, $C_2 \in \mathscr{C}_m^M$ maps the class (8.6.4) into itself, and moreover $(Tf)_n = f_n$.*

Suppose

$$m^{-1} + \frac{M}{m} + n + |\Lambda| \le K.$$

Then

$$N(Tf) \le \text{const } N(f) + \text{const}, \qquad M(Tf) \le \text{const } M(f),$$

with constants which depend only on K.

PROOF. We estimate $\delta c(x)$ in two steps. Express C_1, C_2 as a convex sum of operators of the form $(-\Delta_{B_j} + m_j^2)^{-1}$. First change the boundary conditions in each Δ_{B_j} to the free boundary conditions on the Laplace operator Δ. Second change each mass m_j to a given mass $\bar{m} \in [m, M]$.

For a fixed mass $m_j \ge m$, let $\delta c(x)$ arise from a change in boundary conditions. This can be estimated by the logarithmic bound in (7.6.2) and (7.6.19), along with the assumed bounds on $|\Lambda|$ and on m^{-1}. Then $\|\delta c(x)\|_{L_p(\Lambda)} < R$, for a constant R which depends only on K and on $p < \infty$. The estimates on $N(Tf)$, $M(Tf)$ then follow by (8.6.1) and Hölder's inequality in the form

$$\| (\delta c)^{1/2} h \|_{L_{n/(n-j+1)}} \le \| \delta c \|_{L_{n/2}(\Lambda)}^{1/2} \| h \|_{L_{n/(n-j)}}$$
$$\le R(K, n)^{1/2} \| h \|_{L_{n/(n-j)}}.$$

Here $h = f_j$ or f_j/f_n, and $1 \le j \le n$. Since n is bounded by assumption, the constants can be chosen to depend only on K.

Thus we are reduced to considering a change in mass of the free covariance $C_\varnothing$. In this case $\delta c(x) = (2\pi)^{-1} \ln(m_j/\bar{m})$ which is bounded uniformly using the assumption $\bar{m}$, $m_j \in [m, M]$, so $K^{-1} \le m_j/\bar{m} \le K$. The estimates on $N(Tf)$ and $M(Tf)$ then follow for the change in mass, again by using Hölder's inequality and the assumed bound on $|\Lambda|$.

The main result of this section is

Theorem 8.6.2. *Let f satisfy (8.6.4), and let* $C_1, C_2 \in \mathscr{C}_m^M$. *If*

$$m^{-1} + (M/m) + n + |\Lambda| \leq K,$$

then

$$\int \exp(- :P(\phi, f):_{C_1}) \, d\phi_{C_2}$$

$$\leq \text{const} \exp\{\text{const} \, \|f_n\|_{L_\infty}[N(f) + (\ln(M(f) + 1))^{n/2}]\}. \quad (8.6.7)$$

If

$$\|f_n\|_{L_\infty} + m^{-1} + (M/m) + n + |\Lambda| \leq K,$$

then

$$\int \exp(- :P(\phi, f):_{C_1}) \, d\phi_{C_2} \leq \exp(\text{const}(N(f) + 1)), \quad (8.6.8)$$

and in both cases, the constant depends only on K.

We first prove the case $C_1 = C_2 = C$. Then the proof has two steps. The first step is to establish a semiboundedness property for $:P(\phi, f):_C$. Although $P(\phi, f)$ is bounded below, the Wick ordering destroys this property. Using the approximate delta function δ_κ of (7.1.5), let

$$\phi_\kappa = \phi * \delta_\kappa. \quad (8.6.9)$$

Then $:P_\kappa: \equiv :P(\phi_\kappa, f):_C$ has a κ-dependent lower bound: $-O(\ln \kappa)^{(\deg P)/2} \leq :P_\kappa:$. This is the content of Proposition 8.6.3.

The second step in the proof is to show that the set of configurations $\phi \in \mathscr{D}'$ for which $:P:$ is less than $\inf :P_\kappa:$ has small measure, roughly $O(e^{-\kappa^\delta})$, $\delta > 0$, as follows from the estimates on Gaussian integrals in Section 8.5. This is the content of Proposition 8.6.4.

Proposition 8.6.3. *Let f satisfy (8.6.4), and let* $C \in \mathscr{C}_m$. *Then for* $\kappa \geq 2$,

$$-\text{const}[\|f_n\|_{L_1}(\ln \kappa)^{(\deg P)/2} + \|f_n\|_{L_\infty} N(f)] \leq :P(\phi_\kappa, f):_C. \quad (8.6.10)$$

The constant depends only on $n = \deg P$ *and on* m.

Remark. To illustrate the idea of the proof, consider a simple P of degree $n = 4$. Take $f_0 = \ldots = f_3 = 0$. Note by (8.5.5) that

$$:\phi_\kappa(x)^4: = \phi_\kappa(x)^4 - 6c_\kappa(x)\phi_\kappa(x)^2 + 3c_\kappa(x)^2$$
$$= (\phi_\kappa(x)^2 - 3c_\kappa(x))^2 - 6c_\kappa(x)^2.$$

This is bounded from below by $-6c_\kappa(x)^2 = -O(\ln \kappa)^2$, using Theorem 7.1.1. Integration of this bound yields (8.6.10).

PROOF. We use Hölder's inequality with respect to the measure $f_n(x) \, dx$. Let

$$p_1^{-1} = 0, \quad p_2^{-1} = \frac{j - 2l}{n}, \quad p_3^{-1} = \frac{n - j}{n}, \quad p_4^{-1} = \frac{2l}{n},$$

so that $\sum p_v^{-1} = 1$. Then we bound all terms with powers ϕ^m, $m < n$, by ϕ^n, the leading term in P, as follows:

$$\left| \int c_\kappa^l(x) \phi_\kappa^{j - 2l} f_j(x) \, dx \right| = \left| \int [c_\kappa^l(x)][\phi_\kappa(x)^{j - 2l}] \left[\frac{f_j}{f_n} \right](x) f_n(x) \, dx \right|$$

$$\leq \| c_\kappa \|_{L_\infty}^l \left(\int \phi_\kappa(x)^n f_n(x) \, dx \right)^{(j - 2l)/n}$$

$$\times \left(\int \left| \frac{f_j}{f_n} \right|^{n/(n - j)} f_n \, dx \right)^{(n - j)/n} \| f_n \|_{L_1}^{2l/n}.$$

By the inequality $ab \leq a^p + b^q$ with a, b positive and $p^{-1} + q^{-1} = 1$, the above is bounded by

$$\varepsilon \int \phi_\kappa(x)^n f_n(x) \, dx + a(\varepsilon) \| c_\kappa \|_{L_\infty}^{n/2} \| f_n \|_{L_1} + \int \left| \frac{f_j}{f_n} \right|^{n/(n - j)} f_n \, dx$$

$$\leq \varepsilon \int \phi_\kappa(x)^n f_n(x) \, dx + a(\varepsilon) \| c_\kappa \|_{L_\infty}^{n/2} \| f_n \|_{L_1} + \left\| \frac{f_j}{f_n} \right\|_{L_{n/(n - j)}}^{n/(n - j)} \| f_n \|_{L_\infty}.$$

Furthermore the same bound holds with a new constant $a = a(\varepsilon, n)$ when the left side is multiplied by the combinatorial factor $[j/2]j!/((j - 2l)! \, l! \, 2^l)$. The ε above is arbitrary, but positive. We choose ε small and let $\mathcal{N}$ denote the total number of such terms. For $0 < 1 - \mathcal{N}\varepsilon$, (8.6.10) follows by summation over j and l in (8.5.5). The bound $\| c_\kappa \|_{L_\infty} \leq O(\ln \kappa)$ follows from Property LR3 of Theorem 7.1.1.

Next we come to the second step in the proof of the integrability of e^{-P}, namely to show that the high energy tail δP of P becomes large only on a small set of field configurations. To make this precise, define

$$\delta P_\kappa \equiv :P(\phi, f):_C - :P(\phi_\kappa, f):_C, \tag{8.6.11}$$

so $:P: = :P_\kappa: + \delta P_\kappa$. Let $X(\kappa)$ denote the paths ϕ for which $1 \leq |\delta P_\kappa|$, namely

$$X(\kappa) = \{\phi : 1 \leq |\delta P_\kappa|\}.$$

Proposition 8.6.4. Let $C \in \mathscr{C}_m$, let $m^{-1} + n + |\Lambda| \leq K$, and let $\varepsilon > 0$ be the constant of Theorem 8.5.3. Then there exists a constant α depending only on K such that for $\kappa > 2$,

$$\int_{X(\kappa)} d\phi_C \leq \exp \left[-\alpha \left(\frac{\kappa^\varepsilon}{M(f)} \right)^{2/\deg P} \right]. \tag{8.6.12}$$

PROOF. Since $1 \leq |\delta P_\kappa|$ on $X(\kappa)$, we have for any even integer j,

$$\int_{X(\kappa)} d\phi_C \leq \int_{\mathscr{S}'} (\delta P_\kappa)^j \, d\phi_C.$$

Expand

$$\delta P_\kappa = \sum_{l=1}^n :(\phi^l - \phi_\kappa^l):_C(f_l).$$

Note that the $l = 0$ term does not occur in δP_κ. Thus δP_κ^j is a sum of at most n^j monomial terms, each of the form $\prod_{i=1}^j (R^{(l)} - R_\kappa^{(l)})$, with $R^{(l)} = :\phi^{m_i}(n_{m_i}):_C$ and $1 \le m_i \le n$. Each such term is bounded by Corollary 8.5.4, with the choices $r = 1$, $\sum n_i = m_l \le n$, and $p_l = p_{m_i} = n/(n - m_l)$. Then

$$\|w^{(l)}\|_{L_{p_l}} = \|f_{m_l}\|_{L_{p_l}} \le M(f).$$

Hence summing these bounds,

$$\int_{X(\kappa)} d\phi_C \le (j!)^{n/2}(\text{const } M(f)\kappa^{-\varepsilon})^j, \tag{8.6.13}$$

with the constant depending only on K. The right side of this inequality is minimized by choosing j to be the largest even integer which is less than $(\kappa^\varepsilon/\text{const } M(f))^{2/n}$, with the same constant as in (8.6.13). Stirling's formula then yields (8.6.12).

In the proof of the theorem, we use the following elementary identity.

Lemma 8.6.5. *Let $g(\phi)$ be an L_p function defined on a probability space $\{X, d\mu\}$, and let*

$$h(a) \equiv \mu\{\phi : a \le |g(\phi)|\}. \tag{8.6.14}$$

Then

$$\int |g|^p \, d\mu = p \int_0^\infty a^{p-1} h(a) \, da. \tag{8.6.15}$$

PROOF. This is obtained from

$$\int |g|^p \, d\mu = -\int_0^\infty a^p \, dh(a)$$

by integration of parts.

We apply (8.6.15) in the special case $p = 1$. To prove $g \in L_1(d\mu)$, we need only prove that $h(a)$ is integrable at $a = \infty$. Furthermore, if $a(\kappa)$ is a continuously differentiable monotonically increasing function of κ, going to ∞ with κ, then $g \in L_1(d\mu)$ if

$$h(a(\kappa))\frac{da(\kappa)}{d\kappa}$$

is integrable at $\kappa = \infty$. We assume $da(\kappa)/d\kappa \le a(\kappa)^2$. For $e \le \kappa_0 \le \kappa$,

$$\int |g| \, d\mu \le a(\kappa_0) + \sup_{\kappa \ge \kappa_0} \kappa^2 (a(\kappa)^2 h(a(\kappa))). \tag{8.6.16}$$

PROOF OF THEOREM 8.6.2. For $g(\phi)$ we choose $\exp[- :P(\phi,f):_C]$. For $a(\kappa)$ we exponentiate the bound in (8.6.10), namely

$$a(\kappa) \equiv \exp\{1 + \text{const}[\|f_n\|_{L_1}(\ln \kappa)^{(\deg P)/2} + \|f_n\|_{L_\infty}N(f)]\}.$$

This definition satisfies $da(\kappa)/d\kappa \leq a(\kappa)^2$. Furthermore, by writing

$$e^{- :P:} = e^{- :P_\kappa:}e^{-\delta P_\kappa},$$

we have by Proposition 8.6.3 that

$$h(a(\kappa)) \leq \int_{X(\kappa)} d\phi_C \leq \exp\left[-\alpha\left(\frac{\kappa^\varepsilon}{M(f)}\right)^{2/\deg P}\right],$$

where the second inequality follows by Proposition 8.6.4. Thus

$\ln\{\kappa^2 a(\kappa)^2 h(a(\kappa))\}$

$$\leq \text{const}\left\{\ln \kappa + \|f_n\|_{L_\infty}((\ln \kappa)^{n/2} + N(f)) - \left(\frac{\kappa^\varepsilon}{M(f)}\right)^{2/n}\right\} + \text{const}. \qquad (8.6.17)$$

We claim (8.6.17) is bounded by

$$\text{const}\|f_n\|_{L_\infty}[N(f) + \{\ln(1 + M(f))\}^{n/2}] + \text{const}, \qquad (8.6.18)$$

from which (8.6.7) follows. Furthermore, since

$$M(f) \leq \|f_n\|_{L_\infty}(N(f) + n), \qquad (8.6.19)$$

we infer (8.6.8).

We now establish (8.6.18). Choose κ_0 sufficiently large so that if $\kappa_0 \leq \kappa$, then $1 \leq (\ln \kappa)^{n^2} \leq \kappa^\varepsilon$.

Case 1: $\kappa^\varepsilon \leq (1 + M(f))^{n+1}$. In this case, neglect the negative term in (8.6.17). Note $\|f_n\|_{L_\infty} \leq M(f)$, and thus

$$\|f_n\|_{L_\infty} \leq \text{const}[1 + \|f_n\|_{L_\infty}\{\ln(1 + M(f))\}^{n/2}]. \qquad (8.6.20)$$

Using (8.6.19–20), it follows that

$$\ln \kappa \leq (n + 1)\varepsilon^{-1}\ln(1 + M(f)) \leq \text{const } M(f)$$

$$\leq \text{const}\|f_n\|_{L_\infty}(N(f) + n)$$

$$\leq \text{const}\|f_n\|_{L_\infty}[N(f) + \{\ln(1 + M(f))\}^{n/2}] + \text{const}. \qquad (8.6.21)$$

Furthermore, by our assumption we infer

$$\|f_n\|_{L_\infty}(\ln \kappa)^{n/2} \leq \text{const}\|f_n\|_{L_\infty}\{\ln(1 + M(f))\}^{n/2}. \qquad (8.6.22)$$

Adding (8.6.21–22) yields the desired bound (8.6.18).

Case 2: $\kappa^\varepsilon > (1 + M(f))^{n+1}$. In this case

$$\kappa^\varepsilon = \kappa^{3\varepsilon/4}\kappa^{\varepsilon/4} \geq (1 + M(f))^{3(n+1)/4}(\ln \kappa)^{n^2/4}.$$

Using $\{(3(n + 1)/4) - 1\}(2/n) = 1 + \frac{1}{2}(1 - n^{-1}) \geq 1$, and $M(f) \geq \|f_n\|_{L_\infty}$, it follows that

$$\left(\frac{\kappa^\varepsilon}{M(f)}\right)^{2/n} \geq (1 + M(f))(\ln \kappa)^{n/2}$$

$$\geq \ln \kappa + \|f_n\|_{L_\infty}(\ln \kappa)^{n/2}.$$

Thus (8.6.17) is bounded from above by $\text{const}\|f_n\|_{L_\infty} N(f)$, and (8.6.18) holds.

This completes the proof when $C_1 = C_2$. Finally we reduce the general case $C_1 \neq C_2$ to the special case already proved. It is sufficient to establish (8.6.7). Let $:P(\phi, f):_{C_1} = :P(\phi, Tf):_{C_2}$. By the special case, the bound (8.6.7) holds with Tf replacing f. Using Proposition 8.6.1 and the bound (8.6.20),

$$\|f_n\|_{L_\infty}[N(Tf) + \{\ln(1 + M(Tf))\}^{n/2}]$$
$$\leq \text{const}\|f_n\|_{L_\infty}[N(f) + \{\ln(1 + M(f))\}^{n/2}] + \text{const}.$$

Hence (8.6.7) holds in the desired form.

Remark 1. The proof of Theorem 8.6.2 yields another bound which is useful in the phase transition analysis of Chapter 16. Assume $m^{-1} + (M/m) + n + |\Lambda| \leq K$. Also suppose there exists a function $L(f)$ and a constant depending only on K such that

$$-\text{const}[(1 + M(f))(\ln \kappa)^{n/2} + L(f)] \leq :P(\phi_\kappa, f):_{C_1}.$$

Then

$$\int \exp(- :P(\phi, f):_{C_1}) \, d\phi_{C_2}$$

$$\leq \text{const} \exp\{\text{const}[L(f) + \{1 + M(f)\}\{\ln(1 + M(f))\}^{n/2}]\}, \qquad (8.6.23)$$

where the constants in (8.6.23) depend only on K.

Remark 2. Scale transformation identities extend the usefulness of these bounds. For $\alpha > 0$, consider the transformation of masses $m \to m/\alpha$ and distances $x \to \alpha x$. Let $C_B(m) = (-\Delta_B + m^2)^{-1}$, where B denotes classical boundary conditions on Γ. Define the scaled covariance operator $C_{\alpha B}(\alpha m) = (-\Delta_{\alpha B} + (m/\alpha)^2)^{-1}$, where $\Delta_{\alpha B}$ denotes boundary conditions on $\alpha \Gamma$, the set of points αx for $x \in \Gamma$. For $d = 2$,

$$C_B(m; x, y) = C_{\alpha B}(m/\alpha; \alpha x, \alpha y), \qquad (8.6.24)$$

as in (7.2.2) for the free covariance.

We let R_α denote the scale transformation defined (for $d = 2$) by

$$R_\alpha C_B(m) = C_{\alpha B}(m/\alpha) \text{ and } (R_\alpha f)_j(x) = \alpha^{-2} f_j(x/\alpha). \qquad (8.6.25)$$

Then the scaling identity is

$$\int \exp(- :P(\phi, f):_{C_1} \, d\phi_{C_2} = \int \exp(- :P(\phi, R_\alpha f):_{R_\alpha C_1}) \, d\phi_{R_\alpha C_2}. \qquad (8.6.26)$$

This identity follows in the Gaussian case (P linear) from (8.6.24–25) and the definition of the Gaussian integral. The general case follows from the Gaussian identity and the definition (8.5.5) of Wick order.

8.7 Finite Dimensional Approximations

In order to prove integration by parts identities in Chapter 9, we need an approximation to the factor $e^{- :P(\phi, f):}$ by a function of a finite number of variables. The essential step is the construction of Section 8.6, in which $e^{-:P(\phi,f):}$ is approximated by $e^{-:P(\phi_\kappa,f):}$. In two further steps, we approximate f by a C_0^∞ test function and we approximate the Riemann integral in $:P(\phi_\kappa,f):$ by a Riemann sum. A distinct finite dimensional approximation is developed in Sections 9.5–6, based on the finite difference (lattice) approximation to the Laplace operator. The finite difference approximation preserves the ferromagnetic property of the Laplace operator, and is useful for the proof of correlation inequalities.

Let χ_Λ be the characteristic function of $\Lambda = \text{suppt } f_n$, and let

$$f_{j,\lambda} = \chi_\Lambda(f_j * \delta_\lambda), \qquad f_\lambda = \{f_{j,\lambda} : j = 0, \ldots, n\}. \tag{8.7.1}$$

Then $f_\lambda \to f$ in L_p for some $p > 1$, and also $N(f_\lambda) \to N(f)$; see (8.6.5). Let $P^{(\delta,\lambda,\kappa)}$ be a Riemann approximating sum to the integral $:P(\phi_\kappa,f_\lambda):_C$, namely

$$P^{(\delta,\lambda,\kappa)} = \delta^2 \sum_{j=0}^{n} \sum_{x \in \delta Z^2} :\phi_\kappa(x)^j:_C \ f_{j,\lambda}(x). \tag{8.7.2}$$

Proposition 8.7.1. *For each $C \in \mathscr{C}$, the double limit*

$$\lim_{\lambda\to\infty} \lim_{\delta\to\infty} P^{(\delta,\lambda,\kappa)} = :P(\phi_\kappa,f):_C$$

converges in $L_p(d\phi_C)$ for all $p < \infty$.

Remark. The $\kappa \to \infty$ convergence was established in Theorem 8.5.3.

PROOF. Let $f_{\lambda,\delta}$ be a collection of δ functions chosen as in (8.7.2) so that

$$P^{(\delta,\lambda,\kappa)} = :P(\phi_\kappa,f_{\lambda,\delta}):_C \ .$$

As in the proof of Theorem 8.5.3, the L_p convergence of $P^{(\delta,\lambda,\kappa)}$ reduces to the convergence

$$\langle f_{j,\lambda,\delta}, C_\kappa^j f_{j,\lambda,\delta}\rangle \to \langle f_j, C_\kappa^j f_j\rangle. \tag{8.7.3}$$

The $\delta \to 0$ limit in (8.7.3) is the definition of the Riemann integral, and the $\lambda \to \infty$ convergence follows from the L_p convergence of f_j, since C_κ and C_κ^j are in L_∞.

Proposition 8.7.2. *With $P^{(\delta,\lambda,\kappa)}$ as above,*

$$\lim_{\kappa\to\infty} \lim_{\lambda\to\infty} \lim_{\delta\to 0} \exp(-P^{(\delta,\lambda,\kappa)}) = \exp(-:P(\phi,f):_C) \tag{8.7.4}$$

in $L_p(d\phi_C)$ for all $p < \infty$.

PROOF. By Theorem 8.6.2 and its proof, $\exp(-P^{(\delta,\lambda,\kappa)})$ is bounded in L_p uniformly in δ for fixed λ and κ. The same is true for the interpolating exponentials

$$\exp(-P^{(t)}) = \exp[-tP^{(\delta,\lambda,\kappa)} - (1-t)P^{(0,\lambda,\kappa)}]$$

for $0 \leq t \leq 1$, by a Schwarz inequality. The L_p convergence follows from the inequalities

$$\|e^{-P(0)} - e^{-P(1)}\|_{L_p} \leq \int_0^1 \left\| \frac{d}{dt} e^{-P(t)} \right\|_{L_p}$$

$$\leq \sup_t \left\| \frac{d}{dt} e^{-P(t)} \right\|_{L_p}$$

$$= \sup_t \|(P^{(0)} - P^{(1)})e^{-P(t)}\|_{L_p}$$

$$\leq \|P^{(0)} - P^{(1)}\|_{L_{2p}} \sup_t \|e^{-P(t)}\|_{L_{2p}} .$$

The second factor is bounded, uniformly in δ, by the above remarks, and the first factor converges by Proposition 8.7.1. The λ and κ limits are similar, and the proof is complete.

Proposition 8.7.3. *Let $P^{(\delta, \lambda, \kappa)}$ and $:P(\phi, f):_C \equiv :P:$ be as above, and define derivatives*

$$\left\langle v, \frac{\delta}{\delta\phi} \right\rangle = \int v(x) \frac{\delta}{\delta\phi(x)} \, dx$$

and

$$\Delta_w = \left\langle \frac{\delta}{\delta\phi}, w \frac{\delta}{\delta\phi} \right\rangle = \int w(x, y) \frac{\delta^2}{\delta\phi(x)\,\delta\phi(y)} \, dx \, dy$$

acting algebraically on polynomials $R(\phi)$. Then for v, w continuous and compactly supported,

$$\lim_{\kappa \to \infty} \lim_{\lambda \to \infty} \lim_{\delta \to 0} \left\langle v, \frac{\delta}{\delta\phi} \right\rangle P^{(\delta, \lambda, \kappa)} = \left\langle v, \frac{\delta}{\delta\phi} \right\rangle :P: \quad \text{in } L_p, \text{ all } p < \infty, \quad (8.7.5)$$

$$\lim_{\kappa \to \infty} \lim_{\lambda \to \infty} \lim_{\delta \to 0} \Delta_w P^{(\delta, \lambda, \kappa)} = \Delta_w :P: \quad \text{in } L_p, \text{ all } p < \infty. \quad (8.7.6)$$

PROOF. The proof of Proposition 8.7.1 and Theorem 8.5.3 applies to this case.

Remark. For Proposition 8.7.1 and Proposition 8.7.3, P does not have to be semibounded, nor do the P's in the three propositions have to coincide. Thus in summary, we have

$$(\partial R^{(\delta, \lambda, \kappa)})e^{-P(\delta, \lambda, \kappa)} \to (\partial R)e^{-:P:},$$

where $\partial = I$, $\delta/\delta\phi$, or $\delta^2/\delta\phi\,\delta\phi$.

CHAPTER 9

Calculus and Renormalization on Function Space

9.1 A Compilation of Useful Formulas

In this section we list the main formulas and identities, with references to sections where they are established. Let $R(\phi)$ denote a polynomial in ϕ, or a polynomial in $\phi, \ldots, :\phi(x)^j:$, as discussed in Section 8.5. Let

$$A(\phi) = R(\phi)e^{i\phi(f)}. \tag{9.1.1}$$

(i) Wick Product Identities

Ordering of an Exponential (Section 6.3)

$$:e^{\phi(f)}:_C = e^{\phi(f)}e^{-\langle f, Cf \rangle/2} \tag{9.1.2}$$

Here $\langle \, , \, \rangle$ denotes the real inner product on L_2.

Derivative of a Functional. The directional derivative of a functional $A(\phi)$ on $\mathscr{D}'(R^d)$ in the direction ψ is defined by the formula

$$(D_\psi A)(\phi) = \lim_{\varepsilon \to 0} \frac{A(\phi + \varepsilon\psi) - A(\phi)}{\varepsilon} \tag{9.1.3}$$

Here $\psi \in \mathscr{D}'$, and the derivative may exist for a particular $A(\phi)$ pointwise, in L_p with respect to a measure $d\mu(\phi)$, etc. Some examples are derivatives of functions of the form (9.1.1),

$$D_\psi \phi(f)^n = n\psi(f)\phi(f)^{n-1},$$

$$D_\psi e^{\phi(f)} = \psi(f)e^{\phi(f)}.$$

The special case when $\psi = \delta_x$, the Dirac measure at x, arises often and so is given a special notation:

$$\frac{\delta}{\delta\phi(x)} A(\phi) \equiv D_{\delta_x} A(\phi). \tag{9.1.4a}$$

For A a polynomial with smooth coefficients, i.e. a linear combination of monomials of the form

$$\phi(f_1) \cdots \phi(f_n), \qquad f_j \in \mathcal{D},$$

the definition (9.1.3) coincides with the algebraic definition used in Sections 1.5 and 6.3. For nonsmooth coefficients, e.g. the A in (9.1.1), which may include the local Wick products $:\phi(x)^j:$, the definition (9.1.3) must be extended by continuity. This extension then depends both on the coefficients in A and on the function space measure for which the extension is defined. Since $\phi(y)$ and $:\phi(y)^j:$ are not functions on $\mathcal{D}'$, and not multiplication operators, but are bilinear forms on a suitable domain, either (9.1.4a) must be regarded as an identity between bilinear forms, or $\delta/\delta\phi(x)$ must be regarded as a distribution in the variable x, which simply means that (9.1.4a) must be integrated over x to become meaningful.

Another example is

$$\frac{\delta}{\delta\phi(x)} \phi(y) = \delta(x - y). \tag{9.1.4b}$$

Definition of Wick Order in Covariance C (Section 8.5).   Let $c_\kappa = \delta_\kappa * C * \delta_\kappa$. Then

$$:\phi^n(x):_C = \lim_{\kappa \to \infty} \sum_{j=0}^{[n/2]} \frac{(-1)^j n!}{(n - 2j)! j! 2^j} c_\kappa(x)^j \phi_\kappa(x)^{n-2j}. \tag{9.1.5}$$

Derivative of a Wick Product

$$\frac{\delta : A :_C}{\delta\phi(x)} = : \frac{\delta A}{\delta\phi(x)} :_C \tag{9.1.6}$$

Example:

$$\frac{\delta}{\delta\phi(x)} : \phi(y)^n :_C = n\delta(x - y) : \phi(y)^{n-1} :_C.$$

Multiplication of a Wick Product

$$\phi(f) : A(\phi) :_C = : \phi(f)A :_C + \left(\left\langle Cf, \frac{\delta}{\delta\phi} \right\rangle : A :_C \right). \tag{9.1.7}$$

Example:

$$\phi(x) : \phi(y)^n :_C = : \phi(x)\phi(y)^n :_C + nC(x, y) : \phi^{n-1}(y) :_C. \qquad (9.1.8)$$

Infinitesimal Change of Wick Order. Let $C(t)$ be a family of covariance operators depending smoothly on t. Then

$$\frac{d}{dt} : A :_{C(t)} = -\tfrac{1}{2} \Delta_{\dot{C}} : A :_{C(t)}. \qquad (9.1.9)$$

Here $\dot{C} = (d/dt)C(t)$, and

$$\Delta_C \equiv \left\langle C \frac{\delta}{\delta\phi}, \frac{\delta}{\delta\phi} \right\rangle \equiv \int C(x, y) \frac{\delta}{\delta\phi(x)} \frac{\delta}{\delta\phi(y)} \, dx \, dy \qquad (9.1.10)$$

denotes the second order differential operator on ϕ space with respect to the operator C.

An example of (9.1.9) is

$$\frac{d}{dt} : \phi(x)^n :_{C(t)} = -\binom{n}{2} \dot{C}(x, x) : \phi(x)^{n-2} :_{C(t)}. \qquad (9.1.11)$$

Finite Change in Wick Order. With C_1 and C_2 covariance operators in $\mathscr{C}$, and $\delta c(x) = \lim_{y \to x} [C_2(x, y) - C_1(x, y)]$,

$$: \phi(x)^n :_{C_1} = \sum_{j=0}^{[n/2]} \frac{n!}{(n-2j)! \, j! \, 2^j} \, \delta c(x)^j : \phi(x)^{n-2j} :_{C_2}. \qquad (9.1.12)$$

This follows from (9.1.2) and the formula (1.5.12) for Hermite polynomials.

PROOF OF (9.1.6).   Let $h(\lambda, \phi) = \exp[\lambda\phi(f) - \tfrac{1}{2}\lambda^2\langle f, Cf \rangle]$, so that

$$: \phi(f)^n :_C = \frac{d^n}{d\lambda^n} h(\lambda, \phi) \bigg|_{\lambda=0}.$$

Then

$$\frac{\delta}{\delta\phi(x)} : \phi(f)^n :_C = \frac{\delta}{\delta\phi(x)} \frac{d^n}{d\lambda^n} h(\lambda, \phi) \bigg|_{\lambda=0}$$

$$= \frac{d^n}{d\lambda^n} [\lambda f(x) h(\lambda, \phi)]_{\lambda=0}$$

$$= nf(x) \left[\frac{d^{n-1}}{d\lambda^{n-1}} h(\lambda, \phi) \right]_{\lambda=0}$$

$$= nf(x) : \phi(f)^{n-1} :_C$$

$$= : \frac{\delta}{\delta\phi(x)} \phi(f)^n :_C.$$

Multilinearity of Wick polynomials, and the polarization identity

$$x_1 \cdots x_n = 2^{-n}(n!)^{-1} \sum_{\varepsilon_j = \pm 1} \varepsilon_1 \cdots \varepsilon_n (\varepsilon_1 x_1 + \cdots + \varepsilon_n x_n)^n \qquad (9.1.13)$$

yield (9.1.6) in case $A = \phi(f_1) \cdots \phi(f_n)$. For A of the form (9.1.1) with smooth coefficients, the identity follows by summing the convergent series which defines the exponential. To allow the Wick polynomials of Section 8.5 in A, we must restrict the function space measure and extend the definition of D_f on functionals $A(\phi)$.

For the $d = 2$ Gaussian or finite volume $P(\phi)$ measures of Chapter 8, we introduce an ultraviolet cutoff in A. The result has smooth coefficients and hence satisfies (9.1.6). After integrating over x, as in (9.1.14), the formula is continuous and converges to a limit as the cutoff is removed, by estimates from Section 8.5. This extends by continuity the definition of D_f, so that (9.1.6) remains valid.

PROOF OF (9.1.7).   With $h(\lambda, \phi)$ as before, recall that

$$\left\langle Cf, \frac{\delta}{\delta\phi} \right\rangle h(\lambda, \phi) = \lambda\langle f, Cf \rangle h(\lambda, \phi).$$

Thus
$$:\phi(f)^n:_C = \frac{d^n}{d\lambda^n} h(\lambda, \phi)\bigg|_{\lambda=0}$$

$$= \frac{d^{n-1}}{d\lambda^{n-1}} [(\phi(f) - \lambda\langle f, Cf \rangle)h(\lambda, \phi)]_{\lambda=0} \qquad (9.1.14)$$

$$= \phi(f):\phi(f)^{n-1}:_C - \left\langle Cf, \frac{\delta}{\delta\phi} \right\rangle:\phi(f)^{n-1}:_C.$$

By (9.1.13) and summation of the convergent series for $e^{\phi(f)}$, this proves (9.1.7) in the case of smooth coefficients. The general case follows by introducing and removing an ultraviolet cutoff, as in Section 8.5.

PROOF OF (9.1.9).   For $A(\phi) = e^{i\phi(f)}$,

$$\frac{d}{dt} :A:_{C(t)} = \frac{d}{dt} \exp(\tfrac{1}{2}\langle f, C(t)f \rangle + i\phi(f))$$

$$= \tfrac{1}{2}\langle f, \dot{C}(t)f \rangle :A:_{C(t)} = -\tfrac{1}{2}\Delta_{\dot{C}}:A:_{C(t)}, \qquad (9.1.15)$$

using notation from (9.1.10). Polynomials are included by substituting λf for f and taking λ derivatives evaluated at $\lambda = 0$. Nonsmooth A are included by use of an ultraviolet cutoff, as above. For the above proof, it is sufficient to assume $C(t) \in \mathscr{C}$ (see Section 7.9) and that $\dot{C}(t)$ has a kernel $\dot{C}(t, x, y)$ for which

$$|\dot{C}(t, x, y)| \leq \text{const } C(x, y), \quad \text{some } C \in \mathscr{C}.$$

We only need the simple case:

$$C(t) = tC_1 + (1 - t)C_2,$$
$$\dot{C} = \delta C = C_1 - C_2.$$

(ii) Gaussian Integrals (Section 6.3)

Fourier Transform (Characteristic Function. Mean 0, Covariance **C)**

$$S(f) = \int e^{i\phi(f)} \, d\phi_C = e^{-\langle Cf, f \rangle/2}. \qquad (9.1.16)$$

Moments

$$\int \phi(f)^{2n+1}\, d\phi_C = 0, \qquad \int \phi(f)^{2n}\, d\phi_C = \frac{(2n)!}{2^n n!} \langle Cf, f \rangle^n. \quad (9.1.17)$$

Scalar Product of Exponentials. As a consequence of (9.1.16), (9.1.2),

$$\int :e^{-i\phi(f)}:_C :e^{i\phi(g)}:_C\, d\phi_C = e^{\langle f, Cg \rangle}. \qquad (9.1.18)$$

Orthogonality of Hermite (Wick) Polynomials. Expanding (9.1.18) in a power series yields

$$\int :\phi(f)^n:_C :\phi(g)^m:_C\, d\phi_C = \delta_{nm} n! \langle f, Cg \rangle^n. \qquad (9.1.19)$$

Examples: For $n \geq 1$,

$$\int :\phi(f)^n:\, d\phi_C = 0, \qquad (9.1.20)$$

$$\int :A(\phi):_C\, d\phi_C = A(\phi = 0). \qquad (9.1.21)$$

In the following formulas, let v denote an integral operator on $L_2(R^d, dx)$ with kernel $v(x, y)$. Furthermore, we assume $I + C^{1/2}vC^{1/2} > 0$. Let

$$:V: = \frac{1}{2} \int :\phi(x)v(x, y)\phi(y):\, dx\, dy. \qquad (9.1.22)$$

Wick constant. If vC is trace class,

$$:V:_C = V - \tfrac{1}{2}\mathrm{Tr}(vC). \qquad (9.1.23)$$

This identity follows from $:\phi(x)\phi(y):_C = \phi(x)\phi(y) - C(x, y)$, and $\mathrm{Tr}(vC) = \int v(x, y)C(y, x)\, dx\, dy$.

Functional Determinant (Section 9.3)

$$Z \equiv \int e^{-:V:_C}\, d\phi_C = \exp\{-\tfrac{1}{2}\, \mathrm{Tr}[\ln(I + C^{1/2}vC^{1/2}) - C^{1/2}vC^{1/2}]\}. \quad (9.1.24)$$

Gaussian Perturbation Identity (Section 9.3)

$$d\phi_{(C^{-1}+v)^{-1}} = Z^{-1}e^{-:V:_C}\, d\phi_C. \qquad (9.1.25)$$

Example: let v denote multiplication by $m(x)^2$, $C = (-\Delta + I)^{-1}$. Then

$$:V: = \frac{1}{2} \int m(x)^2 :\phi(x)^2:_C\, dx \qquad (9.1.26a)$$

and

$$d\phi_{(-\Delta + I + m(x)^2)^{-1}} = Z^{-1} e^{-:V:} d\phi_{(-\Delta + I)^{-1}} ;$$

$$Z = \exp\{-\tfrac{1}{2} \text{Tr}[\ln(I + C^{1/2} m^2 C^{1/2}) - C^{1/2} m^2 C^{1/2}]\}. \quad (9.1.26b)$$

Scale Transformation (Section 8.6). Let $\alpha > 0$ and let R_α denote the transformation which changes distance scales by the factor α. In particular with $C = (-\Delta_\Gamma + m^2)^{-1}$, define $R_\alpha C = (-\Delta_{\alpha\Gamma} + (m/\alpha)^2)^{-1}$. For $f \in C_0^\infty$ let $(R_\alpha f)(x) = \alpha^{-(d+2)/2} f(x/\alpha)$. Then the scaling identity is

$$\int e^{i\phi(R_\alpha f)} d\phi_{R_\alpha C} = \int e^{i\phi(f)} d\phi_C .$$

See (8.6.26) for a non-Gaussian scaling identity.

Translation of a Gaussian Measure. Let $g \in \mathscr{S}$ and $\psi \equiv \phi - g$. We replace integration over ϕ by integration over the translated variable ψ. Then

$$d\phi_C = \exp[-\tfrac{1}{2}\langle g, C^{-1} g \rangle - \langle C^{-1} g, \psi \rangle] d\psi_C. \quad (9.1.27)$$

PROOF OF (9.1.27). We show the identity of the Fourier transform. Integrate each side of (9.1.27) with $\exp(i\phi(f)) = \exp(i\psi(f)) \exp(i\langle g, f \rangle)$. The integrals can be evaluated using (9.1.16), and both sides equal $\exp(-\tfrac{1}{2}\langle f, Cf \rangle)$. Since a measure is uniquely determined by its Fourier transform, (9.1.27) is an identity.

(iii) Integration by Parts

The Gaussian integration by parts formula

$$\int \phi(f) A \, d\phi_C = \int \left\langle Cf, \frac{\delta}{\delta\phi} \right\rangle A \, d\phi_C = \int \frac{\delta}{\delta\phi(Cf)} A \, d\phi_C = \int D_{Cf} A \, d\phi_C \quad (9.1.28)$$

was proved in Section 6.3, and it also follows from (9.1.6), (9.1.7) and (9.1.20). To extend this identity to non-Gaussian measures, let

$$V = \int_\Lambda :P(\phi):_{C_1} dx \quad (9.1.29)$$

$$d\mu = d\mu_\Lambda = \frac{e^{-V} d\phi_{C_2}}{\int e^{-V} d\phi_{C_2}}, \quad (9.1.30)$$

and let

$$B = A e^{-V}. \quad (9.1.31)$$

Integration by Parts Identity (General Case: Section 12.2)

$$\int \phi(f) A \, d\mu = \int \left\langle C_2 f, \frac{\delta A}{\delta\phi} - A \frac{\delta V}{\delta\phi} \right\rangle d\mu. \quad (9.1.32)$$

Note that in case the limit $\Lambda \uparrow R^d$ exists in (9.1.32), the corresponding formula (9.1.32) for the infinite volume measure $d\mu$ also holds.

Infinitesimal Change of Covariance (Gaussian Case: Section 9.2). Let $C(t)$ be a smooth function of t. Then

$$\frac{d}{dt} \int B \, d\phi_{C(t)} = \frac{1}{2} \int \Delta_{\dot{C}} B \, d\phi_{C(t)},$$ (9.1.33)

where $\dot{C} = dC/dt$ and Δ_C is defined in (9.1.10).

Infinitesimal Change of Covariance (General Case: Sections 9.2, 12.2). Here $C_1(t)$, $C_2(t)$ are smooth functions of t. Then

$$\frac{d}{dt} \int A e^{-V} \, d\phi_{C_2(t)}$$

$$= \frac{1}{2} \int \left\{ \Delta_{\dot{C}_2} A + A \, \Delta_{\dot{C}_2 - \dot{C}_1} V \right.$$

$$\left. + A \left\langle \dot{C}_2 \frac{\delta V}{\delta \phi}, \frac{\delta V}{\delta \phi} \right\rangle - \left\langle \dot{C}_2 \frac{\delta V}{\delta \phi}, \frac{\delta A}{\delta \phi} \right\rangle \right\} e^{-V} \, d\phi_{C_2(t)}.$$ (9.1.34)

A formula for $(d/dt) \int A \, d\mu$, with $d\mu$ of (9.1.30), follows. When defined, the latter was an infinite volume limit.

Conditioning. Here we set $A = I$, $C_1 = C_2 = C(t)$ in (9.1.31, 34):

$$\frac{d}{dt} \int e^{-V} \, d\phi_{C(t)} = \frac{1}{2} \int \left\langle \dot{C} \frac{\delta V}{\delta \phi}, \frac{\delta V}{\delta \phi} \right\rangle e^{-V} \, d\phi_{C(t)}.$$ (9.1.35)

This identity is most useful when $\dot{C} \geq 0$ or $\dot{C} \leq 0$, as in the case $C = tC_1 + (1 - t)C_2$, with $\dot{C} = C_1 - C_2$ and $C_1 \geq C_2$ or $C_1 \leq C_2$.

Translation. With a finite volume interaction, (9.1.27) becomes

$$e^{-V(\phi)} \, d\phi_{C_2} = e^{-W(\psi)} \, d\psi_{C_2}$$

where $W(\psi) = \int_{\Lambda} :P(\psi + g):_{C_1} dx + \langle C_2^{-1} g, \psi \rangle + \frac{1}{2} \langle g, C_2^{-1} g \rangle$. (9.1.27')

PROOF OF (9.1.32). Observe that substituting B for A in (9.1.28) reduces (9.1.32) to (9.1.28). Thus we only need to justify the chain rule in evaluating the derivative $\delta B/\delta \phi$. We introduce ultraviolet cutoffs in V, to approximate it by a sequence of polynomials $Q^{(l)}$ of the form (8.7.2), using Proposition 8.7.2. With $B^{(l)} = A e^{-Q^{(l)}}$ the chain rule and integration by parts is valid, because $Q^{(l)}$ is a cylinder function (depends on only a finite number of degrees of freedom) and so the functional derivatives of $e^{-Q^{(l)}}$ reduce to ordinary derivatives. Thus the identity holds for $B^{(l)}$. However, by Proposition 8.7.3, both sides of the identity converge as $l \to \infty$, so the proof is complete.

(iv) Limits of Measures

Let C_n be a sequence of covariance operators. If $C_n \to C$ weakly as bilinear forms on $\mathscr{S} \times \mathscr{S}$, then the Gaussian measures converge, $d\phi_{C_n} \to d\phi_C$, in the sense of convergent characteristic functions and convergent moments, by (9.1.16), (9.1.17). Let $\mathscr{W} \subset \mathscr{S}$ be a finite dimensional subspace. A cylinder function f, based on $\mathscr{W}$, is a function of the form

$$f(\phi) = F(\langle \phi, w_1 \rangle, \ldots, \langle \phi, w_n \rangle)$$

for a finite set $w_1, \ldots, w_n \in \mathscr{W}$. The integral $\int f \, d\phi_C$ can be written as a finite dimensional Gaussian integral with Gaussian exponent $\frac{1}{2} C_w^{-1}$, where C_w is the restriction of C to $\mathscr{W} \times \mathscr{W}$. Convergence of C_n to C implies convergence of C_{nw} to C_w and (assuming nondegeneracy) of C_{nw}^{-1} to C_w^{-1}. The degenerate case gives rise to a δ function in the Gaussian measure. Taking this into account, one can show that for f a bounded continuous cylinder function,

$$\int f(\phi) \, d\phi_{C_n} \rightarrow \int f(\phi) \, d\phi_C. \qquad (9.1.36)$$

It is important to extend the convergence, in certain cases, to the $P(\phi)_2$ measure (9.1.30).

Removal of lattice cutoff (Sections 9.5, 6). Let $C \in \mathscr{C}$, and C_δ be the lattice approximation to C with lattice spacing δ. Then $C_\delta \rightarrow C$ weakly as L_2 operators, and convergence extends to $d\mu$, in the sense of convergence of characteristic functions.

The Dirichlet limit (Section 7.8). Let $C \in \mathscr{C}$, let Λ be a rectangle, and let $\chi_{\sim \Lambda}$ be the characteristic function of $\sim \Lambda$. Let

$$C_\lambda = (C^{-1} + \lambda \chi_{\sim \Lambda})^{-1}.$$

Then

$$\lim_{\lambda \to \infty} C_\lambda = C_\infty$$

exists, and $C_\infty f = 0$ for $\operatorname{supp} f \subset \sim \Lambda$. Furthermore suppose $C^{-1} = -\Delta_\Gamma + m^2 I$, where Δ_Γ is the Dirichlet Laplacian, with zero data on $\Gamma \subset \Lambda$. Then on $L_2(\Lambda)$, $C_\infty^{-1} = -\Delta_{\Gamma \cup \partial \Lambda} + m^2 I$ is the Dirichlet Laplacian with zero boundary data on $\partial \Lambda$ in addition to the data of C^{-1} on Γ.

The Neumann limit. Let

$$\langle f, C_\lambda^{-1} f \rangle = \langle f, C^{-1} f \rangle + \lambda \int_{\sim \Lambda} |\nabla f(x)|^2 \, dx.$$

Then

$$\lim_{\lambda \to \infty} C_\lambda = C_\infty$$

exists and (at least formally) is associated with the introduction of Neumann data on $\partial \Lambda$. We do not make use of this limit.

9.2 Infinitesimal Change of Covariance

In this section we consider the formulas for interpolation of measures $d\mu$ of the form (9.1.30). First we consider the Gaussian case, and establish (9.1.33).

Proposition 9.2.1. *Let $C(t)$ be a smooth, one parameter family of covariance operators in $\mathscr{C}$, and let A be of the form (9.1.1). Then*

$$\frac{d}{dt} \int A \, d\phi_{C(t)} = \frac{1}{2} \int (\Delta_{\dot{C}} A) \, d\phi_{C(t)}. \qquad (9.2.1)$$

PROOF I.   Suppose $A = e^{i\phi(g)}$. For simplicity let $C(t) = tC_1 + (1-t)C_2$. Then

$$\frac{d}{dt} \int A \, d\phi_{C(t)} = \frac{d}{dt} \int e^{i\phi(g)} \, d\phi_{C(t)}$$

$$= \frac{d}{dt} \exp(-\tfrac{1}{2}\langle g, C(t)g\rangle)$$

$$= -\tfrac{1}{2}\langle g, \dot{C}(t)g\rangle \exp(-\tfrac{1}{2}\langle g, C(t)g\rangle)$$

$$= \frac{1}{2} \int (\Delta_{\dot{C}} A) \, d\phi_{C(t)}. \qquad (9.2.2)$$

Extension to general A's of the form (9.1.1) follows as in the proof of (9.1.6, 7).

OUTLINE OF PROOF II.   This alternative proof illustrates that (9.1.33) is an integration by parts formula. We approximate (9.1.33) by a finite dimensional integral with respect to measures of the form

$$d\phi_n = Z_n^{-1} \exp(-\tfrac{1}{2} : \langle \phi, C(t)_n^{-1}\phi\rangle :) \, d\phi_{\text{Leb}} \to d\phi_{C(t)}, \qquad (9.2.3)$$

where $d\phi_{\text{Leb}}$ is Lebesgue measure, and Wick ordering is with respect to $C(t)_n$. Then

$$\frac{d}{dt} \int A \, d\phi_n = -\frac{1}{2} \int : \langle \phi, \frac{d}{dt} C_n(t)^{-1}\phi\rangle : A \, d\phi_n$$

$$= \frac{1}{2} \int : \langle C_n(t)^{-1}\phi, \dot{C}_n(t)C_n(t)^{-1}\phi\rangle : A \, d\phi_n. \qquad (9.2.4)$$

We next integrate each ϕ by parts in the expression $: \langle C^{-1}\phi, \dot{C}C^{-1}\phi\rangle :$, yielding

$$\frac{d}{dt} \int A \, d\phi_n = \frac{1}{2} \int (\Delta_{\dot{C}_n} A) \, d\phi_n.$$

Passing to the limit $n \to \infty$ yields (9.1.33). We do not justify here the convergence of the $n \to \infty$ limit, which relies e.g. on material in Section 9.5.

Corollary 9.2.2. *Let C_1, C_2 be covariances, and let A be as in the proposition. Then*

$$\int A \, d\phi_{C_2} = \int A \, d\phi_{C_1} + \frac{1}{2} \int_0^1 dt \int (\Delta_{C_2 - C_1} A) \, d\phi_{tC_2 + (1-t)C_1}. \qquad (9.2.5)$$

Corollary 9.2.3. *The formula* (9.1.34) *holds in a finite volume measure.*

PROOF. We include $\exp(-:V:_{C_{(t)}})$ in A and apply Propositions 8.7.1–3 to justify the application of the chain rule in this case.

9.3 Quadratic Perturbations

In this section we study the quadratic perturbation

$$:V:_C = \frac{1}{2} \int :\phi(x)v(x, y)\phi(y):_C \, dx \, dy, \qquad (9.3.1)$$

(cf. (9.1.22)) and the Gaussian measure

$$d\mu = \frac{e^{-:V:_C} \, d\phi_C}{\int e^{-:V:_C} \, d\phi_C} = Z^{-1} e^{-:V:_C} \, d\phi_C. \qquad (9.3.2)$$

We assume C is a bounded positive self-adjoint operator on $L_2(R^d)$. For example $C \in \mathscr{C}$ satisfies this condition. A fundamental restriction is

$$0 < C^{-1} + v. \qquad (9.3.3)$$

We also require v to be real and symmetric. To avoid technical difficulties, we suppose that v is a bounded operator. Let

$$\hat{v} = C^{1/2}vC^{1/2}. \qquad (9.3.4)$$

Proposition 9.3.1. *Assume* (9.3.3). *Let C be bounded, positive, and self-adjoint, let v be bounded, and let $\hat{v}$ be Hilbert–Schmidt. Then $:V:_C$ and $e^{-:V:_C}$ belong to $L_p(d\phi_C)$ for all $p < \infty$ and*

$$Z = \int e^{-:V:_C} \, d\phi_C = \exp(-\tfrac{1}{2} \operatorname{Tr}[\ln(I + \hat{v}) - \hat{v}]). \qquad (9.3.5)$$

If $\|\hat{v}\|_{HS} < 1$, then also

$$Z = \sum_{n=0}^{\infty} \frac{(-1)^n}{n!} \int (:V:_C)^n \, d\phi_C, \qquad (9.3.6)$$

with absolute convergence of the series.

Remark 1. Z is called the functional determinant, since without Wick ordering

$$\int e^{-V} \, d\phi_C = \det(I + \hat{v})^{-1/2} \qquad (9.3.7)$$

when (9.3.7) exists. Note that for symmetric, positive A, $\det A =$

$\exp(\text{Tr} \ln A)$. Formally (and exactly for finite dimensional Gaussian integrals),

$$\det(I + \hat{v})^{-1/2} = \det C^{-1/2} \det(C^{-1} + v)^{-1/2} = \exp[-\tfrac{1}{2} \text{Tr} \ln(I + \hat{v})].$$

In particular, Z is defined by (9.3.7) only when $\hat{v}$ is trace class.

Remark 2. The Wick ordering of V, namely the replacement of V by $:V:_C$ in (9.3.7), removes from the exponent the terms linear in $\hat{v}$. This is just the factor $\exp(\tfrac{1}{2} \text{Tr } \hat{v})$, since $:V:_C = V - \tfrac{1}{2} \text{Tr } \hat{v}$. It defines the renormalized determinant $\det_1$ as follows:

$$\det_1(I + \hat{v})^{-1/2} = \det(I + \hat{v})^{-1/2} \det(\exp \hat{v})^{1/2}$$

$$= \exp[-\tfrac{1}{2} \text{Tr}(\ln(I + \hat{v}) - \hat{v})].$$

In this case Z is defined although $\hat{v}$ may not be trace class.

Remark 3. We distinguish several cases: (1) If $\|\hat{v}\|_{\text{HS}} < 1$, and if V is Wick-ordered, then Z is given by a convergent series expansion in v. (2) If $\hat{v}$ is Hilbert–Schmidt, and if (9.3.3) holds, then Z is defined by (9.3.5) and equals $\det_1(I + \hat{v})$. (3) Even though Z may equal 0 or ∞, the measure $d\mu$ defined by (9.3.2) may exist as a limit of approximating measures. Proposition 9.3.3 below deals with the possibility that $Z = 0$. If $d\mu$ exists but Z is not defined, we say that there is a vacuum vector renormalization, produced by division by Z. The same phenomenon occurs for non-Gaussian measures, as in our discussion of renormalization of the ϕ^4 model in Section 14.3. (In Chapters 8–12, we follow the terminology of statistical mechanics, in which the multiplicative renormalization of the measure is called a partition function and denoted Z; in Chapter 14, the notation Z is reserved for the field strength renormalization, a multiplicative renormalization of the field ϕ.)

PROOF. One computes

$$\int :V:_C^2 d\phi_C = \|\hat{v}\|_{\text{HS}}^2$$

and the higher powers are similar, so $:V:_C \in L_p$ and is continuous in its dependence on $\hat{v}$. Since $\ln(1 + x) - x = O(x^2)$ for $x \to 0$, we see that the right side of (9.3.5) is finite and continuous in $\hat{v}$ for $\hat{v}$ Hilbert–Schmidt.

    Suppose (9.3.5) holds for $\hat{v}$ in a dense linear subspace of the Hilbert–Schmidt operators. We approximate a general $\hat{v}$ by a sequence $\hat{v}_j$ from this subspace. Then $:V_{j}:_C \to :V:_C$ in L_p, and after passing to a subsequence, the same is true pointwise a.e. Thus $e^{-:V_j:_C} \to e^{-:V:_C}$ pointwise a.e. Moreover

$$\|e^{-:V_i:_C} - e^{-:V_j:_C}\|_{L_2}^2 = \exp(-\tfrac{1}{2} \text{Tr}[\ln(I + 2\hat{v}_i) - 2\hat{v}_i])$$

$$+ \exp(-\tfrac{1}{2} \text{Tr}[\ln(I + 2\hat{v}_j) - 2\hat{v}_j])$$

$$- 2 \exp(-\tfrac{1}{2} \text{Tr}[\ln(I + \hat{v}_i + \hat{v}_j) - \hat{v}_i - \hat{v}_j])$$

$$\to 0 \quad \text{as } i, j \to \infty.$$

It follows that $e^{-:V:_C} \in L_2$ and that (9.3.5) holds. Applying this argument to $e^{-(p/2):V:_C}$, we have $e^{-:V:_C} \in L_p$, all $p < \infty$. Finally the right side of (9.3.5) has a convergent power series for $\|\hat{v}\|_{HS} < 1$, so the convergence of (9.3.6) follows. (As an exercise in the manipulation of Feynman diagrams, we give a separate proof of (9.3.6) after the next proposition.)

Now we restrict to smooth v of finite rank. Then $:V:_C$ and $e^{-:V:_C}$ are cylinder functions, and in this case Z can be evaluated explicitly as a finite dimensional Gaussian integral. By assumption the kernel $v(x, y)$ of v has a representation

$$v(x, y) = \sum_{j=1}^{J} \lambda_j f_j(x) f_j(y)$$

with $f_j \in \mathcal{S}$. Let $g_j = C^{1/2} f_j$, so that

$$\hat{v}(x, y) = \sum_{j=1}^{J} \lambda_j g_j(x) g_j(y).$$

Without loss of generality we take the g_j to be orthonormal. By (9.3.3), $\lambda_j > -1$ for all j. Then

$$Z = e^{1/2\, \mathrm{Tr}\, \hat{v}} (2\pi)^{-J/2} \int \prod_j e^{-(1/2)(\lambda_j + 1)x_j^2} \, dx_j$$

$$= e^{1/2\, \mathrm{Tr}\, \hat{v}} \det(1 + \hat{v})^{-1/2}$$

$$= e^{(-1/2\, \mathrm{Tr}[\ln(1 + \hat{v}) - \hat{v}])},$$

and the proof is complete.

Proposition 9.3.2. *The measure $d\mu$ defined by (9.3.2) and Proposition 9.3.1 has covariance*

$$C_v \equiv (C^{-1} + v)^{-1}. \qquad (9.3.8)$$

PROOF. Since $\hat{v}$ is Hilbert–Schmidt and $-1 < \hat{v}$ by (9.3.3), $-1 + \varepsilon < \hat{v}$ for some $\varepsilon > 0$. Thus $(1 + \hat{v})^{-1}$ is bounded, as is

$$C^{1/2}(1 + \hat{v})^{-1} C^{1/2} = C_v,$$

and C_v is characterized by the identity

$$C_v = C - CvC_v, \qquad (9.3.9)$$

namely, the integration by parts formula applied to the two point function

$$\int \phi(x)\phi(y) \, d\mu.$$

DIRECT PROOF OF (9.3.6), ASSUMING $\|\hat{v}\|_{HS} < 1$. We use the Feynman graph expansions of Corollary 8.3.2 to evaluate each integral in (9.3.6), and to show the series (9.3.6) converges. The graphs which contribute to $\int (:V:_C)^n \, d\phi_C$ have exacly n v-vertices. Each v-vertex has two legs, and the corresponding graphs G are a union of connected components G_i, each G_i a loop with $n_i \geq 2$ vertices; cf. Figure 9.1. Because of the form of these graphs, the expansion (9.3.6) is called the loop expansion.

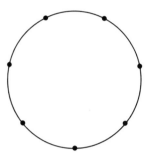

Figure 9.1. A connected component of a graph contributing to $\int (:V_C:)^n \, d\phi_C$.

A graph contributing to $(-1)^n \int (:V:_C)^n \, d\phi_C$ has r connected components with $n_1 + \cdots + n_r = n$. The ith component has the value

$$I(G_i) = (-\tfrac{1}{2})^{n_i} \, \mathrm{Tr}(\hat{v}^{n_i}),$$

which we obtain by noting that

$$\int dx_1 \cdots dx_n \, dy_1 \cdots dy_n \; v(x_1, y_1)C(y_1, x_2)v(x_2, y_2)C(y_2, x_3) \cdots v(x_n, y_n)C(y_n, x_1)$$

$$= \mathrm{Tr}((vC)^n) = \mathrm{Tr}((C^{1/2}vC^{1/2})^n) = \mathrm{Tr}(\hat{v}^n).$$

Thus we write

$$(-1)^n \int (:V:_C)^n \, d\phi_C = \sum_G I(G) = \sum_G \prod_i I(G_i)$$

$$= \sum_G \prod_i (-\tfrac{1}{2})^{n_i} \, \mathrm{Tr}(\hat{v}^{n_i}),$$

where $\{G_i\}$ denotes the set of connected components of G.

Substitution in (9.3.6) gives

$$Z = 1 + \sum_n \frac{1}{n!} \sum_r \sum_{\{n_1, \ldots, n_r : \, \sum n_i = n, \, n_i \geq 2\}} \Sigma'_G \prod_{j=1}^r (-\tfrac{1}{2})^{n_j} \, \mathrm{Tr}(\hat{v}^{n_j}),$$

where Σ'_G is the sum over all G's with the same loop sizes. To be explicit, the vertices of G are labeled, and we let n_1 be the size of the loop containing the first vertex, n_2 the size of the loop containing the first vertex remaining after the first loop, etc. The above expression can be simplified by combining the sum over n with the sum over n_i; this merely removes the constraint on the sum over n_i, so that

$$Z = 1 + \sum_r \sum_{\{n_1, \ldots, n_r : \, n_i \geq 2\}} \left(\left(\sum n_i \right)! \right)^{-1} \Sigma'_G \prod_j (-\tfrac{1}{2})^{n_j} \, \mathrm{Tr}(\hat{v}^{n_j}).$$

However Σ'_G can be evaluated explicitly:

$$\Sigma'_G = \binom{\sum n_i}{n_1, \ldots, n_r} \frac{1}{r!} \prod_j 2^{n_j - 1}(n_j - 1)!.$$

Here the binomial factor counts the number of ways vertices can be assigned to loops. Division by $r!$ then gives the same number in the case of ordered loops.

Also $\frac{1}{2}(n_j - 1)!$ is the number of cyclic orderings of vertices along the loop, without distinguishing between the two orientations of the loop. Finally there is a factor 2 per vertex because of the choice between two legs for the initial contraction at each vertex. Combining the above with an interchange of summation and product yields

$$Z = \sum_{r=0}^{\infty} (r!)^{-1} \left(\sum_{n=2}^{\infty} \frac{1}{2n} \operatorname{Tr}(-\hat{v}^n) \right)^r$$

$$= \exp(-\tfrac{1}{2} \operatorname{Tr}[\ln(1 + \hat{v}) - \hat{v}])$$

to prove (9.3.5), (9.3.6) for $\|\hat{v}\|_{HS} < 1$.

The preceding applies to local mass perturbations in $d = 3$ as well as to $d = 2$. The next result applies to more singular perturbations.

Proposition 9.3.3. *The definition of $d\mu$ in (9.3.2) extends by continuity to the case $\hat{v}$ bounded, $0 < \varepsilon \le I + \hat{v}$.*

PROOF. With $\hat{v}_j \to \hat{v}$ in the strong operator topology, we have $(1 + \hat{v}_j)^{-1} \to (1 + \hat{v})^{-1}$ and $C_{v_j} \to C_v$ also. Convergence of measures in the sense of convergent characteristic functions and moments follows.

A typical approximating kernel is defined by a momentum and space cutoff:

$$v_{\kappa, \Lambda}(x, y) = \chi_\Lambda(x) v_\kappa(x, y) \chi_\Lambda(y),$$

where

$$v_\kappa(x, y) = \int v(x - x', y - y') \delta_\kappa(x') \delta_\kappa(y') \, dx' \, dy'.$$

9.4 Perturbative Renormalization

Renormalization is a reparametrization of the Lagrangian. The parameters in the Lagrangian are not directly measurable, and the purpose of the reparametrization is to replace the perturbation expansion of Section 8.4 by an expansion in terms of measurable quantities: the particle masses and coupling constants. The masses are defined as eigenvalues in the spectrum of the mass operator $M = (H^2 - P^2)$ which labels the Lorentz-invariant hyperbolae in the energy–momentum spectrum. Equivalently masses are defined by exponential decay rates of correlations (see Chapter 14). The coupling constants are defined in terms of scattering theory. In particular, in the nonrelativistic limit of scattering, the nature of the long range force between particles defines a coupling constant. The boson theories studied in this section characteristically have short range forces of a Yukawa type. The large distance asymptotic behavior of the potential is typically

$$V(r) = \frac{\lambda}{4\pi r} e^{-mr} + O(e^{-(m+\varepsilon)r}), \tag{9.4.1}$$

where λ is a coupling constant and $0 < m$ is the mass. These parameters λ and m are functions of the parameters λ_b and m_b which occur in the Lagrangian $\mathcal{L}$. Since the field theory describes observable particles, we wish to adjust λ_b and m_b so that λ and m have predetermined values. In other words, we invert the functional relation $\lambda = \lambda(\lambda_b, m_b)$, $m = m(\lambda_b, m_b)$. We can then regard the Lagrangian as a (not necessarily single valued) function of λ and m through the inverse relation $\lambda_b = \lambda_b(\lambda, m)$, $m_b = m_b(\lambda, m)$. Having made this renormalization, the predictive power of the theory rests in providing bound states (other eigenvalues of M) and the differential cross sections, resonances, production and decay rates, branching ratios, etc. In other words, λ and m fix a particular field theory out of a two parameter family. Any observation independent of λ and m is a prediction.

A second, more technical use of the concept of renormalization refers to the cancellation of divergences which occur in perturbation calculations. Mathematically, these divergences arise because the functional $V(\phi)$ with V nonlinear does not exist on the distribution space $\mathcal{S}'$. This aspect of renormalization is an extension of the reparametrization described above. Again we regard λ_b and m_b as functions of λ and m. In order for λ and m to be finite, λ_b and m_b may be forced to be infinite. Thus the reparametrization modifies the potential in such a way as to require possibly divergent coefficients λ_b or m_b in $\mathcal{L}$. These divergences then cancel in the calculation of m or λ. The resulting renormalized theory gives finite values to observable quantities.

Let us explain the reparametrization in more detail, for the particular case of the ϕ^4 field theory. The bare Euclidean Lagrangian function is

$$\mathcal{L}_b(\phi) = \int \left(\tfrac{1}{2}(\nabla\phi)^2 + \tfrac{1}{2}m_b^2\phi^2 + \lambda_b\phi^4\right) dx. \qquad (9.4.2)$$

Here m_b is the bare mass and λ_b is the bare coupling constant. With λ, m, and the field strength renormalization constant Z_ϕ to be chosen later, we write $\phi = \phi_b$ and

$$m^2 = m_b^2 + \delta m^2,$$
$$\phi_r = Z_\phi^{-1/2}\phi, \qquad (9.4.3)$$
$$\lambda = \lambda_b Z_\lambda^{-1} Z_\phi^2 = (\lambda_b + \delta\lambda)Z_\phi^2.$$

Then the renormalized Lagrangian $\mathcal{L}_r$ is defined by

$$\mathcal{L}_r(\phi_r) = \mathcal{L}_b(\phi)$$

$$= \int \{\tfrac{1}{2}(\nabla\phi_r)^2 + \tfrac{1}{2}m^2\phi_r^2 + \lambda\phi_r^4 + \tfrac{1}{2}(Z_\phi - 1)(\nabla\phi_r)^2$$

$$+ \tfrac{1}{2}(m^2(Z_\phi - 1) - \delta m^2 Z_\phi)\phi_r^2 - Z_\phi^2 \,\delta\lambda\phi_r^4\} \, dx. \qquad (9.4.4)$$

We now regard λ and m as the basic parameters of the theory, and δm^2, Z_ϕ, and $\delta\lambda$ as functions of λ and m, defined by a power series in λ, depending on m. These power series will be determined by the definition of

λ and m as the physical coupling constant and the mass, respectively. The constant Z_ϕ will be determined by a convention.

The effect of this reparametrization on the perturbation expansion of Section 8.4 is to generate new vertices, connected by new lines. The terms $\frac{1}{2}(\nabla\phi_r)^2 + \frac{1}{2}m^2\phi_r^2$ in $\mathcal{L}_r(\phi_r)$ define the covariance and the Gaussian measure in which the expansion is performed. This covariance defines the meaning of the lines in the graphs of the renormalized expansion. For each ϕ^4 vertex in the Section 8.4 expansion, there are now two terms, one with a $\lambda\phi_r^4$ vertex and one with a $\delta\lambda\phi_r^4$ vertex. The remaining quadratic terms $-\frac{1}{2}\langle\phi_r, \delta Q\phi_r\rangle$ are expanded also. Here

$$\frac{1}{2}\langle\phi_r, \delta Q\phi_r\rangle = \frac{1}{2}(1 - Z_\phi)(\nabla\phi_r)^2 + \frac{1}{2}(\delta m^2 Z_\phi + m^2(1 - Z_\phi))\phi_r^2. \quad (9.4.5)$$

The expansion of the quadratic terms can be related to the analysis of Section 9.3. In fact the covariance for $\mathcal{L}_r$ is

$$C_r = (-\Delta + m^2)^{-1},$$

and that for $\mathcal{L}_b$ is

$$C_b = (-\Delta + m_b^2)^{-1}Z_\phi^{-1}$$

(if we take ϕ_r as the integration variable in both cases). Then

$$C_r^{-1} = C_b^{-1} + \delta Q,$$

by Section 9.3, where δQ now denotes the kernel of quadratic form δQ of (9.4.5).

The Neumann series relating C_r to C_b can be written in terms of Feynman graphs (see Figure 9.2). Substituting this graphical version of the Neumann series in the unrenormalized Section 8.4 expansion generates the renormalized C_r lines (propagators) and the δQ vertices of the Section 9.4 renormalized expansion.

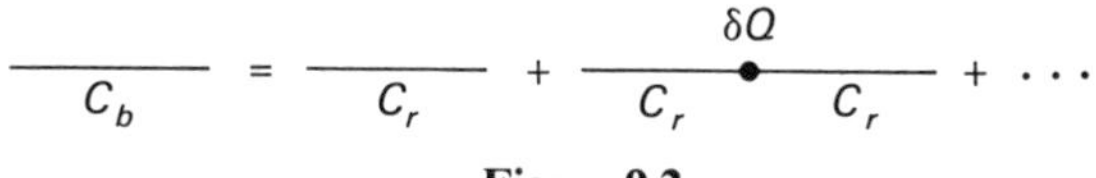

Figure 9.2

We argue that the combinatoric structure of the δQ vertices matches exactly the combinatoric structure of a class of subdiagrams of the expansion. This combinatoric match allows an identification between similar terms. The terms are then compared, and found to cancel partially. The cancellation is the infinite cancellation of renormalized perturbation theory. In fact δQ and the terms which match all contribute to the definition of δm^2 and Z_ϕ. The definition of δm^2 and Z_ϕ then fixes δQ and forces the partial cancellation.

The diagrams which match with δQ in combinatoric structure are mass diagrams. These are diagrams which have two external legs and which cannot be disconnected by the removal of one line (called one particle irreducible, or 1PI).

We define a mass skeleton diagram to be a diagram (with an arbitrary number of external legs) but whose only mass subdiagrams are the single lines (propagators) in it. Associated with a general diagram is its mass skeleton, obtained by collapsing every linear chain of mass diagrams into a single propagator. The allowed linear chains have the structure of the Neumann series of Figure 9.2, and so conversely a general diagram can be reconstructed from its mass skeleton by arbitrary insertion of mass diagrams. Each δQ vertex is also a mass diagram, chosen to cancel the effect of the other mass diagrams to shift the mass spectrum and field strength of the theory.

There is a similar theory of vertex diagrams. In a ϕ^4 theory, a vertex diagram has four external legs and cannot be disconnected by removal of two or fewer internal lines (called two particle irreducible, or 2PI). The $\delta\lambda$ vertices are included among the vertex diagrams. All vertex diagrams occur with the same combinatorial structure. A vertex skeleton diagram has no vertex subdiagrams except the single vertex $\lambda\phi^4$ vertices, and general diagrams can be constructed from vertex skeletons by inserting vertex subdiagrams. There is a partial cancellation among vertex diagrams (including the $\delta\lambda$ vertex), enforced by the definition of $\delta\lambda$.

A skeleton diagram is a diagram which is both a mass skeleton and a vertex skeleton diagram. Because it is necessary to do mass insertions into vertex renormalization and vice versa, both mass and vertex renormalizations must be done simultaneously, as mass and vertex insertions in the class of full skeleton diagrams.

The next stage of renormalization is to define δm^2, $\delta\lambda$, and Z_ϕ explicitly, and to relate these definitions to first principles (particle masses and scattering processes). This will be postponed to Section 14.3, at which point more theoretical tools will have been developed. The result of this definition is to give δm^2, $\delta\lambda$, and Z_ϕ as power series in λ. The definition of the coefficient of λ^n ensures a cancellation in the mass and vertex diagrams of order λ^n. If such a diagram includes a mass or vertex subdiagram, then this subdiagram can be inserted in canceled form, i.e. as the finite sum of two partially canceling subdiagrams. The low order mass and vertex diagrams for the ϕ^4 interaction are illustrated in Section 14.3.

Finally, the statement that a theory is renormalizable is the statement that each skeleton graph, with the insertion of arbitrary renormalized (i.e. partially cancelled) mass and vertex subdiagrams, defines an integral over some R^n (as in Section 8.2) and that this integral is absolutely convergent. The ϕ^4, Yukawa, Dirac, Maxwell, QED, and Yang–Mills fields are renormalizable for $d \leq 4$ dimensions: see [Bogoliubou and Parasiuk, 1957], [Hepp, 1971], ['t Hooft, 1971a, b], [Abers and Lee, 1973] and [Becchi, Rouet and Stora, 1976]. As a general rule renormalizability occurs when the coupling constant has dimensions (length)^j with $j \leq 0$. Superrenormalizability means that the infinities occur only to a finite order in λ. This situation occurs when the coupling constant has dimensions (length)^j with $j < 0$. The examples of superrenormalizable fields in $d \leq 3$ dimensions are fields such as ϕ^4, electrodynamics, or Yang–Mills, which are renormalizable for $d = 4$ dimensions. The renormalizable but not superrenormalizable theories have dimensionless coupling constants, e.g. the fine structure constant $\alpha = e^2/4\pi\hbar c$ in electrodynamics or λ in the ϕ^4 model. They are also characterized by the fact that the Sobolev inequality which dominates the interaction part of the Euclidean action by the kinetic (free) part of the action becomes borderline.

Zero-mass theories result from special values of m_b and λ_b; see Chapter 17. The renormalization of such theories is more complicated.

9.5 Lattice Laplace and Covariance Operators

In this section we discuss the lattice approximation C_δ to the covariance operators C introduced in Chapter 7. As a consequence of norm convergence ($C_\delta \to C$ as the lattice spacing $\delta \to 0$), we obtain a lattice approximation to the corresponding Gaussian measures $d\phi_{C_\delta} \to d\phi_C$, in the sense of convergence of characteristic functionals (Fourier transforms). This convergence holds in arbitrary dimension d.

Detailed proofs are given only in the case of Dirichlet boundary conditions. The results, however, are valid more generally and extend to classical boundary conditions for covariance operators C in the class $\mathscr{C}$ defined in Chapter 7.

For simplicity, let Λ denote a unit cube in R^d,

$$\Lambda = \{x = (x_1, \ldots, x_d) : 0 \le x_\alpha \le 1, \alpha = 1, 2, \ldots, d\}. \tag{9.5.1}$$

We may obtain an $l_1 \times \cdots \times l_d$ rectangular region by scaling the cube sides differently. The Dirichlet Laplace operator Δ_D is diagonalized by the Fourier sine series, defined by the orthonormal basis set

$$e_k(x) = 2^{d/2} \prod_{\alpha=1}^{d} \sin(k_\alpha x_\alpha), \qquad k_\alpha \in \pi Z_+. \tag{9.5.2}$$

(Repeated indices are not summed.) Then

$$-\Delta_D e_k = k^2 e_k, \tag{9.5.3}$$

where $k^2 = \sum_\alpha k_\alpha^2$. Recall that $C_D = (-\Delta_D + m^2)^{-1}$.

We consider the finite lattice Λ_δ with spacing δ. Let

$$\delta Z^d = \{\delta z : z \in Z^d\},$$

$$\text{Int } \Lambda_\delta = \text{Int } \Lambda \cap \delta Z^d, \qquad \partial\Lambda_\delta = \partial\Lambda \cap \delta Z^d, \tag{9.5.4}$$

$$\Lambda_\delta = \text{Int } \Lambda_\delta \cup \partial\Lambda_\delta = \Lambda \cap \delta Z^d.$$

For compatibility, we require that $\delta = 2^{-\nu}$, $\nu \in Z_+$. Then $l_2(\text{Int } \Lambda_\delta)$ is the Hilbert space with inner product

$$\langle f, f \rangle_{\text{Int } \Lambda_\delta} = \sum_{x \in \text{Int } \Lambda_\delta} \delta^d |f(x)|^2. \tag{9.5.5}$$

Regard $l_2(\text{Int } \Lambda_\delta)$ as a subspace of $l_2(\Lambda_\delta)$.

The forward lattice gradient on $l_2(\delta Z^d)$ is defined by

$$(\partial_{\delta, \alpha} f)(x) = \delta^{-1}[f(x + \delta e_\alpha) - f(x)], \qquad (9.5.6)$$

where e_α is the unit vector in the αth coordinate direction. The backward gradient is the adjoint $\partial^*_{\delta, \alpha}$ of $\partial_{\delta, \alpha}$ relative to the $l_2(\delta Z^d)$ inner product, and

$$-\Delta_\delta = \partial^*_\delta \partial_\delta = \partial_\delta \partial^*_\delta = \sum_\alpha \partial^*_{\delta, \alpha} \partial_{\delta, \alpha},$$

so that

$$(\Delta_\delta f)(x) = \delta^{-2}[-2d\, f(x) + \sum_{nn\, x} f(x')]. \qquad (9.5.7)$$

For simplicity, we write below ∂_α for $\partial_{\delta, \alpha}$. In (9.5.7), $\sum_{nn\, x}$ denotes the sum over the $2d$ lattice sites x' which are nearest neighbors to x. Let $\Pi_{\text{Int } \Lambda_\delta}$ be the orthogonal projection of $l_2(\delta Z^d)$ onto $l_2(\text{Int } \Lambda_\delta)$. The Dirichlet difference Laplacian is defined to be

$$\Delta_{\delta, D} = \Pi_{\text{Int } \Lambda_\delta} \Delta_\delta \Pi_{\text{Int } \Lambda_\delta}. \qquad (9.5.8)$$

In particular, for $x \in \text{Int } \Lambda_\delta$ and $f \in l_2(\text{Int } \Lambda_\delta)$, $(\Delta_{\delta, D} f)(x)$ is given by (9.5.7). Summation by parts shows that

$$\langle f, -\Delta_\delta f \rangle_{\delta Z^d} = \delta^d \sum_{\delta Z^d} |\partial_\delta f(x)|^2. \qquad (9.5.9)$$

The gradient is better thought of as a function defined on bonds connecting nearest neighbor lattice sites. Let $B(\delta Z^d)$ be the set of δZ^d bonds, and $B(\Lambda_\delta)$ those bonds contained in Λ_δ. Then for $f \in l_2(\text{Int } \Lambda_\delta)$,

$$\langle f, -\Delta_{\delta, D} f \rangle_{\text{Int } \Lambda_\delta} = \langle f, -\Delta_\delta f \rangle_{\delta Z^d} = \delta^d \sum_{b \in B(\Lambda_\delta)} |\partial_\delta f(b)|^2. \qquad (9.5.10)$$

Thus we see that the Dirichlet difference Laplacian, as with the continuum case, results from a restriction in the domain of the associated bilinear form.

Proposition 9.5.1. *The Dirichlet Laplace operator* $-\Delta_{\delta, D}$ *defines a ferromagnetic interaction for which correlation inequalities were proved in Chapter 4.*

PROOF. The off diagonal terms in $\langle f, -\Delta_{\delta, D} f \rangle_{\Lambda, \delta}$ arise from the $\sum_{nn\, x}$ term in (9.5.7) and are negative multiples of the product $f(x)f(x')$, i.e. ferromagnetic. The diagonal terms (which have the opposite sign and are not ferromagnetic) are regarded as contributing to the measure $d\mu_i(\xi_i)$ of Section 4.1, and not to the interaction H of Section 4.1. Thus the interaction is ferromagnetic.

Let $C_{\delta, D} = (-\Delta_{\delta, D} + m^2)^{-1}$ be the lattice Dirichlet covariance operator, and let $C_\delta = C_{\delta, \varnothing} = (-\Delta_\delta + m^2)^{-1}$ be the free lattice covariance operator. Let us consider $C_{\delta, D}$ as an operator on $l_2(\delta Z^d)$ with Dirichlet boundary conditions on space-filling translates of Λ. Then $C_{\delta, D}$ and C_δ are related by the odd reflection formula (7.5.1).

Proposition 9.5.2. *The following l_2 operator and pointwise inequalities are uniform in δ:*

$$0 \le C_{\delta, D} \le m^{-2} I,$$

$$0 \le C_{\delta, D}(x, y), \qquad x, y \in \text{Int } \Lambda_\delta.$$

PROOF. Pointwise positivity follows from the lattice maximum principle [Bers, John, and Schechter, 1964], as in the proof of Corollary 7.5.2. Operator monotonicity follows from (9.5.10), which implies that $0 \le -\Delta_{\delta, D}$.

For the detailed analysis of $C_{\delta, D}$, we diagonalize $\Delta_{\delta, D}$ by choosing as a basis for $l_2(\text{Int } \Lambda_\delta)$ the $(\delta^{-1} - 1)^d$ functions

$$\{e_k^\delta(x) = e_k(x) : x \in \text{Int } \Lambda_\delta, \ k_\alpha = \pi, 2\pi, \ldots, (\delta^{-1} - 1)\pi\}. \quad (9.5.11)$$

Here e_k is defined in (9.5.2). Observe that e_k vanishes on $\partial \Lambda_\delta$, so $e_k \in l_2(\text{Int } \Lambda_\delta)$.

Proposition 9.5.3. *The vectors e_k^δ diagonalize $-\Delta_{\delta, D}$, and*

$$-\Delta_{\delta, D} e_k^\delta = \lambda_k^\delta e_k^\delta, \qquad \lambda_k^\delta = 4\delta^{-2} \sum_{\alpha=1}^d \sin^2\left(\frac{\delta k_\alpha}{2}\right). \quad (9.5.12)$$

Furthermore

$$\langle e_k^\delta, e_l^\delta \rangle_{\text{Int } \Lambda, \delta} = \delta_{k, l}. \quad (9.5.13)$$

PROOF. Since $\Delta_{\delta, D} = \sum_{\alpha=1}^d \Pi_{\text{Int } \Lambda_\delta} \partial_\alpha^* \partial_\alpha \Pi_{\text{Int } \Lambda_\delta}$ and since the summands $\Pi_{\text{Int } \Lambda_\delta} \partial_\alpha^* \partial_\alpha \Pi_{\text{Int } \Lambda_\delta}$ are commuting self-adjoint operators, we can diagonalize each $\Pi_{\text{Int } \Lambda_\delta} \partial_\alpha^* \partial_\alpha \Pi_{\text{Int } \Lambda_\delta}$ independently. We verify that $\sin(k_\alpha x_\alpha)$ is an eigenfunction of $\Pi_{\text{Int } \Lambda_\delta} \partial_\alpha^* \partial_\alpha \Pi_{\text{Int } \Lambda_\delta}$. In fact $\Pi_{\text{Int } \Lambda_\delta} \sin(k_\alpha x) = \sin(k_\alpha x)$, $x \in \text{Int } \Lambda_\delta$, and

$$-\partial_\alpha^* \partial_\alpha \sin(k_\alpha x_\alpha) = 2\delta^{-2}\{1 - \cos(k_\alpha \delta)\}\sin(k_\alpha x_\alpha), \quad (9.5.14)$$

so (9.5.12) holds. Furthermore, $\sin(k_\alpha x_\alpha)$ is a simple eigenfunction of $\partial_\alpha^* \partial_\alpha$, from which we infer that for $k \ne l$, $\langle e_k^\delta, e_l^\delta \rangle = 0$.

In order to calculate the normalization of e_k, we note that the norm squared factorizes,

$$\|e_k^\delta\|_{\text{Int } \Lambda_\delta}^2 = (2\delta)^d \sum_{\substack{j_\alpha = 0 \\ \alpha = 1, \ldots, d}}^{\delta^{-1}} \prod_{\alpha=1}^d \sin^2(k_\alpha \, \delta j_\alpha), \quad (9.5.15)$$

so we need only perform the $\alpha = 1$ sum. But for $l = 1, 2, \ldots, \delta^{-1} - 1$ and $k = \pi l$,

$$2\delta \sum_{j=0}^{1/\delta} \sin^2(k\delta j) = \delta \sum_{j=0}^{(1/\delta)-1} \{1 - \text{Re } e^{2ik\delta j}\}$$

$$= 1 - \delta \, \text{Re } \frac{1 - e^{2ik}}{1 - e^{2ik\delta}} = 1, \quad (9.5.16)$$

since $2k = 2\pi l$, $e^{2ik} = 1$, but $0 < 2k\delta < 2\pi$. Hence (9.5.13) holds.

Corollary 9.5.4. *The map*

$$i_\delta : e_k^\delta \to e_k \tag{9.5.17}$$

defines an isometric imbedding of $l_2(\text{Int } \Lambda_\delta)$ into $L_2(\Lambda)$.

Let Π_δ be the projection operator acting on $L_2(\Lambda)$ which truncates Fourier series at $k_\alpha/\pi = \delta^{-1}$, so that

$$\Pi_\delta \sum \alpha_k e_k = \sum_{1 \le k_\alpha/\pi \le (1/\delta)-1} \alpha_k e_k. \tag{9.5.18}$$

Then i_δ^* is computed as

$$(i_\delta)^* f = (\Pi_\delta f)|\Lambda_\delta, \tag{9.5.19}$$

i.e. the projection Π_δ followed by restriction to lattice points.

We now may regard $C_{\delta, D}$ as acting on $L_2(\Lambda)$ via the isometry of Corollary 9.5.4. In other words,

$$C_{\delta, D} = i_\delta C_{\delta, D} i_\delta^*, \tag{9.5.20}$$

where the right hand $C_{\delta, D}$ acts on $l_2(\text{Int } \Lambda_\delta)$ and the left hand $C_{\delta, D}$ acts on $L_2(\Lambda)$. As an $L_2(\Lambda)$ operator, $C_{\delta, D}$ has the kernel

$$C_{\delta, D}(x, y) = \sum_{1 \le k_\alpha/\pi \le (1/\delta)-1} (\lambda_k^\delta + m^2)^{-1} e_k(x) e_k(y). \tag{9.5.21}$$

The restriction of this kernel to lattice points $x, y \in \text{Int } \Lambda_\delta$ coincides with the matrix entries of $C_{\delta, D}$ as an $l_2(\text{Int } \Lambda_\delta)$ operator.

Proposition 9.5.5. *The operators $C_{\delta, D}$ converge in norm as $\delta \to 0$ to C_D, the Dirichlet covariance on $L_2(\Lambda)$. Here all operators are considered as acting on $L_2(\Lambda)$, as in Corollary 9.5.4. In fact $\|C_{\delta, D} - C_D\| \le O(\delta^2)$.*

PROOF. Consider

$$D_\delta \equiv \sum_{1 \le k_\alpha/\pi \le (1/\delta)-1} (k^2 + m^2)^{-1} e_k(x) e_k(y).$$

We claim that $\|C_{\delta, D} - D_\delta\| \le O(\delta^2)$. In fact,

$$(\lambda_k^\delta + m^2)^{-1} - (k^2 + m^2)^{-1} = (k^2 - \lambda_k^\delta)(\lambda_k^\delta + m^2)^{-1}(k^2 + m^2)^{-1}. \tag{9.5.22}$$

However,

$$\lambda_k^\delta = \sum_{\alpha=1}^d \left| 2\delta^{-1} \sin \frac{k_\alpha \delta}{2} \right|^2 = \sum_{\alpha=1}^d k_\alpha^2 \left| \frac{\sin \theta_\alpha^\delta}{\theta_\alpha^\delta} \right|^2, \tag{9.5.23}$$

where $\theta_\alpha^\delta \equiv k_\alpha \delta/2$. Hence $0 \le \theta_\alpha^\delta \le \pi/2$, and so

$$4k^2/\pi^2 \le \lambda_k^\delta \le k^2. \tag{9.5.24}$$

Thus

$$0 \le k^2 - \lambda_k^\delta = \sum_{\alpha=1}^d k_\alpha^2 \left(1 - \left| \frac{\sin \theta_\alpha^\delta}{\theta_\alpha^\delta} \right|^2 \right)$$

$$\le O(1) \sum_{\alpha=1}^d k_\alpha^2 (\theta_\alpha^\delta)^2 \le O(1) |k|^4 \delta^2. \tag{9.5.25}$$

since $0 \leq 1 - x^{-2}(\sin x)^2 \leq O(x^2)$. Hence (9.5.22) is bounded by $0(1)\delta^2$. Since $e_k(x)e_k(y)$ is the kernel of the orthogonal projection onto e_k, it follows that $\|C_{\delta, D} - D_\delta\| \leq O(\delta^2)$. Finally

$$C_D(x, y) = \sum_{1 \leq k_\alpha/\pi < \infty} (k^2 + m^2)^{-1} e_k(x)e_k(y).$$

Thus

$$\|D_\delta - C_D\| \leq \inf\{(k^2 + m^2)^{-1} : k_\alpha/\pi \geq \delta^{-1} \text{ for some } \alpha\},$$

and so $C_{\delta, D} \to C_D$ in norm, as claimed.

We now introduce $C_{m, \Lambda}$ as the convex set generated by all $C_{\delta, D}$ operators (m and Λ fixed), with $\delta = 0$ and $\delta = 2^{-\nu}$, $\nu \in Z^+$. The set $\mathscr{C}_{m, \Lambda}$ is technically useful in studying the $\delta \to 0$ limit of $P(\phi)_2$ measures. Each $C \in \mathscr{C}_{m, \Lambda}$ is diagonalized by the basis $\{e_k\}$, and thus the equation

$$Ce_k = (\lambda_k^{(C)} + m^2)^{-1} e_k \tag{9.5.26}$$

defines $\lambda_k^{(C)}$. By (9.5.24) and the convex sum property,

$$4k^2/\pi^2 \leq \lambda_k^{(C)}. \tag{9.5.27}$$

Observe that $\Pi_\delta C = C\Pi_\delta = \Pi_\delta C\Pi_\delta$.

Proposition 9.5.6. *Define the Wick ordering constant*

$$c_\delta(x) = (\Pi_\delta C)(x, x), \qquad C \in \mathscr{C}_{m, \Lambda}.$$

Then

$$0 \leq c_\delta(x) \leq \begin{cases} \text{const} \ln \delta^{-1}, & \text{for } d = 2, \\ \text{const } \delta^{-d+2} & \text{for } d \geq 3, \end{cases}$$

and the constant is independent of $C \in \mathscr{C}_{m, \Lambda}$.

PROOF. By (9.5.2), $|e_k(x)| \leq 1$. Thus

$$0 \leq c_\delta(x) = \sum_{1 \leq k_\alpha/\pi \leq (1/\delta) - 1} (\lambda_k^{(C)} + m^2)^{-1} e_k(x)^2$$

$$\leq \sum_{1 \leq k_\alpha/\pi \leq (1/\delta) - 1} (\lambda_k^{(C)} + m^2)^{-1}.$$

Substitution of (9.5.27) gives

$$c_\delta(x) \leq \sum_{1 \leq k_\alpha/\pi \leq (1/\delta) - 1} (4k^2\pi^{-2} + m^2)^{-1},$$

which is bounded as in the statement of the proposition.

Proposition 9.5.7. *Let $d = 2$, $p < \infty$. Then*

$$C(x, \cdot) \in L_p(\Lambda), \qquad (\Pi_\delta C)(x, \cdot) \in L_p(\Lambda),$$

with a bound uniform in x and $C \in \mathscr{C}_{m, \Lambda}$. As $\delta \to 0$,

$$\sup_{x \in \Lambda} \| C(x, \cdot) - (\Pi_\delta C)(x, \cdot) \|_{L_p(\Lambda)} \leq O(\delta^\varepsilon),$$

$$\sup_{x \in \Lambda} \| C_{\delta, D}(x, \cdot) - C_D(x, \cdot) \|_{L_p(\Lambda)} \leq O(\delta^\varepsilon),$$

for $\varepsilon < \min\{2/p, 1\}$, uniformly in C.

PROOF.   Because Λ is bounded, we may suppose $2 \leq p$. The Hausdorff–Young inequality for the Fourier sine series $\{e_k\}$ follows from the usual (periodic) Fourier series case, provided we consider separately the contribution of even and odd k_α/π. See [Zygmund, 1959] for the periodic case. Thus

$$\| f \|_{L_p(\Lambda)} \leq \| \tilde{f} \|_{l_{p'}(\pi Z^d)},$$

where $p' = p/(p - 1) \in [1, 2]$. Note that

$$\| C_{\delta, D}(x, \cdot) \|_{L_p(\Lambda)} = \left\| \sum_{k/\pi} (\lambda_k^\delta + m^2)^{-1} e_k(x) e_k(\cdot) \right\|_{L_p(\Lambda)}. \tag{9.5.28}$$

The Fourier coefficients for $C_{\delta, D}(x, \cdot)$ are $(\lambda_k^\delta + m^2)^{-1} e_k(x)$. Thus the $l_{p'}$ norm of this series is bounded (using (9.5.24)) by

$$\| (\lambda_k^\delta + m^2)^{-1} 2^{d/2} \|_{l_{p'}} \leq \text{const} \| (k^2 + 1)^{-1} \|_{l_{p'}}$$
$$= \text{const} \| (k^2 + 1)^{-p'} \|_{l_1}^{1/p'} \leq \text{const}, \tag{9.5.29}$$

since $p' > 1$. Furthermore with $C_{\delta, D} - C_D$ replacing $C_{\delta, D}$ above, the Fourier coefficients are

$$[(\lambda_k^\delta + m^2)^{-1} - (k^2 + m^2)^{-1}] e_k(x) \equiv \alpha(k).$$

By (9.5.25), and $1 - (x^{-1} \sin x)^2 \leq O(x^\varepsilon)$ for $0 \leq \varepsilon \leq 2$, we have as from (9.5.22)

$$|\alpha(k)| \leq \text{const } \delta^\varepsilon (k^2 + 1)^{-1 + \varepsilon/2}$$

for any $0 \leq \varepsilon \leq 2$. Now choose $\varepsilon < 2/p$. Then

$$(-2 + \varepsilon)p' < \left(-2 + 2\left(1 - \frac{1}{p'}\right) \right) p' = -2,$$

so that $(k^2 + 1)^{-1 + \varepsilon/2} \in l_{p'}$. Thus

$$\| \alpha(\cdot) \|_{l_{p'}} \leq O(\delta^\varepsilon).$$

The proofs for general $C \in \mathscr{C}_{m, \Lambda}$ are similar.

To conclude this section, we discuss the interpolation between Dirichlet boundary conditions on the boundary of two regions $\Lambda_1 \subset \Lambda_2$. We write $C_{\delta, \Lambda} \equiv C_{\delta, D}$ to make the Λ dependence explicit in C. It is convenient to consider C_{δ, Λ_1} and C_{δ, Λ_2} as matrix operators on $l_2(\Lambda_{2, \delta})$ with $C_{\delta, \Lambda_1} = 0$ on $l_2(\Lambda_{2, \delta} \backslash \text{Int } \Lambda_{1, \delta})$.

Then $C_{\delta, \Lambda}$ is monotonic increasing in Λ, both in the operator sense and pointwise (matrix elements). This is the lattice analog of monotonicity in Section 7.8, and yields the monotonicity of that section in the limit $\delta \to 0$.

Proposition 9.5.8. *The lattice covariance operators $C_{\delta,\Lambda}$ are monotonic in Λ. For $\Lambda_1 \subset \Lambda_2$*

$$C_{\delta,\Lambda_1} \leq C_{\delta,\Lambda_2} \tag{9.5.30}$$

and

$$0 \leq C_{\delta,\Lambda_1}(x, y) \leq C_{\delta,\Lambda_2}(x, y), \qquad x, y \in \Lambda_{2,\delta}. \tag{9.5.31}$$

PROOF. As in Section 7.8, we interpolate between boundary conditions on $\partial\Lambda_2$ and $\partial\Lambda_1$ by introducing a local mass perturbation of $C_{\delta,\Lambda_2}^{-1}$ by $m(x)^2 \to \infty$ for $x \in \Lambda_{2,\delta} \backslash \text{Int } \Lambda_{1,\delta}$. The proof that

$$C_{\delta,\Lambda_1} = \lim_{\lambda \to \infty} C(\lambda) = \lim_{\lambda \to \infty} (C_{\delta,\Lambda_2}^{-1} + \lambda\chi_{\Lambda_2\backslash\text{Int } \Lambda_1})^{-1} \tag{9.5.32}$$

follows the proof of Proposition 7.8.1, but it is elementary, since the C_{δ,Λ_1} is a finite matrix. Operator monotonicity follows from

$$\frac{dC}{d\lambda} = -C(\lambda)\chi_{\Lambda_2\backslash\text{Int } \Lambda_1}C(\lambda) \leq 0. \tag{9.5.33}$$

Pointwise positivity, $C_{\delta,\Lambda}(x, y) \geq 0$, at lattice points $x, y \in \Lambda_\delta$ follows from the lattice maximum principle; cf. Proposition 9.5.2. Pointwise monotonicity in Λ follows from pointwise positivity and from (9.5.33).

In Chapter 18 we establish a lattice analog of the Wiener integral formula (7.8.3). The diffusion process is replaced by a random walk on the lattice δZ^d. The pointwise and operator monotonicity could also be established by (7.8.14) applied in the lattice case to (9.5.33).

9.6 Lattice Approximation of $P(\phi)_2$ Measures

In this section we give a lattice approximation to the $P(\phi)_2$ measures of Chapter 8. In fact we have already introduced a discrete approximation in Section 8.7 based on approximation by Riemann sums. The present approximation is based on truncation of Fourier series. While both methods are useful, the method here preserves OS positivity and also the ferromagnetic nature of the measure. Thus the lattice approximation is suited to the proof of correlation inequalities. The convergence of the lattice approximation for the ϕ^4 model in $d = 3$ dimensions has also been established, but we do not present it here.

For simplicity we choose Λ to be the unit square of Section 9.5. Let

$$f(x) = \sum_{k_\alpha/\pi \in Z_+} f^\sim(k)e_k(x) \tag{9.6.1}$$

be the Fourier sine series for f. Here $e_k(x)$ is given in (9.5.2). Let $C_{\delta,D}$ be the operator (9.5.21) acting on $L_2(\Lambda)$. Then

$$(C_{\delta,D}f)(x) = \sum_{1 \leq k_\alpha/\pi \leq (1/\delta)-1} (\lambda_k^\delta + m^2)^{-1}f^\sim(k)e_k(x). \tag{9.6.2}$$

Recall that $i_\delta : l_2(\text{Int } \Lambda_\delta) \to L_2(\Lambda)$ defined in (9.5.17) is an isometric imbedding, and that $i_\delta^* : L_2(\Lambda) \to l_2(\text{Int } \Lambda_\delta)$ is defined as truncation of the Fourier series followed by restriction to lattice points.

If $\phi(y)$ is a Gaussian process with covariance C, then

$$\phi_\delta(x) \equiv (i_\delta^* \phi)(x), \qquad x \in \text{Int } \Lambda_\delta \tag{9.6.3}$$

defines a Gaussian lattice field, with covariance

$$C_\delta = i_\delta^* C i_\delta = i_\delta^* \Pi_\delta C i_\delta . \tag{9.6.4}$$

Here we choose

$$C \in \mathscr{C}_{m,\Lambda}, \tag{9.6.5}$$

where $\mathscr{C}_{m,\Lambda}$ is the convex set of covariance operators generated by the $C_{\delta,D}$, $\delta = 0$ or $\delta = 2^{-\nu}$, as defined in Section 9.5. The field ϕ_δ can be realized equivalently by a Gaussian measure on $L_2(R^{|\text{Int } \Lambda_\delta|})$ or by a Gaussian measure on $\mathscr{S}'$. Both measures have covariance C_δ, which via the isometry i_δ is either an $l_2(\text{Int } \Lambda_\delta)$ operator or an $L_2(\Lambda)$ operator. Explicitly, let

$$\prod_{x \in \text{Int } \Lambda_\delta} d\phi_\delta(x)$$

denote Lebesgue measure on $R^{|\text{Int } \Lambda_\delta|}$. Then

$$d\phi_{\delta,D} = (\det C_{\delta,D})^{-1/2} \pi^{-|\text{Int } \Lambda_\delta|/2}$$

$$\times \exp\left[-\frac{\delta^{2d}}{2} \sum_{x,y \in \text{Int } \Lambda_\delta} \phi_\delta(x) C_{\delta,D}^{-1}(x,y) \phi_\delta(y) \right] \prod_x d\phi_\delta(x)$$

$$= (\det C_{\delta,D})^{-1/2} \pi^{-|\text{Int } \Lambda_\delta|/2}$$

$$\times \exp\left[-\tfrac{1}{2}(\|\nabla \phi_\delta\|_{l_2}^2 + m^2 \|\phi_\delta\|^2) \right] \prod_x d\phi_\delta(x) \tag{9.6.6}$$

is the Gaussian measure on $R^{|\text{Int } \Lambda_\delta|}$ with covariance $C_{\delta,D}$.

To study convergence as $\delta \to 0$, it is convenient to regard $d\phi_{\delta,D}$ as a measure on $\mathscr{S}'(R^d)$, with covariance $C_{\delta,D}$. Convergence of the Gaussian measures is a direct consequence of convergence of the covariance operators; here convergence means convergence of the moments and of the generating function. This assertion follows from the formula (6.2.2) which expresses the generating function in terms of $C_{\delta,D}$.

The lattice approximation to the $:P(\phi, f):$ interaction polynomial of Section 8.6 is given by

$$:P(\phi_\delta, f):_{C_{\delta,D}} = \delta^d \sum_{x \in \Lambda_\delta} \sum_{j=0}^n :\phi_\delta(x)^j:_{C_{\delta,D}} f_j(x). \tag{9.6.7}$$

In this section we restrict the f_j satisfying (8.6.4) to be smooth functions times the characteristic function of a rectangle, the important point being that f_j must be Riemann integrable for the lattice approximation to be convergent. For brevity we write

$$:P_\delta:_C = :P(\phi_\delta, f):_C$$

and now consider convergence of the interaction measure

$$d\mu_{\delta, D} = \frac{\exp(-:P_\delta:_{C_{\delta, D}}) \, d\phi_{\delta, D}}{\int \exp(-:P_\delta:_{C_{\delta, D}}) \, d\phi_{\delta, D}}. \tag{9.6.8}$$

Convergence in (9.6.8) is not a corollary of the convergence of the $C_{\delta, D}$; it is thus necessary to reexamine the proof of Theorem 8.6.2. Let

$$C(t) = tC_D + (1 - t)C_{\delta, D},$$

$$d\mu_{\delta, D, t} = \frac{\exp(-:P_\delta:_{C(t)}) \, d\phi_{C(t)}}{\int \exp(-:P_\delta:_{C(t)}) \, d\phi_{C(t)}}.$$

Proposition 9.6.1. *Let $d = 2$, and let $Q(\phi_{\delta'})$ be a polynomial in the field $\phi_{\delta'}$. Then as $\delta \to 0$,*

$$\left| \int Q(\phi_{\delta'}) \, d\mu_{\delta, D, 1} - \int Q(\phi_{\delta'}) \, d\mu_{\delta, D, 0} \right| \le O(\delta).$$

Remark. We take $\delta \to 0$ in two steps, first in the covariance in the measure and in the Wick order in P, and second in the ϕ_δ and the discrete sum $\sum_{x \in \text{Int} \Lambda_\delta}$ in P.

PROOF. We use the lattice version of the change of covariance formula (9.1.34). In fact, using the finite dimensional realization of $d\phi_{\delta, t}$, as in (9.6.6), the formula (9.1.34) reduces to the chain rule for evaluation of derivatives. Since

$$\|\dot{C}\| = \|C_{\delta, D} - C_D\| = O(\delta^2)$$

by Proposition 9.5.5, the same rate of convergence holds for the matrix operator

$$\dot{C}_\delta \equiv i_\delta \dot{C} i_\delta^* : l_2(\text{Int} \Lambda_\delta) \to l_2(\text{Int} \Lambda_\delta).$$

We use the Schwarz inequality relative to $d\phi_{C(t)}$ to separate the polynomial Q and exponential factors:

$$\left| \int Q e^{-:P_\delta:_{C(t)}} \, d\phi_{C(t)} \right| \le \left(\int |Q|^2 \, d\phi_{C(t)} \right)^{1/2} \left(\int e^{-2:P_\delta:_{C(t)}} \, d\phi_{C(t)} \right)^{1/2}.$$

Since $\|\dot{C}_\delta\| \le O(\delta^2)$, the $\int |Q|^2 \, d\phi$ factor converges as $O(\delta^2)$. To complete the proof of the first step in convergence, we show that the exponential factor is bounded uniformly in δ and t. The uniformity is established in the next lemma, and follows the proof of Theorem 8.6.2.

Lemma 9.6.2. *Let $d = 2$, $0 \le t \le 1$. Then*

$$\int e^{-:P_\delta:_{C(t)}} \, d\phi_{C(t)}$$

is bounded and bounded away from zero uniformly in δ and in t.

SKETCH OF PROOF. We use $\delta_1 \ge \delta$ to play the role of κ^{-1}, i.e. an inverse ultraviolet cutoff. There are two essential bounds. The first is Proposition 9.5.6, which establishes the lower bound of $:P_{\delta_1}\!:_{C(t)}$ as $O(|\ln \delta_1| + 1)^{(\deg P)/2}$. The second is the bound analogous to Theorem 8.5.3:

$$\left| \int \left(:P_{\delta_1}\!:_{C(t)} - :P_\delta\!:_{C(t)} \right)^j d\phi_{C(t)} \right| \leq j!^{(\deg P)/2} O(\delta_1^{j\varepsilon}), \quad j \text{ even}. \qquad (9.6.9)$$

Proposition 9.5.7 provides these factors $O(\delta_1^\varepsilon)$. The upper bound now follows the proof of Theorem 8.6.2. The lower bound results from a uniform upper bound on

$$\int :P_\delta\!:_{C(t)}^2 d\phi_{C(t)}.$$

For the second step in the $\delta \to 0$ limit, we define

$$P(t) = t:P_{\delta=0}\!:_{C_D} + (1 - t):P_\delta\!:_{C_D},$$

$$d\mu_t = \frac{e^{-P(t)} d\phi_{C_D}}{\int e^{-P(t)} d\phi_{C_D}}.$$

Proposition 9.6.3. *Let $d = 2$. Let $Q(\phi_{\delta'})$ be a polynomial in the field $\phi_{\delta'}$. Then as $\delta \to 0$, with δ' fixed,*

$$\left| \int Q(\phi_{\delta'}) d\mu_1 - \int Q(\phi_{\delta'}) d\mu_0 \right| \leq O(\delta^\varepsilon).$$

PROOF. The Gaussian measures are now defined on $\mathscr{S}'(R^2)$. The difference above is bounded by

$$\sup_t \left| \int Q(\phi_{\delta'}) \frac{d}{dt} d\mu_t \right|. \qquad (9.6.10)$$

Again we separate the polynomial and exponential factors; a second Schwarz inequality separates the $:P_{\delta=0}:$ and $:P_\delta:$ exponential factors. The factor containing $\exp(-:P_{\delta=0}\!:_{C_D})$ was bounded in Section 8.6. It is independent of δ, and hence bounded uniformly in δ. The $:P_\delta:$ factor was bounded in Lemma 9.6.2.

Finally we consider the polynomial factors bounding (9.6.10). These involve the Gaussian integral

$$\int Q^2 \left(\frac{dP(t)}{dt} \right)^2 d\mu_{C_D},$$

which is convergent to zero, as in Theorem 8.5.3.

We summarize these two propositions in a single theorem, which gives convergence of the infinite volume lattice $P(\phi)_2$ measures to the corresponding continuum measures.

Theorem 9.6.4. *Let $d = 2$, and let $Q(\delta')$ be a polynomial in the field $\phi_{\delta'}$. Then with δ' fixed*

$$\lim_{\delta \to 0} \int Q(\delta') d\mu_{\delta, D} = \int Q(\delta') d\mu_D.$$

Remark. The corresponding result holds for the boundary conditions $B = \emptyset, P, N$, with a similar proof.

CHAPTER 10

Estimates Independent of Dimension

10.1 Introduction

The basic bounds for control over the infinite volume limit are given here. These tools are formally independent of dimension. We give proofs for the finite volume $P(\phi)_2$ measures of Chapter 8, but the same methods apply more generally. For example, applied to lattice fields of arbitrary dimension, the bounds of this chapter do not depend on the lattice spacing. This is in contrast to the ultraviolet bounds of Chapter 8, which depend strongly on dimension. Section 12.2, i.e. integration by parts, is similarly independent of dimension.

10.2 Correlation Inequalities for $P(\phi)_2$ Fields

In this section we extend to $P(\phi)_2$ fields the correlation inequalities and the Lee–Yang theorem established in Chapter 4 for a finite lattice. We recall the restrictions on P required by the various inequalities:

FKG inequalities (Section 4.4):

$$\text{arbitrary semibounded } P. \tag{10.2.1a}$$

Griffiths inequalities (Section 4.1):

$$P = \text{even} - \mu\phi, \qquad \mu \geq 0. \tag{10.2.1b}$$

Lebowitz inequalities (Section 4.3):

$$P = \lambda\phi^4 + \sigma\phi^2 - \mu\phi, \qquad \lambda > 0, \quad \mu \geq 0. \qquad (10.2.1c)$$

Lee–Yang theorem (Section 4.5):

$$P = \lambda\phi^4 + \sigma\phi^2 - \mu\phi, \qquad \lambda > 0, \quad \text{Re } \mu > 0. \qquad (10.2.1d)$$

The correlation inequalities (10.2.1a–c) remain valid in any measure obtained by a limit of measures for which the moments converge. The Lee–Yang theorem remains true provided the free energy

$$f \equiv \frac{\ln Z}{\text{volume}}$$

is uniformly bounded and convergent for μ in a complex domain. By Theorem 9.6.4, the convergence of the lattice field to a continuum field for Dirichlet boundary conditions and finite interaction volume satisfies these restrictions. Similarly, the lattice approximation converges for Neumann and periodic boundary conditions. Thus we obtain

Theorem 10.2.1. *The correlation inequalities and the Lee–Yang theorem are true for finite volume continuum $P(\phi)_2$ fields under the restrictions* (10.2.1).

Remark. After construction of the infinite volume limit in Chapter 11 (or in Chapter 18), we conclude that the correlation inequalities and the Lee–Yang theorem are also valid for the resulting infinite volume theories.

Our first application of a correlation inequality is to prove that the Schwinger functions depend monotonically on the location of the Dirichlet boundary conditions, in analogy with statistical mechanics results of Section 4.2.

Theorem 10.2.2. *Let $\langle \cdot \rangle_{\partial\Lambda, D}$ denote the expectation in a $P(\phi)_2$ measure* (9.1.30) *defined in a rectangle $\Lambda \subset R^2$ with Dirichlet boundary conditions on $\partial\Lambda$. In this formula, take $C_2 = (-\Delta_{\partial\Lambda} + m^2)^{-1}$ and C_1 independent of Λ. Assume* (10.2.1b). *Then the Schwinger functions*

$$S^{(n)}(x_1, \ldots, x_n) \equiv \langle \phi(x_1) \cdots \phi(x_n) \rangle_{\partial\Lambda, D} \qquad (10.2.2)$$

are nonnegative and monotone increasing in Λ. Also, for $0 \leq f \in C_0^\infty$, the generating functional

$$\langle e^{\phi(f)} \rangle_{\partial\Lambda, D} = S\{-if\}_{\partial\Lambda, D} \qquad (10.2.3)$$

is positive and monotone increasing in Λ.

Remark. The Wick order of P cannot depend on Λ in this comparison. For example, $C_\varnothing$, the free covariance, is allowed for Wick order.

PROOF. Let $\Lambda_1 \subset \Lambda$. We introduce the lattice approximation of Section 9.5.6, with spacing δ. Let $\langle \cdot \rangle_{\delta, \partial\Lambda, D} = \langle \cdot \rangle$ denote the expectation in the lattice Dirichlet measure $d\mu_{\delta, D}$ of (9.6.8). For δ fixed, we show that $\langle \cdot \rangle$ satisfies the monotonicity in Λ claimed in (10.2.2–3). Then the covergence of the lattice approximation as $\delta \to 0$ (see Theorem 9.6.4) completes the proof.

In the lattice approximation, monotonicity is a consequence of the Griffiths inequality, as we now indicate. In fact, consider a local mass perturbation $m(x)^2 = \lambda \chi_{\Lambda \setminus \mathrm{Int} \, \Lambda_1}(x)$, as in Proposition 9.5.8. For δ fixed, the limit $\lambda \to \infty$ introduces Dirichlet boundary conditions on $\partial\Lambda_1$ in the convariance and hence in $d\mu_{\delta, D}$. However, any increase in λ decreases the Schwinger functions and generating functional, because

$$-\frac{d}{d\lambda} S^{(n)}_{\delta, \partial\Lambda, D}(x_1, \ldots, x_n)$$

$$= \tfrac{1}{2} \delta^d \sum_{x \in \Lambda_\delta \setminus \Lambda_{1\delta}} [\langle \phi_\delta(x_1) \cdots \phi_\delta(x_n) : \phi_\delta(x)^2 : \rangle$$

$$\qquad\qquad - \langle \phi_\delta(x_1) \cdots \phi_\delta(x_n) \rangle \langle : \phi_\delta(x)^2 : \rangle], \qquad (10.2.4)$$

Without the Wick dots on $: \phi(x)^2 :$, the right side is positive by the second Griffiths inequality. To deal with this extra complication, we note that

$$: \phi_\delta(x)^2 : = \phi_\delta(x)^2 - c_\delta(x),$$

where $c_\delta(x)$ is a constant, independent of ϕ. Thus $c_\delta(x)$ cancels between the two terms on the right of (10.2.4). This completes the proof.

Corollary 10.2.3. *Assume* (10.2.1b). *Then the Schwinger functions and the generating functional* $S\{-if\}$ *for* $0 \le f$ *are monotone increasing in* μ.

PROOF. The derivative of S is a Schwinger function, and the derivative of a Schwinger function is a truncated Schwinger function. Thus the corollary follows from the first and second Griffiths inequalities.

10.3 Dirichlet or Neumann Monotonicity and Decoupling

The partition function

$$Z_B(\Lambda) = \int \exp(-: V:_{C_B}) \, d\phi_{C_B}$$

and the free energy

$$\alpha^B(\Lambda) = \frac{\ln Z_B(\Lambda)}{|\Lambda|} \qquad (10.3.1)$$

are the basic quantities for this section. Here $B = \varnothing$, N, Γ, D or p denotes free, Neumann, Dirichlet, or periodic boundary conditions in the covariance C_B, and $V = \int_\Lambda P(\phi(x)) \, dx$, as in Chapters 7, 8. In particular, the coupling constants are restricted by (8.6.2, 4). The N, D, and p cases have boundary conditions on all lines of a lattice in R^2.

Proposition 10.3.1 (Conditioning).   *For fixed Λ and for $\Gamma_1 \subseteq \Gamma_2$,*

$$Z_D \leq Z_{\Gamma_2} \leq Z_{\Gamma_1} \leq Z_\emptyset \leq Z_N,$$

$$Z_D \leq Z_p \leq Z_N. \tag{10.3.2}$$

PROOF.   We establish the second inequality; the first and third are special cases, and the rest have a similar proof. Let $C(t) = tC_{\Gamma_1} + (1 - t)C_{\Gamma_2}$ and $Z(t) = \int \exp(- :V:_{C(t)}) \, d\phi_{C(t)}$. Then $\dot{C}(t) = C_{\Gamma_1} - C_{\Gamma_2} \geq 0$ by (7.7.4), and so by (9.1.35), $0 \leq dZ(t)/dt$.

Note that both Z_D and Z_N factor into products, i.e. decouple, with each factor corresponding to a lattice square $\Delta \subset \Lambda$:

$$Z_D(\Lambda) = \prod_{\Delta \subset \Lambda} Z_D(\Delta), \tag{10.3.3}$$

$$Z_N(\Lambda) = \prod_{\Delta \subset \Lambda} Z_N(\Delta). \tag{10.3.4}$$

Corollary 10.3.2.   *Assume that the coefficient functions f of the interaction polynomial P have finite norms $N(f)$ and $M(f)$ in each lattice square Δ, as defined in (8.6.5–6), and bounded uniformly in Δ. Then*

$$e^{-O(|\Lambda|)} \leq Z_B(\Lambda) \leq e^{O(|\Lambda|)}, \tag{10.3.5}$$

and (10.3.1) is bounded uniformly in Λ.

PROOF.   The upper bound follows from the upper bound on Z_N in Theorem 8.6.2 and (10.3.4). If f_0 in (8.6.2) is the constant coupling constant, then $|\int f_0 \, dx| \leq$ const $|\Lambda|$, by (8.6.4) and the uniformity assumption. Let $V_1 = V - \int f_0 \, dx$, so $\int :V_1:_{C_B} d\phi_{C_B} = 0$. Hence with

$$F_B(s) = \int \exp(-s :V_1:_{C_B}) \, d\phi_{C_B},$$

we have $F_B(0) = 1$, $F'_B(0) = 0$, $0 \leq F''_B(s)$. Thus $1 \leq F_B(s)$, and

$$\exp(-\textstyle\int f_0 \, dx) \leq \exp(-\textstyle\int f_0 \, dx)F_D(1) = Z_D(\Lambda).$$

Remark 1.   Let $:V:_{C_B} = \sum_{j=0}^n :\phi^j(f_j):_{C_B} = :P(\phi, f):_{C_B}$. The proof of the corollary shows that

$$\exp\left(-\int f_0 \, dx\right) \leq \int \exp(- :P(\phi, f):_{C_B}) \, d\phi_{C_B} \equiv Z_B(f). \tag{10.3.6}$$

Remark 2.   Finally, we allow a change in the Wick ordering of P. Let T be the Wick reordering transformation defined by $:P(\phi, f):_{C_{B'}} = :P(\phi, Tf):_{C_B}$. Let χ_Δ denote the characteristic function of a lattice square Δ and let

$f\chi_\Delta \equiv \{f_j \chi_\Delta\}$. Since T is a local transformation, $T(f\chi_\Delta) = (Tf)\chi_\Delta$. By the proposition,

$$\prod_\Delta Z_D(T\chi_\Delta f) \le \int \exp(-:P(\phi, f):_{C_{B'}})\, d\phi_{C_B} \le \prod_\Delta Z_N(T\chi_\Delta f). \quad (10.3.7)$$

Suppose furthermore that C_B, $C_{B'} \in C_m^M$, as defined in Section 7.9, and that

$$m + (M/m) + n + \sup_\Delta N(f\chi_\Delta) + \sup_\Delta M(f\chi_\Delta) \le K.$$

Then

$$\exp(-\mathrm{const}\,|\Lambda|) \le \int \exp(-:P(\phi, f):_{C_B})\, d\phi_{C_B} \le \exp(\mathrm{const}\,|\Lambda|) \quad (10.3.8)$$

where the constants depend only on K.

We next consider the free energy (10.3.1). Let $B = \partial\Lambda$, D and $B = \partial\Lambda$, N denote Dirichlet and Neumann boundary conditions on the boundary of a rectangle Λ.

Proposition 10.3.3. *The free energies for a $P(\phi)_2$ model defined by Dirichlet, free, and Neumann boundary conditions on the boundary of a rectangle Λ satisfy*

$$\alpha^{\partial\Lambda,\, D}(\Lambda) \le \alpha^\varnothing(\Lambda) \le \alpha^{\partial\Lambda,\, N}(\Lambda). \quad (10.3.9)$$

Furthermore $\alpha^{\partial\Lambda,\, D}(\Lambda)$ and $\alpha^{\partial\Lambda,\, N}(\Lambda)$ both converge as $\Lambda \uparrow R^2$.

Remark. The limits actually agree. For the limiting free energies α^D, $\alpha^\varnothing$, and α^N, we have $\alpha^D = \alpha^\varnothing = \alpha^N$ [Guerra, Rosen, and Simon, 1976].

PROOF. We consider first the case where Λ_1 is the union of n nonoverlapping translates of Λ. Then by conditioning, (10.3.2),

$$Z^{\partial\Lambda,\, D}(\Lambda)^n = Z^{\partial\Lambda_1,\, D}(\Lambda_1) \le Z_\varnothing(\Lambda_1) \le Z^{\partial\Lambda_1,\, N}(\Lambda_1) = Z^{\partial\Lambda,\, N}(\Lambda)^n. \quad (10.3.10)$$

Taking logarithms,

$$\alpha^{\partial\Lambda,\, D}(\Lambda) = \alpha^{\partial\Lambda_1,\, D}(\Lambda_1) \le \alpha^\varnothing(\Lambda_1) \le \alpha^{\partial\Lambda_1,\, N}(\Lambda_1) = \alpha^{\partial\Lambda,\, N}(\Lambda). \quad (10.3.11)$$

More generally let $\bar\alpha = \limsup_\Lambda \alpha^{\partial\Lambda,\, D}(\Lambda)$, and for $0 \le \varepsilon$ choose Λ so that $\bar\alpha \le \alpha^{\partial\Lambda,\, D}(\Lambda) + \varepsilon$. If Λ_1 is the union of $n = n_x n_y$ nonoverlapping translates of Λ and of a boundary region covered by $n_x + n_y + 1$ translates of Λ, then as above

$$\alpha^{\partial\Lambda,\, D}(\Lambda) \le \alpha^{\partial\Lambda_1,\, D}(\Lambda_1) + O(1)\,\frac{n_x + n_y + 1}{n_x n_y}.$$

As $\Lambda_1 \to R^2$ with Λ fixed, $n_x \to \infty$, $n_y \to \infty$, and $(n_x + n_y + 1)/n_x n_y \to 0$. Thus

$$\bar\alpha \le \alpha^{\partial\Lambda,\, D}(\Lambda) + \varepsilon \le \alpha^{\partial\Lambda_1,\, D}(\Lambda_1) + 2\varepsilon,$$

and convergence follows. The case of $\alpha^{\partial\Lambda,\, N}(\Lambda)$ is similar.

10.4 Reflection Positivity

We show here that θ-invariant measures $d\mu$ are reflection-positive with respect to θ. The noninvariant case will be discussed in Section 10.6.

The bilinear form

$$b(A, B) = \langle \theta A, B \rangle_{L_2(d\mu)} = \int (\theta A)^- B \, d\mu = \langle A^\wedge, B^\wedge \rangle_{\mathcal{H}} \quad (10.4.1)$$

defines the inner product in the Hilbert space $\mathcal{H}$. The condition of reflection positivity with respect to θ is $0 \leq b(A, A)$ for all $A \in \mathcal{E}_+$. In the case of quantum fields, $\mathcal{H}$ is the space of quantum states. In the case of classical statistical mechanics $\mathcal{H}$ is the space on which the transfer matrix acts as a self-adjoint operator. In either case $A, B \in \mathcal{E}_+$ and θ is reflection about a hyperplane Π. The space $\mathcal{E}_+$ is generated by functionals $e^{i\phi(f)}$ with suppt $f \subset \Pi_+$; $\Pi_\pm$ are the two connected components of $R^d \backslash \Pi$. We study here $P(\phi)_2$ measures, lattice fields, and Ising type models with Dirichlet, Neumann, and/or periodic boundary conditions. For simplicity we use $P(\phi)$ notation.

The basic consequence of reflection positivity is the Schwarz inequality

$$|b(A, B)| = |\langle A^\wedge, B^\wedge \rangle_{\mathcal{H}}| \leq \|A^\wedge\|_{\mathcal{H}} \|B^\wedge\|_{\mathcal{H}} = b(A, A)^{1/2} b(B, B)^{1/2}, \quad (10.4.2)$$

valid for $A, B \in \mathcal{E}_+$.

In Chapter 6 we introduced the Fourier transform S_μ of a measure $d\mu$ on $\mathcal{D}'$,

$$S_\mu\{f\} = \int e^{i\phi(f)} \, d\mu(\phi).$$

A continuous isomorphism of $\mathcal{D}'$, such as θ, defines a transformed measure $\theta \, d\mu$ through the relation

$$S_\mu\{\theta^{-1}f\} = S_{\theta\mu}\{f\}. \quad (10.4.3)$$

Definition 10.4.1. The measure $d\mu$ is θ-invariant if

$$\theta \, d\mu = d\mu; \quad \text{i.e.,} \quad S\{\theta f\} \quad \text{for all } f.$$

Recall also that in Chapter 6 the Gaussian functional with mean zero and covariance C was defined by the formula

$$S_C\{f\} = \exp(-\tfrac{1}{2}\langle f, Cf \rangle) = \int e^{i\phi(f)} \, d\phi_C.$$

(In the Gaussian case, the measure is uniquely determined by its covariance.) Thus the transformed measure has the Fourier transform

$$S_C\{\theta^{-1}f\} = \exp(-\tfrac{1}{2}\langle f, \theta C \theta^{-1}f \rangle) = S_{\theta C \theta^{-1}}\{f\}.$$

Then $\theta \, d\phi_C$ is the Gaussian measure with mean zero and covariance $\theta C \theta^{-1}$, so $d\phi_C$ is θ-invariant if and only if $[\theta, C] = 0$.

Theorem 10.4.2. *Let $d\phi_C$ be θ-invariant, where $C = (-\Delta_B + I)^{-1} = C_B$ is a covariance operator with the classical boundary conditions studied in Section 7.10. Then $d\phi_C$ satisfies reflection positivity with respect to θ.*

Remark. Since C may be defined on R^d, on T^d, or on Z^d, we infer that quantum or lattice fields, Ising models, Heisenberg models, etc., are special cases.

PROOF. By the remarks above, $[\theta, C] = 0$. By Theorems 7.10.1–3, the covariance operator C is reflection positive. By Theorem 6.2.2, $d\phi_C$ is reflection positive.

We consider a finite volume $P(\phi)_2$ measure (or lattice $P(\phi)_d$ measure) with classical boundary conditions. Let C_B denote a covariance operator with boundary conditions on $\Gamma \supset \partial\Lambda$, where Λ is a bounded region of R^2, and Γ is a union of lattice line segments as in Section 7.10. Define

$$d\mu = d\mu(V, \Lambda, C_B) = Z^{-1} e^{-V(\Lambda)} \, d\phi_{C_B}, \qquad (10.4.4)$$

where

$$V(\Lambda) = \int_\Lambda :P(\phi(x)):_{C_\varnothing} dx \qquad (10.4.5)$$

and

$$Z = Z(V, \Lambda, C_B) = \int e^{-V(\Lambda)} \, d\phi_{C_B}. \qquad (10.4.6)$$

Note that

$$\theta \, d\mu(V, \Lambda, C_B) = d\mu(\theta V, \theta\Lambda, \theta C_B \theta^{-1}).$$

Here θ invariance entails $\theta\Lambda = \Lambda$ and $\theta V = V$, as well as $[\theta, C_B] = 0$.

Theorem 10.4.3. *If $d\mu$ defined in (10.4.4) is θ-invariant, then it satisfies reflection positivity with respect to θ.*

Remark. Since reflection positivity is preserved under limits, an infinite volume measure $d\mu(V, C_B)$, whenever it exists, satisfies reflection positivity.

PROOF. Write $V(\Lambda) = V_+ + V_-$, where

$$V_\pm = V(\Lambda_\pm) \quad \text{and} \quad \Lambda_\pm = \Lambda \cap \Pi_\pm. \qquad (10.4.7)$$

Thus $\theta V_\pm = V_\mp$ and $V = V_+ + \theta V_+$. By Theorem 10.4.2, $d\phi_C$ is reflection positive with respect to θ. Hence writing

$$Z \, d\mu = (\theta e^{-V_+}) e^{-V_+} \, d\phi_C,$$

we conclude that $d\mu$ is reflection positive with respect to θ.

10.5 Multiple Reflections

The Schwarz inequality in (10.4.2) is useful for estimating function space integrals. In this section we develop systematically some multiple reflection bounds. These bounds arise from a sequence of Schwarz inequalities defined by a sequence of hyperplanes Π and reflection operators θ_Π.

We consider here a general $\mathcal{D}'$ probability measure $d\mu$ satisfying the axioms OS0,2,3 (Fourier space $d\tilde{\mu}$ analyticity, invariance, and reflection positivity) of Section 6.1, as well as the finite volume $P(\phi)_2$ measure $d\mu_\Lambda$ constructed in Section 8.6. In each case let $\mathscr{E} = L_2(\mathcal{D}', d\mu)$, and for open sets $\Lambda \subset R^d$ define $\mathscr{E}(\Lambda)$ as the subspace of $\mathscr{E}$ generated by $e^{i\phi(f)}$ with suppt $f \subset \Lambda$. Thus $\mathscr{E}_+ = \mathscr{E}(R^{d-1} \times (0, \infty))$, and for $\Lambda = R^{d-1} \times (s_1, s_2)$ we introduce the special notation

$$\mathscr{E}(s_1, s_2) = \mathscr{E}(R^{d-1} \times (s_1, s_2)). \tag{10.5.1}$$

There are three geometric configurations in which we apply multiple reflection bounds. The first configuration involves reflections θ_{Π_v} about the $v = 1, 2, \ldots, d$ orthogonal coordinate hyperplanes through the origin. In this case suppose $(R_+)^d$ is the first orthant, i.e. $\{x : 0 < x_v, v = 1, 2, \ldots, d\}$. The appropriate reflected function $R(k)$ for $k \in \mathscr{E}((R_+)^d)$ is

$$R(k) = \prod_{I \subset \{1, 2, \ldots, d\}} \left[\left(\prod_{v \in I} \theta_{\Pi_v} \right) k^{(-)} \right]. \tag{10.5.2}$$

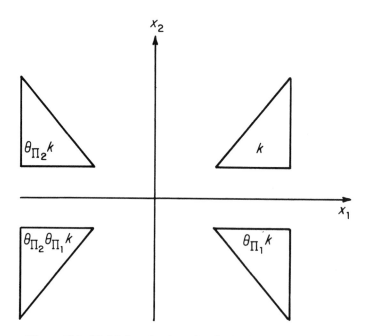

Figure 10.1. Multiple reflections for the lattice reflection group.

Here $(-)$ denotes complex conjugate for $|I|$ odd and the identity for $|I|$ even. In Figure 10.1 we show the effect of R on subsets of R^d thought of as supports for k and its reflections for $d = 2$. The reflection operators $\prod_{v \in I} \theta_{\Pi_v}$ form a group, called the lattice reflection group.

Proposition 10.5.1. *Let the measure $d\mu$ be reflection positive and invariant under the generators θ_{Π_v} of the lattice reflection group. Let $k \in \mathscr{E}((R_+)^d)$. Then*

$$\left| \int k\, d\mu \right| \le \left(\int R(k)\, d\mu \right)^{2^{-d}} \tag{10.5.3}$$

PROOF. The inequality (10.5.3) follows from d successive applications of the Schwarz inequality with respect to the inner products

$$(A,\, B)_v = \int (\theta_{\Pi_v} A)^- B\, d\mu$$

for $v = 1, 2, \ldots, d$ with $A,\, B \in \mathscr{E}(R^{v-1} \times (R_+)^{d-v+1})$. The positive definiteness of the inner product results from Theorem 10.4.3.

The second geometric configuration of reflections will be reflections in a lattice of translates of a given hyperplane Π. For Π we can take the $t = 0$ hyperplane. In this case we study the reflections illustrated in Figure 10.2 indexed by the integer time translation group. Suppose $k \in \mathscr{E}(0,\, t)$.

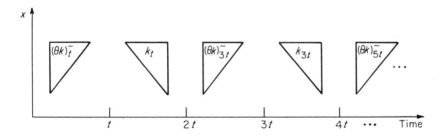

Figure 10.2. Multiple reflections for the integer time translation group.

Define

$$M_n(k) = \prod_{j=1}^{n} (\theta k)^-_{(2j-1)t}\, k_{(2j-1)t}. \tag{10.5.4}$$

Here θ denotes reflection in the $t = 0$ hyperplane, and $k_s = T(s)kT(s)^{-1}$ denotes k translated by time s. Furthermore let

$$M(k) = \limsup_{n \to \infty} \left(\int M_n(k)\, d\mu \right)^{1/2n}. \tag{10.5.5}$$

Proposition 10.5.2. *Let $d\mu$ satisfy axioms OS2–3 of invariance and reflection positivity. Then for $k \in \mathscr{E}(0, t)$,*

$$\left| \int \dot{k} \, d\mu \right| \leq M(k). \tag{10.5.6}$$

PROOF. By reflection positivity, the Schwarz inequality, and translation invariance of $d\mu$,

$$\left| \int \dot{k} \, d\mu \right| = |\langle 1, k \rangle_{\mathscr{E}}| = |\langle \theta 1, k \rangle_{\mathscr{E}}|$$

$$\leq \langle \theta k, k \rangle_{\mathscr{E}}^{1/2} = \langle 1, (\theta k)_t^- k_t \rangle_{\mathscr{E}}^{1/2} = \langle \theta 1, (\theta k)_t^- k_t \rangle_{\mathscr{E}}^{1/2}.$$

We continue to apply the Schwarz inequality and translation invariance. After r applications we obtain (with the substitution $2^r = n$)

$$\left| \int \dot{k} \, d\mu \right| \leq \langle 1, M_{2^r}(k) \rangle_{\mathscr{E}}^{2^{-r-1}} = \left(\int M_{2^r}(k) \, d\mu \right)^{2^{-r-1}}$$

and $|\int \dot{k} \, d\mu| \leq M(k)$, as desired.

The third geometric configuration is useful when k is supported in a cube Δ (or more generally in a rectangle $X \equiv [0, a_i] \times [0, a_2] \times \cdots \times [0, a_d]$). In this case we use the full lattice translation group, generalizing (10.5.4) above. Let $\mathscr{L}_n^{(\nu)}$ denote the product (10.5.4) constructed along the νth coordinate direction, with $\theta_\nu = \theta_{\Pi_\nu}$ denoting reflection in the $x_\nu = 0$ hyperplane:

$$\mathscr{L}_n^{(\nu)}(k) = \prod_{j=1}^{n} (\theta_\nu k)_{(2j-1)a_\nu}^- k_{(2j-1)a_\nu}. \tag{10.5.7}$$

Note that the unit cube Δ is the case $a_\nu = 1$ for $\nu = 1, 2, \ldots, d$. Let

$$\mathscr{L}(k) = \lim_{n_1, \ldots, n_d \to \infty} \sup \left[\int \mathscr{L}_{n_1}^{(1)} (\mathscr{L}_{n_2}^{(2)} \cdots (\mathscr{L}_{n_d}^{(d)}(k) \cdots)) \, d\mu \right]^{1/(2n_1 \cdots 2n_d)}. \tag{10.5.8}$$

The lattice of reflections for $d = 2$ is illustrated in Figure 10.3.

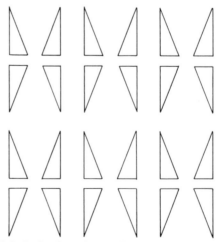

Figure 10.3. Reflections for the full lattice translation group.

Proposition 10.5.3. *For $k \in \mathscr{E}(X)$, as above, and $d\mu$ satisfying axioms OS2–3 of translation invariance and reflection positivity,*

$$\left| \int k \, d\mu \right| \leq \mathscr{L}(k). \tag{10.5.9}$$

PROOF. The proof follows that of the previous proposition in each coordinate direction.

In the case of a translation invariant measure, it is convenient to work with the operator $k_{\mathscr{H}} = k^\wedge$ that multiplication by k defines on the Hilbert space of quantum mechanics, $\mathscr{H}$. Recall from Chapter 6 that

$$\wedge : \mathscr{E}_+ \rightarrow \mathscr{E}_+ / \mathscr{N} \subset \mathscr{H}$$

is the canonical imbedding of the path space $\mathscr{E}_+$ into $\mathscr{H}$, where $\mathscr{N} \subset \mathscr{E}_+$ is the null space of the reflection positive form on $\mathscr{E}_+$. By (6.1.12), $\wedge$ allows the transfer of certain operators S on $\mathscr{E}_+$ onto operators $S_{\mathscr{H}} = S^\wedge$ on $\mathscr{H}$. For example, $T(t)^\wedge = T(t)_{\mathscr{H}} = e^{-tH}$, by Theorem 6.1.3. Note also that Theorem 6.1.3 is proved using a multiple reflection bound.

We now generalize the proof of Theorem 6.1.3 to construct an operator $k^\wedge$ on $\mathscr{H}$ from functions $k \in \mathscr{E}(0, t)$. Since $k \in L_2(d\mu)$, multiplication by k defines an operator on $\mathscr{E}_+$ with domain $\mathscr{E}_+ \cap L_\infty$. To construct $k^\wedge$, we restrict the domain of k. Multiple reflections bound the norm of $k^\wedge$ as an operator of $\mathscr{H}$.

Proposition 10.5.4. *Consider a $\mathscr{D}'$ probability measure $d\mu$ which satisfies axioms OS0,2,3 of analyticity, invariance, and reflection positivity. Consider $k \in \mathscr{E}(0, t)$, $0 < t$, as a multiplication operator on $\mathscr{E}_+$ with domain $\mathscr{D}(k) = T(t)(\mathscr{E}_+ \cap L_\infty)$. Then $k^\wedge$ defined by (6.1.12) is densely defined on $\mathscr{H}$.*

PROOF. First we note that the domain $\mathscr{D}(k)^\wedge = (T(t)(L_\infty \cap \mathscr{E}_+))^\wedge$ is dense in $\mathscr{H}$. By Theorem 6.1.3, especially the relation $\lim_{t\to 0} e^{-tH} = I$, it follows that e^{-tH} has no nonzero null vectors. For if $e^{-tH}\theta = 0$ for some t, then $e^{-tH/2}\theta = 0$ etc., so that $\lim_{t\to 0} e^{-tH}\theta = 0$, ensuring $\|\theta\| = 0$. Let ψ in $\mathscr{H}$ be orthogonal to $\mathscr{D}(k)^\wedge$. Then $e^{-tH}\psi \perp (L_\infty \cap \mathscr{E}_+)^\wedge$, and since $(L_\infty \cap \mathscr{E}_+)^\wedge$ is dense in $\mathscr{H}$, $e^{-tH}\psi = 0$, and by the above, $\psi = 0$. Thus $\mathscr{D}(k)^\wedge$ is dense in $\mathscr{H}$.

To show that k is defined, we verify (6.1.13), namely

$$A \in \mathscr{D}(k) \cap \mathscr{N} \quad \Rightarrow \quad b(kA, kA) = \langle \theta kA, kA \rangle_\mathscr{E} = 0. \tag{10.5.10}$$

First consider the case that k is bounded, i.e. $k \in L_\infty$. Consider A of the form $A = T(t)B$, $B \in \mathscr{E}_+$. Since $\theta(kT(t)B) = (\theta k)T(-t)\theta B$ and $T(t)^* = T(-t)$,

$$b(kA, kA) = \langle \theta kT(t)B, kT(t)B \rangle_\mathscr{E}$$

$$= \langle \theta B, T(t)(\theta k)^- kT(t)B \rangle_\mathscr{E}$$

$$= b(B, (\theta k)_t^- k_t T(2t)B)$$

$$\leq b(B, B)^{1/2} b(C, C)^{1/2},$$

where $C = (\theta k)_t^- k_t T(2t)B \in \mathscr{E}_+$. By assumption, $A^\wedge = 0$ and $A^\wedge = e^{-tH}B^\wedge$. As observed, this implies $B^\wedge = 0$, i.e. $B \in \mathcal{N}$. Thus $b(B, B) = 0$ and (10.5.10) holds.

In general, k is not bounded, but we use the above to replace k by

$$k_j = \begin{cases} k & \text{if } |k| \leq j, \\ 0 & \text{if } |k| > j. \end{cases}$$

Then $\|k - k_j\|_{L_2(d\mu)} \to 0$ as $j \to \infty$. Since $A \in \mathscr{D}(k)$, $A \in T(t)L_\infty \subset L_\infty$. Also $k_j A \in \mathcal{N}$ by the above, and so the $k_j A$ terms (including cross terms) do not contribute in the inequality:

$$b(kA, kA) = b((k - k_j)A, (k - k_j)A) \leq \|(k - k_j)A\|^2_{L_2(d\mu)}$$

$$\leq \|k - k_j\|^2_{L_2} \|A\|^2_{L_\infty} \to 0.$$

This completes the proof.

Theorem 10.5.5. *Let $d\mu$ be a $\mathscr{D}'$ probability measure satisfying the axioms OS0,2,3 of analyticity, invariance, and reflection positivity. Let $k \in \mathscr{E}(0, t)$, $t > 0$. Then*

$$\|k^\wedge e^{-tH}\|_{\mathscr{H}} \leq M(k). \tag{10.5.11}$$

PROOF. Let $Q = e^{-tH}k^\wedge {}^* k^\wedge e^{-tH}$, and let $A \in \mathscr{E}_+ \cap L_\infty$. Then by the Schwarz inequality,

$$\|k^\wedge e^{-tH} A^\wedge\| = \langle A^\wedge, QA^\wedge \rangle^{1/2} \leq \|A^\wedge\|_{\mathscr{H}}^{1/2} \|QA^\wedge\|_{\mathscr{H}}^{1/2}.$$

We continue to apply the Schwarz inequality and after n applications obtain

$$\|k^\wedge e^{-tH} A^\wedge\|_{\mathscr{H}} \leq \|A^\wedge\|_{\mathscr{H}}^{1 - 2^{-n}} \|Q^{2^{n-1}} A^\wedge\|_{\mathscr{H}}^{2^{-n}}. \tag{10.5.12}$$

By definition $(k^\wedge e^{-tH})^* A^\wedge = (T(t)(\theta k)^- A)^\wedge = ((\theta k)_t^- T(t)A)^\wedge$, or

$$e^{-tH}k^\wedge {}^* = ((\theta k)_t^-)^\wedge e^{-tH}.$$

Thus with M_n defined in (10.5.4),

$$\|Q^{2^{n-1}} A^\wedge\|^2_{\mathscr{H}} = \langle \theta M_{2^{n-1}}(k)T(2^n t)A, M_{2^{n-1}}(k)T(2^n t)A \rangle_{\mathscr{E}}$$

$$= \langle \theta A, M_{2^n}(k)T(2^{n+1}t)A \rangle_{\mathscr{E}} \leq \|A\|^2_{L_\infty}\left(\int |M_{2^n}(k)| \, d\mu \right).$$

Thus by (10.5.5),

$$\limsup_n \|Q^{2^{n-1}} A^\wedge\|^{2^{-n}}_{\mathscr{H}} \leq \limsup_n \|A\|^{2^{-n}}_{L_\infty} M(k).$$

Substituting in (10.5.12), we obtain

$$\|k^\wedge e^{-tH} A^\wedge\|_{\mathscr{H}} \leq M(k) \|A^\wedge\|_{\mathscr{H}}$$

to complete the proof.

Let us now define

$$\mathscr{H}_\delta = e^{-\delta H}\mathscr{H}. \tag{10.5.13}$$

$\mathscr{H}_\delta$ is a domain of analytic vectors for H. It is useful in constructing the analytic continuation from Euclidean to Minkowski space and in the proof of a Euclidean Feynman–Kac formula in Chapter 19. The first step of these constructions is given below.

Corollary 10.5.6.   *Let* $\delta > 0$ *and* $0 \leq \tau < \delta/4$, $\delta' > t + \delta/2$, $M(k) < \infty$. *Then* $(k_\tau)^\wedge$ *as a bilinear form on* $\mathcal{H}_\delta \times \mathcal{H}_{\delta'}$ *is analytic in* τ *and extends to a complex analytic function in the circle* $|\tau| < \delta/8$. *Also*

$$\left\| e^{-\delta H} \frac{d^n}{d\tau^n} (k_\tau)^\wedge e^{-\delta' H} \right\| \leq \left(\frac{8}{\delta} \right)^n n! \, \| k^\wedge e^{-(t+\delta/4)H} \|. \qquad (10.5.14)$$

PROOF.   Since

$$e^{-\delta H} \frac{d}{d\tau} (k_\tau)^\wedge e^{-\delta'H} = e^{-(\delta+\tau)H}[k^\wedge, H] \, e^{-(\delta'-\tau)H},$$

we use the relations

$$C_m = \| H^m e^{-\delta H/4} \| \leq (4/\delta)^m m!,$$

$$C_m C_{n-m} \leq (4/\delta)^n n! \binom{n}{m}^{-1}$$

and the theorem to establish (10.5.14). The stated analyticity then follows.

The operator norm estimates on $\mathcal{H}$ can be reformulated as estimates on $d\mu$ integrals.

Corollary 10.5.7.   *Let* $k^{(0)}, k^{(1)}, \ldots, k^{(r)} \in \mathscr{E}(0, t)$. *Then*

$$\left| \int \prod_{j=0}^{r} k_{jt}^{(j)} \, d\mu \right| \leq \prod_{j=0}^{r} M(k^{(j)}). \qquad (10.5.15)$$

PROOF.   Observe that $k_{jt} \in \mathscr{E}(jt, (j+1)t)$. Also

$$\int \prod_{j=0}^{r} k_{jt}^{(j)} \, d\mu = \left\langle \Omega, \left(\prod_{j=0}^{r} k_{jt}^{(j)} \right)^\wedge \Omega \right\rangle$$

$$= \left\langle \Omega, \prod_{j=0}^{r} (k^{(j)\wedge} e^{-tH}) \Omega \right\rangle. \qquad (10.5.16)$$

The ordered product on the right side of (10.5.16) extends from $k^{(0)\wedge} e^{-tH}$ on the left to $k^{(r)\wedge} e^{-tH}$ on the right. We apply the inequality for operator norms $\| \prod_j A_j \|_{\mathscr{H}} \leq \prod_j \| A_j \|_{\mathscr{H}}$ and the theorem to (10.5.16) to obtain (10.5.15).

Corollary 10.5.8.   *Let* $J \subset Z^d$ *be a finite set, and for* $j \in J$ *let* $k^{(j)} \in \mathscr{E}(\Delta_j)$ *be given. Then*

$$\left| \int \prod_{j \in J} k_{jt}^{(j)} \, d\mu \right| \leq \prod_{j \in J} \mathscr{L}(k^{(j)}), \qquad (10.5.17)$$

with $\mathscr{L}(k)$ *defined in* (10.5.8).

PROOF. Formally, apply Corollary 10.5.7 successively in each coordinate direction. The actual proof requires following the proof of the theorem and corollary so that $n_v \to \infty$ only after reflections in all directions have been performed.

Remark. A nonsymmetric multiple reflection bound is given in Theorem 12.4.2. In this case, $d\mu$ is not assumed to be reflection invariant. The Schwarz inequality above is replaced by (10.6.8) in the nonsymmetric case.

10.6 Nonsymmetric Reflections

The reflection-positive Schwarz inequality can be extended to measures $d\mu$ which are not symmetric under the reflection θ about some hyperplane Π. In the nonsymmetric case, the Schwarz inequality has the form

$$|b(A, B)| \leq \text{const } b_1(A, A)^{1/2} b_2(B, B)^{1/2}, \qquad (10.6.1)$$

where b_1 and b_2 are θ-invariant and reflection-positive with respect to θ. Since this section is more technical than the one for the symmetric reflection case (Section 10.4), we point out that it is needed only for the regularity of $P(\phi)_2$ fields, but not in the proof of their existence.

Let $d\mu$ be as in (10.4.4) with classical boundary conditions on Γ. We first define measures $d\mu_\pm$ which are related to $d\mu$ and which are θ-invariant. We assume that if $\Gamma \cap \Pi \neq \varnothing$, the intersection is transversal, namely $\dim(\Gamma \cap \Pi) \leq d - 2$. Let $\Pi_\pm$ denote the two half spaces $R^d \backslash \Pi$, and let

$$\Gamma_+ = (\Gamma \cap \Pi_+) \bigcup (\theta\Gamma \cap \Pi_-), \qquad (10.6.2)$$

$$\Gamma_- = (\Gamma \cap \Pi_-) \bigcup (\theta\Gamma \cap \Pi_+). \qquad (10.6.3)$$

Then Γ_+ and Γ_- are both θ-invariant. Let θB denote the boundary data B reflected by Π and let

$$B_+ \equiv B \text{ data on } \Gamma \cap \Pi_+ \text{ and } \theta B \text{ data on } \theta\Gamma \cap \Pi_-,$$

$$B_- \equiv B \text{ data on } \Gamma \cap \Pi_- \text{ and } \theta B \text{ data on } \theta\Gamma \cap \Pi_+, \qquad (10.6.4)$$

$$C_+ = C_{B_+}, \qquad C_- = C_{B_-}.$$

By construction, C_+ and C_- are θ-invariant.

To be explicit, introduce three variables: interaction V, volume Λ, and Gaussian covariance C, which specify $d\mu$. With $V_\pm = V(\Lambda_\pm)$ and $\Lambda_\pm = \Lambda \cap \Pi_\pm$, define

$$d\mu_+ = d\mu(V_+ + \theta V_+, \Lambda_+ \bigcup \theta\Lambda_+, C_{B_+})$$

$$= Z_+^{-1} e^{-(V_+ + \theta V_+)} d\phi_{C_+}, \qquad (10.6.5)$$

$$d\mu_- = d\mu(V_- + \theta V_-, \Lambda_- \bigcup \theta\Lambda_-, C_{B_-})$$

$$= Z_-^{-1} e^{-(V_- + \theta V_-)} d\phi_{C_-}.$$

Here $Z_\pm$ are the customary normalizing factors, chosen so that $\int d\mu_\pm = 1$.

Since the V's, Λ's, and C's which define $d\mu_\pm$ are θ-invariant,

$$\theta \, d\mu_\pm = d\mu_\pm . \qquad (10.6.6)$$

By Theorem 10.4.2, the measures $d\mu_\pm$ satisfy reflection positivity with re-spect to θ. Note that if $d\mu$ is θ-invariant, then $C = C_+ = C_-$, $d\mu = d\mu_+ = d\mu_-$, and $Z = Z_+ = Z_-$.

The generalization of reflection positivity is a Schwarz inequality, which reduces, in case $d\mu$ is θ-invariant, to

$$|\langle A, B\rangle_\mu|^2 \leq \langle \theta A, A\rangle_\mu \langle \theta B, B\rangle_\mu, \qquad A \in \mathscr{E}_-, \quad B \in \mathscr{E}_+. \quad (10.6.7)$$

The generalization of (10.6.7) is

$$|\langle A, B\rangle_\mu|^2 \leq \frac{Z_+ Z_-}{Z^2} \det(C^{-2}C_+ C_-)^{1/2}\langle \theta A, A\rangle_{\mu_-}\langle \theta B, B\rangle_{\mu_+} (10.6.8)$$

Note that the bilinear form $\langle A, B\rangle_\mu$ is not necessarily positive definite.

Proposition 10.6.1. *The inequality (10.6.8) is valid if it is valid in the Gaussian case.*

PROOF. The factor $Z_+ Z_- Z^{-2}$ can be removed if we replace the measures $d\mu$, $d\mu_+$, and $d\mu_-$ by unnormalized measures $d\tilde\mu = Z \, d\mu$ and $d\tilde\mu_\pm = Z_\pm \, d\mu_\pm$. Fur-thermore the factor $e^{-V} = e^{-V_+} e^{-V_-}$ can be included in A and B. Having done this, (10.6.8) reduces to

$$|\langle A, B\rangle_{d\phi_C}|^2 \leq \det(C^{-2}C_+ C_-)^{1/2}\langle \theta A, A\rangle_{d\phi_{C_-}} \langle \theta B, B\rangle_{d\phi_{C_+}}, \qquad (10.6.9)$$

which is the Gaussian case ($V = 0$) of (10.6.8).

The proof in the Gaussian case follows the same pattern: we reduce to a fixed θ-invariant covariance C_0, namely the case of Theorem 10.4.2. We discuss the formal aspects of this argument, and later give mathematical details in the special cases needed for Chapter 12: Theorem 10.6.2 and Corollary 10.6.3.

The inequality (10.6.9) can be understood by writing

$$\det(C^{-2}C_+ C_-)^{1/2} = \frac{Z_{C_+} Z_{C_-}}{Z_C^2} \qquad (10.6.10)$$

as a ratio of partition functions. We think of Z_{C_+}/Z_C as normalizing the Gaussian measure $d\phi_{C_+}$ relative to $d\phi_C$. However, as these are both prob-ability measures, we must be more precise. In particular, let us express $d\phi_C$ and $d\phi_{C_+}$ relative to a chosen reference Gaussian $d\phi_{C_0}$. Define the integral operators v and $v_\pm$ by

$$v = C^{-1} - C_0^{-1}, \qquad v_+ = C_+^{-1} - C_0^{-1}, \qquad v_- = C_-^{-1} - C_0^{-1}. \quad (10.6.11)$$

They have kernels $v(x, y)$, $v_\pm(x, y)$. Also define

$$V_C = \frac{1}{2} \int \phi(x)v(x, y)\phi(y) \, dx \, dy, \qquad (10.6.12)$$

$V_{C_\pm}$, etc. Then by (9.3.8), formally

$$d\phi_C = Z_C^{-1} \, e^{-V_C} \, d\phi_{C_0} \tag{10.6.13}$$

with

$$Z_C = \int e^{-V_C} \, d\phi_{C_0}. \tag{10.6.14}$$

Similarly, define $d\phi_{C_\pm}$, using $v_\pm$ in place of v. Using the formula (9.3.7) for Gaussian functional integrals, $Z_C = \det(I + C_0^{1/2} v C_0^{1/2})^{-1/2}$. Thus the ratio

$$Z_{C_+}/Z_C = \det(C^{-1} C_+)^{1/2}$$

is independent of C_0. Likewise (10.6.10) follows. In this calculation we could take $C_0 = C$, yielding $Z_C = 1$. However, it is convenient to choose C_0 to satisfy θ reflection positivity. The inequality (10.6.9) is then a consequence of the θ reflection positivity of $d\phi_{C_0}$ (Theorem 10.4.2) and an application of the Schwarz inequality in the associated inner product. Namely, for $A \in \mathscr{E}_-$, $B \in \mathscr{E}_+$,

$$|\langle A, B \rangle_{d\phi_{C_0}}|^2 \le \langle \theta A, A \rangle_{d\phi_{C_0}} \langle \theta B, B \rangle_{d\phi_{C_0}}. \tag{10.6.15}$$

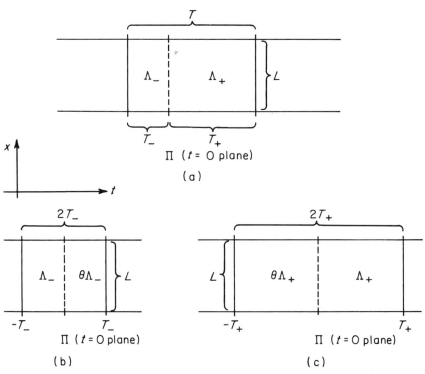

Figure 10.4. (a) In the measure $d\mu$, assume V is supported in $\Lambda = \Lambda_+ \bigcup \Lambda_-$, with $\Pi \subset \Lambda_+ \cap \Lambda_-$. The solid lines indicate Γ, which consists of $\partial\Lambda$ and the lines extending two sides of $\partial\Lambda$. The dashed line indicates the plane Π, which we choose as the $t = 0$ plane. (b, c) The second and third diagrams similarly represent $d\mu_-$, $d\mu_+$ with $C_{B_\pm}$ having Dirichlet data on the solid lines $\Gamma_\pm$.

In terms of $d\phi_C$, etc., we write (10.6.15) as

$$|\langle A, B \rangle_{d\phi_C}|^2 = Z_C^{-2}|\langle AB\, e^{-Vc} \rangle_{d\phi_{C_0}}|^2$$
$$\leq Z_C^{-2}\langle (\theta A)A \exp(-V_{C_-}) \rangle_{d\phi_{C_0}} \langle (\theta B)B \exp(-V_{C_+}) \rangle_{d\phi_{C_0}}$$
$$= Z_{C_+} Z_{C_-} Z_C^{-2} \langle \theta A, A \rangle_{d\phi_{C_-}} \langle \theta B, B \rangle_{d\phi_{C_+}}$$
$$= \det(C^{-2}C_+ C_-)^{1/2} \langle \theta A, A \rangle_{d\phi_{C_-}} \langle \theta B, B \rangle_{d\phi_{C_+}}, \qquad ((10.6.16)$$

which is the desired inequality. Here we use $v = v_1 + v_2$, where $v_- = v_1 + \theta v_1 \theta^{-1}$, $v_+ = v_2 + \theta v_2 \theta^{-1}$, and where v_1, v_2 are localized in Π_-, Π_+ respectively.

A technical complication in the above formal argument arises from the fact that the Gaussian measures $d\phi_{C_B}$ are mutually singular for distinct boundary conditions B. Hence the operators v, $v_\pm$ are singular, and Z_{C_+}, Z_{C_-}, Z_C do not exist. This complication is avoided by regularizing (10.6.11–12) and passing to the limit in (10.6.16). In this limit Z_C diverges, but the ratio $\det(C^{-2}C_+ C_-)$ converges, as do $\langle \cdot \rangle_{d\phi_C}$ and $\langle \cdot \rangle_{d\phi_{C_+}}$.

In order to use (10.6.8), it is necessary to bound both $Z_+ Z_- /Z^2$ and $\det(C^{-2}C_+ C_-)^{1/2}$, as well as to obtain estimates of their volume dependence. We postpone the bound on $Z_+ Z_- /Z^2$ to Section 12.4, and here we investigate the ratio $\det(C^{-2}C_+ C_-)^{1/2}$ of functional determinants. To simplify the estimates, we restrict attention to some special cases of Dirichlet boundary conditions on $\Gamma \supset \partial\Lambda$.

In the following, we assume that $d = 2$ and that Λ is an $L \times T$ rectangle oriented along (x, t) axes, as in Figure 10.4. Assume Π is the $t = 0$ plane, and that Π divides Λ into $\Lambda_- \cup \Lambda_+$, two rectangles of size $L \times T_-$ and $L \times T_+$ respectively. Then $T = T_+ + T_-$. Assume that x axis is chosen so that the lines $x = 0$, $x = L$ pass through $\partial\Lambda$. We choose

$$\Gamma_0 = \{(x, t) : x = 0 \text{ or } x = L\}, \qquad \Gamma = \partial\Lambda \cup \Gamma_0,$$
$$C_0 = (-\Delta_{\Gamma_0} + m^2)^{-1}, \qquad C_B = (-\Delta_\Gamma + m^2)^{-1}.$$

Theorem 10.6.2. *Under the above restrictions on B, (10.6.8) is valid. Let Γ, C_B be as above, and suppose $m^{-1} \leq T_- \leq T_+$, where m is the mass in C_B. Then*

$$1 \leq \det(C^{-2}C_+ C_-)^{1/2} \leq \text{const } e^{\text{const } L/T_-}, \qquad (10.6.17)$$

and the constant is uniform in the choice of L, $T_\pm$.

PROOF. The lower bound in (10.6.17) follows from (10.6.9) with $A = B = 1$. To establish (10.6.9) and the upper bound, we evaluate the ratio of determinants. Let H_0 denote the Hamiltonian determined by $d\phi_{C_0}$ on the Hilbert space $\mathcal{H}_0$ constructed in the canonical manner of Chapter 6 using θ-reflection in the ($t = 0$) hyperplane Π. Since $d\phi_{C_0}$ is Gaussian $\mathcal{H}_0$ is a Fock space and H_0 is the free field Hamiltonian on $\mathcal{H}_0$. In particular, $\mathcal{H}_0$ is the symmetric tensor algebra over the one-particle space

$$L_2([0, L], dx) \oplus L_2((-\infty, 0], dx) \oplus L_2([L, \infty), dx).$$

Thus $\mathcal{H}_0$ is a tensor product of the Fock spaces defined by these three subspaces. Only the first of these subspaces contributes to the determinant. Since e^{-itH_0} preserves the tensor product structure, we can restrict our attention to the factor $\mathcal{H}_0([0, L])$, which is the symmetric tensor algebra over $L_2([0, L], dx)$. Let h_0 denote H_0 restricted to this one-particle space. The eigenvalues of h_0 are $(k^2 + m^2)^{1/2} = \mu(k)$, where $k \in (\pi/L)Z$, and the corresponding eigenfunction is $\sin kx$.

$\mathcal{H}_0$ has a Schrödinger representation, in which it is an infinite tensor product

$$\mathcal{H}_0 = \bigotimes_{k \in (\pi/L)Z} L_2(R, dv_k(q_k)) = \bigotimes_{k \in (\pi/L)Z} \mathcal{H}_0^{(k)}. \tag{10.6.18}$$

Here $dv_k(q_k) = (\mu(k)/\pi)^{1/2} \exp(-q_k^2 \mu(k)) \, dq_k$ and $\mathcal{H}_0^{(k)}$ is the state space of a harmonic oscillator with position variable q_k; see Sections 1.5, 6.2, and 6.4. The operator e^{-tH_0} also factors:

$$e^{-tH_0} = \prod_k e^{-tH_0^{(k)}}.$$

Then $H_0^{(k)}$ is the harmonic oscillator Hamiltonian for the kth oscillator, and $e^{-tH_0^{(k)}}$ has a kernel given by Mehler's formula (1.5.26):

$$e^{-tH_0^{(k)}}(q_k, q_k') \equiv p_t^{(k)}(q_k, q_k')$$

$$= \pi^{-1/2}(1 - e^{-2\mu(k)t})^{-1/2}$$

$$\cdot \exp\left\{-(q_k^2 - q_k'^2)\frac{\mu(k)}{2} - \frac{\mu(k)(e^{-\mu(k)t}q_k - q_k')^2}{1 - e^{-2\mu(k)t}}\right\}. \tag{10.6.19}$$

For a product state $\psi = \prod_k f_k(q_k)$,

$$\langle \psi, e^{-tH_0}\psi \rangle_{\mathcal{H}_0} = \prod_k \langle f_k, e^{-tH_0^{(k)}} f_k \rangle_{\mathcal{H}_0^{(k)}}. \tag{10.6.20}$$

For clarity, we first give a formal calculation, only justifying the calculation at the end. On $\mathcal{H}_0$, use the Feynman–Kac formula to write

$$Z_C = \int e^{-V_C} \, d\phi_{C_0} = \langle \psi, e^{-TH_0}\psi \rangle_{\mathcal{H}_0}. \tag{10.6.21}$$

Here $V_C = V_1 + V_2$, where $v_C = \Delta_{C_0} - \Delta_C$ is a local expression, localized on $\Gamma \bigcup \{x = 0, L\} \equiv \Gamma_1 \bigcup \Gamma_2$, where $\Gamma_1 \cap \Pi_+ = \Gamma_1$, $\Gamma_2 \cap \Pi_- = \Gamma_2$. Let v_1, v_2 denote v_C on $L_2(\Pi_+)$. Then ψ is the state e^{-V_1} expressed in the $t = T_+$ Hilbert space $\mathcal{H}_0$, or e^{-V_2} in the $t = -T_-$ space. Note further that ψ is a product state over modes, proportional to a δ function in each mode, $\psi = \prod_k c\delta(q(k))$. (In each mode, ψ introduces Dirichlet data.) Thus the normalizing factors $\prod_k c$ cancel in the ratio

$$\frac{Z_{C_+}Z_{C_-}}{Z_C^2} = \frac{\langle \psi, e^{-2T_+ H_0}\psi \rangle \langle \psi, e^{-2T_- H_0}\psi \rangle}{\langle \psi \, e^{-TH_0}\psi \rangle^2}$$

$$= \prod_{k, k_+, k_-} \frac{\langle \delta, \exp(-2T_+ H_0^{(k_+)}) \, \delta \rangle \langle \delta, \exp(-2T_- H_0^{(k_-)}) \, \delta \rangle}{\langle \delta, \exp(-TH_0^{(k)}) \, \delta \rangle^2}$$

$$= \det(C^{-2}C_+ C_-)^{1/2}. \tag{10.6.22}$$

In this product, the momenta k, $k_\pm$ all belong to the same lattice:

$$k, k_+, k_- \in (\pi/L)Z. \tag{10.6.23}$$

Using (10.6.19), evaluate (10.6.22) as

$$\det(C^{-2}C_+ C_-)^{1/2} = \prod_{k, k_\pm} \frac{1 - e^{-2\mu(k)T}}{(1 - e^{-4\mu(k_+)T_+})^{1/2}(1 - e^{-4\mu(k_-)T_-})^{1/2}}. \quad (10.6.24)$$

In (10.6.24) the products over k, k_+, and k_- converge individually, for fixed T_+, T_-, L, since for $\alpha t > 0$, each $e^{-\alpha\mu(k)t}$ converges exponentially to zero.

Since each factor in the numerator is bounded by 1 and since $T_- \le T_+$ by hypothesis,

$$\det(C^{-2}C_+ C_-)^{1/2} \le \prod_k (1 - e^{-4\mu(k)T_-})^{-1}$$

$$= \exp \sum_k \ln (1 - e^{-4\mu(k)T_-})^{-1}. \quad (10.6.25)$$

Since $m^{-1} \le T_-$ by hypothesis, and since $m \le \mu(k)$ because the covariance $C_B \in \mathscr{C}_m$ has mass at least m by hypothesis, it follows that $4 \le 4\mu(k)T_-$. Thus $4\mu(k)T_-$ is bounded away from zero in (10.6.25). Also

$$e^{-4\mu(k)T_-} \le e^{-4mT_-} \le e^{-4} < 1$$

is bounded away from 1. Using the inequality

$$\ln (1 - \varepsilon)^{-1} \le \text{const } \varepsilon$$

for $0 \le \varepsilon \le \varepsilon_0 < 1$ with a constant depending on ε_0, we have

$$\det(C^{-2}C_+ C_-)^{1/2} \le \exp (\text{const} \sum_k e^{-4\mu(k)T_-})$$

$$\le \text{const} \exp O(L/T_-).$$

To justify the above formal argument, return to (10.6.21). Now compute a regularized Z_C, obtained by replacing $e^{-V_{1, 2}}$ with a smoothed-out product wave function $\psi^k = \prod_{|k| \le \kappa_1} c\delta_{\kappa_2}(q(k))$. Since we proceed by limits, note that in (10.6.9) it is sufficient to choose A, B from a dense subspace of $\mathscr{E}_\pm$. We take A, B to be bounded continuous functions of a finite number of normal modes, and depending on a finite number of times (cylinder functions). For such A, B, convergence of the Gaussian integral, as $\kappa_1 \to \infty$, follows from convergence of the generating functional, and hence from weak convergence of the covariance operators as L_2 operators. Since the normal modes diagonalize the covariance operators, weak convergence as the number of modes increases is obvious. Thus it is sufficient to consider a fixed finite value of κ_1.

The inequality (10.6.9) follows from its validity in each factor, as one sees by applying the Schwarz inequality iteratively in each factor $\mathscr{H}_0^{(k)}$. Thus it is sufficient to consider a single mode, namely ordinary Wiener measure, with paths taking values in R. In Section 7.8 it was shown that a locally infinite mass perturbation in the covariance operator generates Dirichlet data. This analysis applies in each normal mode, i.e. to a covariance which is a function of t alone. This Dirichlet covariance $C^{(k)}$ defines the Gaussian measure which describes the path space of the kth oscillator. This oscillator has a t-dependent mass. Consider an approximation $C^{(k, \kappa_2)}$ with a t-dependent mass labeled by κ_2; as $\kappa_2 \to \infty$, the mass becomes infinite about the points (e.g. $t = \pm T$) where the Dirichlet data are located, and $C^{(k, \kappa_2)}$ converges to the Dirichlet covariance.

By Section 9.3, the mass perturbation in the covariance $C^{(k, \kappa_2)}$ can be realized by a Feynman–Kac exponent in the Gaussian measure defined by $C^{(k, \kappa_2)}$. The exponent is time-dependent, and has the form

$$V(q_k, t) = \begin{cases} 0 \\ \kappa_2 q_k^2 \end{cases}$$

with the choice between the two values for V depending on t. In this way, we get an explicit sequence of approximate δ functions δ_{κ_2} in $\mathscr{H}_0^{(k)}$, each of which gives rise to a Gaussian measure, whose covariance converges to the Dirichlet covariance $C^{(k)}$ of the kth mode.

Estimates which show the Hölder continuity of the Wiener sample paths also show $\delta_{\kappa_2} \to \delta$ in this limit. This gives $q = q' = 0$ in (10.6.19) and justifies the evaluation of (10.6.22) by (10.6.24).

The convergence of the integrals as in (10.6.9) follows from the convergence of the covariance, for $\kappa_2 \to \infty$, where A and B are restricted to be bounded continuous cylinder functions as above. This completes the proof.

As a corollary to the theorem, we state a bound for measures with Dirichlet data on $\partial\Lambda$ (but not on the infinite lines $x = 0$, L as required above). We now restrict A, B to be localized in Λ_-, Λ_+.

Corollary 10.6.3. *L be an $L \times T$ rectangle as in the theorem. Let $C_B = C_{\partial\Lambda}$ be the Dirichlet covariance with data on $\partial\Lambda$. Let $A \in \mathscr{E}(\Lambda_-)$, $B \in \mathscr{E}(\Lambda_+)$. Then*

$$|\langle A, B \rangle_\mu| \leq \text{const} \frac{Z_+ Z_-}{Z^2} \langle \theta A, A \rangle_{\mu_-} \langle \theta B, B \rangle_{\mu_+} \exp\left(\frac{L}{T_+} + \frac{L}{T_-}\right),$$
$$(10.6.26)$$

where the constant is uniform in the choice of L, $T_\pm \geq 1$.

PROOF. Since A, B, V are all localized in Λ (i.e., they are elements of $\mathscr{E}(\Lambda)$), and since $d\phi_{C_{\partial\Lambda}}$ factorizes on $\mathscr{E}(\Lambda) \otimes \mathscr{E}(R^2 \backslash \Lambda)$, we can write

$$\langle A, B \rangle_\mu = \langle A, B \rangle_{\tilde{\mu}},$$

where in $\tilde{\mu}$ we replace $C_{\partial\Lambda}$ by the C of Theorem 10.6.2. Likewise $\mu_\pm$ can be replaced by $\tilde{\mu}_\pm$ for expectations on $\mathscr{E}(\Lambda_\pm \bigcup \theta\Lambda_\pm)$. We apply the theorem to the measures $\tilde{\mu}$, $\tilde{\mu}_\pm$.

CHAPTER 11

Fields Without Cutoffs

11.1 Introduction

The construction of $P(\phi)_2$ quantum fields given here is valid for semi-bounded interaction polynomials P of the form $P = \text{even} + \text{linear}$. Other constructions apply to general semibounded P; see also Chapter 18. In this chapter, the problem of existence is separated from regularity. We prove the existence of a Euclidean measure $d\mu$, obtained as an infinite volume limit of the finite volume measures of Chapter 8. The proof uses monotone convergence and uniform upper bounds.

11.2 Monotone Convergence

The general strategy of this section is monotone convergence. Monotonicity of the Schwinger functions follows by correlation inequalities for the case $P = \text{even} + \text{linear}$. Uniform upper bounds are proved in Section 11.3 by combining the finite volume estimates of Chapter 8 with the multiple reflection bounds of Chapter 10.

Let us define the finite volume $P(\phi)_2$ measure $d\mu_\Lambda$ with Dirichlet boundary conditions on $\partial\Lambda$ by

$$d\mu_\Lambda = Z^{-1} e^{-V(\Lambda)} \, d\phi_{C_{\partial\Lambda}}. \tag{11.2.1}$$

Here $\Delta_{\partial\Lambda}$ is the Laplace operator on R^2 with Dirichlet data on $\partial\Lambda$, and $C_{\partial\Lambda} = (-\Delta_{\partial\Lambda} + m^2)^{-1}$. Also for $P(\xi)$ a polynomial bounded from below,

$$V(\Lambda) = \int_\Lambda :P(\phi(x)):_{C_\varnothing} dx, \tag{11.2.2}$$

where the Wick ordering is with respect to the free covariance $C_\varnothing = (-\Delta + m^2)^{-1}$. The normalization factor is

$$Z = Z(\Lambda) = \int e^{-V(\Lambda)} d\phi_{C_{\partial\Lambda}}. \tag{11.2.3}$$

Finally let

$$S_\Lambda\{f\} = \int e^{i\phi(f)} d\mu_\Lambda. \tag{11.2.4}$$

Theorem 11.2.1 (Existence). *Let $P = even + linear$ and $f \in C_0^\infty$. Then*

$$S\{f\} = \lim_{\Lambda \uparrow R^2} S_\Lambda\{f\} \tag{11.2.5}$$

exists and satisfies the Euclidean axioms OS0 and OS2–3 of Section 6.1.

PROOF.   Assume the existence of the limit (11.2.5). Reflection positivity OS3 holds for the measure $d\mu_\Lambda$ with respect to reflection θ in a hyperplane Π such that $\theta\Lambda = \Lambda$; cf. Theorem 10.4.2. Since positivity is preserved under limits, positivity holds for the limiting functional $S\{f\}$.

Euclidean invariance OS2 also follows from the existence of the limit (11.2.5). Let E be an arbitrary Euclidean transformation. By assumption both $S_\Lambda\{f\}$ and $S_{E\Lambda}\{f\}$ have limits as $\Lambda \uparrow R^2$, and these limits coincide. Since $S_{E\Lambda}\{f\} = S_\Lambda\{Ef\}$, it follows that $S\{f\} = S\{Ef\}$.

Thus the proof of both reflection positivity and Euclidean invariance is reduced to the proof of the existence of the limit (11.2.5). The basic upper bound is stated as Theorem 11.3.1 below. Assuming this bound, we establish (11.2.5).

Without loss of generality, we assume the linear term in P is negative, for otherwise substitute $-\phi$ for ϕ. Now choose $g \in C_0^\infty$, $0 \le g$. By Theorem 10.2.2, $S_\Lambda\{-ig\}$ is positive and increases monotonically in Λ. Furthermore, the uniform bound of Theorem 11.3.1 ensures that the limit $S\{-ig\}$ exists.

Next consider a set $\{g_j\}$ of n such functions, $j = 1, 2, \ldots, n$, in C_0^∞, and n complex numbers z_j. Let $zg \equiv z_1 g_1 + \cdots + z_n g_n$. The finite volume generating function $S_\Lambda\{zg\}$ is an entire function on C^n, satisfying the bound, uniform in Λ (but not in n),

$$|S_\Lambda\{zg\}| \le \prod_{i=1}^n \exp\{c(|K| + \|nz_i g_i\|_{L_p}^p)\}, \tag{11.2.6}$$

by Theorem 11.3.1. Hence by Vitali's theorem, as $\Lambda \uparrow R^2$,

$$S_\Lambda\{zg\} \to S\{zg\} \tag{11.2.7}$$

with uniform convergence on compact sets of z, and the limiting functional is entire. In particular, the Schwinger functions also converge for $0 \le g$:

$$\int \phi(g_1) \cdots \phi(g_n) d\mu_\Lambda \to \int \phi(g_1) \cdots \phi(g_n) d\mu. \tag{11.2.8}$$

By continuity, the bounds (11.2.6) and convergence (11.2.7) extends to $g_i \in L_1 \cap L_p$ with compact support. Now let $f_j \in C_0^\infty$, and let

$$f_{j\pm}(x) = \begin{cases} \pm f_j(x) & \text{if } 0 \le \pm f_j(x), \\ 0 & \text{otherwise,} \end{cases}$$

so that $f_j = f_{j+} - f_{j-}$ and $0 \le f_{j\pm}$. Then $f_{j\pm}$ is not in general in C_0^∞, but $f_{j\pm} \in L \cap L_p$. Thus the convergence (11.2.8) holds with g_j replaced by $f_{j\pm}$; and by a finite summation, with g_j replaced by f_j, we have

$$\int \phi(f_1) \cdots \phi(f_n) \, d\mu_\Lambda \to \int \phi(f_1) \cdots \phi(f_n) \, d\mu. \tag{11.2.9}$$

By summation,

$$S_\Lambda \left\{ \sum_{i=1}^n z_i f_i \right\} \to S \left\{ \sum_{i=1}^n z_i f_i \right\} \tag{11.2.10}$$

for $f_i \in C_0^\infty$, $z_i \in C$, and the limit in (11.2.10) is an entire function of $(z_1, \ldots, z_n) \in C^n$. Hence $S\{f\}$ is an entire function on C_0^∞. This completes the proof of the theorem, assuming Theorem 11.3.1.

11.3 Upper Bounds

We now give the main estimate of the existence proof.

Theorem 11.3.1. *Let $0 < m$, and let $d\mu_\Lambda$ be given by (11.2.1). Let $P(\xi) =$ even + linear be of degree n and bounded from below. Let $p = n/(n-1)$. Let $f \in L_1 \cap L_p$ be supported in a rectangle K of area $|K|$. Then there exists a constant $c < \infty$ which is independent of Λ such that*

$$\left| \int e^{\phi(f)} \, d\mu_\Lambda \right| \le \exp\{c(|K| + \|f\|_{L_p}^p)\}. \tag{11.3.1}$$

Remark. In Chapter 12 we eliminate the K dependence from the bound (11.3.1), as required by OS1.

PROOF. It is no loss of generality to assume $0 \le f$. In fact, by the positivity of the Schwinger functions (Theorem 10.2.2),

$$\left| \int e^{\phi(f)} \, d\mu_\Lambda \right| \le \int e^{\phi(|f|)} \, d\mu_\Lambda.$$

Theorem 10.2.2 also shows that for $0 \le f$,

$$\langle e^{\phi(f)} \rangle_\Lambda = S_\Lambda\{-if\} \tag{11.3.2}$$

is monotonic increasing in Λ.

We now begin a sequence of enlargements and reflections. Let $K \subset \Lambda$ be the rectangle containing supp f, and let $\Lambda^{(1)} \supset \Lambda$ be a rectangle whose center is at one corner of K and whose axes are parallel to those of K, as in Figure 11.1(b).

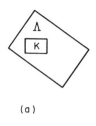

(a)

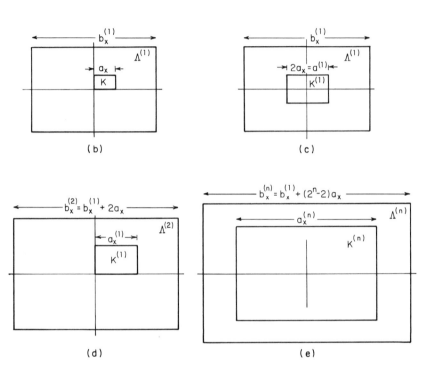

(b)   (c)   (d)   (e)

Figure 11.1. The rectangles K, Λ for multiple reflection bounds: (a) initially; (b) after enlargement of Λ to $\Lambda^{(1)}$; (c) after one reflection in each axis; (d) after enlargement of $\Lambda^{(1)}$ to $\Lambda^{(2)}$; (e) after n reflections and enlargements (not drawn to scale).

Let us label the axes x and y. Assume the side lengths of K are $\{a_x, a_y\}$, and those of $\Lambda^{(1)}$ are $\{b_x^{(1)}, b_y^{(1)}\}$. The first reflections are in the axes of $\Lambda^{(1)}$. Since $\Lambda^{(1)}$ is invariant under these reflections, we use reflection positivity of $d\mu_{\Lambda^{(1)}}$ to derive an upper bound.

Let

$$f^{(1)} \equiv f + \theta_{\Pi_x} f + \theta_{\Pi_y} f + \theta_{\Pi_x}\theta_{\Pi_y} f$$
$$= (1 + \theta_{\Pi_x})(1 + \theta_{\Pi_y})f,$$

and

$$K^{(1)} = K \cup \theta_{\Pi_x} K \cup \theta_{\Pi_y} K \cup \theta_{\Pi_x}\theta_{\Pi_y} K.$$

Thus suppt $f^{(1)} \subset K^{(1)}$. With R the reflection operator of (10.5.2),

$$R(e^{\phi(f)}) = e^{\phi(f^{(1)})}.$$

By Proposition 10.5.1 and Theorem 10.2.2,

$$\int e^{\phi(f)} \, d\mu_\Lambda \leq \int e^{\phi(f)} \, d\mu_{\Lambda^{(1)}} \leq \left(\int e^{\phi(f^{(1)})} \, d\mu_{\Lambda^{(1)}} \right)^{1/4}. \tag{11.3.3}$$

This reflection is shown in Figure 11.1(c).

We now iterate the above process. Enlarge $\Lambda^{(j)}$ to $\Lambda^{(j+1)}$, chosen so that $K^{(j)}$ lies in the first quadrant of $\Lambda^{(j+1)}$. Then reflect in the axes of $\Lambda^{(j+1)}$, producing $f^{(j+1)}$ supported in $K^{(j+1)}$, as in Figure 11.1(d–e). After n steps we find that $K^{(n)}$ and $\Lambda^{(n)}$ have comparable size and we terminate the enlargement–reflection process. Iteration of (11.3.3) yields

$$\langle e^{\phi(f)} \rangle_\Lambda \equiv \langle e^{\phi(f)} \rangle_{\Lambda, \, C_{\partial\Lambda}} \leq \langle e^{\phi(f)} \rangle_{\Lambda^{(1)}} \leq \langle e^{\phi(f^{(n)})} \rangle_{\Lambda^{(n)}}^{4^{-n}}. \tag{11.3.4}$$

To estimate (11.3.4), note that $K^{(n)}$ is a rectangle with sides of length

$$a_x^{(n)} = 2^n a_x, \qquad a_y^{(n)} = 2^n a_y, \tag{11.3.5}$$

and has area $|K^{(n)}| = 4^n |K|$. Furthermore, $\Lambda^{(n)}$ has sides of length

$$b_x^{(n)} = b_x^{(1)} + (2^n - 2)a_x \leq b_x^{(1)} + 2^n a_x,$$
$$b_y^{(n)} = b_y^{(1)} + (2^n - 2)a_y \leq b_y^{(1)} + 2^n a_y. \tag{11.3.6}$$

This follows from

$$b_x^{(n)} = b_x^{(n-1)} + a_x^{(n-1)} = (2^{n-1} + \cdots + 2)a_x + b_x^{(1)} = (2^n - 2)a_x + b_x^{(1)}.$$

For n sufficiently large,

$$b_x^{(1)} \leq 2^n a_x, \qquad b_y^{(1)} \leq 2^n a_y. \tag{11.3.7}$$

Then

$$|\Lambda^{(n)}| \leq 4(2^n)(2^n)|K| = 4|K^{(n)}|, \tag{11.3.8}$$

so $K^{(n)}$ covers at least one-quarter the area of $\Lambda^{(n)}$; see Figure 11.1(e)

Writing (11.3.4) as a ratio,

$$\langle e^{\phi(f^{(n)})} \rangle_{\Lambda^{(n)}} = \frac{\int e^{\phi(f^{(n)}) - V(\Lambda^{(n)})} \, d\phi_{\partial\Lambda^{(n)}}}{\int e^{-V(\Lambda^{(n)})} \, d\phi_{\partial\Lambda^{(n)}}},$$

we bound the numerator and denominator separately. By Proposition 10.3.1 and by (10.3.8),

$$\langle e^{\phi(f)} \rangle_{\Lambda, \, C} \leq \left(O(1)^{|\Lambda^{(n)}|} \int e^{\phi(f^{(n)}) - V(\Lambda^{(n)})} \, d\phi_{\partial\Lambda^{(n)}}^N \right)^{4^{-n}}, \tag{11.3.9}$$

where $d\phi_{\partial\Lambda^{(n)}}^N$ denotes the Gaussian measure with Neumann boundary data on all unit squares in $\Lambda^{(n)}$. Since the right side of (11.3.9) factorizes, we estimate it using Theorem 8.6.2. In fact, (8.6.8) gives an upper bound $\exp(0(N(g) + |\Lambda|)$, where for $j \neq 1$, $g_j = f_j$ is a coefficient of the polynomial P and $g_1 = f_1 + f^{(n)}$, as in (8.6.2, 4). With a constant depending on P, i.e., in the f_j, but not on f, nor on $f^{(n)}$,

$$N(g) \leq O(|\Lambda|)(1 + \|f^{(n)}\|_{L_p}^p).$$

Thus

$$\langle e^{\phi(f)} \rangle_{\Lambda, C} \leq [O(1)^{\Lambda^{(n)}} \exp\{c\|f^{(n)}\|_{L_p}^p\}]^{4^{-n}}. \tag{11.3.10}$$

Note that $f^{(n)}$ is a sum of reflections of f into nonoverlapping regions. Hence

$$\|f^{(n)}\|_{L_p}^p = 4^n \|f\|_{L_p}^p.$$

Moreover, using (11.3.8), (11.3.10),

$$\langle e^{\phi(f)} \rangle_{\Lambda, C} \leq \exp\{c(|K| + \|f\|_{L_p}^p)\}. \tag{11.3.11}$$

This concludes the proof.

Historical Remark. The use of multiple reflections based on the full lattice group Z^d to reduce estimates of local perturbations to free energy bounds originated in [Glimm, Jaffe, and Spencer, 1975]. This method has considerably simplified the analysis of the $V \to \infty$ limit. For general semibounded P, weak coupling boundary conditions [Glimm and Jaffe, 1975b] are defined by use of a cluster expansion with a large external field [Spencer, 1974b]; see also Chapter 18. Monotonicity in the external field follows from the FKG inequalities of Section 10.2. Monotonicity and multiple reflection bounds allow removal of the large external field. See [Fröhlich and Simon, 1977].

CHAPTER 12

Regularity and Axioms

12.1 Introduction

The calculus of integration by parts, developed in Chapter 9 for finite volume theories, extends to the infinite volume measures $d\mu$ constructed in Chapter 11. These basic identities generate series which exhibit regularity and other detailed properties of the quantum field models. The integration by parts identities generate the perturbation expansion of Sections 8.4, 9.4 as well as the high and low temperature expansions studied in Part III and in the literature. These tools allow a detailed investigation of the local (ultraviolet) singularities of the models on the one hand and the large distance (infrared) decoupling on the other.

Because the $P(\phi)_2$ models are superrenormalizable, the ultraviolet behavior is governed by low order terms in the perturbation series. The dominant singularity is the free field behavior and is exhibited in closed form in terms of the Green's functions of Chapter 7. The remainder is seen to be regular. In Section 12.5 we use integration by parts to exhibit infrared decoupling from the upper bound on the generating functional $S\{if\}$ and to eliminate the dependence on the area of K (suppt $f \subset K$) in Theorem 12.4.1.

Another way to view the integration by parts identity (9.1.32) is as a Euclidean equation of motion—namely, for $P(\phi)$ models,

$$(-\Delta + m^2)\phi(x) + P'(\phi(x)) = \frac{\delta}{\delta\phi(x)} + \left(\frac{\delta}{\delta\phi(x)}\right)^*. \qquad (12.1.1)$$

219

In particular, after analytic continuation to real time, the right side of (12.1.1) vanishes and the left side becomes the nonlinear equation for ϕ,

$$(-\Box + m^2)\phi(x) + P'(\phi(x)) = 0. \tag{12.1.2}$$

The difficulty in establishing (12.1.1) has already been studied for finite volume models in Section 9.1; it revolves about the definition and regularity of the renormalized (in this case Wick-ordered) polynomial $P(\phi)$. In this chapter we treat these same issues for the infinite volume theory, establishing estimates on $:\phi^j:$ uniform in the volume. A consequence of the infinite volume convergence of Chapter 11 is the identification of $:\phi^j:$, in the infinite volume theory, as both the limit of the finite volume expression and the Wick-ordered polynomial constructed directly in infinite volume. With this control to justify the interchange of limits ($V \to \infty$ and Wick ordering), we recover the integration by parts identity in finite volume. The remainder of the chapter then deals with the application described above and completes the proof of the Euclidean axiom OS1 and hence verifies the hypotheses of the reconstruction Theorem 6.1.5. We study the generating functional

$$S\{f\} = \int e^{i\phi(f)} \, d\mu(\phi)$$

for a $P(\xi)$ even + linear polynomial self interaction with $P(\xi)$ bounded from below.

Theorem 12.1.1 (Axioms). *The functional $S\{f\}$ exists (Theorem 11.2.1) and satisfies the Euclidean axioms OS0–3 of Section 6.1. Hence it yields a quantum field which satisfies the Wightman axioms W1–3.*

PROOF. This is a direct consequence of Theorems 11.2.1 and 12.5.1.

12.2 Integration by Parts

Since the integration by parts formulas such as (12.1.1) involve derivatives and Wick polynomials $:P'(\phi(x)):$, we must begin by establishing the existence and differentiability of an appropriate class of functions $A(\phi)$ on the Euclidean space $\mathscr{E} = L_2(\mathscr{D}', d\mu)$. The construction of Chapter 11 yields the existence of fields $\phi(f)$ and their products $A(\phi) = \phi(f_1) \cdots \phi(f_n)$, but not the Wick powers $:\phi^j:$. We now show that the Wick powers are limits of cutoff powers in the infinite volume theory:

$$:\phi(x)^j: = \lim_{\kappa \to \infty} :\phi_\kappa(x)^j:. \tag{12.2.1}$$

Here $\phi_\kappa = \delta_\kappa * \phi$ with δ_κ an approximate delta function,

$$\delta_\kappa(x) = \kappa^2 h(\kappa x) \in C_0^\infty, \qquad \int h \, dx = 1.$$

Furthermore, the $\kappa \to \infty$ limit will be shown to be uniform in the volume Λ, and it follows that the $\Lambda \uparrow R^2$ and $\kappa \to \infty$ limits can be interchanged. Thus $:\phi^j:$ is a function of ϕ on $\mathscr{E}$, and

$$:\phi^j: = \lim_{\Lambda \uparrow R^2} \lim_{\kappa \to \infty} :\phi^j_\kappa: = \lim_{\kappa \to \infty} \lim_{\Lambda \uparrow R^2} :\phi^j_\kappa:. \qquad (12.2.2)$$

The estimates required to establish (12.2.2) are somewhat technical and necessitate the extension of Section 8.6 bounds to a class of nonlocal perturbations. However, these estimates then allow the extension of the results of Chapter 11 to Wick monomials of arbitrary degree. In particular, we define generalized Schwinger functions

$$\int \phi(x_1) \cdots \phi(x_k) :\phi^2(y_1): \cdots :\phi^2(y_l): \cdots :\phi^{n-1}(z_1): \cdots :\phi^{n-1}(z_m): d\mu,$$

and we prove estimates on their regularity. The arbitrary powers $:\phi^r:$, $n < r$, are controlled by integration by parts which reduce them to their present case.

In the following theorem we consider the coefficient functions $g = \{g_1, g_2, \ldots, g_{n-1}\}$ in C_0^∞; these coefficients define a lower order perturbation $\delta P = \sum :\phi^j(g_j):$. Here $n = \deg P$. We consider $:\phi^j: \equiv :\phi^j:_C$, where $C \in \mathscr{C}_m$ is a covariance analyzed in Chapter 7.

Theorem 12.2.1.   *Let $P = even + linear$. Then*

$$S\{g\} = \lim_{\Lambda \uparrow R^2} \int \exp\left(i \sum_{j=1}^{n-1} :\phi^j(g_j):\right) d\mu_\Lambda \qquad (12.2.3)$$

exists and defines Wick ordered powers $:\phi^j:$ in the measure $d\mu$. The κ and Λ limits can be interchanged, so that $:\phi^j: = \lim_{\kappa \to \infty} :\phi^j_\kappa:$ in $L_2(d\mu)$, and $:\phi^j:$ is a function of the field ϕ. Furthermore $S\{g\}$ is an entire analytic function of $g_1, \ldots, g_{n-1}$.

PROOF.   See Section 12.4.

Theorem 12.2.2.   *There is a constant $c < \infty$ such that for any $g \in L_1 \cap L_p$ with $p = n/(n-j)$, $j < n$ and supp $g \subset \Lambda \subseteq R^2$,*

$$\left| \int \exp(:\phi^j(g):) d\mu_\Lambda \right| \le \exp\{c(\|g\|_{L_1} + \|g\|_{L_p}^p)\}. \qquad (12.2.4)$$

In Section 12.5, the special case $j = 1$ of this theorem (Theorem 12.5.1) is proved, and modifications to cover the general case are indicated briefly.

Let $\mathfrak{A}$ be the algebra of functions $A = A(\phi)$ generated by the $:\phi^j(g_j):$ and their exponentials. As in Chapter 7, let $C_\varnothing = (-\Delta + m^2)^{-1}$.

Corollary 12.2.3. *The integration by parts formulas* (9.1.32) *and* (12.1.1) *hold for* $A \in \mathfrak{A}$; *namely, for* $f \in C_0^\infty$,

$$\int \phi(f)A(\phi) \, d\mu = \int \left(\left\langle C_\varnothing f, \frac{\delta A}{\delta \phi} \right\rangle - A(\phi) \left\langle C_\varnothing f, \frac{\delta V}{\delta \phi} \right\rangle \right) d\mu. \quad (12.2.5)$$

PROOF.   For bounded Λ, integration by parts was established in Section 9.1. In the Λ bounded version of (12.2.5), the covariance $C_\varnothing$ is replaced by $C_{\partial\Lambda}$. In the formula, we substitute $C_{\partial\Lambda}^{-1} f = C_\varnothing^{-1} f$ for f. Then both sides of the equation have compactly supported test functions which are independent of Λ. By Theorem 12.2.1, both sides converge to the corresponding $\Lambda = R^2$ expressions. Thus the identity is established for $f \in C_\varnothing^{-1} C_0^\infty$. By Theorem 12.2.2 and the Cauchy integral theorem, the identity extends by continuity to $f \in L_1 \cap L_{n/(n-1)}$.

Remark.   This identity can be extended to the case $f = \delta_x$ provided each term on the right side extends by continuity. The resulting expression $\phi(x)$ on the left side is then defined as a bilinear form; it is neither a function nor an operator.

We allow g to be supported in a nonoverlapping union of unit squares Δ, namely suppt $g \subset \bigcup_{\Delta \in \mathcal{N}} \Delta$. Also let $g_\Delta = \chi_\Delta g$, where χ_Δ is the characteristic function of Δ, so that

$$g = \sum_{\Delta \in \mathcal{N}} g_\Delta. \quad (12.2.6)$$

The Cauchy integral theorem gives an estimate on generalized Schwinger functions.

Corollary 12.2.4. *Let* g *be defined as above. With the above hypothesis,* $p = n/(n-j)$, $j < n$ *and* $h_{\Delta, i}$ *supported in* Δ,

$$\left| \int \prod_{\Delta \in \mathcal{N}} \left[\left(\prod_{i=1}^{n_\Delta} :\phi^j(h_{\Delta, i}): \right) \exp(:\phi^j(g_\Delta):) \right] d\mu \right|$$

$$\leq \prod_{\Delta \in \mathcal{N}} \left[(n_\Delta !)^{1 - 1/p} \left(\prod_{i=1}^{n_\Delta} c \|h_{\Delta, i}\|_{L_p} \right) \exp\{c(1 + \|g_\Delta\|_{L_p}^p)\} \right]. \quad (12.2.7)$$

PROOF.   By a Schwarz inequality, we can separate the exponential from the polynomial and restrict $\mathcal{N}$ to those Δ for which $n_\Delta \neq 0$. In view of Theorem 12.2.2, the exponential factor is bounded as desired, and we thus take $g = 0$. By polarization identity, we reduce to the special case of a single $h_{\Delta, i}$ for each fixed Δ, as follows. In fact assuming (12.2.7) in this case, write

$$2^{n-1}n! \prod_{i=1}^{n} a_i = \sum_{\varepsilon_i = \pm 1} \varepsilon_2 \cdots \varepsilon_n (a_1 + \varepsilon_2 a_2 + \cdots + \varepsilon_n a_n)^n. \quad (12.2.8)$$

Since we may assume all norms $\|h_{\Delta, i}\|_{L_p}$ are equal by scaling, substitution of (12.2.8) and the triangle inequality

$$\|h_{\Delta, 1} + \varepsilon_2 h_{\Delta, 2} + \cdots \|_{L_p} \leq n_\Delta \|h_{\Delta, i}\|_{L_p}$$

in (12.2.7) gives

$$\left| \int \prod_{\Delta \in \mathcal{N}} \prod_{i=1}^{n_\Delta} : \phi^j(h_{\Delta,i}): d\mu \right| \le \prod_{\Delta \in \mathcal{N}} c(n_\Delta!)^{-1/p} n_\Delta^{n_\Delta} \|h_{\Delta,i}\|_{L_p}^{n_\Delta}.$$

Stirling's formula then completes the proof, assuming the special case.

Now we prove the corollary in the special case. Let $z_\Delta \in C^{|\mathcal{N}|}$ denote a family of complex variables indexed by $\Delta \in \mathcal{N}$. Let

$$zh(x) = \sum_{\Delta \in \mathcal{N}} z_\Delta h_\Delta(x).$$

Then the left side of (12.2.7) is equal to

$$\mathcal{M} = \left(\prod_{\Delta \in \mathcal{N}} \left(\frac{d}{dz_\Delta} \right)^{n_\Delta} \right) \int \exp(:\phi^j(zh):) \, d\mu \Big|_{z(\Delta)=0}.$$

By the theorem, $S\{-izh\} = \int \exp(:\phi^j(zh):) \, d\mu$ is entire analytic. Thus by the Cauchy integral formula,

$$\mathcal{M} = \int S\{-izh\} \prod_{\Delta \in \mathcal{N}} \frac{n_\Delta! \, dz_\Delta}{(2\pi i) z^{n_\Delta+1}}, \tag{12.2.9}$$

where the integrals extend over a product of circles centered at $z_\Delta = 0$ and of radius r_Δ. Hence by (12.2.4),

$$|\mathcal{M}| \le \prod_{\Delta \in \mathcal{N}} n_\Delta! r_\Delta^{-n_\Delta} \exp\{c(r_\Delta \|h_\Delta\|_{L_1} + r_\Delta^p \|h_\Delta\|_{L_p}^p)\}. \tag{12.2.10}$$

Now choose

$$r_\Delta = \tfrac{1}{2} n_\Delta^{1/p} \|h_\Delta\|_{L_p}^{-1}.$$

Since $\|h_\Delta\|_{L_1} \le \|h_\Delta\|_{L_p}$,

$$r_\Delta \|h_\Delta\|_{L_1} + r_\Delta^p \|h_\Delta\|_{L_p}^p \le n_\Delta.$$

Using (12.2.10),

$$|\mathcal{M}| \le \prod_{\Delta \in \mathcal{N}} (n_\Delta!)^{1-1/p} (c\|h_\Delta\|_{L_p})^{n_\Delta},$$

which proves (12.2.7) as desired.

12.3 Nonlocal ϕ^j Bounds

The Chapter 8 estimates on Wick polynomials $:\phi^j:$ are derived from weakly divergent bounds on the cutoff fields $:\phi_\kappa^j:$, as in Proposition 8.6.3. To obtain uniform bounds (convergent as $\kappa \to \infty$) for $:\phi_\kappa^j:$, we introduce the doubly momentum-cut-off field

$$\phi_{\kappa,\kappa'} = \delta_\kappa * \delta_{\kappa'} * \phi \tag{12.3.1}$$

and use weakly divergent bounds on $\phi_{\kappa,\kappa'}$. The double cutoff in (12.3.1) also determines a doubly cutoff Wick ordering constant $c_{\kappa,\kappa'}$ used below. The section title refers to the fact that ϕ_κ is nonlocal.

For technical reasons we suppose that $\bigcup_{j=1}^{n-1}$ suppt g_j is contained in the interior of $\Lambda = $ suppt f_n. We use the notation of (8.6.4). The g_j's define perturbations of the polynomial P, considered first in doubly cut-off form. A g_n perturbation could also be allowed, provided $\|g_n\|_{L_\infty}$ is sufficiently small. Let

$$: Q(\phi_{\kappa, \kappa'}, g) := \sum_{j=1}^{n-1} \int : \phi_{\kappa, \kappa'}(x)^j : g_j(x)\, dx. \tag{12.3.2}$$

In analogy with (8.6.5), we define

$$N'(g) = \sum_{j=1}^{n-1} \|g_j\|_{L_{n/(n-j)}}^{n/(n-j)}$$

Proposition 12.3.1. *There is a constant c depending on m, n, and $\|f_n\|_{L_\infty} + \|f_n^{-1}\|_{L_\infty(\Lambda)}$ such that*

$$-c\|f_n\|_{L_1}(\ln \kappa)^{(\deg P)/2} + c(N(f) + N'(g)) \le :P(\phi_\kappa, f): + :Q(\phi_{\kappa, \kappa'}, g):. \tag{12.3.3}$$

PROOF. Because $\phi_{\kappa, \kappa'}$ is a nonlocal function of ϕ_κ, a modification is required in the proof of Proposition 8.6.3. By definition

$$\phi_{\kappa, \kappa'}(x)^j = \prod_{i=1}^{j} \int \phi_\kappa(y_i)\, \delta_{\kappa'}(y_i - x)\, dy_i$$

$$= \int \left(\prod_{i=1}^{j} \phi_\kappa(y_i) \right) \left(\prod_{l=1}^{j} \delta_{\kappa'}(y_l - x)\, dy_l. \right)$$

The inequality $\prod_{i=1}^{j} a_i \le \sum_{i=1}^{j} a_i^j$ for $0 \le a_i$ gives

$$|\phi_{\kappa, \kappa'}(x)^j| \le \int \sum_{i=1}^{j} |\phi_\kappa(y_i)^j| \prod_{l=1}^{j} \delta_{\kappa'}(y_l - x)\, dy_l.$$

Since $\int \delta_\kappa(y)\, dy = 1$,

$$|\phi_{\kappa, \kappa'}(x)^j| \le \int \sum_{i=1}^{j} |\phi_\kappa(y_i)^j| \, \delta_{\kappa'}(y_i - x)\, dy_i$$

and

$$\left| \int \phi_{\kappa, \kappa'}(x)^j g_j(x)\, dx \right| \le j \int |\phi_\kappa(x)^j| (\delta_{\kappa'} * |g_j|)(x)\, dx.$$

By Hölder's inequality, as in Proposition 8.6.3,

$$\left| \int c_{\kappa, \kappa'}^l \phi_{\kappa, \kappa'}(x)^{j-2l} g_j(x)\, dx \right|$$

$$\le \varepsilon \int \phi_\kappa(x)^n f_n(x)\, dx + a(\varepsilon) c_{\kappa, \kappa'}^{n/2} \int f_n(x)\, dx$$

$$+ \int \left(\frac{(\delta_{\kappa'} * |g_j|)(x)}{f_n(x)} \right)^{n/(n-j)} f_n(x)\, dx. \tag{12.3.4}$$

The proposition follows from (12.3.4), as in the proof of Proposition 8.6.3, since

$$\|\delta_{\kappa'} * g\|_{L_p} \leq \|\delta_{\kappa'}\|_{L_1} \|g\|_{L_p} = \|g\|_{L_p},$$

and for $g = |g_j|$,

$$\int \left(\frac{\delta_{\kappa'} * g}{f_n}\right)^{n/(n-j)} f_n \, dx = \int (\delta_{\kappa'} * g)^{n/(n-j)} f_n^{-j/(n-j)} \, dx$$

$$\leq \text{const} \int g^{n/(n-j)} \, dx = \text{const } N'(g).$$

We now follow the proof of Theorem 8.6.2 to establish

Proposition 12.3.2.   *Let $\kappa \leq \kappa' \leq \infty$ and*

$$\|f_n\|_{L_\infty} + \|f_n^{-1}\|_{L_\infty(\Lambda)} + m^{-1} + n + |\Lambda| \leq L.$$

Then

$$\int \exp(:Q(\phi_\kappa, g):) \, d\mu_\Lambda \leq \exp(\text{const}(N(f) + N'(g) + 1))$$

and

$$\int \exp(:Q(\phi_\kappa, g) - Q(\phi_{\kappa'}, g):) \, d\mu_\Lambda$$

$$\leq \kappa^{-\varepsilon} M(g) \exp(\text{const}(N(f) + N'(g) + 1)).$$

In both cases, the constant depends only on L.

12.4 Uniformity in the Volume

We show that the Section 12.3 estimates are uniform in the volume $|\Lambda|$, in analogy with Theorem 11.3.1. Because the polynomial function Q being estimated is not of the form even + linear, we cannot use the monotonicity of S_Λ in Λ. Thus a different sequence of multiple reflection bounds is required. We use the asymmetric reflection method developed in Section 10.6. A benefit of this more general method is that the interaction polynomial P must be semibounded, but is otherwise unrestricted. In particular P is not restricted to the class even + linear in this section.

Below we restrict K and Λ to be rectangles with $K \subset \Lambda$. Furthermore Λ is restricted to be chosen from a specific sequence $\Lambda_v \uparrow \infty$; the sequence Λ_v depends on K. These restrictions arise for technical reasons, but still allow conclusions to be drawn about the infinite volume measure.

Theorem 12.4.1.   *There are constants c and ε such that for any $\kappa < \kappa' \leq \infty$ and any rectangle K with width and length bounded away from* *zero, a sequence of rectangles* $\Lambda_v \uparrow R^2$ *exists with*

$$\int \exp(:\phi_\kappa^j(g):)\, d\mu_{\Lambda_l} \le \exp(c\{|K| + \|g\|_{L_p}^p\}),$$

$$\int \exp(:\phi_\kappa^j(g) - \phi_{\kappa'}^j(g):)\, d\mu_{\Lambda_l} \le \kappa^{-\varepsilon} M(g) \exp(c\{|K| + \|g\|_{L_p}^p\})$$

provided suppt $g \subset K$.

PROOF. To begin, we take Λ_v and K to have parallel sides and a common center. Let K be an $l \times t$ rectangle and choose Λ_v to be a $2L_v + l$ by $2T_v + t$ rectangle. We restrict Λ_v below so that T_v is much larger than $L_v + l$, and in this way obtain a sequence $\Lambda_v \uparrow R^2$. Now we establish estimates for a given v, and thus suppress the v-dependence of Λ_v, L_v, T_v, etc. These estimates are uniform in v. Also, without loss of generality, we assume $l \le L$. The axis of the first reflection is the line through one of the sides of K. This axis is not a line of symmetry of Λ, nor of the boundary conditions B. Thus we use the nonsymmetric form of reflection positivity, Proposition 10.6.1. Let Λ_+ be the part of Λ containing K. Then $A = I$ in (10.6.8), and the factor in (10.6.8) containing A reduces to one, because of normalization. The basic reflection step is indicated in Figure 12.1.

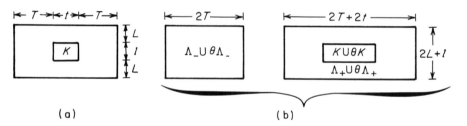

(a)   (b)

Figure 12.1. (a) Before and (b) after reflection.

The basic reflection step is iterated, as in the proof of Theorem 11.3.1, and after n steps produces regions $K^{(n)}$, $\Lambda^{(n)}$ and a function $B^{(n)}$ such that for large n

$$|\Lambda^{(n)}| \le 4|K^{(n)}|.$$

Also $B = B^{(0)} = \exp(:\phi_\kappa^j(g):)$, etc. After these n reflections we obtain

$$\left| \int B\, d\mu_\Lambda \right| \le \prod_{j=1}^{n} \left[\left(\frac{Z_+^{(j)} Z_-^{(j)}}{Z^{(j)\,2}} \right)^{2^{-j}} \det(C^{(j)-2} C_+^{(j)} C_-^{(j)})^{2^{-j-1}} \right]$$
$$\cdot \left(\int (\theta B^{(n-1)})^- B^{(n-1)}\, d\mu_{\Lambda^{(n)}} \right)^{2^{-n}}. \qquad (12.4.1)$$

The above restriction on Λ is that $L + l \ll T(2L + l) \le T$, in the notation of Figure 12.1, for some number $T(2L + l)$ sufficiently large. This restriction on Λ makes Λ very oblong. We reflect first about the vertical side of K, as follows. With j_0 and j_1 chosen so that

$$2^{j_0} t < 2T + t \le 2^{j_0 + 1} t,$$

$$2^{j_1} l < 2L + l \le 2^{j_1 + 1} l,$$

we use j_0 reflections about a vertical axis and then j_1 reflections about a horizontal axis.

Consider the following bounds on the individual terms in (12.4.1): First, the determinants are bounded using Theorem 10.6.2. The j_0 reflections about the vertical axis satisfy

$$L_j = 2L + l, \qquad (T_j)_- = T, \qquad (T_j)_+ = T + 2^{j-1}t.$$

Thus $L_j/(T_j)_\pm \le (2L + l)/ \le 1$, and these determinants are bounded by

$$\prod_{j=1}^{j_0} O(1)^{2^{-j}} \le O(1).$$

The reflections about the horizontal axis satisfy

$$(L_{j_0+k})_- = L, \qquad (L_{j_0+k})_+ = L + 2^{k-1}l,$$
$$T_{j_0+k} = 2T + 2^{j_0}t.$$

Thus as $1 \le L$,

$$\frac{T_{j_0+k}}{(L_{j_0+k})_\pm} \le \frac{2T + 2^{j_0}t}{L} \le 2^{j_0+2}t.$$

The determinants arising from $j_0 < j \le j_0 + j_1$ are therefore bounded by

$$\prod_{k=1}^{j_1} \exp[0(2^{j_0}t)\, 2^{-j_0-k}] \le \exp(O(t)) \le \exp(O(|K|)).$$

Next consider the product of Z's. This product partially telescopes, because of the relation $Z_+^{(j)} = Z^{(j+1)}$. It equals

$$\left(\prod_{j=1}^{j_0+j_1} \left(\frac{Z_-^{(j)}}{Z^{(1)}} \right)^{2^{-j}} \right) \left(\frac{Z_+^{(j_0+j_1)}}{Z^{(1)}} \right)^{2^{-(j_0+j_1)}}. \tag{12.4.2}$$

The Z's are bounded above and below by conditioning (10.3.8). Thus the Z_+ factor in (12.4.2) can be bounded using the above relations on the areas of $\Lambda^{(n)}$, $n = j_0 + j_1$. It satisfies

$$(Z_+^{(n)}/Z^{(1)})^{2^{-n}} \le \exp(O(TL)2^{-n})$$
$$\le \exp(O(tl)) = \exp(O(|K|)).$$

A similar bound applies to the $B^{(n)}$ integral in (12.4.1), using also Proposition 12.3.2, and introduces a factor $\exp[\text{const } N'(g)]$. (We fix f in this estimate.)

Let $Z(a, b)$ denote the normalizing partition function for an $a \times b$ rectangle. Then

$$Z^{(1)} = Z(2T + t, 2L + l),$$
$$Z_-^{(j)} = Z(2T, 2L + l), \qquad 1 \le j \le j_0,$$
$$Z_-^{(j)} = Z(2T + 2^{j_0}t, 2L), \qquad j_0 + 1 \le j \le j_0 + j_1 = n.$$

Thus the remaining ratios in (12.4.1) reduce to

$$\left(\frac{Z(2T, 2L + l)}{Z(2T + t, 2L + l)} \right)^{1-2^{-j_0}} \left(\frac{Z(2T + 2^{j_0}t, 2L)}{Z(2T + t, 2L + l)} \right)^{2^{-j_0}(1 - 2^{-j_1})}. \tag{12.4.3}$$

The second factor in (12.4.3) is bounded by further nonsymmetric reflections about a horizontal axis. In fact, by (10.6.8) and the preceding arguments,

$$Z(2T + 2^{j_0}t, 2L)$$

$$\leq e^{O(T)} Z(2T + 2^{j_0}t, 2L - l)^{1/2} Z(2T + 2^{j_0}t, 2L + l)^{1/2}. \tag{12.4.4}$$

Iterating this inequality j_1 times bounds the second factor in (12.4.3) by

$$e^{O(|K|)} \left(\frac{Z(2T + 2^{j_0}t, 2L - (2^{j_1} - 1)l)}{Z(2T + t, 2L + l)} \right)^{2^{-j_0}(1 - 2^{-j_1})2^{-j_1}}$$

$$\times \left(\frac{Z(2T + 2^{j_0}t, 2L + l)}{Z(2T + t, 2L + l)} \right)^{2^{-j_0}(1 - 2^{-j_1})(1 - 2^{-j_1})}.$$

Using (10.3.8), as before, this is bounded by

$$e^{O(|K|)} \left(\frac{Z(2T + 2^{j_0}t, 2L + l}{Z(2T + t, 2L + l} \right)^{2^{-j_0}(1 - 2^{-j_1})^2}. \tag{12.4.5}$$

The remainder of the proof concerns the asymptotic analysis of $Z(t, l)$ as $t \to \infty$ with l fixed. Using the construction of Chapter 6 and the infinite volume limit of Chapter 11 adapted to a $t \times l$ rectangle, with $t \to \infty$ while l is held fixed, we obtain a Hamiltonian H_l and semigroup e^{-tH_l} associated with a space interval of length l and Dirichlet boundary conditions. For a choice of vector ψ_l, we have

$$Z(t, l) = \langle \psi_l, e^{-tH_l}\psi_l \rangle = \int e^{-\lambda t} \, d\rho_l(\lambda),$$

where $d\rho_l$ is the spectral measure determined by ψ_l and H_l. Let $E_l = \inf \operatorname{suppt} d\rho_l$. Then

$$Z(t, l) = O(e^{-tE_l}) \quad \text{as } t \to \infty. \tag{12.4.6}$$

By the normalization of Chapter 6, $0 \leq H$, so that $0 \leq E_l$. In particular, (12.4.5) is bounded by

$$e^{O(|K|)} \exp[-(2^{j_0} - 1)tE_{2L+l}2^{-j_0}(1 - 2^{-j_1})(1 - 2^{-j_1})]. \tag{12.4.7}$$

However, by use of the spectral theorem, we can strengthen (12.4.6) to obtain a bound on the first factor in (12.4.3),

$$\frac{Z(2T, 2L + l)}{Z(2T + t, 2L + l)} \leq 2e^{tE_{2L+l}} \quad \text{as } T \to \infty, \tag{12.4.8}$$

with l, t, and L fixed. $T(2L + l)$ is defined as the minimum size of T for (12.4.8) to hold. Note that a priori $T(2L + l)$ may also depend on t, but this possibility is excluded by the spectral theorem. Thus

$$\exp[-\text{const } N'(g)] \left| \int B \, d\mu_\Lambda \right| \leq e^{O(|K|)} \exp[tE_{2L+l}(1 - 2^{-j_0})]$$

$$\exp[-tE_{2L+l}(1 - 2^{-j_0})(1 - 2^{-j_1})^2]$$

$$\leq e^{O(|K|)} \exp[tE_{2L+l}(O(2^{-j_1}))].$$

However by (10.3.8), $|E_{2L+l}| \leq O(2L + l) \leq O(2^{j_1}l)$, so

$$\exp[-\text{const } N'(g)] \left| \int B \, d\mu_\Lambda \right| \leq e^{O(|K|)} e^{O(tl)} = e^{O(|K|)}.$$

This proves the first inequality of the theorem. The second is similar, but uses the second inequality from Proposition 12.3.2 in place of the first.

PROOF OF THEOREM 12.2.1.   By Theorem 11.2.1, the $:\phi_\kappa^j:$ Schwinger functions converge to define a field $:\phi_\kappa^j:$ in $L_2(d\mu)$. By Theorem 12.4.1 and the Cauchy theorem (as in the proof of Corollary 12.2.4) these Schwinger functions are bounded in a manner which allows summation of the exponential. These bounds are uniform in Λ and also hold for $\Lambda = R^2$. Thus for $\kappa < \infty$, $\int \exp(:\phi_\kappa^j(g_j):)\,d\mu_\Lambda$ is convergent. By the uniform bounds of Theorem 12.4.1 and a 3 epsilon argument, $\int \exp(:\phi^j(g):)\,d\mu_\Lambda$ is also convergent. The other bounds follow similarly.

We formulate the reflection bound as a basic tool.

Theorem 12.4.2. *For Λ an $L \times T$ rectangle, let $d\mu_\Lambda$ be given by* (11.2.1–3). *Assume for some constant a the uniform bound*

$$\exp(-a|\Lambda|) \le Z(\Lambda) \le \exp(a|\Lambda|). \qquad (12.4.9)$$

Let B be a function of ϕ localized in a rectangle $K \subset \Lambda$, where K and Λ have parallel sides and the width and length of K are bounded away from zero. Let $B^{(n)}$ be the n-fold reflection of B, which is localized in $K^{(n)} \subset \Lambda^{(n)}$ as above. For T sufficiently large (i.e., $T_0 = T_0(L, P, m) \le T$),

$$\left| \int B\,d\mu_\Lambda \right| \le e^{\text{const}|K|} \left(\int B^{(n)}\,d\mu_{\Lambda^{(n)}} \right)^{2^{-n}} \qquad (12.4.10)$$

where the constant depends only on the constant a.

PROOF.   Follow the proof of Theorem 12.4.1, but use (12.4.9) in place of the conditioning inequality (10.3.8).

12.5 Regularity of the $P(\phi)_2$ Field

In this section we study the Euclidean $P = \text{even} + \text{linear}$ models constructed in Chapter 11. We complete the verification of the axioms OS0–3 of Chapter 6, by proving the regularity bound OS1. This involves the elimination of $|K|$ from the bound (11.3.1). The existence of a $P(\phi)_2$ field theory satisfying the Wightman and Haag–Kastler axioms then follows from the reconstruction theorem of Chapters 6, 19. The regularity we require is the following:

Theorem 12.5.1. *Let $P = \text{even} + \text{linear}$. Then there exists $c < \infty$ such that for all $f \in C_0^\infty$,*

$$S\{-if\} = \int \exp(\phi(f))\,d\mu \le \exp\{c(\|f\|_{L_1} + \|f\|_{L_p}^p)\}. \qquad (12.5.1)$$

Here $p = n/(n-1)$ and $n = \deg P$. The measure $d\mu$ satisfies axiom OS1.

A first step in the elimination of $|K|$ from Theorem 12.4.1 is given by multiple reflections.

Lemma 12.5.2. *Let* $p = n/(n - j)$ *with* $j < n$. *Let* $g \in L_1 \cap L_p$ *and let* g *have compact support. Define* g_Δ *by* (12.2.6). *Then*

$$\left| \int \exp(:\phi^j(g):) \, d\mu \right| \leq \prod_{\Delta \in \mathcal{N}} \exp\{c(1 + \|g_\Delta\|_{L_p}^p)\}$$

$$= \exp\{c(|\mathcal{N}| + \|g\|_{L_p}^p)\}. \tag{12.5.2}$$

PROOF. Use the multiple reflection bound of Corollary 10.5.8 with $k^{(l)} = \exp:\phi^j(g_{\Delta_l}):$. Estimate each factor $\mathcal{L}(k^{(l)})$ by Theorem 12.4.1.

Lemma 12.5.3. *Let* g *be defined as above. With the above hypothesis,* $p = n/(n - j)$, $j < n$ *and* $h_{\Delta, i}$ *supported in* Δ,

$$\left| \int \prod_{\Delta \in \mathcal{N}} \left[\left(\prod_{i=1}^{n_\Delta} :\phi^j(h_{\Delta, i}): \right) \exp(:\phi^j(g_\Delta):) \right] d\mu \right|$$

$$\leq \prod_{\Delta \in \mathcal{N}} \left[(n_\Delta !)^{1 - 1/p} \left(\prod_{i=1}^{n_\Delta} c \|h_{\Delta, i}\|_{L_p} \right) \exp\{c(1 + \|g_\Delta\|_{L_p}^p)\} \right].$$

PROOF. Follow the proof of Corollary 12.2.4, but use Theorem 12.4.1 in place of Theorem 12.2.2.

PROOF OF THEOREM 12.5.1. First decompose f into its large and small parts as follows. Let $f = g + h$, where

$$h = \sum_{i \in \mathcal{M}} \chi_{\Delta_i} f, \qquad g = \sum_{i \in Z^2 \setminus \mathcal{M}} \chi_{\Delta_i} f, \tag{12.5.3}$$

where χ_{Δ_i} is the characteristic function for the unit square Δ_i in a cover $\{\Delta_i\}$ of R^2, and where $\mathcal{M}$ is the set of indices such that

$$\|f\|_{L_1(\Delta_i)} + \|f\|_{L_p(\Delta_i)} \leq 1. \tag{12.5.4}$$

Thus g is the "large" part of f and h is the "small" part of f. Also

$$1 < \|f\|_{L_1(\Delta_i)} + \|f\|_{L_p(\Delta_i)}^p \quad \text{for } i \in Z^2 \setminus \mathcal{M}. \tag{12.5.5}$$

As a preliminary step, we use the Schwarz inequality

$$\int e^{\phi(f)} \, d\mu \leq \left(\int e^{\phi(2g)} \, d\mu \int e^{\phi(2h)} \, d\mu \right)^{1/2} \tag{12.5.6}$$

to separate g from h. Using Lemma 12.5.2 and (12.5.5), we bound the g factor as desired. Hence we may without loss of generality assume $g = 0$ and $f = h$. Similarly, we use a Schwarz inequality to separate h into a sum of four parts, so that for each part, $\mathcal{M}$ labels squares with no common edges or corners. In other words either $\chi_{\Delta_i} h$ or $\chi_{\Delta_j} h$ vanishes if $i \neq j$ but $\text{dist}(\Delta_i, \Delta_j) \leq 1$.

For convenience, we now let a subscript j index the support of f. Let $f_j = h_j = \chi_{\Delta_j} f$. Then

$$e^{\phi(f_i)} = 1 + F_i + G_i, \tag{12.5.7}$$

where

$$F_i = \phi(f_i), \qquad G_i = \phi(f_i)^2 \int_0^1 \int_0^1 \lambda \, e^{\lambda \gamma \phi(f_i)} \, d\lambda \, d\gamma.$$

Also

$$e^{\phi(f)} = \sum_{I \subset Z^2} \sum_{J \subset I} F_J G_{I \setminus J}, \qquad (12.5.8)$$

where

$$F_J = \prod_{j \in J} F_j, \qquad G_J = \prod_{j \in J} G_j.$$

The sums in (12.5.8) run over finite subsets of Z^2. The sums and products are finite, because with a finite number of exceptions $f_i = 0$. We claim there exists $c < \infty$ such that

$$\left| \int F_J G_{I \setminus J} \, d\mu \right| \le \prod_{i \in I} c\{\|f_i\|_{L_1} + \|f_i\|_{L_p}^p\}. \qquad (12.5.9)$$

Hence

$$\int e^{\phi(f)} \, d\mu \le \sum_{I \subset Z^2} \sum_{J \subset I} \prod_{i \in I} (c\{\|f_i\|_{L_1} + \|f_i\|_{L_p}^p\})$$

$$\le \sum_{I \subset Z^2} \prod_{i \in I} (2c\{\|f_i\|_{L_1} + \|f_i\|_{L_p}^p\})$$

$$= \prod_{i \in Z^2} (1 + 2c\{\|f_i\|_{L_1} + \|f_i\|_{L_p}^p\}), \qquad (12.5.10)$$

where the second inequality follows from the identity $\sum_{J \subset I} 1 = 2^{|I|}$. Now use $1 + \alpha x \le e^{\alpha x}$ to conclude

$$\int e^{\phi(f)} \, d\mu \le \exp(2c\{\|f\|_{L_1} + \|f\|_{L_p}^p\}), \qquad (12.5.11)$$

as desired.

To complete the proof, we establish (12.5.9). We begin by integrating each factor $\phi(f_j)$ in F_J by parts; cf. (12.2.5). The $\phi(f_i)$ contract (i) to each other, (ii) to the factors $\phi(f_j)^2 \exp(\lambda\mu \, \phi(f_j))$ in $G_{I \setminus J}$, and (iii) to the exponent. We shall bound these factors using Lemma 12.5.3, explicit bounds on covariance operators, and elementary combinatoric estimates.

The terms of type (i) with contractions between distinct linear factors in F or the contractions of type (ii) between linear factors in F and G occur in nonadjacent squares. These contractions yield factors $\langle f_i, C f_j \rangle$, bounded by

$$|\langle f_i, C f_j \rangle| \le O(1) \, e^{-m|i - j|} \|f_i\|_{L_1} \|f_j\|_{L_1}. \qquad (12.5.12)$$

Here $C = C_\varnothing$ is bounded by Proposition 7.2.1.

The contractions from the $\phi(f_j)$ to the exponent produce factors of the form

$$(P^{(r)}(\phi))(\bar{f}_{j_1} \times \cdots \times \bar{f}_{j_r}),$$

where $r \le n$ and $j_1, \ldots, j_r$ are distinct localization indices. Here $P^{(r)}$ denotes the rth derivative of P, and $\bar{f}_j \equiv C f_j$. The monomials in $P^{(r)}$ are of degree at most $n - r$, so if we use Lemma 12.5.3, we are interested in the local $L_{p(r)}$ norms of the kernel with $p(r) \le n/r$. Since the local L_p norms increase with p, we set $p(r) = n/r$ without loss of generality.

Lemma 12.5.4. *Given* $\alpha < m$, *there exists* $c < \infty$ *such that for all* $r \leq n$,

$$\sum_{i \in Z^2} e^{\alpha(|i-j_1| + \cdots + |i-j_r|)} \|\bar{f}_{j_1} \cdots \bar{f}_{j_r}\|_{L_{p(r)}}(\Delta_i)$$

$$\leq c \|f_{j_1}\|_{L_1} \cdots \|f_{j_r}\|_{L_1}. \qquad (12.5.13)$$

PROOF. By Hölder's inequality

$$\|\bar{f}_{j_1} \cdots \bar{f}_{j_r}\|_{L_{p(r)}(\Delta_i)} \leq \prod_{s=1}^{r} \|\bar{f}_{j_s}\|_{L_n(\Delta_i)}. \qquad (12.5.14)$$

We claim that

$$\|\bar{f}_j\|_{L_n(\Delta_i)} \leq O(1)e^{-m|i-j|} \|f_j\|_{L_1}, \qquad (12.5.15)$$

Inserting this bound in (12.5.14) and summing yields (12.5.13). For $|i - j| > 1$, the bound (12.5.15) follows from

$$|\bar{f}_j(x)| \leq O(1)e^{-m|x-j|} \|f_j\|_{L_1},$$

as in (12.5.12). For $|i - j| \leq 1$, we use the Hausdorff–Young and Hölder inequalities to establish

$$\|\bar{f}_j\|_{L_n(\Delta_i)} \leq \|\bar{f}_j\|_{L_n(R^2)} \leq \|\tilde{C}_\varnothing f_j\|_{L_{n'}}$$

$$\leq \|\tilde{C}_\varnothing\|_{L_{n'}} \|\hat{f}_j\|_{L_\infty} \leq \|\tilde{C}_\varnothing\|_{L_{n'}} \|f_j\|_{L_1}$$

$$\leq O(1)\|f_j\|_{L_1}. \qquad (12.5.16)$$

Here $n' = n/(n-1)$ and $\tilde{C} = \tilde{C}_\varnothing = \text{const} (p^2 + m^2)^{-1} \in L_p$ for $p > 1$. Since $n < \infty$, it follows that $1 < n'$ and (12.5.16) holds. This completes the proof of the lemma.

Returning to the proof of the theorem, let $n(\Delta)$ denote the number of contractions to $P^{(r)}$ vertices in Δ, where $r = 1, 2, \ldots, n$. Let us enumerate these $n(\Delta)$ contractions $k = 1, 2, \ldots, n(\Delta)$ in order of increasing distance to Δ. Since each square Δ' contains at most one F_J vertex, the distance d_k of the kth contraction satisfies

$$k \leq \text{const } d_k^2. \qquad (12.5.17)$$

Each contraction to Δ is accompanied by an exponential decay $O(1)e^{(m-\varepsilon) d_k}$, as seen in (12.5.12) and the lemma above. The product of these factors is thus bounded by

$$\prod_{k=1}^{n(\Delta)} (O(1)e^{-(m-\varepsilon) d_k}) \leq \prod_{k=1}^{n(\Delta)} O(1)e^{-\text{const } k^{1/2}}$$

$$\leq e^{\text{const } n(\Delta) - \text{const } n(\Delta)^{3/2}}$$

$$\leq \text{const } e^{-\text{const } n(\Delta)^{3/2}}. \qquad (12.5.18)$$

Furthermore, the number of terms with the same set of contraction indices is bounded by $\prod_\Delta n(\Delta)!^{\text{const}}$. Also, use of Lemma 12.5.3 to bound the $P^{(r)}$ monomials gives rise to a further factor $\prod_\Delta n(\Delta)!^{\text{const}}$. Since $n(\Delta)! \leq \exp(n(\Delta) \ln n(\Delta))$, both factors are dominated by the convergence from (12.5.18). Let $J_e \equiv \{j : \phi(f_j)$

contracts to exponent}. The above estimates yield for contractions to the exponent

$$\sum_{\text{contractions}} \sum_{\text{localizations}} \|\text{kernels}\|_{L_{p(r)}} \prod_{\Delta} n(\Delta)!^{\text{const}} \leq \prod_{j \in J_e} \text{const} \|f_j\|_{L_1}. \quad (12.5.19)$$

We estimate G vertices which remain after contraction of F vertices using Lemma 12.5.3. Since each $I\backslash J$ square has at least two G vertices,

$$\left| \int F_J \, G_{I\backslash J} \, d\mu \right| \leq \prod_{j \in J} (\text{const} \|f_j\|_{L_1}) \times \prod_{i \in I\backslash J} (\text{const} \|f_i\|_{L_p}^2). \quad (12.5.20)$$

Since $p \leq 2$ and $\|f_i\|_{L_p} \leq 1$, $\|f_i\|_{L_p}^2 \leq \|f_i\|_{L_p}^p$. Increasing the bound (12.5.20) using

$$\|f_j\|_{L_1}, \|f_j\|_{L_p}^p \leq \|f_j\|_{L_1} + \|f_j\|_{L_p}^p,$$

we obtain (12.5.9) as desired. This completes the proof of Theorem 12.5.1.

The proof of Theorem 12.2.2 is similar. For $n/2 < j < n$, i.e., for $2 < p$, it is necessary to integrate the G vertices by parts also.

PART III

THE PHYSICS OF QUANTUM FIELDS

Part III concentrates on properties of the solutions, i.e., on the physics of quantum fields and of statistical mechanics. The individual chapters may be read independently, though the material here is less complete and is written at a more advanced level than Parts I and II. We begin with a discussion of the particle interpretation of field theory: the scattering matrix, the bound state spectrum, and the asymptotic completeness question. In Chapter 16 we present a complete proof of the existence of phase transitions for ϕ^4 quantum fields. We follow this with a report in Chapter 17 on the state of knowledge of the ϕ^4 critical point. In Chapter 18 we give a second existence proof for $d = 2$ quantum fields. Here we use expansion methods (cluster expansions) to analyze the infinite volume limit. Within their domain of convergence these expansions not only provide an alternative to the multiple reflection bounds of Part II, but in addition they establish exponential clustering of correlations and thereby reveal properties of the mass spectrum. In principle these expansions allow a complete understanding of the low-lying energy states of the theory. In Chapter 19 we establish the connection between Minkowski and Euclidean fields which was stated in Chapter 6. One aspect of this discussion shows the relation between multiple vacuum states in quantum theory and the decomposition of an equilibrium statistical mechanics state into its pure phases. In fact, the results of Chapters 16–19 also pertain to properties of equilibrium states and the transfer matrix in statistical physics, as described in these chapters. Chapter 20 provides an introduction to the literature for major issues outside the scope of this book.

CHAPTER 13

Scattering Theory: Time-Dependent Methods

13.1 Introduction

Scattering is concerned with the asymptotic behavior, as $t \to \pm\infty$, of the solutions

$$\theta(t) = e^{-itH}\theta(0) \qquad (13.1.1)$$

of the Schrödinger equation

$$i\dot{\theta} = H\theta.$$

Because $\theta(t) = e^{it\omega}\theta(0)$ for θ an eigenvector or generalized eigenvector of H with eigenvalue ω, the problem reduces to spectral analysis of H. For a translation-invariant problem, H and $\mathbf{P}$ commute, and we seek a joint spectral resolution of H and $\mathbf{P}$. However, the joint spectrum is not simple, and so to obtain a well-defined problem, we return to (13.1.1). At large $|t|$, the wavefunction $\theta(t)$ separates into isolated noninteracting single particles and clusters (bound states). The separation occurs in x space, and implies the existence of a free energy operator H_0, to describe the noninteracting motion of the isolated particles and bound states. For $t \to \pm\infty$, we find generalized eigenvectors θ of H so that $\theta(t)$ is asymptotic to the free dynamics e^{-itH_0}. These eigenvectors define the in/out spectral resolution of H, and the unitary operator which interchanges these two spectral resolutions has the property of mapping $t \to -\infty$ asymptotes onto $t \to +\infty$ asymptotes. It is called the S matrix.

Specializing now to a relativistic field theory, the in/out asymptotes are described by free fields (see Chapter 6), labeled, for example, $\phi_{\text{in/out}}$. The in/out fields act on a Fock space $\mathscr{F}$, and vectors $\theta_{\text{in/out}} \in \mathscr{F}$ label the large $|t|$ asymptotes of the quantum field. In case the theory contains several types of particles (elementary particles and bound states), there will be several in/out fields, acting on a tensor product Fock space $\mathscr{F}_1 \otimes \mathscr{F}_2 \otimes \cdots = \mathscr{F}$. Because of Lorentz invariance, the energy–momentum spectrum lies on hyperboloids of constant

$$M = (P_0^2 - \mathbf{P}^2)^{1/2}$$

(see Figure 13.1). $\mathscr{H}_\mu$ denotes the eigenspace of M for eigenvalue μ. In the multiparticle space $\mathscr{H}_{\geq 2m}$, we have drawn the particle–bound-state threshold $M = m + m_b$ and the three particle threshold $M = 3m$.

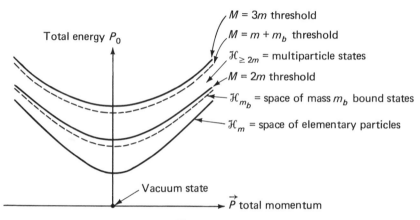

Figure 13.1

Asymptotic completeness is then the statement that all the states of the quantum field are labeled by the free field asymptotes in $\mathscr{F}$. To be more precise, the discrete spectrum of M (and of the spin operator and other "internal" quantum numbers of the theory) determine the single particle subspaces of the free field factors $\mathscr{F}_1$, $\mathscr{F}_2$, ..., in $\mathscr{F}$, and then $\mathscr{F}$ describes multiparticle states formed with these masses, spins, and internal quantum numbers. If asymptotic completeness holds, then there are no other states in the theory beyond these multiparticle states.

In Section 13.2 we orient our discussion by presenting multiparticle potential scattering. In particular, we construct the wave operators and S matrix, in order to present scattering theory in its best-known form. For the remainder of the chapter we describe time-dependent methods for quantum field theory. The generalization of the methods to yield a wave operator and an S matrix for field theory comprises the Haag–Ruelle construction.

As input to the Haag–Ruelle theory we assume the existence of an isolated eigenvalue[1] m for the mass operator M. In addition we assume

---

[1] This assumption is unsuitable in the presence of zero mass particles. While much progress has been made in understanding the scattering of zero mass particles, an adequate generalization of the time-dependent methods still needs to be developed.

that the field ϕ satisfies the Wightman axioms. The output is the construction of multiparticle states. These states $\mathcal{H}_{\text{in/out}}$ actually represent the asymptotic incoming or outgoing multiparticle states for scattering of the mass m particles. In case the field theory describes many different species of particles or bound states, the Haag–Ruelle construction can be applied to yield multiparticle states with n_1 particles of type 1, n_2 particles of type 2, etc. The only additional input required is the existence of an isolated eigenvalue m_i of M for each such particle or bound state.

The spaces $\mathcal{H}_{\text{in/out}}$ constructed by the Haag–Ruelle theory are subspaces of $\mathcal{H}$. They have a natural Fock space structure, and are isomorphic to the space $\mathcal{F}$ of scattering state labels. In fact we can identify $\mathcal{F}$ with $\mathcal{H}_{\text{in}}$ or with $\mathcal{H}_{\text{out}}$. A consequence of the assumptions is that $\mathcal{H}_{\text{in}} = \mathcal{H}_{\text{out}}(=\mathcal{F})$. Thus the Haag–Ruelle theory gives a concrete representation for $\mathcal{H}_{\text{in/out}}$.

The question of asymptotic completeness alluded to above is the question whether $\mathcal{H}_{\text{in/out}} = \mathcal{H}$. On physical grounds we expect completeness. In other words, we expect that every state of a physical system can be regarded either as composed of particles (including bound states) or else as decaying into such a state as time progresses. An example of a situation where asymptotic completeness fails is a Fock space restricted to vectors with an even number of particles. Thus single particle states never occur in such a theory. Were such states (say soliton pair states) to occur in the $\mathcal{H}_{\text{in}}$ of a field theory on $\mathcal{H}$, we might conclude that the original $\mathcal{H}$ needed to be enlarged to describe the physics adequately. From the point of view of Haag–Ruelle theory (or physics) such examples are pathological. On the other hand, it is a very deep (and open) mathematical question to establish $\mathcal{H}_{\text{in/out}} = \mathcal{H}$, even for the concrete examples of quantum fields constructed in Part II. Partial results (at low energies and weak coupling) are described in Chapter 14.

The S matrix itself can be expressed in terms of time-ordered Wightman functions (the Lehmann–Symanzik–Zimmermann formalism) and thus is in some sense effectively computable. The study of the particle and bound state spectrum (i.e., of the $\mathcal{H}_m$), as well as asymptotic completeness, uses the Bethe–Salpeter equation. This equation is the field theory counterpart to time-independent methods in potential scattering. In this analogy, the Bethe–Salpeter kernel K corresponds to the potential V in potential scattering. The analogy becomes precise in the nonrelativistic limit, $c \to \infty$, in which K converges to V, after appropriate rescaling.

In field theory, K is regarded as a derived quantity, since it does not occur in the energy operator, nor in the equations of motion. In keeping with this tradition, one is able to derive, at least for weak coupling, the required properties of K, starting from the (bare) interaction, e.g. $:\phi^4:$. This derivation uses high temperature cluster expansion methods; cf. Chapter 18.

13.2 Multiparticle Potential Scattering

Consider as a total energy operator

$$H = H_0 + V = \sum_{j=1}^{n} \frac{1}{2m_j} p_j^2 + \sum_{i \neq j} V_{ij}(q_i - q_j)$$

acting on $L_2(R^{3n})$. We introduce a cluster decomposition

$$\mathcal{D} = \{C_1, \ldots, C_m\}.$$

$\mathcal{D}$ is by definition a partition of $\{1, \ldots, n\}$ into disjoint subsets. Let us define the total cluster momentum, mass, energy, center of mass etc., by

$$P_K = \sum_{j \in C_K} p_j, \qquad M_K = \sum_{j \in C_k} m_j,$$

$$H_K = \sum_{j \in C_K} \frac{1}{2m_j} p_j^2 + \sum_{i \neq j \in C_K} V_{ij}(q_i - q_j) = H_{0,K} + V_K,$$

$$H_{\mathcal{D}} = \sum_{K=1}^{m} H_K, \qquad Q_K = \frac{\sum_{j \in C_K} m_j q_j}{M_K}.$$

In order to specify an asymptote, i.e., an in/out scattering state, we need to know

1. the cluster decomposition $\mathcal{D}$,
2. the free motion of the center of mass of each cluster,
3. the internal motion of each cluster.

To separate the internal motion from the center of mass motion, we make a change of coordinates within each cluster. Q_K is the center of mass coordinate for the cluster C_K. Let $q_{K,\text{rel}}$ denote $|C_K| - 1$ independent coordinates selected from the span of $\{q_i - q_j : i, j \in C_K\}$.

The linear transformation $A: \{q_i\} \to \{Q_K, q_{K,\text{rel}}\}$ induces a linear map A^{*-1} from p_i to momentum coordinates conjugate to $\{Q_K, q_{K,\text{rel}}\}$. We can avoid some tedious linear algebra by observing that A^{*-1} is determined by the fact that the p's and q's transform canonically, i.e., with preservation of commutator or Poisson brackets. Since

$$\{Q_K, P_K\} \equiv \sum \frac{\partial Q_K}{\partial q_i} \frac{\partial P_K}{\partial p_i} - \frac{\partial Q_K}{\partial p_i} \frac{\partial P_K}{\partial q_i} = 1,$$

$$\{q_i - q_j, P_K\} = 0,$$

P_K is conjugate to the center of mass coordinate Q_K. Let $p_{K,\text{rel}}$ denote the momentum variables conjugate to $q_{K,\text{rel}}$. Then H_0 is a bilinear form in P_K, $p_{K,\text{rel}}$, and the dependence on P_K is also computed using Poisson brackets, in the two coordinate systems $\{Q_K, q_{K, \text{rel}}\}$ and $\{q_i\}$. Since

$$\frac{\partial H_{0, K}}{\partial P_K} = \{Q_K, H_{0, K}\}$$

$$= \frac{\sum m_i \dfrac{\partial H_{0, K}}{\partial p_i}}{\sum m_i} = \frac{P_K}{M_K},$$

it follows that

$$H_{0, K} = \frac{1}{2M_K} P_K^2 + h_{0, K}(p_{K, \text{rel}})$$

for some quadratic form $h_{0, K}$. Choosing a sequence of refinements of cluster decompositions, with separation of center of mass at each stage, leads to what are known as Jacobi coordinates.

In summary, we have shown that in the coordinates $Q_K, q_{K, \text{rel}}$,

$$H_K = \tfrac{1}{2} M_K P_K^2 + h_{0, K}(p_{K, \text{rel}}) + V_K(q_{K, \text{rel}}),$$

and so in the tensor product decomposition

$$\mathscr{H}_K = L_2(R^{3|C_K|}, dq) = L_2(R^3, dQ_K) \otimes L_2(R^{3(|C_K|-1)}, dq_{\text{rel}}),$$

$(2M_K)^{-1} P_K^2$ acts on the first factor and $h_K \equiv h_{0, K} + V_K$ acts on the second factor. We now define a bound state for the cluster C_K as a discrete eigenvector ϕ_K for h_K:

$$h_K \phi_K = E_K \phi_K.$$

The choice of ϕ_K specifies the internal motion of the cluster. If the eigenvalue E_K has multiplicity greater than one, then we choose ϕ_K from some orthonormal basis of this eigenspace. A channel

$$\alpha = \{\mathscr{D}, \phi_1, \ldots, \phi_m\}$$

is defined as a cluster decomposition together with a bound state ϕ_K for each cluster C_K having two or more elements.

For each channel α, we define an isometric mapping U_α,

$$U_\alpha : \mathscr{H}_{\mathscr{D}} \equiv L_2(R^{3m}) \to L_2(R^{3n})$$

as follows:

$$U_\alpha f = f(Q_{K_1}, \ldots, Q_{K_m}) \prod_{K=1}^m \phi_K(q_{K, \text{rel}}).$$

Here f specifies the center of mass state for each cluster. $\sum_\alpha U_\alpha$ is a crude approximation to the wave operator, which we now define. Let

$$W_{\mathscr{D}}(t) = e^{itH} e^{-itH_{\mathscr{D}}},$$

$$W_{\mathscr{D}}^{\pm} = \operatorname*{st\,lim}_{t \to \pm\infty} W_{\mathscr{D}}(t),$$

$$W^{\pm} = \sum_\alpha W_{\mathscr{D}}^{\pm} U_\alpha.$$

Theorem 13.2.1. $W^{\pm}$ *is an isometric map from* $\mathscr{H}' = \sum_{\alpha} \oplus \mathscr{H}_{\mathscr{D}}$ *into* $\mathscr{H} = L_2(R^{3n}, dq)$, *provided each* $V_{ij} \in L_2 + L_{3-\varepsilon}$. *In particular the strong limit defining* $W_{\mathscr{D}}^{\pm}$ *exists, and the ranges of the operators* $W_{\mathscr{D}}^{\pm} U_{\alpha}$, *for distinct* α, *are orthogonal.*

Lemma 13.2.2. *On* $L_2(R^d)$, *the kernel of* $e^{-it\Delta}$ *is*

$$(-4\pi it)^{-d/2} e^{-ix^2/4t}.$$

PROOF. This follows by analytic continuation, starting with the solution operator $e^{t\Delta}$ for the heat equation.

Lemma 13.2.3. *As a map from* $L_1(R^d)$ *to* $L_{\infty}(R^d)$,

$$\|e^{-itP^2}\| \leq t^{-d/2}.$$

PROOF. $\|e^{-itP^2}\| = \|(-4\pi it)^{-d/2} e^{-x^2/4it}\|_{L_{\infty}}.$

Lemma 13.2.4. *As an operator on* $L_2(R^d)$, $e^{-itP^2} \to 0$ *weakly as* $t \to \pm\infty$.

PROOF. For θ_1, $\theta_2 \in L_1 \cap L_2$, $\langle \theta_1, e^{-itP^2}\theta_2 \rangle \to 0$ by Lemma 13.2.3. Since e^{-itP^2} is unitary, weak convergence on a dense set of vectors is sufficient.

Lemma 13.2.5. $W_{\mathscr{D}}^{\pm} U_{\alpha} \mathscr{H}_{\alpha} \perp W_{\mathscr{D}'}^{\pm} U_{\beta} \mathscr{H}_{\beta}$ *if* $\alpha = \{\mathscr{D}, \phi\} \neq \beta = \{\mathscr{D}', \phi'\}$.

PROOF. We assume here the existence of the strong limit $W_{\mathscr{D}}^{\pm}$. Let $\psi_{\alpha} \in \mathscr{H}_{\mathscr{D}}$, $\psi_{\beta} \in \mathscr{H}_{\mathscr{D}'}$. If $\mathscr{D} = \mathscr{D}'$ then

$$\langle W_{\mathscr{D}}^{\pm} U_{\alpha}\psi_{\alpha}, W_{\mathscr{D}'}^{\pm} U_{\beta}\psi_{\beta} \rangle = \langle U_{\alpha}\psi_{\alpha}, U_{\beta}\psi_{\beta} \rangle$$
$$= \langle \psi_{\alpha}, \psi_{\beta} \rangle \langle \phi, \phi' \rangle = 0,$$

since in this case ϕ and ϕ' are orthogonal eigenvectors of the relative energy portion of $H_{\mathscr{D}}$. If $\mathscr{D} \neq \mathscr{D}'$ then

$$|\langle W_{\mathscr{D}}^{\pm} U_{\alpha}\psi_{\alpha}, W_{\mathscr{D}'}^{\pm} U_{\beta}\psi_{\beta} \rangle| = \lim_{t \to \pm\infty} |\langle e^{itH} e^{-itH_{\mathscr{D}}} U_{\alpha}\psi_{\alpha}, e^{itH} e^{-itH_{\mathscr{D}'}} U_{\beta}\psi_{\beta} \rangle|$$
$$= \lim_{t \to \pm\infty} |\langle e^{-itH_{\mathscr{D}}} U_{\alpha}\psi_{\alpha}, e^{-itH_{\mathscr{D}'}} U_{\beta}\psi_{\beta} \rangle|$$
$$= \lim_{t \to \pm\infty} |\langle e^{-it\sum_{\kappa} P_{\kappa}^2} U_{\alpha}\psi_{\alpha}, e^{-it\sum_{\kappa'} P_{\kappa'}^2} U_{\beta}\psi_{\beta} \rangle|$$
$$= \lim_{t \to \pm\infty} |\langle U_{\alpha}\psi_{\alpha}, e^{it(\sum_{\kappa} P_{\kappa}^2 - \sum_{\kappa'} P_{\kappa'}^2)} U_{\beta}\psi_{\beta} \rangle|.$$

A change of variables diagonalizes the exponent, so that

$$\sum_{\kappa} P_{\kappa}^2 - \sum_{\kappa'} P_{\kappa'}^2 = \lambda \frac{\partial^2}{\partial w_1^2} + \cdots$$

with $\lambda \neq 0$. Then

$$e^{it(\sum_{\kappa} P_{\kappa}^2 - \sum_{\kappa'} P_{\kappa'}^2)} = e^{-it\lambda \partial^2/\partial w_1^2} \otimes U_1(t),$$

with $U_1(t)$ unitary. By Lemma 13.2.4, the limit above is zero, and the proof is complete.

Lemma 13.2.6. Let $V_{ij} \in L_2 + L_\infty$, and $\theta \in \mathscr{D}(H)$. Then $W_{\mathscr{D}}(t)\theta$ is strongly differentiable and

$$\frac{d}{dt} W_{\mathscr{D}}(t)\theta = ie^{itH} V'_{\mathscr{D}} e^{-itH_{\mathscr{D}}} \dot{\theta},$$

where $V'_{\mathscr{D}} = H - H_{\mathscr{D}}$ is the sum of all intercluster potentials.

PROOF. V is a Kato perturbation of H_0, and similarly $V'_{\mathscr{D}}$ is a Kato perturbation of H and $H_{\mathscr{D}}$. Thus $\mathscr{D}(H) = \mathscr{D}(H_{\mathscr{D}}) = \mathscr{D}(H_0)$. The rest is standard operator theory.

Lemma 13.2.7. Let $\mathscr{M} \subset \mathscr{H}$ be the set of states for which

$$\|V'_{\mathscr{D}} e^{-itH_{\mathscr{D}}}\theta\| \le O(|t|^{-3/2})$$

as $t \to \pm\infty$. Then $\mathscr{M} \cap \mathscr{D}(H)$ is dense in $\mathscr{H}$.

PROOF OF THEOREM 13.2.1. Since $W_{\mathscr{D}}(t)$ is unitary, it is sufficient to prove convergence for θ in the dense set $\mathscr{M} \cap \mathscr{H}$. For such θ,

$$\|[W_{\mathscr{D}}(t_1) - W_{\mathscr{D}}(t_2)]\theta\| \le \int_{t_1}^{t_2} \|V'e^{-itH}\theta\| \, dt$$

$$\le \int_{t_1}^{t_2} O(|t|^{-3/2}) \, dt,$$

which tends to zero as $t_1, t_2 \to \pm\infty$.

PROOF OF LEMMA 13.2.7. To simplify the proof, we only consider the case $V_{ij} \in L_2$. Let

$$\theta(q) = \prod_{K=1}^{m} f_K(Q_K)g_K(q_{K,\,\mathrm{rel}}).$$

With f_K, $g_K \in \mathscr{S}$, we assert $\theta \in \mathscr{M}$, to complete the proof.

Let V_{jl} be a single term contributing to $V'_{\mathscr{D}}$, so that $j \in C_K$, $l \in C_{K'}$ with $K \ne K'$. Let

$$Q_{KK'} = \frac{M_K Q_K + M_{K'} Q_{K'}}{M_K + M_{K'}}, \qquad q_{KK',\,\mathrm{rel}} = Q_K - Q_{K'}.$$

As before,

$$\frac{1}{2M_K} P_K^2 + \frac{1}{2M_{K'}} P_{K'}^2 = \frac{1}{2(M_K + M_{K'})} \frac{\partial^2}{\partial Q_{KK'}^2} + \frac{1}{2m} \frac{\partial^2}{\partial q_{KK',\,\mathrm{rel}}^2}$$

with $m = M_1 M_2/(M_1 + M_2)$ and

$$V_{jl} = V_{jl}(q_{KK',\,\mathrm{rel}} + L(q_{K,\,\mathrm{rel}}, q_{K',\,\mathrm{rel}})),$$

where L is some linear functional. Then

$$\| V_{jl} e^{-itH_{\mathscr{D}}} \theta \|^2 = \text{const} \| V_{jl} e^{-it(H_K + H_{K'})} f_K f_{K'} g_K g_{K'} \|^2$$

$$= \text{const} \int |V_{jl}|^2 |\psi_1(Q_{KK'}, q_{KK', \text{rel}})|^2 |\psi_2(q_{K, \text{rel}}, q_{K', \text{rel}})|^2 \, dq$$

where const is independent of t and

$$\psi_1 = e^{-i(t/2m)\partial^2/\partial q_{KK', \text{rel}}^2} f_K f_{K'},$$

$$\psi_2 = e^{-it(h_K + h_{K'})} g_K g_{K'}.$$

Thus making a change of variables and applying Hölder's inequality,

$$\| V_{jl} e^{-itH_{\mathscr{D}}} \theta \|^2 \le \text{const} \| V_{jl} \|_2^2 \| \psi_2 \|_2^2 \left\| \sup_{q_{KK', \text{rel}}} \psi_1 \right\|_{L_2(Q_{KK'})}^2.$$

Since the Schrödinger evolution operator in ψ_1 has a norm $t^{-3/2}$ as an operator from L_1 to L_∞,

$$\left\| \sup_{q_{KK', \text{rel}}} \psi_1 \right\|_{L_2(Q_{KK'})}^2 \le O(|t|^{-3}) \int \left[\int |\chi(Q_{KK'}, q_{KK', \text{rel}})| \, dq_{KK', \text{rel}} \right]^2 dQ_{KK'},$$

where

$$\chi(Q_{KK'}, q_{KK', \text{rel}}) = f_K(q_K) f_{K'}(q_{K'}) \in \mathscr{S}.$$

This completes the proof.

13.3 The Wave Operator for Quantum Fields

In this section we define a wave operator appropriate for quantum fields. As in Section 13.1, let $\mathscr{H}_{m_1}$, $\mathscr{H}_{m_2}$, ..., be the eigenspaces of the mass operator M with eigenvalues m_1, m_2, ... Each $\mathscr{H}_{m_i}$ is a representation space for a representation of the Lorentz group. Let $\mathscr{F}_i$ be the free field Fock space with single particle space $\mathscr{H}_{m_i}$, and let $\mathscr{F} = \otimes_i \mathscr{F}_i$. Then $\mathscr{F}$ can be considered the space of labels of the asymptotic states.

The wave operator we construct can be considered as a mapping from the set of labels $\mathscr{F}$ to $\mathscr{H}$. The S matrix it defines acts on $\mathscr{H}$ (and in particular maps $\mathscr{H}_{\text{out}}$ onto $\mathscr{H}_{\text{in}}$). At the end of the construction we will identify $\mathscr{F}$ with $\mathscr{H}_{\text{in/out}}$, so the scattering matrix S is also defined on the label space $\mathscr{F}$.

To construct a wave operator we begin with a crude approximation $U : \mathscr{F} \to \mathscr{H}$, as in Section 13.2. Let H_0 be the free field energy operator on $\mathscr{F}$, and define

$$W(t) = e^{itH} U e^{-itH_0}. \tag{13.3.1}$$

We study below a map U and conditions for which

$$W^{\pm} = \lim_{t \to \pm\infty} W(t) \tag{13.3.2}$$

exists. We find that $W^{\pm}$ intertwine H and H_0, and that

$$S = W^+(W^-)* \qquad (13.3.3)$$

is a unitary transformation on the space $\mathscr{H}_{\text{in/out}} = \text{Range } W^+ = \text{Range } W^-$.

The operator U is called a solution to the one body problem. Namely, we find a polynomial ψ_m in the (physical) field ϕ such that

$$0 \neq \psi_m \Omega \in \mathscr{H}_m.$$

For simplicity suppose that $\mathscr{H}_m$ is an irreducible spin zero representation space for the Lorentz group. (See [Hepp, 1966a] for the general case.) Let

$$\psi_m(\mathbf{x}) = e^{i\mathbf{x} \cdot \mathbf{P}} \psi_m e^{-i\mathbf{x} \cdot \mathbf{P}}. \qquad (13.3.4)$$

Let ϕ_{m_i} denote the free field in $\mathscr{F}_i$, and let $^*\phi_{m_i}$ denote a possible time derivative of ϕ_{m_i}. The definition of U is given on polynomials of $^*\phi_{m_i}$, applied to Ω. In particular

$$U^*\phi_{m_1}(0, \mathbf{x}_1) \cdots {}^*\phi_{m_n}(0, \mathbf{x}_n)\Omega = {}^*\psi_{m_1}(0, \mathbf{x}_1) \cdots {}^*\psi_{m_n}(0, \mathbf{x}_n)\Omega, \qquad (13.3.5)$$

where Ω denotes the Fock vacuum on the left, and on the right it denotes the physical vacuum in $\mathscr{H}$.

The solution to the one body problem comes in several steps. It is often the case that the particle and bound state hyperboloids $M = \mu_i$ are isolated eigenvalues. In the case of some symmetry (e.g., $\phi \to -\phi$ for an even theory) or a superselection rule, the $M = \mu_i$ hyperboloids may be isolated only with respect to spectra with the same symmetry or superselection value. The physical basis for this idea is that otherwise the particle would be energetically unstable relative to decay into subparticles. Such unstable objects do occur in physics; they are resonances, and the corresponding spectra occur off the physical sheet and not as eigenvalues of M. At worst the theory contains massless particles, or more generally a spectrum which fills the entire forward cone. In this case the scattering theory is much more difficult, and in fact is incompletely understood. We rule out that possibility at the start, in order to avoid technical details. (It is necessary, of course, to understand scattering in the presence of photons and neutrinos, both of which are believed to be massless.)

In superrenormalizable field theories far from critical points (at weak coupling, i.e., in the approximately Gaussian region) it can be proved that the particle and bound state spectra m_i are isolated, in the above sense.

Because the vectors $P(\phi)\Omega$ are dense in $\mathscr{H}$ for P a polynomial in the field, we can choose a polynomial ψ_m^{prelim} with the property

$$\psi_m^{\text{prelim}}\Omega \perp \mathscr{H}_m.$$

Convolute ψ_m^{prelim} with a function h_m whose Fourier transform $h_{\tilde{m}}$ satisfies $h_{\tilde{m}} = h_{\tilde{m}}^- = f(p^2)$. Suppose further for $\sigma(M)$, the spectrum of the mass operator $M = (H^2 - P^2)^{1/2}$, that

$$\text{suppt } \tilde{h}_m \cap \sigma(M) = \{M = m\}. \tag{13.3.6}$$

This defines

$$\psi_m = h_m * \psi_m^{\text{prelim}}. \tag{13.3.7}$$

For simplicity, let $\psi_m = \psi_m^*$ (neutral particles). The existence of h_m follows from the assumption that m is an isolated point in the spectrum of M, and ψ_m solves the one body problem. There is an arbitrary multiplicative constant in h_m, which is chosen so that $\phi_{\text{in/out}}$ (defined below) equals a canonically renormalized free field, rather than just being proportional to one.

Proposition 13.3.1. *Suppose $\psi_m^{\text{prelim}}(t, \mathbf{x})$ is a Schwartz distribution in t, $\mathbf{x}$, and an unbounded operator on $\mathcal{H}$, defined on a ψ_m^{prelim}-invariant and Lorentz-invariant domain. Then $\psi_m(t, \mathbf{x})$ is a Schwartz distribution in $\mathbf{x}$, C^∞ in t, on the same domain, and this domain is ψ_m-invariant. In particular, the ψ_m equal time expectation values exist, satisfy*

$$\langle \Omega, \psi_{m_1}(t, \mathbf{x}) \cdots \psi_{m_n}(t, \mathbf{x})\Omega \rangle \in \mathscr{S}'(R^{n(d-1)}), \tag{13.3.8}$$

and are independent of t.

PROOF. With $f \in \mathscr{S}(R^{d-1})$, a test function for ψ_m, we have

$$f^\sim(\mathbf{p})\tilde{h}_m(p)e^{ip_0t} \in \mathscr{S}(R^d),$$

a test function for ψ_m^{prelim}. The equal time expectations exist, and the same holds for their time derivatives.

Remark. A typical case is $\psi_m^{\text{prelim}} = \phi$, the Wightman field. For simplicity, we assume this below. Then the hypothesis of Proposition 13.3.1 is contained in the Wightman axioms. The normalization hypothesis on h now is a normalization hypothesis on ϕ. It is conventional field strength renormalization, satisfied by $\phi_{\text{ren}} = Z^{-1/2}\phi$ in place of ϕ.

It is useful to represent e^{-itH_0} as an integral operator. This operator will be studied in detail in the next section. Let H_{0, m_i} be the energy operator in $\mathscr{F}_i$, so that $H_0 = \sum_i H_{0, m_i}$. Let $\dot{\phi}_m$ be the derivative of ϕ_m. To simplify notation, let

$$u_m(t, \mathbf{x}) = \begin{pmatrix} \phi_m(t, \mathbf{x}) \\ \dot{\phi}_m(t, \mathbf{x}) \end{pmatrix} = \begin{pmatrix} \phi_m(t, \mathbf{x}) \\ \partial_t\phi_m(t, \mathbf{x}) \end{pmatrix}, \tag{13.3.9}$$

let $v_m(t, \mathbf{x})$ denote the similar vector formed from ψ_m and $\dot{\psi}_m$, and let $G_m(t)$ denote the 2×2 solution matrix for the Cauchy problem,

$$u_m(t, \mathbf{x}) = G_m(t)u_m(0, \mathbf{x}) = e^{itH_0}u_m(0, \mathbf{x})e^{-itH_0}.$$

Thus with tensor product multiplication,

$$W(t)u_{m_1}(0, \mathbf{x}_1) \cdots u_{m_n}(0, \mathbf{x}_n)\Omega$$

$$= e^{itH} U G_{m_1}(-t)u_{m_1}(0, \mathbf{x}_1) \cdots G_{m_n}(-t)u_{m_n}(0, \mathbf{x}_n)\Omega$$

$$= e^{itH} G_{m_1}(-t)v_{m_1}(0, \mathbf{x}_1) \cdots G_{m_n}(-t)v_{m_n}(0, \mathbf{x}_n)\Omega$$

$$= G_{m_1}(-t)v_{m_1}(t, \mathbf{x}_1) \cdots G_{m_n}(-t)v_{m_n}(t, \mathbf{x}_n)\Omega. \qquad (13.3.10)$$

In the following sections we prove

Theorem 13.3.2. *Assume a Wightman field theory with isolated one par-*
ticle spectrum and $\psi_m^{\text{prelim}} = \phi$. *Average* (13.3.10) *with test functions* $f_i(\mathbf{x})$,
$\hat{f}_i(\mathbf{x}) \in \mathscr{S}(R^{d-1})$, *and assume the* f's *are nonoverlapping in velocity space.*
Then (13.3.10) *converges strongly with a rate* $O(t^{-N})$, *for any* N, *as* $t \to \infty$.
The limit operators $W^{\pm}$ *are isometric.*

Remark. By Propositions 13.3.1 and 13.4.1, the operator $W(t)$ is defined
on vectors $u \cdots u\Omega$, as in (13.3.10). The nonoverlapping property is defined
in Section 13.4.

Corollary 13.3.3. *The limit*

$$\phi_{m, \text{in/out}}(t, \mathbf{x}) = W^{\pm} \phi_m(t, \mathbf{x})(W^{\pm})^*$$

is a free field.

Corollary 13.3.4. $HW^{\pm} = W^{\pm}H_0$, *and* H *is the Hamiltonian for the free*
dynamics of $\phi_{m, \text{in/out}}$.

Corollary 13.3.5. *The* S *matrix defined by* (13.3.3) *is unitary on*
$\mathscr{H}_{\text{in}} = \mathscr{H}_{\text{out}} = \text{Range } W^{\pm}$.

PROOF. A deep consequence of the Wightman axioms is the identity $TCP = I$,
where T is the time reversal operator, P is space reversal, and C is charge
conjugation ([Streater and Wightman, 1964], [Jost, 1965]). Because
$C\mathscr{H}_m = \mathscr{H}_m = P\mathscr{H}_m$, T also leaves $\mathscr{H}_m$ invariant. Similarly C and P map the
multiparticle in/out states Range $W^{\pm}$ onto themselves, and thus T does likewise.
However, by definition T interchanges Range W^+ and Range W^-, so these two
spaces coincide. Restricted to $\mathscr{H}_{\text{in}} = \mathscr{H}_{\text{out}} = \text{Range } W^{\pm}$, S is defined by (13.3.3)
as a product of two unitary operators, and thus is unitary.

13.4 Wave Packets for Free Particles

As in Section 13.2, one can prove a $t^{-3/2}$ decay property for solutions of
the free field in $d = 4$ space-time dimensions. However, in the case of
nonoverlapping velocities, there is a decay t^{-N} for all N and all dimensions
$d \geq 2$. It is the latter decay which we establish here. It is used to prove

Theorem 13.3.1 for the superrenormalizable fields in $d = 2$, 3 dimensions, provided the test functions are restricted to have nonoverlapping velocities. The decay takes place outside of cones in x-t space, and thus states that free particles remain essentially within cones defined by their velocities, with the velocities determined by the support in momentum space. Within the velocity cone, we use the following very simple bound.

Proposition 13.4.1. *Let $f(t, \mathbf{x})$ be a solution of the Klein–Gordon equation with data in $\mathscr{S}(R^{d-1})$. Then for any $\mathscr{S}$ norm $\|\cdot\|_{\mathscr{S}}$ and any t derivative ∂_t^j, $j \geq 0$, the function $\mathbf{x} \to \partial_t^j f(t, \mathbf{x})$ has a finite norm $\|\partial_t^j f(t, \cdot)\|_{\mathscr{S}}$ with at most polynomial growth in t.*

PROOF.   The partial Fourier transform, $f^{\sim}(t, \mathbf{p})$, has the form

$$f^{\sim}(t, \mathbf{p}) = (2\pi)^{-(d-1)/2}[e^{-i\mu(\mathbf{p})t}g_+(\mathbf{p}) + e^{i\mu(\mathbf{p})t}g_-(\mathbf{p})]$$

with $\mu = (\mathbf{p}^2 + m^2)^{1/2}$, $g_\pm \in \mathscr{S}$. Since $\mathscr{F}\mathscr{S} = \mathscr{S}$, the proposition follows.

The velocities and momenta are related by the relativistic formulas

$$\mathbf{p} = \pm m \frac{\mathbf{v}}{(1 - v^2)^{1/2}}, \qquad \text{suppt } \mathbf{p} \in g_\pm, \tag{13.4.1}$$

$$\mathbf{v} = \pm \mathbf{p}/\mu(\mathbf{p}). \tag{13.4.2}$$

Let $\mathscr{V}$ be a set of velocities. The forward velocity cone $\mathscr{C}_{\mathscr{V}}$ is defined to be

$$\mathscr{C}_{\mathscr{V}} = \{t, \mathbf{x} \in R^d \,|\, t \geq 0, \ \mathbf{x}/t \in \mathscr{V}\} = \{t, t\mathscr{V} \,|\, t \geq 0\}.$$

We choose $\mathscr{V}$ closed, and to contain a neighborhood of the velocities defined by $\mathbf{p} \in \text{supp } g_\pm$.

Proposition 13.4.2. *With f and $\mathscr{V}$ as above, f is rapidly decreasing in t, outside of $\mathscr{C}_{\mathscr{V}}$. In other words*

$$\sup_{\{\mathbf{x} : \mathbf{x}/t \notin \mathscr{V}\}} (1 + |\mathbf{x}|)^L |f(t, \mathbf{x})| \leq O(t^{-N})$$

for all L, N, as $t \to \infty$, and the same bound holds for arbitrary derivatives of f.

PROOF.   It is sufficient to consider g_+ alone. Then

$$f(t, t\mathbf{v}) = (2\pi)^{(d-1)/2} \int e^{it(-\mu + \mathbf{v} \cdot \mathbf{p})} g_+(\mathbf{p}) \, d\mathbf{p}$$

$$= \int e^{its} h_v(s) \, ds$$

where

$$h_v(s) = (2\pi)^{(d-1)/2} \int \delta(s + \mu(\mathbf{p}) - \mathbf{v} \cdot \mathbf{p}) g_+(\mathbf{p}) \, d\mathbf{p}.$$

The geometry is easiest to visualize in energy-momentum space, where the integration can be regarded as lying on the intersection of the hyperboloid $p_0 = \mu(\mathbf{p})$ with the hyperplane $s = -p_0 + \mathbf{v} \cdot \mathbf{p} = ((1, \mathbf{v}), p)$ (Lorentz inner product). The only possible loss of regularity in $h_v(s)$ occurs at s for which the hyperplane and hyperboloid are tangent. We assert that this condition of tangency determines exactly the relativistic relation between $\mathbf{v}$ and $\mathbf{p}$ above. Consequently for $\mathbf{v} \notin \mathscr{V}$, the $\mathbf{p}$ integration, restricted to suppt g_+, is bounded away from tangency, uniformly in s. It follows that h and its s derivatives are in $L_1(ds)$, and thus $f(t, t\mathbf{v})$ has the desired decay rate. For $\mathbf{v}$ bounded, this proves the proposition. For $\mathbf{v}$ unbounded (e.g., $|\mathbf{v}| > 1$), we use finite propagation speed and the fact that the data are in $\mathscr{S}$.

To prove the assertion, note that tangency requires $v^2 < 1$. Assuming this, we check that $\hat{p} = m\mathbf{v}/(1 - v^2)^{1/2}$, $s = -m(1 - v^2)^{1/2}$ defines a point $(\mu(\hat{p}), \hat{p})$ of intersection for the hyperplane and the hyperboloid. A general point on the hyperplane is $p = (\mu(\hat{p}), \hat{p}) + p^\perp$ with $(p^\perp, (1, \mathbf{v})) = 0$. Then $p^\perp$, being Lorentz-orthogonal to the timelike vector $(1, \mathbf{v})$, must be spacelike, i.e., $(p^\perp, p^\perp) \geq 0$. Using (13.4.2), we show that $-(p, p) < m^2$ if $p^\perp \neq 0$, so that $p^\perp = 0$ is the unique point of intersection of the hyperplane and the hyperboloid. This completes the proof.

We also consider a one parameter family of functions $f^{(t)} \in \mathscr{S}(R^d)$, defined for $f \in \mathscr{S}(R^d)$ as follows:

$$f^{(t)\sim}(p) = f^\sim(p)e^{i(p_0 - \varepsilon(p_0)\mu(\mathbf{p}))t}, \tag{13.4.3}$$

where

$$\varepsilon(p_0) = \varepsilon(p) = \operatorname{sgn} p_0. \tag{13.4.4}$$

Let $f = 0$ in a neighborhood of $p_0 = 0$.

Proposition 13.4.3. *For any $\mathscr{S}$ norm $\|\cdot\|_{\mathscr{S}}$ on $\mathscr{S}(R^{d-1})$ and any t derivative ∂_t^j, there is an L such that for any K we have*

$$\|\partial_t^j f^{(t)}(x_0, \cdot)\|_{\mathscr{S}} \leq C_K(1 + |x_0 - t|)^{-K}(1 + |t|)^L.$$

Also

$$\sup_{\{\mathbf{x} : \mathbf{x}/t \notin \mathscr{V}\}} (1 + |\mathbf{x}|)^L |f(t, \mathbf{x})| \leq O(t^{-N})$$

for all N, as $t \to \infty$, and the same bound holds for arbitrary derivatives of f.

PROOF. Regard $x_0 - t$, t, and $\mathbf{x}$ as independent variables. Since

$$f^{(t)}(x) = (2\pi)^{-d/2} \int g(p)e^{i\mathbf{p} \cdot \mathbf{x}} e^{it(p_0 - \varepsilon(p_0)\mu(\mathbf{p}))}\, dp,$$

where

$$g(p) = f^\sim(p)e^{-ip_0(x_0 - t)},$$

we see that $f^{(t)}(x)$ is in $\mathscr{S}(R^1)$ in $x_0 - t$ and a smooth solution of the Klein–Gordon equation in $(t, \mathbf{x})$. Thus the proposition follows from Proposition 13.4.1.

13.5 The Haag–Ruelle Theory

In proving convergence of the wave operators in field theory, we follow the same method as in potential scattering, but we encounter one new difficulty. The falloff of the potential, e.g., $V_{ij} \in L_2$, as assumed in Section 13.2, is replaced in field theory by decay properties of truncated vacuum expectation values. These correlations reflect interparticle potentials. The difficulty is that these decay properties are not assumed, but are derived from first principles (axioms, or the choice of interaction Lagrangian). To begin this section we bypass this difficulty, and assume the required decay properties.

Given any set of n-point functions (for example the Wightman functions $\mathscr{W}_n(x_1, \ldots, x_n)$), we can define truncated n-point functions as follows:

$$\mathscr{W}_n = \sum_{\pi \in \mathscr{P}} \prod_{P \in \pi} \mathscr{W}^T_{|P|}(x_{i_1}, \ldots, x_{i_{|P|}}), \tag{13.5.1}$$

$$\mathscr{W}^T_n = \sum_{\pi \in \mathscr{P}} (-1)^{|\pi|+1}(|\pi| - 1)! \prod_{P \in \pi} \mathscr{W}_{|P|}(x_{i_1}, \ldots, x_{i_{|P|}}). \tag{13.5.2}$$

Here $\mathscr{P}$ is the set of partitions of $\{1, \ldots, n\}$, $\pi = \{P_1, \ldots, P_{|\pi|}\}$ denotes an element of $\mathscr{P}$, and $\{i_1, \ldots, i_{|P|}\}$ denotes an element P of π. A combinatorial argument known as Moebius' theorem shows that both formulas give the same definition of $\mathscr{W}^T$ in terms of $\mathscr{W}$.

Let $\#$, as previously, denote a possible time derivative, and

$$\mathscr{F}_n(x_1, \ldots, x_n) = \langle \Omega, {}^{\#}\psi_{m_1}(0, x_1) \cdots {}^{\#}\psi_{m_n}(0, x_n)\Omega \rangle. \tag{13.5.3}$$

Theorem 13.5.1. *Assume the conditions of Theorem 13.3.2. As a function of the difference variables* $x_1 - x_j$, $\mathscr{F}^T_n$ *is a Schwartz distribution of rapid decrease.*

PROOF OF THEOREM 13.3.2. Let $\theta(t)$ denote the left side of (13.3.10), averaged with test functions $f, \hat{f}$. Assume that the distinct f's have nonoverlapping support in velocity space. Then

$$\|\theta(t_1) - \theta(t_2)\| \le \int_{t_1}^{t_2} \left\| \frac{d\theta(t)}{dt} \right\| dt$$

and $\|d\theta(t)/dt\|^2$ has an expression in terms of the inner product (13.5.3); (13.5.3) in turn is expanded in sums of products of truncated expectation values $\mathscr{F}^T_n$.

Any term with an $\mathscr{F}^T_1$ factor is zero, because $\psi_m\Omega \in \mathscr{H}_m \perp \Omega$, and so $\langle \Omega, \psi_m\Omega \rangle = 0$. Any term with only $\mathscr{F}^T_2$ factors is also zero. In fact such a term contains factors with time derivatives, namely

$$\left\langle e^{-itH} {}^{\#}\Big|G_{m_i}(-t)\Big(\frac{\psi_{m_i}(0, x_i)\Omega}{\dot\psi_{m_i}(0, x_i)\Omega}\Big)\Big|, \frac{d}{dt} e^{-itH} {}^{\#}\Big|G_{m_j}(-t)\Big(\frac{\psi_{m_j}(0, x_j)\Omega}{\dot\psi_{m_j}(0, x_j)\Omega}\Big)\Big| \right\rangle$$

However, for the vectors $\psi_m\Omega$ and $\dot\psi_m\Omega$, the free dynamics (defined by $G_m(t)$) and the physical dynamics (defined by e^{-itH}) coincide, since ${}^{\#}\psi_m\Omega \in \mathscr{H}_m$. Thus allowing for the minus in each $G_m(-t)$, each of the vectors in the inner product is time-independent, without the d/dt. Including the time derivative, the inner product is zero.

Next we consider a term with an $\mathscr{F}_j^T$ factor, $j \geq 3$. Now the time derivative d/dt plays no role. In $\mathscr{F}_j^T$, some of the j points come from fields ψ_m in the left variable $\langle \cdot \, \Omega, \cdot \, \Omega \rangle$ of the inner product and some from the right. Because $j \geq 3$, at least two points come from the same side. The test functions for these two points have nonoverlapping velocities. Thus the corresponding velocity cones do not overlap. Outside the velocity cones, the free field $G_m(-t)f$ has rapid decay, while inside velocity cones, we get rapid decay from the decay of $\mathscr{F}_j^T$ and the relative separation $\mathbf{x}_{i_1} - \mathbf{x}_{i_2}$ of the arguments of $\mathscr{F}_j^T$. It follows that terms with an $\mathscr{F}_j^T$ factor, $j \geq 3$, rapidly decreasing as $t \to \infty$.

From the proof, we see that the $t \to \pm\infty$ limit is determined by the time-independent products of $\mathscr{F}_2^T$ factors. This limit is by inspection a free field. From this fact, it follows that $W^\pm$ is isometric.

We now turn to the proof of Theorem 13.5.1. The key is a rapid decrease property of $\mathscr{F}^T$ under spacelike separation of the variables, as a consequence of the assumed mass gap. Now $\mathscr{F}^T$ is a distribution of rapid decrease in the relative variable $\mathbf{x}_{\mathrm{rel}}$ if and only if $\mathscr{F}^{T\sim}(\mathbf{p}_{\mathrm{rel}})$ is polynomially bounded and C^∞ in $\mathbf{p}_{\mathrm{rel}}$. We write

$$\mathscr{F}_n^{T\sim}(\mathbf{p}_{\mathrm{rel}}) = \int \cdots \int \prod_{i=1}^{n} h^\sim(p_i \cdot p_i) \mathscr{W}^{T\sim}(\mathbf{p}_{\mathrm{rel}}) \prod_{i=1}^{n} dp_{0,\,i},$$

where $p_i = (p_{0,\,i}, \mathbf{p}_i)$, and then observe that it is sufficient to have

$$\int \cdots \int f^\sim(p) \mathscr{W}^{T\sim}(p) \prod_{i=1}^{n} dp_{0,\,i} \in \mathscr{S}(\mathbf{p}_{\mathrm{rel}})$$

for all $f \in \mathscr{S}(R^{dn})$; hence it is also sufficient to have

$$(f * \mathscr{W}^T)(\mathbf{y}_{\mathrm{rel}}) = \int f(x) \mathscr{W}^T(\mathbf{y}_{\mathrm{rel}} - x) \, dx \in \mathscr{S}(R^{(d-1)(n-1)}) \quad (13.5.4)$$

for all $f \in \mathscr{S}(R^{dn})$. Since smoothness is ensured by the convolution, the key property is rapid decrease in $\mathbf{y}_{\mathrm{rel}}$.

At this point, we remove the region of x integration where $\|x\| \geq \sum \|\mathbf{y}_{\mathrm{rel}}\|/n$. Let g be a C^∞ function, equal to 1 in a small neighborhood of the origin and supported in a larger but still small neighborhood of the origin. Then

$$f * (1 - g)\mathscr{W}^T \equiv \int f(x) \left[1 - g\left(\frac{x^2}{x^2 + y_{\mathrm{rel}}^2} \right) \right] \mathscr{W}^T(\mathbf{y}_{\mathrm{rel}} - x) \, dx \in \mathscr{S}(\mathbf{y}_{\mathrm{rel}}),$$

and so we are left with the analysis of $f * g\mathscr{W}^T$.

Regard $\mathbf{y} \equiv (0, \mathbf{y}_{\mathrm{rel}})$ as a collection of n points in R^{d-1}. A simple geometrical argument shows that there are parallel hyperplanes in R^{d-1}, of separation at least $\|\mathbf{y}_{\mathrm{rel}}\|/n$, which partition these n points into two disjoint sets, located on opposite sides of the parallel hyperplanes. Furthermore for

$\|x\| \leq \sum \|\mathbf{y}_{\text{rel}}\|/n$, the n points in R^d defined by $\mathbf{y} - x$ are similarly partitioned into two sets, with the points of the first set (say $\mathbf{y}_i - x$, $i \in X$) spacelike separated from the points of the second set ($\mathbf{y}_j - x$, $j \in X'$).

Let $\pi \in \mathfrak{S}_n$ be the permutation on $\{1, \ldots, n\}$ which orders all indices $i \in X$ ahead of all $j \in X'$, but does not interchange relative positions within X, or within X'; similarly let $\pi' \in \mathfrak{S}_n$ order X' ahead of X. If $\mathscr{W}_\pi$ denotes the action of π on $\mathscr{W}$ by permutation of variables, then the locality axiom combined with our above analysis implies that

$$g\mathscr{W}^T = g\mathscr{W}_\pi^T = g\mathscr{W}_{\pi'}^T.$$

Let $\{\chi_X\}$ denote a partition of unity in $\mathbf{y}_{\text{rel}}$, where each χ_X is supported in the $\mathbf{y}_{\text{rel}}$'s which can lead as above to that given X. Since

$$\chi_X f * (\mathscr{W}^T - \mathscr{W}_\pi^T), \qquad \chi_X f * (\mathscr{W}_\pi^T - \mathscr{W}_{\pi'}^T) \in \mathscr{S}(\mathbf{y}_{\text{rel}}),$$

the key use of the mass gap comes in the following lemma.

Lemma 13.5.2. *There is a distribution $h_X \in \mathscr{S}'(R^{dn})$ of rapid decrease such that*

$$h_X * \mathscr{W}_\pi^T = \mathscr{W}_\pi^T, \qquad h_X * \mathscr{W}_{\pi'}^T = 0.$$

Using the lemma, we have

$$f * \mathscr{W}^T = \sum_X \chi_X f * \mathscr{W}^T$$
$$= \sum_X \chi_X f * (\mathscr{W}^T - \mathscr{W}_\pi^T) + \sum_X \chi_X f * h_X * (\mathscr{W}_\pi^T - \mathscr{W}_{\pi'}^T),$$

and since $f * h_X \in \mathscr{S}(R^{dn})$, we have $f * \mathscr{W}^T \in \mathscr{S}(\mathbf{y}_{\text{rel}})$ to complete the proof of Theorem 13.5.1.

Let

$$V_+^m = \{p \in R^d : -p \cdot p \geq m^2, p_0 > 0\}.$$

Proposition 13.5.3. *In a Wightman theory with a mass $m > 0$, the support of $\mathscr{W}^{T\sim}(p_1, \ldots, p_n)$ is contained in*

$$\sum_{l=s}^n p_l \in V_+^m, \quad s \neq 1; \qquad \sum_{l=1}^n p_l = 0.$$

PROOF OF LEMMA 13.5.2. Let $P_X = \sum_{i \in X} p_i$ and $P_{X'} = \sum_{j \in X'} p_j$. Then the support of $\mathscr{W}_\pi^{T\sim}$ is contained in

$$-P_X, P_{X'} \in V_+^m,$$

and the support of $\mathscr{W}_{\pi'}^{T\sim}$ is contained in

$$P_X, -P_{X'} \in V_+^m.$$

We choose $h_{\tilde{X}}$ to be a bounded C^∞ function of $P_{X,0}$ equal to one for $P_{X,0} < -m/2$ and equal to zero for $P_{X,0} > m/2$.

PROOF OF PROPOSITION 13.5.3. The restriction $\sum_{l=1}^{n} p_l = 0$ comes from translation invariance. Let $\hat{V} = \{0\} \cup V_+^m$; $\hat{V}$ contains the support of the spectral measure for the translation group and consequently $\mathscr{W}^\sim(p_1, \ldots, p_n)$ has support contained in the sets

$$P_s \equiv \sum_{l=s}^{n} p_l \in \hat{V}; \qquad \sum_{l=1}^{n} p_l = 0. \tag{13.5.5}$$

Note that $\hat{V}$ is a semigroup, i.e., closed under addition. From (13.5.2) or an induction on n, from (13.5.1), we see that $\mathscr{W}^{T\sim}$ is also supported in (13.5.5). The whole point of the truncation is to remove the origin $p = 0$ from the supports, i.e., to replace $\hat{V}$ by V_+^m.

We prove the proposition by induction on n. For $n = 1$, the statement $P_s \in V_+^m$, $s \neq 1$, is vacuous and hence true. Now assume the result for $l \leq n - 1$. Let $g \in \mathscr{S}(R^d)$ have the property

$$\text{suppt } g^\sim \cap \hat{V} \subset \{0\}.$$

Then

$$\int g(a)\mathscr{W}_l(x_1, \ldots, x_{j-1}, x_j + a, \ldots, x_l + a)\, da$$

$$= \left\langle \phi(x_{j-1}) \cdots \phi(x_1)\Omega, \int e^{-ia \cdot P} g(a)\, da\, \phi(x_j) \cdots \phi(x_l)\Omega \right\rangle$$

$$= g^\sim(0)\langle \Omega, \phi(x_1) \cdots \phi(x_{j-1})\Omega \rangle \langle \Omega, \phi(x_j) \cdots \phi(x_l)\Omega \rangle$$

$$= g^\sim(0)\mathscr{W}_{j-1}(x_1, \ldots, x_{j-1})\mathscr{W}_{l-j+1}(x_j, \ldots, x_l).$$

Now set $l = n$ and substitute (13.5.1). The result is

$$g^\sim(0)\mathscr{W}_{j-1}\mathscr{W}_{n-j+1} = \int g(a)\mathscr{W}_n^T(x_1, \ldots, x_{j-1}, x_j + a, \ldots, x_n + a)\, da$$

$$+ \sum_{\{1, \ldots, n\} \neq \pi \in \mathscr{P}} (\cdots).$$

By the induction hypothesis, each nontrivial partition π gives zero unless it is a refinement of the partition

$$\pi_j = \{\{1, \ldots, j - 1\}, \{j, \ldots, n\}\}.$$

The partitions which do refine π_j sum to $g^\sim(0)\mathscr{W}_{j-1}\mathscr{W}_{n-j+1}$ by (13.5.1). Thus

$$0 = \int g(a)\mathscr{W}_n^T(x_1, \ldots, x_{j-1}, x_j + a, \ldots, x_n + a)\, da$$

$$= g^\sim(P_j)\mathscr{W}^{T\sim}\Big|_{P_j = 0}.$$

This completes the induction and the proof.

We now reformulate and extend Theorem 13.3.2. The approach $t \to \pm\infty$ to the asymptotic limit can be expressed entirely as an evolution of the test function f, for $f \in \mathscr{S}(R^d)$. In fact under the physical dynamics, $\phi(x_0, \mathbf{x}) \to \phi(x_0 + t, \mathbf{x})$, and after change of variables, this is equivalent to $f(x_0, \mathbf{x}) \to f(x_0 - t, \mathbf{x})$, or in Fourier transform space to $f^\sim \to e^{itp_0}f^\sim$ subject to the sign conventions

$$f(x) = (2\pi)^{-d/2} \int e^{ipx} f\tilde{}(p) \, dp, \qquad (13.5.6)$$

$$p \cdot x = -p_0 x_0 + \mathbf{p} \cdot \mathbf{x}. \qquad (13.5.7)$$

With Cauchy data $f_0, \dot{f}_0 \in \mathscr{S}(R^{d-1})$ given, the construction of ψ from $\psi^{\text{prelim}} = \phi$ in Section 13.3 is equivalent to the change of test function from

$$f(y) = f_0(\mathbf{x})\delta(y_0) + \dot{f}_0(\mathbf{x})\delta'(y_0)$$

for integration against ψ to

$$f(x) = \int \left(h(x-y)f_0(\mathbf{y}) - \partial_{x_0} h(x-y)\dot{f}_0(\mathbf{y}) \right) dy$$

or

$$f\tilde{}(p) = (f\tilde{}_0(\mathbf{p}) + ip_0 \, \dot{f}\tilde{}_0(\mathbf{p}))h\tilde{}(p) \qquad (13.5.8)$$

for integration against ϕ. The combined physical and (inverse) free dynamics, as in (13.3.10), is expressed in Fourier space as

$$f\tilde{}(p) \to e^{it(p_0 - \varepsilon(p_0)\mu(\mathbf{p}))} f\tilde{}(\mathbf{p}) \equiv f^{(t)}\tilde{}(p), \qquad (13.5.9)$$

where ε is defined in (13.4.4). The key estimate on (13.5.9) is Proposition 13.4.3.

For the proof of Theorem 13.3.2, the important property of f is

$$\text{supp} \tilde{f} \subset \{p : |p^2 + m^2| \le \varepsilon\} \qquad (13.5.10)$$

where $\varepsilon > 0$ is small enough that (13.5.10) implies

$$\text{supp} \tilde{f} \cap \sigma(H, \mathbf{p}) = \{p : -p^2 = m^2\}.$$

The dynamics $f \to f^{(t)}$ extends to all f satisfying (13.5.10). Theorem 13.3.2 also extends to all such f, and states that

$$\lim_{t \to \pm\infty} \prod_{i=1}^{n} \phi(f_i^{(t)})\Omega = \prod_{i=1}^{n} \phi_{\text{in/out}}(f_i)\Omega,$$

with convergence $O(t^{-N})$, for nonoverlapping velocities.

The extension of this result is to allow the $t = t_i$ to go to their limits separately.

Theorem 13.5.4. *Let* $\{f_i\}$ *have nonoverlapping velocities, and assume* (13.5.10) *for each* f_i. *Let* $\theta_{\text{in/out}} = \prod_{i=m+1}^{n} \phi_{\text{in/out}}(f_i)\Omega$. *Then*

$$\lim_{\substack{t_1 \le t_2 \le \cdots \le t_m \\ t_i \to +\infty}} \phi(f_1^{(t_1)}) \cdots \phi(f_m^{(t_m)})\theta_{\text{out}} = \prod_{i=1}^{n} \phi_{\text{out}}(f_i)\Omega.$$

A similar statement holds for $t_i \to -\infty$ *and* θ_{in}, *provided* $t_m \le \cdots \le t_2 \le t_1$ *in the limit process.*

PROOF. We use induction on m, starting with $m = 0$, for which the theorem is trivially true. We first change the time of $t_{m+1} = \cdots = t_n$ from ∞ to t_m, then change $t_m = t_{m+1} = \cdots$ back to ∞. The first change introduces the error term bounded in norm by

$$\int_{t_m}^{\infty} \left\| \phi(f_1^{(t_1)}) \cdots (f_m^{(t_m)}) \frac{d}{ds} \prod_{i=m+1}^{n} \phi(f_i^{(s)})\Omega \right\| ds.$$

The integrand squared is an inner product and can be expanded in truncated vacuum expectation values. As in the proof of Theorem 13.3.2, we focus on the fields, $\partial_s \phi(f_i^{(s)})$. If this field acts directly on Ω, the term is zero, since

$$\partial_s \phi(f_i^{(s)})\Omega = 0.$$

If not, there must be two $t = s$ fields on the same side of the inner product. These two fields have velocities which do not overlap with each other, leading to convergence at the rate $O(s^{-N})$. After integration over s, the convergence is $O(t_m^{-N})$.

The second change of times is bounded in norm by

$$\int_{t_{m-1}}^{\infty} \left\| \phi(f_1^{(t_1)}) \cdots \phi(f_{m-1}^{(t_{m-1})}) \frac{d}{dt} \prod_{i=m}^{n} \phi(f_i^{(s)})\Omega \right\| ds \leq O(t_{m-1}^{-N})$$

by the same reasoning.

REFERENCES

[Jost, 1965], [Hepp, 1966a], [Reed and Simon, 1972–9].

CHAPTER 14

Scattering Theory: Time-Independent Methods

14.1 Time-Ordered Correlation Functions

The S matrix is the central object of interest in field theory because it expresses the observed interactions between particles. The Chapter 13 definition of the S matrix does not provide a convenient framework for further analysis of its properties. However, we now show that the S matrix has a simple representation in terms of time-ordered correlation functions. This representation is convenient, and it allows, e.g., the construction of perturbation series, the decomposition into connected components, and the study of momentum space analyticity.

To define the time-ordered product, let

$$\theta(x) = \theta(x_0, \mathbf{x}) = \begin{cases} 1 & \text{if } x_0 > 0, \\ 0 & \text{if } x_0 < 0, \end{cases}$$

and

$$T\phi(x_1) \cdots \phi(x_n) = \sum_{\pi \in \mathfrak{S}_n} \theta(x_{\pi(1)} - x_{\pi(2)}) \cdots \theta(x_{\pi(n-1)} - x_{\pi(n)})$$
$$\times \phi(x_{\pi(1)}) \cdots \phi(x_{\pi(n)}) \tag{14.1.1}$$

where $\mathfrak{S}_n$ is the symmetric group on n elements. Our goal is to show that

$$\tau(x_1, \ldots, x_n) = \langle T\phi(x_1) \cdots \phi(x_n) \rangle \tag{14.1.2}$$

is a tempered distribution whose Fourier transform $\tau^\sim(p)$ has sufficient regularity to allow the δ functions in the expression

$$\prod_{i=1}^{n} \delta(p_i^2 + m^2)\theta(p_i^0)\left[\prod_{i=1}^{n}(p_i^2 + m^2)\right]\tau^{\sim}(p_1, \ldots, p_m, -p_{m+1}, \ldots, -p_n).$$
(14.1.3)

This expression is (up to a constant factor) the S matrix element between "out" states of momenta $p_1, \ldots, p_m$ and "in" states of momenta $p_{m+1}, \ldots, p_n$.

The first problem is to show that τ is a tempered distribution, or is defined in any sense at all. Because $\theta \notin C^\infty$, multiplication of the Wightman function $W \in \mathscr{S}'$ by θ is not in general defined. Using the regularity of the axioms of Chapter 6 (which go beyond the regularity implied by the Wightman axioms), and in particular the property that the Schwinger functions have continuous extensions as functionals defined on all of $\mathscr{S}(R^{nd})$ (including coinciding points), it can be shown that $\tau \in \mathscr{S}'(R^{nd})$ [Eckmann and Epstein, 1979a]. However, it is easier to bypass this problem than to solve it, as we now see. Let $\alpha \in C_0^\infty$ be nonnegative with integral 1. Let

$$T_\alpha \phi(x_1) \cdots \phi(x_n) = \sum_{\pi \in \mathfrak{S}_n} \theta * \alpha(x_{\pi(1)} - x_{\pi(2)}) \cdots \theta * \alpha(x_{\pi(n-1)} - x_{\pi(n)})$$
$$\times \phi(x_{\pi(1)}) \cdots \phi(x_{\pi(n)}). \tag{14.1.4}$$

Since $\theta * \alpha \in \mathscr{S}$,

$$\tau_\alpha = \langle T_\alpha \phi(x_1) \cdots \phi(x_n)\rangle \in \mathscr{S}'. \tag{14.1.5}$$

It will turn out that τ_α can be substituted for τ in (14.1.3). The second problem, which is smoothness in p space, cannot be bypassed. It is equivalent to certain decay properties in x space, and extends the results of Section 13.5.

Let $f_i \in \mathscr{S}(R^d)$, $1 \le i \le n$, with nonoverlapping supports in velocity space, and assume (13.5.10). Let $f_i^{(t)}$ be defined by (13.5.9), and let

$$X^{\text{out}} = \prod_{i=1}^{k} \phi^{\text{out}}(f_i)\Omega,$$

$$X^{\text{in}} = \prod_{i=m+1}^{n} \phi^{\text{in}}(f_i)\Omega,$$

$$\dot{f}_i^{(t)} = \partial_t f_i^{(t)}.$$

Theorem 14.1.1. Let all f_i be supported in a small neighborhood of $-p^2 = m^2$. Then as a function of the variables $t_{k+1}, \ldots, t_m$,

$$\int\left(\prod_{i=k+1}^{m} \dot{f}_i^{(t_i)} dx_i\right)\langle X^{\text{out}}, T_\alpha \phi(x_{k+1}) \cdots \phi(x_m)X^{\text{in}}\rangle \tag{14.1.6}$$

lies in $\mathscr{S}(R^{m-k})$.

PROOF. $\partial_t f^{(t)}$ has the same form as $f^{(t)}$, so we only need to show the rapid decrease. The idea is that a block of fields with largest and nearly equal times can be removed from the time ordering operation T_α, and converge to "out" fields acting on X^{out}, according to Theorems 13.3.2 and 13.5.4, at a rate $O(t^{-N})$. Because of the ∂_t in $\dot{f}$, the resulting "out" fields are zero, so the rapid decrease comes from the convergence rate. Similarly fields with smallest and nearly equal time can be removed from T_α and applied to X^{in}, with the same result.

Due to symmetry, we can consider (14.1.6) in the sector

$$t_{k+1} \geq \cdots \geq t_m \quad \text{and} \quad t \equiv t_{k+1} \geq |t_m|. \tag{14.1.7}$$

For some l, $k + 1 \leq l \leq m$, consider as an approximation to (14.1.6)

$$\int \left(\prod_{i=k+1}^{m} \dot{f}^{(t_i)} \, dx_i \right) \langle X^{\text{out}}, \phi(x_{k+1}) \cdots \phi(x_l) T_\alpha(\phi(x_{l+1}) \cdots \phi(x_m)) X^{\text{in}} \rangle. \tag{14.1.8}$$

According to Theorem 13.5.4,

$$\left\| \prod_{i=k+1}^{l} \phi(\dot{f}^{(t_i)}_i) X^{\text{out}} \right\| = O(t^{-N})$$

in the sector (14.1.7), since $\phi_{\text{in/out}}(\dot{f}_i) = 0$. Also

$$\left\| \int \left(\prod_{i=l+1}^{m} \dot{f}^{(t_i)} \, dx_i \right) T_\alpha(\phi(x_{l+1}) \cdots \phi(x_m)) X^{\text{in}} \right\| \leq O(t^L)$$

for some L, by Proposition 13.4.3, since the regularized time-ordered product can be written in terms of ordinary products with smooth coefficients. Thus (14.1.8) is bounded in magnitude by $O(t^{-N})$ for all N.

To show that (14.1.6) and (14.1.8) differ by $O(t^{-N})$, we choose l, depending on $\{t_{k+1}, \ldots, t_m\}$, so that the initial block of times $t_{k+1}, \ldots, t_l$ are nearly equal, say

$$\left. \begin{array}{l} 0 \leq t_i - t_{i+1} \leq \varepsilon t, \quad k + 1 \leq i \leq l - 1, \\[2mm] t_l - t_{l+1} > \varepsilon t. \end{array} \right\} \tag{14.1.9}$$

In the definition of T_α, consider first the permutations $\pi \in \mathfrak{S}_{m-k}$ which map the initial set $\{k + 1, \ldots, l\}$ of indices onto itself, i.e., $\pi = (\pi^{l-k}, \pi^{m-l}) \in \mathfrak{S}_{l-k} \times \mathfrak{S}_{m-l}$. Because of the nonoverlapping velocities, the test functions $\dot{f}^{(t_i)}_i$, $k + 1 \leq i \leq l$, have spacelike separated supports, up to terms which are $O(t^{-N})$, by Proposition 13.4.3. Thus by locality, the contribution of π to T_α is independent of its π^{l-k} factor, up to $O(t^{-N})$. Hence the sum over $\pi^{l-k} \in \mathfrak{S}_{l-k}$ removes time ordering from the initial $l - k$ fields, and the sum over $\pi^{m-l} \in \mathfrak{S}_{m-l}$ gives time ordering T_α of the final $m - l$ fields. We have shown that the sum over $\pi \in \mathfrak{S}_{l-k} \times \mathfrak{S}_{m-l}$ in (14.1.6) coincides with (14.1.8) up to $O(t^{-N})$.

It remains to consider a $\pi \notin \mathfrak{S}_{l-k} \times \mathfrak{S}_{m-l}$, i.e., incompatible with the sector (14.1.9). The corresponding term in T_α has support in

$$x_{0, \pi(j)} - x_{0, \pi(j+1)} \geq -\text{const} \equiv -\max\{|s| : s \in \text{suppt } \alpha\}.$$

By the choice of π, $\pi(j + 1) \leq l < \pi(j)$ for some j. Thus

$$t_{\pi(j)} - t_{\pi(j+1)} \leq -\varepsilon t$$

and so

$$\left| x_{0,\,\pi(j)} - t_{\pi(j)} \right| + \left| x_{0,\,\pi(j+1)} - t_{\pi(j+1)} \right| \geq \varepsilon t/3.$$

By Proposition 13.4.3, the contribution of π to T_α in (14.1.6) is $O(t^{-N})$, and the proof is complete.

14.2 The S Matrix

The Lehmann–Symanzik–Zimmermann reduction formulae express the S matrix in terms of the τ functions of Section 14.1. These formulae are a simple consequence of Theorem 14.1.1.

Proposition 14.2.1.

$$\left\langle S \prod_{i=1}^{n} \phi(f_i)^{\mathrm{in}} \Omega, \prod_{j=1}^{m} \phi(g_j)^{\mathrm{in}} \Omega \right\rangle = \left\langle \prod_{i=1}^{n} \phi(f_i)^{\mathrm{out}} \Omega, \prod_{j=1}^{m} \phi(g_j)^{\mathrm{in}} \Omega \right\rangle.$$

$$(14.2.1)$$

PROOF. $\phi^{\mathrm{out}} = S\phi^{\mathrm{in}}S^{-1}$.

Proposition 14.2.2. *Assume nonoverlapping velocities and (13.5.10). Then*

$$\sum_{X \subset \{1,\,\ldots,\,n\}} (-1)^{|X|} \left\langle \prod_{i \notin X} \phi(f_i)^{\mathrm{out}}, \prod_{j \in X} \phi(f_j)^{\mathrm{in}} \right\rangle = \int \left(\prod_{i=1}^{n} \partial_{t_i} f_i^{(t_i)}(x_i) \right) \tau_\alpha(x) \, dx \, dt.$$

$$(14.2.2)$$

PROOF. The convergence of the dt integral is $O(t^{-N})$ by the proof of Theorem 14.1.1. Thus the identity is just the fundamental theorem of calculus, with the evaluation $\left.\right|_{t_i = -\infty}^{t_i = +\infty}$ also given by the proof of Theorem 14.1.1.

Remark. It is an elementary exercise in free field combinatorics to eliminate the sum over X in (14.2.2) and thus by (14.2.1) to obtain S matrix elements in terms of τ functions. We omit this step, and continue with the evaluation of the right side of (14.2.2).

Proposition 14.2.3. *Assume nonoverlapping velocities and (13.5.10). Then*

$$\int \left(\prod_{j=1}^{n} \partial_{t_j} f_j^{(t_j)} \right) \tau_\alpha(x) \, dx \, dt = \int \prod_{j=1}^{n} [i\Delta_m * (\square - m^2) f_j(x_j)] \tau_\alpha(x) \, dx$$

where

$$\tilde{\Delta_m}(p) = \varepsilon(p) \, \delta(p^2 + m^2).$$

PROOF. The left hand side is evaluated as

$$\int \prod_{j=1}^{n} \partial_{t_j} e^{it_j(p_{0,j} - \varepsilon(p_j)\mu(\mathbf{p}_j))} f_{\tilde{j}}(p_j)\tau_{\tilde{\alpha}}(-p)\, dp\, dt$$

$$= \int \prod_{j=1}^{n} \frac{e^{it_j(p_{0,j} - \varepsilon(p_j)\mu(\mathbf{p}_j))}}{i(p_{0,j} + \varepsilon(p_j)\mu(\mathbf{p}_j))} (p_j^2 + m^2) f_{\tilde{j}}(p_j)\tau_{\tilde{\alpha}}(-p)\, dp\, dt.$$

After the change of variables $dp = d\mathbf{p}\, d(p_0 - \varepsilon(p)\mu(\mathbf{p}))$, the rapid decrease in t allows interchange of the dt and dp integrations, which gives the result

$$\int \prod_{j=1}^{n} (-i\varepsilon(p_j)) \frac{\delta(p_{0,j} - \varepsilon(p_j)\mu(\mathbf{p}_j))}{2\mu(\mathbf{p}_j)} (p_j^2 + m^2) f_{\tilde{j}}(p_j)\tau_{\tilde{\alpha}}(-p)\, dp$$

$$= \int \prod_{j=1}^{n} (-i\varepsilon(p_j))\, \delta(p_j^2 + m^2)(p_j^2 + m^2) f_{\tilde{j}}(p_j)\tau_{\tilde{\alpha}}(-p)\, dp$$

$$= \int \prod_{j=1}^{n} (i)\Delta_m * (\Box - m^2) f(x)\tau_{\tilde{\alpha}}(x)\, dx.$$

The first equality results from substitution of the definition of $\delta(p^2 + m^2)$.

Remark. Perturbation theory is asymptotic for S, at least in the superrenormalizable cases under mathematical control. In four dimensional quantum electrodynamics, coefficients have been computed through sixth order. A correct discussion of time-ordered products is required. It is sufficient to show that the singularities at coinciding points are at worst distributions, so that the Schwinger functions belong to $\mathscr{S}'$. Such bounds follow from the form of the Euclidean axioms given here. See [Eckmann and Epstein, 1979a] for a discussion of these questions in a general framework.

14.3 Renormalization

The simplest of the renormalizations of quantum field theory is the vacuum renormalization. In the Euclidean formulation, the vacuum renormalization is the statement that the measure $d\mu$ of Section 6.1 and Chapter 11 is a probability measure, or in other words that the division by the normalizing factor Z has been performed, as in Section 11.1. In terms of the canonical formalism, the vacuum renormalization can be regarded as two separate effects. We write

$$1 = \frac{Z}{Z} = Z^{-1} \int e^{-\int :P(\phi(x)):\, dx}\, d\phi_C$$

$$= \int \exp[-\ln Z - \int :P(\phi(x)):\, dx]\, d\phi_C.$$

Consider a rectangular interaction region Λ. Then $\ln Z$ has the asymptotic behavior

$$\ln Z = c_1 T + c_2,$$

where T is the length of the time interval over which the interaction occurs and $|\Lambda|/T$ is fixed, cf. Chapter 11. The coefficient $c_1 = \delta E$ can be interpreted as an additive constant contributing to the Hamiltonian H, while e^{-c_2} is a multiplicative constant, which fixes the norm of the vacuum state $\Omega \in \mathcal{H}$ to one.

On the level of formal perturbation theory, the vacuum renormalizations are reflected in the fact that only connected diagrams contribute to the Schwinger functions, as explained in Section 8.4. Thus the perturbative renormalization of the Schwinger functions and the S matrix does not use the vacuum renormalization explicitly. From the definition

$$\delta E = T^{-1} \ln Z + o(1) \quad \text{as } T \to \infty,$$

we can extract a perturbative expansion for δE. In case $P = \lambda \phi^4$, the low order terms are given in Figure 14.1.

Figure 14.1. δE for $P(\phi) = \lambda \phi^4$ through third order.

The constant c_2 (which we call the vacuum wave function renormalization) is given perturbatively by a sum of connected diagrams also. The diagrams are the same as the diagrams for δE, but the numerical values are different. For example δE has one vertex at $t = 0$, to implement the division by T. First order contributions to both δE and c_2 have been removed by Wick ordering.

Next in order of complexity are the mass and field strength renormalizations. These renormalizations are defined in terms of the two point function, which we study using the Dyson equation and its kernel, the proper self-energy operator Σ. To study mass renormalization, we assume that the mass m is given at the outset, for example by measurement, and that we only want to consider interaction polynomials P which produce fields with mass m particles. These restricted (mass m) polynomials lie in a submanifold of codimension 1 in the space of all polynomials. (Of course we may wish to specify $n > 1$ particle masses if more than one eigenvalue of m occurs. In this case P_{ren} lies in a submanifold of polynomials of codimension n.) For simplicity we discuss $n = 1$. Any map

$$P \to P_{\text{ren}} = P + \delta P$$

from the space of all polynomials to the submanifold of mass m polynomials with the property

$$P_{\text{ren}} = P \quad \text{if } P \text{ has mass } m \text{ particles}$$

is a renormalization. This definition is nonunique and too general. We obtain a unique definition in the weak coupling region by the *Ansatz*

$$\delta P(\phi) = \tfrac{1}{2}\delta m^2 \; \phi^2.$$

δm^2 is then a function of m and the coefficients of P, and the renormalization problem is to determine this function.

Let $S^{(2)T\sim}$ be the truncated two point Schwinger function in (Euclidean) momentum space, and let $S_0^{(2)T\sim}$ be the corresponding free field quantity, both with mass m. Then

$$S_0^{(2)T\sim} = (p^2 + m^2)^{-1} \tag{14.3.1}$$

and by the Lehmann spectral formula,

$$S^{(2)T\sim} = \frac{Z}{p^2 + m^2} + \int_M^\infty \frac{dv(a)}{p^2 + a^2} . \tag{14.3.2}$$

Here we assume an isolated one particle hyperboloid, so that $M > m$. Moreover Z, the field strength renormalization constant, is defined by (14.3.2). (This Z is unrelated to the partition function Z of the vacuum renormalization.) The Dyson equation is

$$S^{(2)T\sim} = S_0^{(2)T\sim} - S_0^{(2)T\sim} \; \Sigma^\sim S^{(2)T\sim}, \tag{14.3.3a}$$

and defines Σ. This is just the resolvent equation and can also be written

$$-\Gamma^{(2)\sim} \equiv (S^{(2)T\sim})^{-1} = S_0^{(2)\sim -1} + \Sigma^\sim. \tag{14.3.3b}$$

Here products and inverses are pointwise operations in momentum space.

Continuing analytically to the pole at $p^2 = -m^2$, we see that (14.3.3) is inconsistent unless

$$\Sigma^\sim \Big|_{p^2 = -m^2} = 0. \tag{14.3.4}$$

Thus (14.3.4) defines the submanifold of mass m polynomials.

Diagrammatically, $\Sigma^\sim$ is the sum of all diagrams with two external legs and which are one-particle-irreducible between these legs. The latter condition means that the diagram is connected and that the removal of a single line in the diagram cannot disconnect it into two components, each containing one of the external legs. See Figure 14.2.

We express $\Sigma^\sim|_{p^2 = -m^2}$ as a formal power series in the coefficients of $P(\phi) = \sum_{j=1}^n a_j \phi^j$. Then $a_2 \equiv \tfrac{1}{2} \delta m^2$ and

$$\Sigma^\sim|_{p^2 = -m^2} = \delta m^2 + O(a_1^2, a_3^2, \ldots, a_n^2). \tag{14.3.5}$$

This formula combined with (14.3.4) defines δm^2 as a formal power series in $a_1, \ldots, a_n$; this series is the perturbative definition of mass renormalization.

Comparing (14.3.2) and (14.3.3), we also see that

$$Z^{-1} - 1 = \frac{\partial \Sigma^\sim}{\partial p^2}\Big|_{p^2 = -m^2}, \tag{14.3.6}$$

Figure 14.2. $\Sigma^{\sim}$ for $P(\phi) = \lambda\phi^4$ through third order.

so that $\Sigma^{\sim}$ determines the field strength renormalization constant Z as well. Conventionally we write

$$\phi_{\text{ren}} = Z^{-1/2}\phi, \tag{14.3.7}$$

so that the two point Schwinger function, expressed in terms of the renormalized field ϕ_{ren}, has a unit strength pole at the mass shell $p^2 = -m^2$.

In the form (14.3.3b), the Dyson equation can be used as part of an argument to establish the isolated one particle hyperboloid (e.g., $M > m$ in (14.3.2)) for weak coupling. Let S_0 have a "bare mass" m_0. We need to know that $\Gamma^{(2)\sim}$ is analytic in a strip

$$\text{Re } p^2 \geq -M^2$$

for $M^2 > m^2$, and we need to know that $\Sigma^{\sim}$ is small for weak coupling. Then Rouché's theorem applies and shows that $\Gamma^{(2)\sim}$ has a simple zero in p^2, near $-m_0^2$. This zero defines the (physical) mass m. Alternatively, this fact follows from monotonicity of $\Gamma^{\sim}$; see [Burnap, 1977].

The final renormalization is charge renormalization. There are a variety of conventions, which give different definitions for the renormalized charge. For a $\lambda\phi^4$ theory, all definitions involve the connected four point function, since the latter determines the two particle scattering process, by Section 14.2. With any of these definitions, the physical charge λ_{phys} is given perturbatively as a power series in λ, the coefficients of which are Feynman diagrams (see Figure 14.3). The numerical value to be associated with these

Figure 14.3. λ_{phys} through second order for a $\lambda\phi^4$ theory.

diagrams depends on the renormalization convention. Charge renormalization itself is just the inversion of the functional relation $\lambda_{\text{phys}} = \lambda_{\text{phys}}(\lambda)$, so that $\lambda = \lambda(\lambda_{\text{phys}})$, which is a power series in λ_{phys}, is substituted for λ in $\lambda\phi^4$ in the interaction Lagrangian. As an example of a renormalization convention, λ_{phys} can be taken as the amputated connected four point function of the renormalized fields ϕ_{ren}, evaluated at space momenta $\mathbf{p} = 0$. Amputation means that the operator $-\square + m^2$ has been applied to each variable, as in Proposition 14.2.3.

The results of this section are summarized in Table 14.1.

Table 14.1. Short Distance Divergence of Renormalizations, as a Function of the Ultraviolet Cutoff Parameter κ

	$:P(\phi):_2$	Yukawa$_2$	ϕ_3^4	ϕ_4^4
Vacuum energy	finite	$(\ln \kappa)^2$	κ	κ^4
Vacuum wave function renormalization	finite	finite	$\ln \kappa$	κ^3
Mass	finite	$\ln \kappa$	$\ln \kappa$	κ^2
Z = field strength renormalization	finite	finite	finite	$\ln \kappa$
Charge	finite	finite	finite	$\ln \kappa$

14.4 The Bethe–Salpeter Kernel

The Bethe–Salpeter equation is used to study the four point function, or more generally n-point functions with bounded n, $n \le N$. The equation contains (and can be used to define) a new unknown quantity: the Bethe–Salpeter kernel K. If we assume analyticity properties of K in energy-momentum space, then the Bethe–Salpeter equation can be used to study the mass spectrum and low energy scattering states contained in polynomials of bounded degree:

$$\phi(x_1) \cdots \phi(x_j)\Omega, \qquad j \le J \quad (\text{e.g.}, j = 2). \tag{14.4.1}$$

This method applies to the bound state and resonance problem and to low energy asymptotic completeness. Required analyticity properties of K can be proved for superrenormalizable quantum field models at weak coupling. The results actually proved are representative of this picture, but do not yet establish it at the above level of generality. There has also been progress with the converse program, which is to assume properties of the mass spectrum and derive analytic properties of K ([Bros, 1970], [Bros and LaSalle, 1977]).

The $n = 2$ Bethe–Salpeter equation is the Dyson equation. The $n = 4$ Bethe–Salpeter equation also has the structure of a resolvent equation:

$$R = R_0 - R_0 KR. \tag{14.4.2}$$

To simplify the exposition we assume an even theory, so that all odd order Schwinger functions vanish: $S^{(2j+1)} = 0$. We work in Euclidean space-time. The operator R_0, which plays the role of "free resolvent" is by definition

$$R_0(x, y) = S^{(2)T}(x_1 - y_1)S^{(2)T}(x_2 - y_2)$$
$$+ S^{(2)T}(x_1 - y_2)S^{(2)T}(x_2 - y_1). \tag{14.4.3}$$

Moreover

$$R(x, y) = \langle \phi(x_1)\phi(x_2), (1 - P_\Omega)\phi(y_1)\phi(y_2)\rangle$$
$$= S^{(4)}(x_1, x_2, y_1, y_2) - S^{(2)}(x_1 - x_2)S^{(2)}(y_1 - y_2), \quad (14.4.4)$$

where P_Ω is the orthogonal projection, in the Euclidean Hilbert space, onto the vacuum state $\Omega \equiv 1$. There are both generalizations of (14.4.2) and alternatives to it, which we do not discuss here.

In perturbation theory, K is defined as the sum of all diagrams with four external legs (amputated), which are connected and are two particle channel irreducible. To explain this term, we identify two of the four external legs in a diagram contributing to K with the x-variables (out legs) and two with the y-variables (in legs). Then two particle channel irreducible means that after removing up to two internal lines from the diagram, each connected component of the remainder of the diagram must contain at least one "in" leg and at least one "out" leg. (See Figure 14.4.) Components consisting only of a single external leg are not counted in this definition.

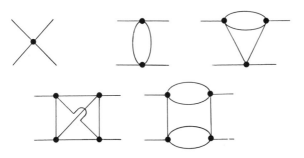

Figure 14.4. Diagrams in K for $P(\phi) = \phi^4$.

Nonperturbatively, K is defined by Equation (14.4.2) [Glimm and Jaffe, 1975e]; however, this fact is of little value unless further information concerning K is available from some other source. We discuss now some general axioms for the analysis of K and R. We distinguish between fairly sharp statements which can be verified at weak coupling and which serve to justify perturbative calculations, and more general statements which should hold for all noncritical theories.

BS1 (Completeness of polynomial states). Given an energy E, there is a $J = J_E$ such that no state of energy $\leq E$ is orthogonal to the span of the states (14.4.1). In (14.4.1), $\Omega \equiv 1$ is the Euclidean vacuum, i.e., $\Omega \in \mathcal{E}$, and $x_v \in R^d$ is a (Euclidean) space-time point, with time $x_{d,v} \geq 0$, and the Euclidean vector (14.4.1) in $\mathcal{E}$ is projected onto its image in the physical Hilbert space $\mathcal{H}$.

In weak coupling $\lambda P(\phi)_2$ fields, this axiom has been verified [Glimm, Jaffe, and Spencer, 1974]. Here we must take $\lambda = \lambda(E)$ sufficiently small.

Specifically, for the analysis of bound states at the two body threshold, we take $E = 4m - \varepsilon$, $\varepsilon > 0$, and $J = 2$ in the subspace of even states.

BS2 (Solution of the one body problem).   The minimum assumption is the existence of an isolated one particle spectrum. A stronger assumption is that $\Sigma^{\sim}(p)$ is bounded and analytic in the regions

$$\text{(Euclidean)} \quad p^2 > -(3m)^2 + \varepsilon \tag{14.4.5a}$$

for an even theory, and

$$\text{(Euclidean)} \quad p^2 > -(2m)^2 + \varepsilon \tag{14.4.5b}$$

otherwise.

(14.4.5) is known for weak coupling $\lambda P(\phi)_2$ fields [Spencer, 1975]. However, this region of analyticity implies small binding energies, and cannot be true for general fields.

BS3 (Analyticity of K).   The minimum assumption is that for given E, the sufficiently irreducible (large n) K must be analytic for $p^2 \geq -E$.

For $n = 2$, we set

$$x_{\text{tot}} = \frac{x_1 + x_2}{2}, \qquad x_{\text{rel}} = \frac{x_1 - x_2}{2},$$

and similarly define y_{tot}, y_{rel}. Let p, q be the Euclidean momenta, dual to x and y, respectively. Then $p_1 + p_2 = p_{\text{tot}}$ and $p_1 - p_2 = p_{\text{rel}}$ are dual to x_{tot} and x_{rel}, and similarly for q_{tot} and q_{rel}. By translation invariance, $K = K(x, y)$ is a function of the difference variables alone. This means that the kernel of the operator K has a Fourier space representation of the form

$$\delta(p_{\text{tot}} - q_{\text{tot}})K^{\sim}(p_{\text{tot}}, p_{\text{rel}}, q_{\text{rel}}).$$

Working in a subspace of constant p_{tot}, $K^{\sim}$ is the kernel of an operator mapping functions of q_{rel} into functions of p_{rel}. By Lorentz invariance, it is no loss of generality in assuming $\mathbf{p}_{\text{tot}} = 0$. We define $E = ip_{\text{tot}}^0$. The region of assumed analyticity is

$$\begin{aligned} &|E| \leq 4m - \varepsilon, \\ &|\operatorname{Im} p_{\text{rel}}| \leq m - \varepsilon, \qquad |\operatorname{Im} q_{\text{rel}}| \leq m - \varepsilon. \end{aligned} \tag{14.4.6}$$

In this region, $K^{\sim}$ is assumed to be bounded. For weak coupling $P(\phi)_2$ fields, such bounds have been established in slightly smaller regions [Spencer, 1975]. The sup norm on $K^{\sim}$ is $O(\lambda)$ as $\lambda \to 0$.

We summarize the results as follows:

I. *Compactness of K, boundedness of R_0 and R.* We introduce below Paley–Wiener–Sobolev spaces A_δ. As operators between suitable A_δ spaces, R_0 and R are bounded and K is compact.

II. *Continuation to the second sheet.* R_0 is analytic in (14.4.6) except for cuts beginning at $E = \pm 2m$. The beginnings of the cuts are square root branch points for R_0, and R_0 has a continuation onto a two sheeted Riemann surface around the branch points. The analytic Fredholm theory shows that R has a meromorphic continuation onto the second sheet. The only poles of R on the first sheet lie on the real E axis and coincide with the point spectrum of the mass operator (i.e. bound states). The poles of R on the second sheet are interpreted as resonances.

III. *Spectral properties.* For an even $P(\phi)_2$ field at weak coupling, there is at most one bound state. Asymptotic completeness for $E \leq 3m - \varepsilon$ is known, and the particle–bound-state S matrix (for all energies) is obtained. Perturbation theory is asymptotic in this region.

The spaces A_δ are functions of two variables: $x_{rel} = x^0_{rel}$, $\mathbf{x}_{rel}$. Here δ is a multiindex,

$$\delta = (\delta_1, \delta_2) = (\delta^0_1, \boldsymbol{\delta}_1, \delta^0_2, \boldsymbol{\delta}_2),$$

and δ is a real parameter (i.e. a scalar, not a vector). To begin, $A_0 = L_2^{sym}$ is the subspace of L_2 symmetric under the map $x_{rel} \to -x_{rel}$. $A_{\delta_1, 0}$ is the Sobolev space

$$A_{\delta_1, 0} = \{f : (-\partial^2_{x_{0, rel}} + (5m)^2)^{\delta^0_1/2} f \in A_0, (-\partial^2_{\mathbf{x}_{rel}} + (5m)^2)^{\delta_1/2} f \in A_0\}.$$

Finally,

$$A_\delta = \{f : e^{\delta^0_2((x_{0, rel})^2 + 1)^{1/2}} e^{\delta_2(\mathbf{x}_{rel}^2 + 1)^{1/2}} f \in A_{\delta_1, 0}\}.$$

Elementary properties of the spaces A_δ are given in [Glimm and Jaffe, 1979c]. (The analysis there, for functions of $\mathbf{x}_{rel}$ alone, easily extends to functions of $x_{0, rel}, \mathbf{x}_{rel}$.)

With $\delta_2 > 0$, A_δ is the Fourier transform of functions analytic in a strip

$$|\operatorname{Im} p_{0, rel}| < \delta^0_2, \qquad |\operatorname{Im} \mathbf{p}_{rel}| < \delta_2$$

and with boundary values at the edge of the strip which after multiplication by

$$(p^2_{0, rel} + (5m)^2)^{\delta^0_1/2} + (\mathbf{p}^2_{rel} + (5m)^2)^{\delta_1/2}$$

are in L_2. For $\gamma \leq \delta$, the inclusion map

$$I(\delta, \gamma) : A_\delta \to A_\gamma$$

is bounded, and for $\gamma < \delta$ (all four components $\gamma^0_1 < \delta^0_1$, etc.), $I(\delta, \gamma)$ is compact.

Consider K as a map from $A_\delta \to A_\gamma$. By the analyticity axiom BS3, K tends to increase the size of δ_2, but because the kernel K is only bounded (and not e.g., L_2), it tends to decrease the size of δ_1. Specifically, K is compact and Hilbert–Schmidt for the following choices of γ, δ:

$$\begin{aligned} &\delta^0_1, \delta_1 > 1, &&\gamma^0_1, \gamma_1 < -1, \\ &\delta^0_2, \delta_2 > -m, &&\gamma^0_2, \gamma_2 < m. \end{aligned} \qquad (14.4.7)$$

R_0, as a map from $A_\delta \to A_\gamma$, has just the opposite effect on the indices δ. From Section 14.3, we can write $S^{(2)T} = ZS_0^{(2)T} + \text{remainder}$, where the remainder improves both analyticity and polynomial decrease in momentum space. Similarly, R_0, regarded as a multiplication operator in the momentum space realization of A, can be written as

$$\frac{Z}{p_1^2 + m^2} \frac{Z}{p_2^2 + m^2} + \text{remainder}, \qquad (14.4.8)$$

where the remainder improves both indices δ_1 and δ_2. Control over the remainder follows from the boundedness and analyticity assumptions BS2. The first term in (14.4.8) can be rewritten as

$$R_{00} = \frac{16Z^2}{[(p_{0,\,\text{rel}} - iE)^2 + \mathbf{p}_{\text{rel}}^2 + 4m^2][(p_{0,\,\text{rel}} + iE)^2 + \mathbf{p}_{\text{rel}}^2 + 4m^2]}.$$

In the L_2 inner product, we identify $A^* = A_{-\gamma}$. Thus we study R_{00} as a bounded bilinear form on $A_{-\gamma} \times A_\delta$. Because of the symmetry restriction on A_δ, we have

$$g^{\sim -}(p) = -g^{-\sim}(p) \in A_\delta,$$

and the inner product

$$\langle g^\sim, R_{00} f^\sim \rangle = \int g^{\sim -}(p_{\text{rel}}) R_{00}(p_{\text{rel}}) f^{\sim -}(p_{\text{rel}}) \, dp_{\text{rel}}$$

has an analytic continuation in a strip. Thus we assume $g \in A_{-\gamma}, f \in A_\delta$, and

$$\delta_2^0, \delta_2 > m - \varepsilon, \qquad \gamma_2^0, \gamma_2 < -m + \varepsilon.$$

The $dp_{0,\,\text{rel}}$ integral can be evaluated as an integral along the upper boundary of the strip plus the residue from the pole in the upper half of the strip. The first of these two terms is bounded and analytic in E (no cut) for

$$\delta_1^0, \delta_1 \geq -2, \qquad \gamma_1^0, \gamma_1 \leq 2.$$

The second term requires no restriction on δ_1^0, γ_1^0, and is bounded for

$$\delta_1 \geq -\tfrac{3}{2}, \qquad \gamma_1 \leq \tfrac{3}{2}.$$

For these same values of δ, γ, the second term has an analytic continuation in $\zeta = (4m^2 - E^2)^{1/2}$ around the cut.

Combining all above bounds, we have a convergent Neumann series

$$R = \sum_{n=0}^{\infty} (-R_0 K)^n R_0$$

which is bounded from A_δ to A_δ for

$$-\tfrac{3}{2} \leq \delta_1^0, \delta_1 < -1, \qquad 0 < \delta_2^0, \delta_2 < m - \varepsilon.$$

and has a meromorphic continuation in ζ, as a bounded operator from A_δ to A_δ, for small λ.

The analysis of the spectrum now proceeds along lines similar to potential scattering (see [Spencer and Zirilli, 1976] and [Dimock and Eckmann, 1976 and 1977]). Subtractions associated with higher irreducible kernels are studied in [Combescure and Dunlop, 1979].

CHAPTER 15

The Magnetic Moment of the Electron

15.1 Classical Magnetic Moments

Consider a classical magnet m in an external constant magnetic field $\mathbf{B}$. The field exerts no net force on m (i.e., the center of mass remains fixed), but the magnet tends to orient itself with respect to the direction of $\mathbf{B}$. The torque to rotate m is linear in $\mathbf{B}$ (assuming that $\mathbf{B}$ does not change the magnetization of m). Thus

$$\text{torque} = \boldsymbol{\mu} \times \mathbf{B}, \qquad (15.1.1)$$

where $\boldsymbol{\mu}$ is a vector along the axis of m. The vector $\boldsymbol{\mu}$ is the magnetic moment of m. Equivalent to (15.1.1) is the statement that m has a potential energy $-\boldsymbol{\mu} \cdot \mathbf{B}$ in the magnetic field. This energy is minimized by aligning $\boldsymbol{\mu}$ with $\mathbf{B}$.

A simple example of such a moment arises from considering a classical loop of current. Let l denote a circular loop of radius r with current $\mathbf{J}$ (units of charge/time); see Figure 15.1. We imagine that $\mathbf{J}$ consists of a density ρ of unit charges/cm traveling around the loop l at a velocity $\mathbf{v}$ (cm/sec). Thus

$$\mathbf{J} = \rho \mathbf{v}. \qquad (15.1.2)$$

Define the moment $\boldsymbol{\mu}$ of $\mathbf{J}$ about the center of the loop as

$$\boldsymbol{\mu} = \frac{1}{2c} \int_{\text{loop}} \mathbf{r} \times \mathbf{J} = \frac{\rho v}{c} \left(\pi r^2 \right) \mathbf{n}$$

$$= \frac{1}{c} (\text{current})(\text{area}) \mathbf{n}. \qquad (15.1.3)$$

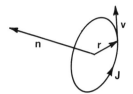

Figure 15.1. A circular current loop l.

Here c is the speed of light, and $\mathbf{n}$ is the unit vector normal to l (oriented by the current).

In general, the magnetic moment of a planar (but not necessarily circular) loop is

$$\boldsymbol{\mu} = \frac{1}{2c} \int_{\text{loop}} \mathbf{r} \times \mathbf{J} = \frac{1}{c} (\text{current})(\text{area})\mathbf{n}, \qquad (15.1.4)$$

with $\mathbf{n}$ normal to the loop.

To justify this definition of $\boldsymbol{\mu}$, we verify that the torque on a circular loop is given by (15.1.1). The force $\mathbf{F}$ on a unit charge is given by the Lorentz formula (for zero electric field)

$$\mathbf{F} = \frac{1}{c} \mathbf{v} \times \mathbf{B}. \qquad (15.1.5)$$

By definition the torque $\mathbf{T}$ is

$$\mathbf{T} = \int_{\text{loop}} \mathbf{r} \times \mathbf{F} = \int_{\text{loop}} \mathbf{r} \times \frac{\rho}{c}(\mathbf{v} \times \mathbf{B}). \qquad (15.1.6)$$

Parametrize the loop by an angle $\theta \in [0, 2\pi)$, where $\theta = 0$ denotes the direction of $\mathbf{B}_l$, the projection of $\mathbf{B}$ onto the plane of the loop l. Since $\mathbf{r} \cdot \mathbf{v} = 0$,

$$\mathbf{T} = \frac{\rho}{c} \int_0^{2\pi} \mathbf{r} \times (\mathbf{v} \times \mathbf{B}) r \, d\theta = \frac{\rho}{c} \int_0^{2\pi} \mathbf{v}(\mathbf{r} \cdot \mathbf{B}_l) r \, d\theta. \qquad (15.1.7)$$

Let 1 and 2 denote the directions in the plane of the loop parallel and perpendicular to $\mathbf{B}_l$, respectively. Then

$$v_1 = -v \sin \theta, \qquad v_2 = v \cos \theta,$$

$$\mathbf{r} \cdot \mathbf{B}_l = rB_l \cos \theta,$$

so $\int \cos \theta \sin \theta \, d\theta = 0$ ensures $T_1 = 0$ and

$$\mathbf{T} = (0, (c^{-1}\rho v)(\pi r^2)B_l, 0) = \boldsymbol{\mu} \times \mathbf{B}.$$

Thus (15.1.1) holds.

Suppose, furthermore, that the charge density ρ is proportional to the mass density,

$$\rho = \left(\frac{e}{m}\right)\rho_{\text{mass}}, \qquad (15.1.8)$$

where e/m is a fixed charge to mass ratio. Then

$$\mathbf{\mu} = \frac{1}{2c} \int \mathbf{r} \times \rho \mathbf{v}$$

$$= \frac{e}{2mc} \int \mathbf{r} \times \rho_{\text{mass}} \mathbf{v}$$

$$= \left(\frac{e}{2mc}\right)\mathbf{L} = g\mu_B \mathbf{L}, \tag{15.1.9}$$

where $\mathbf{L}$ denotes the angular momentum of the charges in the current loop. This standard proportionality between the magnetic moment of a charge distribution and the angular momentum defines the gyromagnetic ratio g. The unit of magnetic moment is the Bohr magneton, by definition

$$\mu_B = \frac{e}{2mc}, \tag{15.1.10}$$

where e/m is the charge to mass ratio of the electron.

The magnetic moment of the electron is $g\mu_B$. According to the classical calculation above, $g = 1$. For the Dirac electron of Section 15.3, $g = 2$. The anomalous magnetic moment of the electron includes field theory effects and gives a value $g = 2.002 \ldots$; see Section 15.4.

15.2 The Fine Structure of the Hydrogen Atom and the Dirac Equation

The nth energy level of the hydrogen atom was calculated to have multiplicity n^2 neglecting spin; see Section 1.7. Here we consider the first excited state, $n = 2$ and $n^2 = 4$. Allowing for two electron spin states and two proton spin states, the degeneracy is $4n^2 = 16$. Associated with the spins are magnetic moments. Each magnetic moment interacts with the other particle, and thus necessitates a change in the Hamiltonian of Section 1.7. The exact form of the perturbation is given below, and is derived in Section 15.3 as a nonrelativistic limit of the Dirac equation. The extra term, called spin–orbit coupling, changes the spectrum of H and in particular partially breaks the $4n^2$ degeneracy of the nth energy level. Splitting due to the spin and magnetic moment of the electron is called fine structure, and is considered here.

Recall from Section 1.3 that the spin space S of the spin $s = \frac{1}{2}$ electron is $S = C^{2s+1} = C^2$. On this space, the angular momentum is defined by the two dimensional representation $\mathscr{D}^{(1/2)}$ of SU(2) (the covering group of the group SO(3) of space rotations). The Hilbert space (center of mass motion removed) is $\mathscr{H} = L_2(R^3) \otimes S$, and the total angular momentum is defined by the representation of SU(2) on $\mathscr{H}$.

In particular the representation of SU(2) on $\mathscr{H}_n \otimes S$ can be reduced using the Clebsch–Gordan formulas

$$\mathscr{D}^{(l)} \otimes \mathscr{D}^{(1/2)} = \mathscr{D}^{(l+1/2)} \oplus \mathscr{D}^{(l-1/2)} \qquad (l \neq 0),$$

$$\mathscr{D}^{(0)} \otimes \mathscr{D}^{(1/2)} = \mathscr{D}^{(1/2)}. \tag{15.2.1}$$

Thus (1.7.10) is replaced by

$$\mathscr{H}_n \otimes S = \mathscr{D}^{(2n-1)/2} \oplus \sum_{l=1}^{n-1} 2\mathscr{D}^{(2l-1)/2}. \tag{15.2.2}$$

We interpret (15.2.2) as saying that the electron in hydrogen has a total angular momentum $J\hbar$, where

$$\mathbf{J} = \mathbf{L} + \mathbf{S},$$

in which $\mathbf{L}$ is an "orbital" angular momentum associated with the binding of the atom, and $\mathbf{S}$ is an intrinsic spin $\frac{1}{2}$. The irreducible representations in (15.2.2) of total angular momentum j arise from the combination of a spin $\frac{1}{2}$ representation with an orbital angular momentum $l = j \pm \frac{1}{2}$.

The fine structure of hydrogen can then be calculated by the inclusion in H of a perturbation

$$W(x_1 - x_2)\mathbf{L} \cdot \mathbf{S} \tag{15.2.3}$$

coupling the orbital and intrinsic electron spin. (A form for W can be derived from the coupling of the magnetic moment of the moving electron to the Coulomb field; see [Baym, 1969].) The appearance of n eigenvalues on $\mathscr{H}_n \otimes S$ is explained by the decomposition (15.2.2), as well as the multiplicities. Note that dim $\mathscr{H}_n \otimes S = 2n^2$. The calculation of the first order energy shift due to (15.2.3) yields the approximate fine structure.

Dirac proposed a Lorentz-covariant wave equation for the electron, acting (for one particle) on the space $L_2(R^3) \otimes \{S \oplus S\}$. (Note that the space $S \oplus S$ arises naturally in the reduction of the spin $\frac{1}{2}$ representation $\mathscr{D}^{(1/2,\, 0)} \oplus \mathscr{D}^{(0,\, 1/2)}$ of the Lorentz group into components irreducible under the special Lorentz group (no reflections).) The Dirac Hamiltonian for a particle of mass μ in a Coulomb field $-e^2/|x|$ is the first order operator

$$H = \alpha_4 \mu - c \sum_{j=1}^{3} \alpha_j \cdot i\hbar \frac{\partial}{\partial x_j} - \frac{e^2}{|x|}. \tag{15.2.4}$$

Here α_i are constant 4×4 matrices acting on $S \oplus S$. Furthermore, the α's represent the Clifford algebra

$$\alpha_i \alpha_j + \alpha_j \alpha_i = 2\delta_{ij} I. \tag{15.2.5}$$

The eigenvalues of (15.2.4) can in fact be found explicitly. They are

$$E_{n,\,j} = \mu \left\{ 1 + \left(\frac{\alpha}{(n - j - \frac{1}{2}) + \sqrt{(j + \frac{1}{2})^2 - \alpha^2}} \right)^2 \right\}^{-1/2}, \tag{15.2.6}$$

where $\alpha = e^2/(\hbar c)$ is the fine structure constant; see (1.7.2). Here $n = 1, 2, \ldots$

is the principal quantum number, while $j = \frac{1}{2}, \frac{3}{2}, \ldots, n - 1/2$. In fact these j-values are exactly the angular momenta which occur in (15.2.2). Furthermore the multiplicities also agree, and j is the angular momentum quantum number.

For $\alpha \ll j + \frac{1}{2}$, (15.2.6) can be expanded as a series in α, yielding

$$E_{n, j} = \mu - \frac{\mu \alpha^2}{2n^2} \left\{ 1 + \frac{\alpha^2}{n^2} \left(\frac{n}{j + \frac{1}{2}} - \frac{3}{4} \right) \right\} + O(\alpha^6). \tag{15.2.7}$$

The term μ is the rest mass of the electron. The $O(\alpha^2)$ term is the nonrelativistic expression (1.7.6) for E_n, and the $O(\alpha^4)$ term is the shift due to the spin–orbit interaction. Evaluation of this term yields the observed fine structure.

15.3 The Dirac Theory

The classical model of the current loop in Section 15.1 yields a magnetic moment

$$\boldsymbol{\mu}_d = \mu_B \mathbf{S}, \tag{15.3.1}$$

where $\mathbf{S} = (\hbar/2)\boldsymbol{\sigma}$ is the intrinsic spin and

$$\mu_B = \frac{e\hbar}{2mc} = 0.578 \cdots \times 10^{-14} \text{ MeV gauss}^{-1}$$

is the Bohr magneton. This moment can be measured by observing the Zeeman effect, i.e., the splitting of energy levels in an applied magnetic field $\mathbf{h}$; see Figure 15.2. The interpretation of that experiment requires a moment equal to $2\mu_{cl}$ of (15.3.1), i.e. to a gyromagnetic ratio $g = 2$.

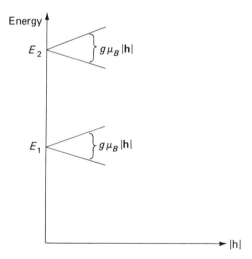

Figure 15.2. Nonrelativistic Zeeman effect, splitting the energy level E_n of Section 1.7 for small magnetic field h.

The best explanation of the origin of g comes from the Dirac theory of the electron. The Hamiltonian (15.2.4) is modified in the presence of a magnetic potential $A(x)$, namely

$$H = \alpha_4 \mu - c \sum_{j=1}^{3} \alpha_j \left(i\hbar \frac{\partial}{\partial x_j} - \frac{e}{c} A_j(x) \right)$$

$$= H_0 + e\boldsymbol{\alpha} \cdot \mathbf{A}, \tag{15.3.2}$$

where H_0 is the Hamiltonian for a free electron.

The natural scalar product in the Dirac theory leads to the expectation of (15.3.2) in a state ψ as

$$\langle H \rangle_\psi = \langle H_0 \rangle_\psi + \int \mathbf{J} \cdot \mathbf{A} \, dx, \tag{15.3.3}$$

where $\mathbf{J} = e\bar{\psi}\gamma_j\psi$, $\gamma_j = \alpha_4\alpha_j$, $j = 1, 2, 3$ and $\bar{\psi} = \psi^*\alpha_4$. The correction term $\mathbf{J} \cdot \mathbf{A}$ has the form of the classical energy of a current density in a magnetic potential, namely the scalar product. We now interpret this result.

Using the Dirac equation for ψ, namely

$$\sum_i \gamma_i \left(\hbar \frac{\partial}{\partial x_i} + i \frac{e}{c} A_i \right) \psi = mc\psi,$$

where $\gamma_4 = \alpha_4$, $x_4 = it$, the current $J_\mu = e\bar{\psi}\gamma_\mu\psi$ can be written

$$J_\mu = e\bar{\psi}\gamma_\mu\psi = \left(\frac{e\hbar}{2mc} \right) \frac{\partial}{\partial x_\nu} (\bar{\psi}\sigma_{\mu\nu}\psi) + \frac{e\hbar}{2mc} \operatorname{Im} \left(\bar{\psi} \frac{\partial}{\partial x_\mu} \psi \right)$$

$$+ i \left(\frac{e^2}{mc^2} \right) A_\mu(\bar{\psi}\psi), \tag{15.3.4}$$

where

$$\sigma_{\mu\nu} = \frac{1}{2i} [\gamma_\mu, \gamma_\nu]. \tag{15.3.5}$$

The second term on the right side of (15.3.4) has the form of a Schrödinger electric current. The last term has an extra factor c^{-1}, and vanishes relative to the others as $c \to \infty$. The first term provides the coupling to the magnetic field. In fact, inserting that term into $\mathbf{J} \cdot \mathbf{A}$ in (15.3.3) yields after integration by parts (neglecting the surface term)

$$-\left(\frac{e\hbar}{2mc} \right) \sum \int \bar{\psi}\sigma_{\mu\nu}\psi \, \partial_\nu A_\mu \, dx = \frac{e\hbar}{4mc} \int \bar{\psi}\sigma_{\mu\nu} \psi F_{\mu\nu} \, dx. \tag{15.3.6}$$

Now we remark that the matrices γ_i form a Clifford algebra

$$\{\gamma_\mu, \gamma_\nu\} = 2\delta_{\mu\nu}, \qquad \mu, \nu = 1, 2, 3, 4.$$

The Dirac representation is by an irreducible set of 4×4 matrices and is, up to unitary equivalence, unique. In this representation, for $\mu, \nu = 1, 2, 3$,

$$\sigma_{\mu\nu} = \varepsilon_{\mu\nu\lambda} \begin{pmatrix} \sigma_\lambda & 0 \\ 0 & \sigma_\lambda \end{pmatrix}, \tag{15.3.7}$$

where σ_λ, $\lambda = 1, 2, 3$, are the Pauli matrices

$$\sigma_1 = \begin{pmatrix} 0 & 1 \\ 1 & 0 \end{pmatrix}, \qquad \sigma_2 = \begin{pmatrix} 0 & -i \\ i & 0 \end{pmatrix}, \qquad \sigma_3 = \begin{pmatrix} 1 & 0 \\ 0 & -1 \end{pmatrix}. \qquad (15.3.8)$$

Thus (15.3.6) equals the magnetic term

$$\left(\frac{e}{mc}\right) \int \bar{\psi} \mathbf{S} \psi \cdot \mathbf{B} \, d\mathbf{x} = 2\mu_B \int \langle \psi, (\mathbf{S} \cdot \mathbf{B})\psi \rangle \, d\mathbf{x},$$

where $F_{ij} = \varepsilon_{ijk} B_k$, plus a term coupling to the electric field (and for which σ_{4j} is off diagonal). This is exactly a magnetic moment coupling with $g = 2$.

15.4 The Anomalous Moment

In fact, accurate experimental observation by Kusch in 1947 showed that g is not exactly 2, but very nearly so, namely $\kappa \equiv \frac{1}{2}(g - 2) = 0.001$. The difference $\kappa = \kappa_{\text{electron}}$ between the actual g factor and the Dirac prediction is known as the anomaly in the magnetic moment. Today, this anomaly provides the most accurate agreement between a calculation and a measured number, in the 8th decimal place of κ and the 11th decimal place of g; thus it merits discussion. The technique of Dehmelt and coworkers [van Dyke et al., 1979] isolates single electrons in a "magnetic bottle" to measure the anomaly as

$$\kappa_{\text{expt}} = 0.001159652200(40). \qquad (15.4.1)$$

The number in parentheses indicates the estimated uncertainty in the final digits.

In fact each elementary Dirac particle with accurately measured g has its own anomaly. The muon anomaly also agrees with calculations, almost as well as κ_{electron}. The proton has a much larger anomaly, nearly 1.8, which has not been calculated accurately. Even the neutron has a magnetic moment (approximately minus the anomalous moment of the proton). These moments reflect the complicated internal structure of these particles.

The theory of the anomaly κ_{electron} requires quantum fields, rather than the Dirac theory of a single electron. In Section 15.3, we calculated g by studying the solutions to the (linear) Dirac equation for the electron wave function ψ and its current J_μ reacting to a given electromagnetic potential A_μ. One understands the anomaly κ as arising from modification to the current J_μ due to its interaction with the electromagnetic field. This leads to the nonlinear coupled Maxwell–Dirac equations, with both A_μ and ψ regarded as unknowns.

Since the previous calculation, with A_μ specified, gives a value of g correct to 0.1%, it is reasonable to calculate g by assuming

$$A_\mu = A_\mu^{\text{ext}} + \delta A_\mu, \qquad (15.4.2)$$

where A_μ^{ext} is specified, as before, and δA_μ is a correction. The correction κ to $g = 2$ is then expanded as a power series in δA_μ, generating a perturbation series. Because the nonlinearity in the Maxwell–Dirac equations arises from the interaction $J_\mu A_\mu = O(e)$, the perturbation series for κ is a power series in the electric charge e. It turns out to depend only on powers of e^2, so the series is generally expressed in terms of the fine structure constant α. While we expect that the series will be divergent, the leading terms turn out to be small and to yield κ close to (15.4.1).

In fact, there is some debate today about whether one even expects the series to be asymptotic to a full theory (like the series in Part II for $\lambda P(\phi)$ models, which are asymptotic but not convergent about $\lambda = 0$). Thus it is a theoretical puzzle whether a *theory* of electrodynamics exists in the sense of a mathematical framework, e.g. as described in Chapters 7–12 for models in dimension $d < 4$, or in other words whether electrodynamics by itself (without strong, or possibly also weak, interactions) can be self-consistent. The same theoretical objection does not apply to the non-Abelian gauge models, and so a theory of the electron and proton, in combination with quarks and gluons, is on a sounder basis than the theory of electrons and photons in isolation from other matter.

Returning to the anomaly, we remark that Schwinger derived κ to order α in 1947, and found $\kappa = \frac{1}{2}(\alpha/\pi) = 0.001159$. Presently, the best theoretical number is given to order α^3 ([Levine and Roskies, 1976] and [Kinoshita, 1979]):

$$\kappa_{\text{calc}} = \frac{1}{2}\left(\frac{\alpha}{\pi}\right) + 0.328478966\left(\frac{\alpha}{\pi}\right)^2 + 1.1835(61)\left(\frac{\alpha}{\pi}\right)^3, \qquad (15.4.3)$$

where the uncertainty arises from numerical integrations. Using $\alpha^{-1} = 137.035963(15)$ [Williams and Olsen, 1979] gives

$$\kappa_{\text{calc}} = 0.001159652566.$$

There appears at present a discrepancy between the κ_{calc} and κ_{expt} which may test the (as yet not known) $O(\alpha^4)$ corrections. (This discrepancy has not been resolved by considering the effects of strong interactions.)

The calculation of g involves the expectation of the interaction energy density $J_\mu(x)A^\mu(x)$, in a state vector for a single electron, evaluated as a series in the electric charge e. As in (15.3.4–6), we can identify a term proportional to a magnetic moment coupling, with the form $\langle\frac{1}{2}\boldsymbol{\sigma}\cdot\mathbf{B}\rangle$. Its coefficient, at zero momentum, is the desired quantity $geh/2mc = g\mu_B$. Thus g is expressed as a series in e, obtained from $\langle\mathbf{J}\cdot\mathbf{A}\rangle$ by successive integration by parts, i.e., by perturbation theory as in Chapter 8 for boson models.

The calculation reduces to a sum of Feynman graphs. It is necessary to distinguish fermion lines (denoted by solid, directed lines) from photon (wavy) lines. The calculation to order α for g is given by the contributions from two graphs, shown in Figure 15.3.

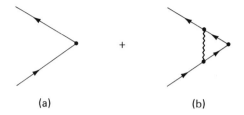

Figure 15.3. The graphs contributing to $g = 2(1 + O(\alpha))$.

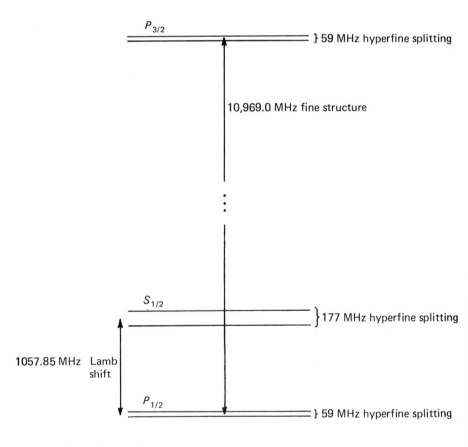

Figure 15.4. The $n = 2$ levels of hydrogen. The S states have $l = 0$, $j = \frac{1}{2}$. The P states have $l = 1$, $j = \frac{3}{2}$ or $j = \frac{1}{2}$. The $n = 2$ levels are approximately 2.5×10^9 MHz above the $n = 1$ levels. The hyperfine splitting of the $n = 1$ levels is 21 cm = 1420 MHz. In the nonrelativistic theory of Chapter 1, all the levels with given n are degenerate. In the Dirac equation, the $S_{j=1/2}$ and $P_{j=1/2}$ levels are exactly degenerate, but the fine structure separates the $P_{j=3/2}$ levels.

The graph (a) gives $g = 2$; cf. Section 15.3. The graph (b) gives a contribution $g = \alpha/\pi$, and together they yield $\kappa = \frac{1}{2}(\alpha/\pi)$, as described above. See [Scadron, 1979].

15.5 The Hyperfine Structure and the Lamb Shift of the Hydrogen Atom

The proton in the hydrogen atom also has spin $\frac{1}{2}$ and a magnetic moment which perturbs the Hamiltonian by a spin–orbital coupling term. The size of this splitting (the "hyperfine structure") is at most $\frac{1}{10}$ the fine structure. It occurs in the proton spin state degeneracy, which was unaffected by the electron magnetic moment. Thus it shows up as a further splitting of degenerate levels.

In the ground state ($n = 1$) the hyperfine splitting corresponds to a wavelength of 21 cm and is a basic frequency observed in radio astronomy.

The $n = 2$ levels are classified by total angular momentum j and orbital angular momentum l, $0 \leq l \leq 1$. By convention $l = 0$ (spherically symmetric) is denoted S, and $l = 1$ is denoted P. Thus by (15.2.2), the states $S_{j=1/2} = S_{1/2}$, $P_{1/2}$, and $P_{3/2}$ occur in the $n = 2$ levels. The fine structure splits the $P_{3/2}$ from the $S_{1/2}$ and $P_{1/2}$ levels. The hyperfine structure splits each of these levels in two, but leaves the $S_{1/2}$ and $P_{1/2}$ levels coinciding.

The Lamb shift, which is a purely field theoretic effect, is the main contribution to the splitting of $S_{1/2}$ from $P_{1/2}$ levels. (See Figure 15.4). The best experimental measurement is 1057.845(9) MHz [Lundeen and Pipkin, 1981].

REFERENCE

[Scadron, 1979].

CHAPTER 16

Phase Transitions

16.1 Introduction

To define the basic concepts, we consider a quantum field determined by a function space measure $d\mu$ defined on $\mathscr{S}'$. Assuming the axioms OS0–3 of Chapter 6.1, $d\mu$ can be decomposed into irreducible components $d\mu$, called *pure phases*, each of which satisfies the axioms OS0–4. The pure phases are characterized by the property OS4 of a unique vacuum, and $d\mu$ is itself a pure phase if and only if its vacuum state is unique.

As in Chapter 11, $d\mu$ is specified by a Euclidean Lagrangian $\mathscr{L}$, which for a $P(\phi)_2$ measure is

$$\mathscr{L}(\phi(x)) = \; :\tfrac{1}{2}(\nabla\phi)^2(x) + \tfrac{1}{2}m^2\phi^2(x) + \lambda P(\phi(x)): . \qquad (16.1.1)$$

The Lagrangian is required as input for the Chapter 11 construction of $d\mu$. Moreover, $\mathscr{L}$ can be recovered from $d\mu$ using the integration by parts formula (12.1.1). Note that this statement survives renormalization in the sense that the renormalized current P' is a well-defined bilinear form. In known superrenormalizable models and formally for ϕ_4^4, P' is explicitly defined by removal of ultraviolet cutoffs; the possibly divergent coefficients in P' determine those in P algebraically, and hence those in $\mathscr{L}$ also.

We call any measure $d\mu$ which satisfies (12.1.1) a *phase* of the $\lambda P(\phi)_2$ Lagrangian.

In addition to $\mathscr{L}$, an explicit choice of boundary conditions was used to construct $d\mu$. (In Chapter 11, we used Dirichlet boundary conditions; the methods of Chapter 18 can be adapted to a wide variety of boundary conditions.) Any distinct measure $d\tilde{\mu}$ constructed by another limit procedure, but using the same Lagrangian $\mathcal{L}$, is defined as an alternative phase or mixture of pure phases. We say that $\mathcal{L}$ has a *first order phase transition* if two or more distinct phases occur for $\mathcal{L}$.

There is a second point of view used to consider first order phase transitions. Instead of varying the boundary conditions, consider a variation of the Lagrangian

$$\mathcal{L} = \mathcal{L}_0 + \delta\mathcal{L}.$$

If the measure

$$d\mu_{\mathcal{L}} = d\mu_{\mathcal{L}_0 + \delta\mathcal{L}}$$

is discontinuous in $\mathcal{L}$ at $\mathcal{L} = \mathcal{L}_0$, then distinct phases

$$d\mu_{\mathcal{L}} = \lim_{j \to \infty} d\mu_{\mathcal{L}_0 + \delta\mathcal{L}_j}$$

can be constructed from the Lagrangian $\mathcal{L}_0$, by distinct choices of a sequence $\delta\mathcal{L}_j \to 0$. The word transition refers to the fact that a continuous variation of the parameters (i.e., $\mathcal{L}$) produces a discontinuous change in the theory. In most examples, the same first order phase transitions and the same multiple phases are produced by either of the two methods suggested above: variation of boundary conditions at $x = \infty$ or infinitesimal variation of $\mathcal{L}$. In quantum field theory, the main significance of phase transitions is that the long distance behavior (i.e., particle states and scattering) of the field may be qualitatively different from that suggested by the parameters in $\mathcal{L}$. The Higgs mechanism and solitons are examples.

In this section, the criteria for the occurrence of a pure phase—i.e., a unique vacuum—is reduced to a study of the two point function for a $P(\phi)_2$ quantum field. Moreover, in the case of a unique vacuum, the two point function is shown to contain the physical mass, that is, the slowest exponential decay rate for correlations of the theory. Finally, the relation of phase transitions to singularities of the thermodynamic functions is discussed.

Theorem 16.1.1. *Let a measure $d\mu$ on $\mathcal{S}'$ satisfy OS0–3 and the FKG inequality (Sections 4.4, 10.2). Then axiom OS4 (unique vacuum) holds if and only if the truncated two point Schwinger function $S_2(x - y)$ vanishes as $|x - y| \to \infty$. In this case, the exponential decay rate of S_2 as $|x - y| \to \infty$ coincides with the gap in the spectrum of the Hamiltonian H between the vacuum energy level (zero) and the rest of the spectrum.*

PROOF   The basic ingredient of the FKG inequality is a set of monotone functions of the fields. Let

$$\sigma(x) = \begin{cases} x & \text{if } |x| \le 1, \\ \text{sgn } x & \text{if } 1 \le |x|, \end{cases}$$

$$\rho(x) = \tfrac{1}{2}(1 + \sigma(x)),$$

and for f a nonnegative function in $\mathscr{S}$, let

$$\sigma(f) = \sigma(\phi(f)), \qquad \rho(f) = \rho(\phi(f)).$$

Note that

$$\langle \rho(f)\rho(g) \rangle - \langle \rho(f) \rangle \langle \rho(g) \rangle = \tfrac{1}{4}[\langle \sigma(f)\sigma(g) \rangle - \langle \sigma(f) \rangle \langle \sigma(g) \rangle].$$

From the identity $\phi(f) = \lim_{\lambda \to \infty} \lambda\sigma(\lambda^{-1}f)$, we see that products of the σ's or of the ρ's span $\mathscr{E}$, and with f restricted to have support in R_+, the same products span $\mathscr{E}_+$, and under the map $\wedge : \mathscr{E}_+ \to \mathscr{E}_+/\mathscr{N} \subset \mathscr{H}$, they span $\mathscr{H}$. Let $\psi \in \mathscr{H}$ be the image of such a product. We assert that there is an $f \in \mathscr{S}$ with $\mathrm{suppt}\, f \subset R^+$ such that

$$\langle \psi, e^{-tH}\psi \rangle - \langle \psi, \Omega \rangle^2 \le \langle \phi(f)^\wedge \Omega, e^{-tH}\phi(f)^\wedge \Omega \rangle - \langle \phi(f)^\wedge \Omega, \Omega \rangle^2. \quad (16.1.2)$$

Let P_Ω be the orthogonal projection onto $\Omega \in \mathscr{H}$. Then (16.1.2) can be rewritten

$$\langle \psi, e^{-tH}(1 - P_\Omega)\psi \rangle \le \langle \phi(f)^\wedge \Omega, e^{-tH}(1 - P_\Omega)\phi(f)^\wedge \Omega \rangle. \quad (16.1.3)$$

If the right hand side vanishes as $t \to \infty$, then $(1 - P_\Omega)\mathscr{H}$ cannot contain eigenvectors for H with zero eigenvalue, or in other words, the vacuum Ω is unique. In this case, the bottom of the spectrum of $H\,|\,(1 - P_\Omega)\mathscr{H}$ is realized by vectors in the span of $(1 - P_\Omega)\phi(f)^\wedge \Omega$, again by (16.1.3). Thus the "if" part of the theorem follows from the assertion.

To prove the assertion, let $A = \{f_1, \ldots, f_n\}$ with $f_i \in \mathscr{S}$, $0 \le f_i$, let

$$\Pi_A = \prod_{f \in A} \rho(f),$$

$$\Sigma_A = \sum_{f \in A} \rho(f),$$

and observe that the following functions are monotone in ϕ:

$$\phi(f_i), \ \sigma(f_i), \ \phi(f_i) - \sigma(f_i), \ \Sigma_A, \ \Pi_A, \ \Sigma_A - \Pi_A. \quad (16.1.4)$$

For the last of these functions, the fact that $0 \le \rho \le 1$ is used to verify monotonicity in each variable $\rho(f_i)$ separately.

From the FKG inequality,

$$0 \le \langle \sigma_1(\phi_2 - \sigma_2) \rangle - \langle \sigma_1 \rangle \langle \phi_2 - \sigma_2 \rangle,$$

$$0 \le \langle (\phi_1 - \sigma_1)\phi_2 \rangle - \langle \phi_1 - \sigma_1 \rangle \langle \phi_2 \rangle.$$

Adding these inequalities gives

$$\langle \sigma_1 \sigma_2 \rangle - \langle \sigma_1 \rangle \langle \sigma_2 \rangle \le \langle \phi_1 \phi_2 \rangle - \langle \phi_1 \rangle \langle \phi_2 \rangle. \quad (16.1.5)$$

By the same reasoning, applied to the monotone functions Π, Σ, $\Sigma - \Pi$,

$$\langle \Pi_A \Pi_B \rangle - \langle \Pi_A \rangle \langle \Pi_B \rangle \le \langle \Sigma_A \Sigma_B \rangle - \langle \Sigma_A \rangle \langle \Sigma_B \rangle. \quad (16.1.6)$$

The left side of (16.1.6) coincides with the left side of (16.1.2), if a Euclidean time translation by $t/2$ is included in the f_i. The right side of (16.1.6) is expanded, and majorized by (16.1.5); it then coincides with the right side of (16.1.2). This completes the proof of the assertion and the "if" part of the theorem. The converse is contained in the following elementary results.

Proposition 16.1.2. *Let H be a positive self-adjoint operator. Then*

$$\text{st lim}_{t \to \infty} e^{-tH}$$

is the projection onto the eigenspace of H for the eigenvalue zero.

Proposition 16.1.3. *Let $d\mu$ be a measure on $\mathscr{S}'$ which satisfies OS0–4. Then for all vectors $\psi, \chi \in \mathscr{H}$,*

$$0 = \lim_{t \to \infty} \langle \psi, e^{-tH} \chi \rangle - \langle \psi, \Omega \rangle \langle \Omega, \chi \rangle.$$

PROOF. $$P_\Omega = |\Omega\rangle\langle\Omega| = \text{st lim}_{t \to \infty} e^{-tH}$$

by Proposition 16.1.2 and the axiom OS4.

We continue the discussion of phase transitions on a formal level. In a finite volume Λ, with a measure $d\mu_\Lambda$, the free-energy per unit volume $a_\Lambda(h)$ is

$$a_\Lambda(h) = |\Lambda|^{-1} \ln\left(\int e^{h \int_\Lambda \phi(x)\, dx} \, d\mu_\Lambda \right), \tag{16.1.7}$$

and by monotone convergence (for the even and linear interaction of Chapter 11 with Dirichlet boundary conditions), $a_\Lambda(h) \to a(h)$ as $\Lambda \uparrow R^2$. By a formal interchange of a derivative and the two $\Lambda \uparrow R^2$ limits—one in the exponent and one in $d\mu_\Lambda$—we can evaluate da/dh and obtain

$$\frac{da}{dh} = \langle \phi(x) \rangle_h, \tag{16.1.8}$$

where $\langle \cdot \rangle_h$ is the expectation in the measure with external field h. Continuing, the nth derivatives are the truncated n point Schwinger functions

$$\frac{d^2 a}{dh^2} = \int \left(\langle \phi(x)\phi(y) \rangle - \langle \phi(x) \rangle \langle \phi(y) \rangle \right) dy \tag{16.1.9}$$

and

$$\frac{d^n a}{dh^n} = \int \langle \phi(x_1), \ldots, \phi(x_n) \rangle^T \, dx_2 \cdots dx_n. \tag{16.1.10}$$

Note that da/dh is the magnetization and $d^2 a/dh^2$ is the susceptibility. In the language of Chapter 5, $a(h)$ is the equation of state, and the derivatives $d^n a/dh^n$ are each thermodynamic functions.

From the FKG or Griffiths inequalities, we see that $0 \le d^2 a/dh^2$. Thus a is convex as a function of h. Moreover the magnetization da/dh is monotone increasing, and hence continuous, except for a countable number of values h. A point h_0 of discontinuity of da/dh is a first order phase transition, and has at least two phases (corresponding to one sided continuity of

da/dh and $\langle\cdot\rangle_h$), namely

$$\langle\cdot\rangle_{h\pm0} = \lim_{\varepsilon\to+0} \langle\cdot\rangle_{h\pm\varepsilon}.$$

As a partial converse, suppose that da/dh is continuously differentiable at $h = h_0$. Then the truncated two point function is integrable, as in (16.1.9). Since $0 \le \langle\phi(x)\phi(y)\rangle^T = \langle\phi(0)\Omega, \ e^{-|x-y|H}(I - P_\Omega)\phi(0)\Omega\rangle$ is monotonic decreasing in $|x - y|$, necessarily $\langle\phi(x)\phi(y)\rangle^T \to 0$ as $|x - y| \to \infty$. Thus by Theorem 16.1.1, multiple phases do not occur in the $h = h_0$ theory. Moreover there is no first order phase transition.

It is also possible for the thermodynamic functions to develop singularities other than jump discontinuities. In this case there will be a unique phase for the given Lagrangian, and the phase transition is said to be *second* or *higher order*. Normally critical points are points of second order phase transition; by definition a critical point is a boundary point (in the space of Lagrangians $\mathscr{L}$) of the manifold of first order phase transitions.

16.2 The Two Phase Region

We consider the simplest case in which the existence of at least two phases and a first order phase transition can be established, namely the polynomial interaction

$$V(\phi) = \lambda : (\phi^2 - \lambda^{-1})^2 :_{\lambda^{1/2}}, \qquad 0 < \lambda \ll 1, \qquad (16.2.1)$$

in $d = 2$ dimensions. Here the subscript $\lambda^{1/2}$ indicates Wick order with respect to the covariance $(-\Delta + \lambda)^{-1}$. The same methods allow ϕ^4 to be replaced by a positive even polynomial. The Wick ordering group maps other parts of the two phase regions for even polynomials onto (16.2.1); see [Glimm, Jaffe and Spencer 1976b]. Extensions of the methods presented here allow the construction of two phases in certain cases where the interaction V is neither even nor symmetric about any value of ϕ; see Chapter 20.

Theorem 16.2.1. *Let $d\mu$ be the measure on $\mathscr{S}'(R^2)$ constructed in Chapter 11, for the interaction polynomial V of (16.2.1) and with $m = \lambda^{1/2}$. Then for $0 < \lambda$ sufficiently small, $d\mu$ has a nonunique vacuum state, i.e., multiple phases.*

The proof of the theorem is motivated by the Ising model proof of Section 5.4. In order to eliminate short distance fluctuations, which are not relevant to the existence of phase transitions, we consider the average of ϕ over a small region. We cover R^2 by a lattice of unit squares Δ_i and consider

$$\phi(\Delta) = \int_\Delta \phi(x)\, dx. \qquad (16.2.2)$$

Then

$$\sigma(\Delta) = \text{sgn } \phi(\Delta) \tag{16.2.3}$$

is an Ising variable, since it takes values $+1$, and is defined on integer lattice points (the centers of the squares Δ_i). Because of the $\phi \to -\phi$ symmetry of $d\mu$,

$$\langle \sigma(\Delta) \rangle = \int \sigma(\Delta) \, d\mu = 0. \tag{16.2.4}$$

The theorem requires an analysis of the probability of phase boundaries. For a.e. classical field configuration $\phi \in \mathscr{S}'(R^2)$, $\sigma(\Delta)$ defines an Ising model configuration. Thus $\sigma \in \{\pm 1\}^{Z^2}$, so that σ is a function from Z^2 (as indices for the set of lattice squares) to the values ± 1. The phase boundary of the configuration ϕ or σ is the boundary of the set $\sigma^{-1}(1)$, or in other words, the set of line segments across which the function $\Delta \to \phi(\Delta)$ changes sign. Because we are studying the infinite volume theory with a translation invariant expectation, the set of configurations which give rise to any specific phase boundary

$$\partial\sigma^{-1}(1) = \text{phase bdry } \sigma = \text{phase bdry } \phi \tag{16.2.5}$$

is a set of measure zero. Thus we analyze as elementary events the sets

$$\{\phi : \Gamma \subset \text{phase bdry } \phi\} \tag{16.2.6}$$

where Γ is a *finite* set of lattice line segments. By abuse of notation, we also let $\Gamma \subset \mathscr{S}'(R^2)$ denote the set (16.2.6).

Theorem 16.2.2. *For λ sufficiently small,*

$$\text{Pr}(\Gamma) = \int_\Gamma d\mu \le e^{-\text{const } \lambda^{-1/2}|\Gamma|},$$

where $|\Gamma|$ is the length of Γ in R^2, and the constant is independent of Γ and λ.

PROOF OF THEOREM 16.2.1, ASSUMING THEOREM 16.2.2. We show that

$$\langle \sigma(\Delta)\sigma(\Delta') \rangle \not\to 0$$

as dist$(\Delta, \Delta') \to \infty$. Since $\langle \sigma(\Delta) \rangle = 0$ by (16.2.4), it follows from Proposition 16.1.3 that the vacuum is not unique.

Let

$$\rho_\pm(\Delta) = \tfrac{1}{2}\{1 \pm \sigma(\Delta)\}$$

denote the characteristic function of the positive and negative values for $\phi(\Delta)$. Since $\langle \phi \rangle = \langle \sigma(\Delta) \rangle = 0$,

$$\langle \sigma(\Delta)\sigma(\Delta') \rangle = 1 - 4\langle \rho_+(\Delta)\rho_-(\Delta') \rangle,$$

and the proof is completed by showing

$$\langle \rho_+(\Delta)\rho_-(\Delta') \rangle \le e^{-O(\lambda^{-1/2})}$$

for λ small.

The essential observation is that $\rho_+(\Delta)\rho_-(\Delta') \neq 0$ only for field configurations ϕ whose phase boundary (16.2.5) contains a curve Γ disconnecting Δ from Δ'. Thus

$$\langle \rho_+(\Delta)\rho_-(\Delta') \rangle \leq \sum_{\{\Gamma \,:\, \Delta \subset \text{Int } \Gamma \text{ or } \Delta' \subset \text{Int } \Gamma\}} \Pr(\Gamma).$$

Clearly, each curve Γ has length $|\Gamma| \geq 4$, and we claim that the number of Γ of length $|\Gamma| = n$ is bounded by $n^2\, 3^n$. This follows by induction. The starting point of Γ can be chosen in at most n^2 ways. Given a curve of length j, it can be enlarged, to length $j + 1$ without change of starting point in at most three ways, by adding one of the three bonds which meet its endpoint. Thus by Theorem 16.2.2,

$$\langle \rho_+(\Delta)\rho_-(\Delta') \rangle \leq \sum_{n=4}^{\infty} n^2\, 3^n e^{-nO(\lambda^{-1/2})}$$

$$\leq e^{-O(\lambda^{-1/2})}$$

to conclude the proof.

In order to prove Theorem 16.2.2, we study perturbations of the Euclidean measure $d\mu$ defined by polynomials of the form

$$Q(\xi, X) = \sum_{\nu=1}^{3} Q_\nu(\xi^{(\nu)}, X),$$

where X is a union of lattice squares Δ and

$$Q_1(\xi^{(1)}, X) = \lambda^{1/2} \sum_{\Delta_j \subset X} \xi_j^{(1)} \int_{\Delta_j} (:\phi^2(x):_{\lambda^{1/2}} - \lambda^{-1})\, dx,$$

$$Q_2(\xi^{(2)}, X) = |\ln \lambda|^{-1} \sum_{\Delta_j \subset X} \xi_j^{(2)} \int_{\Delta_j} :\phi^2(x) - \phi(\Delta_j)^2:_{\lambda^{1/2}}\, dx,$$

$$Q_3(\xi^{(3)}, X) = \sum_{\substack{\Delta_i, \Delta_j \subset X \\ |i-j|=1}} \xi_{ij}^{(3)}(\phi(\Delta_i) - \phi(\Delta_j)).$$

Theorem 16.2.3. *There is a constant $K < \infty$ such that for all $|\xi^{(\nu)}| \leq 1$, and λ sufficiently small,*

$$\langle e^{Q(\xi, X)} \rangle \leq e^{K|\ln \lambda|^2 |X|}.$$

The idea is to use multiple reflections to reduce local, infinite volume estimates to global, finite volume estimates. By Proposition 10.6.1 and Theorem 12.4.2, it is sufficient to bound

$$\langle e^{Q_\nu(\xi^{(\nu)}, X)} \rangle_\Lambda$$

in the case where all $\xi_j^{(\nu)}$ are equal and X is a rectangle which occupies at least a fixed fraction of the area of Λ.

PROOF OF THEOREM 16.2.2, ASSUMING THEOREM 16.2.3. We begin with the identity

$$\left\langle \prod_{(\Delta, \Delta') \in \mathscr{B}} [\rho_+(\Delta)\rho_-(\Delta') + \rho_-(\Delta)\rho_+(\Delta')] \right\rangle = \Pr(\Gamma), \qquad (16.2.7)$$

where $\mathcal{B}$ is the set of nearest neighbor pairs such that the union of the bonds $\Delta \cap \Delta'$ is Γ. Define

$$\rho_+ = \chi_{(0,\ (1/2)\ \lambda^{1/2})} + \chi_{((1/2)\ \lambda^{-1/2},\ \infty)} = \rho_{+,s} + \rho_{+,l},$$

$$\rho_- = \chi_{(-\infty,\ -(1/2)\ \lambda^{-1/2})} + \chi_{(-(1/2)\ \lambda^{-1/2},\ 0)} = \rho_{-,l} + \rho_{-,s}, \tag{16.2.8}$$

where $\chi_{(a,\ b)}(\xi)$ denotes the characteristic function of the interval $a < \xi < b$, and $\xi = \phi(\Delta)$ or $\phi(\Delta')$ in (16.2.8). The subscripts s, l on $\rho_\pm$ in (16.2.8) denote the "small" or "large" range of ϕ. We substitute in (16.2.7) and expand, getting $8^{|\mathcal{B}|} = 8^{|\Gamma|}$ terms; it is sufficient to consider a single term, since each term is positive and we can choose the largest. Thus for each pair $(\Delta_i, \Delta_j) \in \mathcal{B}$, we have a product of a $\rho_{+,l\ or\ s}$ with a $\rho_{-,l\ or\ s}$.

If both ρ's are ρ_l's, we use the bound for any even integer M,

$$\rho_{+,l}(\Delta_i)\rho_{-,l}(\Delta_j) \leq [\lambda^{1/2}(\phi(\Delta_i) - \phi(\Delta_j))]^M$$

$$= \lambda^{M/2}\left(\frac{d}{d\xi_{ij}^{(3)}}\right)^M e^{Q_3}\bigg|_{\xi=0}.$$

The other three types of factors contain either $\rho_{+,s}$ or $\rho_{-,s}$, which is bounded using $1 \leq (4/3)(1 - \lambda\phi(\Delta)^2)$, whenever $2\lambda^{1/2}|\phi(\Delta)| \leq 1$. Note

$$\lambda^{-1} - \phi(\Delta)^2 = \left(\lambda^{-1} - \int_\Delta\ :\phi(x)^2:\ dx\right) + \left(\int_\Delta\ :\phi(x)^2:\ dx\ -\ :\phi(\Delta)^2:\right)$$

$$+ \int_{\Delta\times\Delta}(-\Delta + \lambda)^{-1}(x,\ y)\ dx\ dy.$$

The last term is $0(|\ln \lambda|)$, and hence λ times this term is less than $1/3$ for small λ. Thus for λ small, we bound 1 by a Q_1 ór a Q_2 term, namely

$$1 \leq 2\lambda\left(\lambda^{-1} - \int_\Delta\ :\phi(x)^2:\ dx\right) + 2\lambda\left(\int_\Delta\ :\phi(x)^2:\ dx\ -\ :\phi(\Delta)^2:\right).$$

Furthermore, for M even,

$$\rho_s(\Delta) \leq \left[4\lambda\int_\Delta(:\phi(x)^2:\ -\ \lambda^{-1})\ dx\right]^M + \left[4\lambda\int_\Delta\ :\phi(x)^2\ -\ \phi(\Delta)^2:\ dx\right]^M.$$

Hence

$$\rho_s(\Delta_j) \leq (4\lambda^{1/2})^M\left[\left(\frac{d}{d\xi_j^{(1)}}\right)^M e^{Q_1} + \left(\frac{d}{d\xi_j^{(2)}}\right)^M \lambda^{M/2}(\ln \lambda)^M e^{Q_2}\right]_{\xi=0}.$$

Note $\lambda(\ln \lambda)^2 \leq 1$ for $\lambda \leq 1/2$, so the factor $\lambda^{M/2}(\ln \lambda)^M$ can be omitted. Again expand and choose the largest term from the less than $5^{|\Gamma|}$ terms obtained in this manner. The maximum term has a specific assignment of a Q_v in each square Δ bordering Γ. Let $\mathcal{B}_v$ denote the set of squares (or for $v = 3$, pairs of squares) with a Q_v term assigned, and let X_v denote the union of the squares in $\mathcal{B}_v$. Let $\mathcal{B} = \cup_v \mathcal{B}_v$ and $X = \cup_v X_v$. Thus

$$\Pr(\Gamma) \leq (40)^{|\Gamma|} \prod_{\Delta_j \in X_1}\left(4\lambda^{1/2}\frac{d}{d\xi_j^{(1)}}\right)^M \prod_{\Delta_j \in X_2}\left(4\lambda^{1/2}\frac{d}{d_j^{\nu(2)}}\right)^M$$

$$\times \prod_{(\Delta_i,\ \Delta_j)\ \in\ \mathcal{B}_3}\left(\lambda^{1/2}\frac{d}{d\xi_{ij}^{(3)}}\right)^M \langle e^{Q(\xi,\ X)}\rangle\bigg|_{\xi=0}.$$

Evaluate the derivatives at $\xi = 0$ using the Cauchy integral formula in the variables $\xi_j^{(v)}$, $\xi_{ij}^{(3)}$. Thus extend $\langle \exp Q(\xi, X) \rangle$ to a function of several complex variables, and integrate on the product of circles $|\xi_j^{(v)}| = |\xi_{ij}^{(3)}| = 1$. Because Q is linear in ξ, we have $|\exp(Q(\xi))| \le \exp(Q(\mathrm{Re}\ \xi))$, and therefore the Cauchy theorem yields

$$\Pr(\Gamma) \le (40)^{|\Gamma|} (4\lambda^{1/2})^{M(|\mathscr{B}_1| + |\mathscr{B}_2| + |\mathscr{B}_3|)} (M\,!)^{|\mathscr{B}_1| + |\mathscr{B}_2| + |\mathscr{B}_3|}$$

$$\times \sup_{|\xi^{(1)}| = |\xi^{(2)}| = |\xi_{ij}^{(3)}| = 1} \langle e^{Q(\mathrm{Re}\ \xi,\ X)} \rangle.$$

Let $\mathscr{N} \equiv |\mathscr{B}_1| + |\mathscr{B}_2| + |\mathscr{B}_3|$, and note $|\Gamma| \le \mathscr{N} \le 2|\Gamma|$. Also $|X| \le 2|\Gamma|$. By Theorem 16.2.3

$$\Pr(\Gamma) \le (40)^{|\Gamma|} [\lambda^{1/2} 4M\,!^{(1/M)}]^{\mathscr{N} M} \exp[2K(\ln\lambda)^2 |\Gamma|].$$

Using Stirling's formula, $M\,!^{(1/M)} \sim M/e$

$$\Pr(\Gamma) \le [(5\lambda^{1/2} M/e)^M \exp\{2K(\ln\lambda)^2 + \ln(40)\}]^{|\Gamma|}.$$

Choose M to be the largest even integer for which $5\lambda^{1/2} M \le 1$. Then there is a constant K_1 independent of λ, Γ, and such that for λ sufficiently small,

$$\Pr(\Gamma) \le \exp[\{-2K_1\lambda^{-1/2} + 2K(\ln\lambda)^2 + \ln(40)\}|\Gamma|]$$

$$\le \exp[-K_1\lambda^{-1/2}|\Gamma|].$$

This completes the proof of Theorem 16.2.2 and reduces the existence of phase transitions to Theorem 16.2.3.

As a preliminary step to proving Theorem 16.2.3, we state two related bounds.

Proposition 16.2.4. *Let* $0 < \lambda \le 1/2$ *and let* Λ *be an* $L \times T$ *rectangle with* $L \le T$ *and* $\lambda^{-3/2} \le L$. *For* $\xi(x) \in L_\infty(\Lambda)$, ξ *real, define*

$$W = \int_\Lambda V\,dx + Q_1$$

$$= \int_\Lambda :\lambda(\phi^2 - \lambda^{-1})^2 + \lambda^{1/2}\xi(x)(\phi^2 - \lambda^{-1}):_{\lambda^{1/2}}\,dx.$$

Let $C = (-\Delta_{\partial\Lambda} + \lambda)^{-1}$ *denote the covariance with Dirichlet data on* $\partial\Lambda$. *Then*

$$\exp[-K|\Lambda|] \le \int \exp[-W]\,d\phi_C \le \exp[K(\ln\lambda)^2|\Lambda|], \quad (16.2.9.)$$

where K *is a constant depending only on* $\|\xi\|_{L_\infty}$.

Remark. The $(\ln\lambda)^2$ factor can be eliminated from the upper bound [Glimm, Jaffe, and Spencer, 1976a].

Proposition 16.2.5. *There exists* $K < \infty$ *such that for all* $0 < \lambda \le 1/2$ *and* $|\xi_i^{(2)}|$, $|\xi_{ij}^{(3)}| \le 24$, *and all rectangles* X,

$$\int \exp[Q_2(\xi^{(2)}, X) + Q_3(\xi^{(3)}, X)]\,d\phi_C \le \exp[K|X|]. \quad (16.2.10)$$

PROOF OF THEOREM 16.2.3, ASSUMING PROPOSITIONS 16.2.4–5.   First use
Hölder's equality to give

$$\langle \exp Q(\xi, X) \rangle \leq \sup_{1 \leq v \leq 3} \langle \exp Q_v(3\xi^{(v)}, X) \rangle, \qquad ((16.2.11)$$

so bounds on the individual Q_v are sufficient, but with $\|\xi\|_{L_\infty} \leq 3$. For $v = 1, 2$,
next reduce to the case that X is a rectangle by using the multiple reflection bound
Corollary 10.5.8, with $k^{(j)} = \exp Q_v(\xi^{(v)}, \Delta_j)$. The reflections of such $k^{(j)}$ have the
form $\exp Q_v(\xi^{(v)}, Y)$, where Y is a rectangle. The bound,

$$\langle Q_v(\xi^{(v)}, Y) \rangle \leq \exp[K(\ln \lambda)^2 |Y|] \qquad (16.2.12)$$

which we prove below for rectangles Y, then yields the desired bound on (16.2.11)
for $v = 1, 2$. For $v = 3$, we make a corresponding reduction as follows: First use
Hölder's inequality to reduce Q_3 to a sum over non-overlapping pairs. (In the
process $|\xi| \leq 1$ is replaced by $|\xi| \leq 4$.) Now use Corollary 10.5.8, modified so
that Δ_j is replaced by $\Delta \cup \Delta'$, for $(\Delta, \Delta') \in \mathcal{B}_3$. In this case we are reduced to
bounding expectations by (16.2.12) with $v = 3$, but where Y is a $2L_1 \times L_2$
rectangle and where $\xi_{ij}^{(3)} \neq 0$ only for (Δ_i, Δ_j) in a set $\mathcal{B}$ of $L_1 L_2$ nonoverlapping
2×1 rectangles which cover Y.

The finite volume expectations $\langle \ \rangle_\Lambda$ defined by $d\mu_\Lambda$ of (11.2.1) converge to
$\langle \ \rangle$ as $\Lambda \uparrow R^2$. Thus

$$\langle \exp Q_v(\xi^{(v)}, Y) \rangle \leq 2\langle \exp Q_v(\xi^{(v)}, Y) \rangle_\Lambda \qquad (16.2.13)$$

whenever Λ contains a sufficiently large set depending on ξ, Y, λ and v. We say
$\Lambda \supset \Lambda(\xi, Y, \lambda, v)$. We now bound (16.2.13) for fixed (ξ, Y, λ, v), but obtain bounds
$\exp[K(\ln \lambda)^2 |Y|]$ where K does not depend on (ξ, Y, λ, v). It is no loss of generality
to choose Λ to be an $L \times T$ rectangle with $\lambda^{-3/2} \leq L \ll T$.

The next step in bounding (16.2.13) is to enlarge Y with respect to Λ, until
$|\Lambda| \leq 4Y$. Use the nonsymmetric multiple reflection bound of Theorem 12.4.2 in
the case $B = \exp Q_v(\xi^{(v)}, Y)$, and $K = Y$. Then $|K^{(n)}| \leq 4|Y^{(n)}|$ and $B^{(n)} =
\exp Q_v(\xi^{(v)}, K^{(n)})$. The assumption (12.4.9) on $Z(\Lambda)$ is the bound (16.2.9) of
Proposition 16.2.4, for the case $\xi \equiv 0$. We are now reduced to estimating

$$\langle B^{(n)} \rangle_\Lambda = \langle \exp Q_v(\xi^{(v)}, K^{(n)}) \rangle_\Lambda$$

$$= Z(\Lambda)^{-1} \int \exp[Q_v(\xi^{(v)}, K^{(n)}) - V(\Lambda)] \, d\phi_C$$

with $\Lambda = \Lambda^{(n)}$. It is sufficient to establish

$$\langle B^{(n)} \rangle_{\Lambda^{(n)}} \leq \exp[K(\ln \lambda)^2 |\Lambda^{(n)}|], \qquad (16.2.14)$$

from which (16.2.12) follows.

The lower bound on $Z(\Lambda^{(n)})$ of the form $\exp[-K|\Lambda^{(n)}|]$ is given by Proposition
16.2.4 for the case $\xi = 0$. The upper bound on $\int \exp[Q_1 - V] \, d\phi_C$ follows by the
same proposition. For $v = 2, 3$, apply the Schwartz inequality to give

$$\int \exp[Q_v - V] \, d\phi_C \leq \left\{ \int \exp[2Q_v] \, d\phi_C \int \exp[-2V] \, d\phi_C \right\}^{1/2}.$$

The two factors on the right are bounded from above by Propositions 16.2.5
and 16.2.4, respectively. This completes the proof.

PROOF OF PROPOSITION 16.2.4. Apply the scaling identity (8.6.26) with $\alpha = \lambda^{1/2}$ to yield (in the notation of Section 8.6),

$$\int \exp[-W] \, d\phi_C = \int \exp[-:P(\phi,f):_{C_\varnothing}] \, d\phi_{C_B}. \tag{16.2.15}$$

Here $C_\varnothing = (-\Delta + I)^{-1}$ and $C_B = (-\Delta_{\partial(\lambda^{1/2}\Lambda)} + I)^{-1}$. Let $\chi_X(x)$ denote the characteristic function of X. Then

$$f_4 = \chi_{\lambda^{1/2}\Lambda}, \qquad f_2 = \{-2\lambda^{-1} + \lambda^{-1/2}\xi(x\sqrt{\lambda})\}\chi_{\lambda^{1/2}\Lambda},$$
$$f_0 = \{\lambda^{-2} - \lambda^{-3/2}\xi(x/\sqrt{\lambda})\}\chi_{\lambda^{1/2}\Lambda}.$$

The next step is to re-Wick order P, using the transformation (8.6.1) to calculate g satisfying

$$:P(\phi,f):_{C_\varnothing} = :P(\phi,f+g):_{C_B}.$$

Then

$$g_2 = -6\,\delta c(x)\chi_{\lambda^{1/2}\Lambda}$$
$$g_0 = \{3(\delta c(x))^2 + 2\lambda^{-1}\,\delta c(x) - \lambda^{-1/2}\xi(x/\sqrt{\lambda})\,\delta c(x)\}\chi_{\lambda^{1/2}\Lambda}.$$

As a preliminary step, note that

$$|\delta c(x)| \le O(1)e^{-d}(1 + |\ln d|), \tag{16.2.16}$$

where $d = \mathrm{dist}(x, \partial(\lambda^{1/2}\Lambda))$. Hence

$$\left|\int g_0(x)\,dx\right| \le K|\Lambda|, \tag{16.2.17}$$

for a constant K depending only on $\|\xi\|_{L_\infty}$. Here we use the fact that $|\lambda^{1/2}\Lambda| = \lambda|\Lambda|$. As a consequence of (16.2.17), the contribution of g_0 need not be considered in proving (16.2.9).

(i) *The lower bound.* Let $\psi(x) = \phi(x) + h(x)$ define a translated field variable, with $h \in C_0^\infty(\lambda^{1/2}\Lambda)$ chosen below. Using the translation identity (9.1.27'),

$$\int \exp[-:P(\phi,f+g_2):_{C_B}] \, d\phi_{C_B} = \int \exp[-:P(\psi,l):_{C_B}] \, d\psi_{C_B}, \tag{16.2.18}$$

where $l = \{l_j\}$ is a set of coupling constants defined by (16.2.18). By Remark 1 after Corollary 10.3.2,

$$\exp\left[-\int l_0(x)\,dx\right] \le \int \exp[-:P(\psi,l):_{C_B}] \, d\psi_{C_B},$$

and we are free to choose h to optimize this bound. Hence the lower bound in (16.2.9) is a consequence of the upper bound

$$\int l_0(x)\,dx \le K|\Lambda|, \tag{16.2.19}$$

which we now establish with a particular choice of h. Note

$$\int l_0(x)\,dx = P(h,f+g_2) + \tfrac{1}{2}\langle h, C_B^{-1}h\rangle$$
$$= \int_{\lambda^{1/2}\Lambda} [(h(x)^2 - \lambda^{-1})^2 + \lambda^{-1/2}(h(x)^2 - \lambda^{-1})\xi(x/\sqrt{\lambda})$$
$$- 6\,\delta c(x)h(x)^2]\,dx + \tfrac{1}{2}\langle h, (-\Delta + I)h\rangle \tag{16.2.20}$$

Let $h(x) = \lambda^{-1/2}\chi_s(x)$, where χ_s is a smooth approximate characteristic function of $\lambda^{1/2}\Lambda$ with the following properties:

$$0 \leq \chi_s(x) \leq 1, \qquad \chi_s \in C_0^\infty(\lambda^{1/2}\Lambda),$$

$$\chi_s(x) = 1 \quad \text{whenever} \quad \text{dist}(x, \partial(\lambda^{1/2}\Lambda)) \geq 1,$$

and furthermore

$$|\nabla\chi_s(x)| \leq \text{const}, \tag{16.2.21}$$

with the constant independent of λ, Λ. Note

$$\nabla\chi_s(x) = 0 \quad \text{if} \quad \text{dist}(x, \partial(\lambda^{1/2}\Lambda)) \geq 1. \tag{16.2.22}$$

Using (16.2.21–2) and the assumption $\lambda^{-3/2} \leq L$,

$$\tfrac{1}{2}\langle h, (-\Delta + I)h \rangle = (2\lambda)^{-1}\{\|\nabla\chi_s\|^2 + \|\chi_s\|^2\}$$

$$\leq (2\lambda)^{-1}\{\text{const } \lambda^{1/2}(L + T) + \lambda|\Lambda|\}$$

$$\leq \text{const}\{\lambda^{-1/2}(L + T) + |\Lambda|\} \leq \text{const } |\Lambda|.$$

Thus the second term in (16.2.20) satisfies the bound (16.2.19).

To analyze the first term, note $h(x)^2 - \lambda^{-1} \equiv 0$ unless $\text{dist}(x, \partial(\lambda^{1/2}\Lambda)) \leq 1$. Also $0 \leq \lambda^{-1} - h(x)^2 \leq 2\lambda^{-1}$. Thus

$$\int [(h(x)^2 - \lambda^{-1})^2 + \lambda^{-1/2}(h(x)^2 - \lambda^{-1})\xi(x/\sqrt{\lambda})]\,dx$$

$$\leq 2(4\lambda^{-2})\lambda^{1/2}(L + T) + 4\lambda^{-3/2}\|\xi\|_{L_\infty}\lambda^{1/2}(L + T)$$

$$\leq \text{const } \lambda^{-3/2}(L + T) \leq \text{const } |\Lambda|,$$

with a constant depending only on $\|\xi\|_{L_\infty}$. Finally, using (16.2.16),

$$\left| \int \delta c(x)h(x)^2\,dx \right| \leq \lambda^{-1}\int_{\lambda^{1/2}\Lambda} |\delta c(x)|\,dx$$

$$\leq \text{const } \lambda^{-1}\lambda|\Lambda| \leq \text{const } |\Lambda|.$$

Adding these bounds proves (16.2.19).

(ii) *The upper bound.* We reduce the desired upper bound in (16.2.9) to an estimate in unit squares Δ which cover $\lambda^{1/2}\Lambda$. Use the conditioning inequality (10.3.7) to bound (16.2.15) from above. Taking (16.2.17) into account,

$$\int \exp[-W]\,d\phi_C = \int \exp[-: P(\phi, f + g):_{C_B}]\,d\phi_{C_B}$$

$$= Z_B(f + g) \leq \prod_\Delta Z_N((f + g)\chi_\Delta)$$

$$\leq \exp[K|\Lambda|]\prod_\Delta Z_N((f + g_2)\chi_\Delta).$$

Here Z_N denotes the partition function with Neumann covariance $C_N = (-\Delta_N + I)^{-1}$ used both in Wick ordering and in the covariance of the Gaussian. In order to estimate $Z_N((f + g_2)\chi_\Delta)$ from above, we derive a Wick lower bound on $: P(\phi_\kappa, (f + g_2)\chi_\Delta):_{C_N}$ which can be used with (8.6.23) to obtain

$$Z_N((f + g_2)\chi_\Delta) \leq \exp[K\lambda^{-1}(\ln \lambda)^2], \tag{16.2.23}$$

from which the desired inequality

$$\int \exp[-W]\, d\phi_C \leq \exp[K\lambda^{-1}\{1 + (\ln \lambda)^2\}\lambda |\Lambda|]$$

$$\leq \exp[\text{const}(\ln \lambda)^2 |\Lambda|]$$

follows. (Note that $Z_N((f + g_2)\chi_\Lambda)$ can be bounded using Proposition 8.6.2; however, this bound is too divergent as $\lambda \to 0$.)

Use the Wick ordering identity (8.5.5) to write

$$-\int_\Lambda B(x)\, dx \leq \int_\Lambda [A(x)^2 - B(x)]\, dx =: P(\phi_\kappa, (f + g_2)\chi_\Lambda):_{C_N}.$$

Here A is defined by completing a square and B is the remainder, in which leading order divergences cancel:

$$A(x) = \{\phi_\kappa(x)^2 + \tfrac{1}{2}f_2(x) + \tfrac{1}{2}g_2(x) - 3c_\kappa(x)\}\chi_{\lambda^{1/2}\Lambda}(x)$$

$$B(x) = \{\tfrac{1}{4}(f_2(x) + g_2(x) - 6c_\kappa(x))^2 - f_0(x) + c_\kappa(x)(f_2(x) + g_2(x))$$
$$- 3c_\kappa(x)^2\}\chi_{\lambda^{1/2}\Lambda}(x)$$

$$= \{6\lambda^{-1}\,\delta c(x) - 3\lambda^{-1/2}\xi\,\delta c + 9(\delta c(x))^2 - 2c_\kappa(x)(f_2(x) + g_2(x))$$
$$+ (4\lambda)^{-1}\xi^2 + 6c_\kappa(x)^2\}\chi_{\lambda^{1/2}\Lambda}(x),$$

where $c_\kappa(x) \equiv \delta_\kappa * C_N * \delta_\kappa = O(\ln \kappa)$ and $\delta c(x)$ is defined above. Thus

$$\int_\Lambda B(x)\, dx \leq \text{const}[\lambda^{-1} + (\ln \kappa)^2 + \lambda^{-1}\ln \kappa],$$

where the constant depends only on $\|\xi\|_{L_\infty}$. Note that the norm $M((f + g_2)\chi_\Lambda)$ defined in (8.6.6) satisfies

$$\text{const } \lambda^{-1} \leq M((f + g_2)\chi_\Lambda) \leq \text{const } \lambda^{-1}$$

where the constants depend only on $\|\xi\|_{L_\infty}$. Thus

$$-\{1 + M((f + g_2)\chi_\Lambda)\}(\ln \kappa)^2 \leq :P(\phi_\kappa, (f + g_2)\chi_\Lambda):_{C_N}. \qquad (16.2.24)$$

Use (8.6.23) with $m = M = 1$, $\Lambda = \Delta$, $n = 4$, $L = 0$. Therefore

$$Z_N((f + g_2)\chi_\Lambda) \leq \exp[K\lambda^{-1}(\ln \lambda)^2]$$

as desired, and the proof of the proposition is complete.

PROOF OF PROPOSITION 16.2.5.   By the Schwarz inequality, the cases Q_2 and Q_3 can be bounded separately, with twice the range of ξ. Consider first Q_3. By a further application of Hölder's inequality, Q_3 can be reduced to a sum over nonoverlapping pairs of squares (Δ, Δ'), as in the proof of Theorem 16.2.3. Let $\mathscr{B}$ denote such a set of pairs.

In this proof, let $C_N = (-\Delta_N + \lambda)^{-1}$, where Δ_N has Neumann data on $\partial(\Delta \cup \Delta')$ for all pairs (Δ, Δ') in $\mathscr{B}$. By the conditioning inequality of Proposition 10.3.1,

$$\int \exp[Q_3(\xi^{(3)}, X)]\, d\phi_C \leq \int \exp[Q_3(\xi^{(3)}, X)]\, d\phi_{C_N}. \qquad (16.2.25)$$

Let (Δ, Δ') be one pair in $\mathscr{B}$ and let $h \equiv \chi_\Delta - \chi_{\Delta'}$. Since the right side of (16.2.24) factors,

$$\int \exp[Q_3(\xi^{(3)}, X)] \, d\phi_N \leq \exp[\tfrac{1}{2}\langle h, C_N h\rangle \|\xi^{(3)}\|^2_{L_\infty} |\mathscr{B}|].$$

The function h is perpendicular to functions which are constant on $\Delta \cup \Delta'$. Thus h is perpendicular to the ground state of $-\Delta_N$ on $L_2(\Delta \cup \Delta')$. On this Hilbert space C_N is compact. If E_1 is the smallest nonzero eigenvalue of $-\Delta_N$,

$$\langle h, C_N h\rangle \leq (E_1 + \lambda)^{-1}\|h\|^2_{L_2} \leq 2E_1^{-1},$$

which is a bound independent of λ. Since $|\mathscr{B}| \leq |X|$,

$$\int \exp[Q_3(\xi^{(3)}, X)] \, d\phi_C \leq \exp[K|X|]$$

with a constant independent of λ, X, Λ, as desired.

In the case of Q_2, first Wick reorder Q_2 with respect to the covariance $C = (-\Delta_{\partial\Lambda} + \lambda)^{-1}$ with Dirichlet data on $\partial\Lambda$. Then

$$: Q_2(\xi^{(2)}, X):_{\lambda^{1/2}} = \; : Q_2(\xi^{(2)}, X):_C + \alpha,$$

where α is a constant satisfying $|\alpha| \leq O(|X|)$ with the constant uniform in λ. Thus α need not be considered further in the proof.

Let $C_N = (-\Delta_N + \lambda)^{-1}$ now denote Dirichlet data on the boundaries of all lattice squares Δ. By the conditioning inequality of Proposition 10.3.1,

$$\int \exp[Q_2(\xi^{(2)}, X)] \, d\phi_C \leq \prod_i \int \exp[: Q_2(\xi_i^{(2)}, \Delta_i):_{C_N}] \, d\phi_{C_N}.$$

It is sufficient to bound each term in the product by a constant independent of λ.

Let χ_Δ denote the operator of multiplication by χ_Δ and let P_Δ denote the orthogonal projection in $L_2(\Delta)$ onto the constant functions. Then

$$|\ln \lambda|^{-1}\xi^{(2)} : \phi^2(\Delta) - \phi(\Delta)^2:_{C_N} = \int : \phi(x)\phi(y):_{C_N} v(x, y) \, dx \, dy$$

where $v(x, y)$ is the kernel of the operator $v = |\ln \lambda|^{-1}\xi^{(2)}(\chi_\Delta - P_\Delta)$. It follows by (9.1.26b) that

$$\int \exp\left[: Q_2(\xi^{(2)}, \Delta):_{C_N}\right] d\phi_{C_N} = \exp[-\tfrac{1}{2} \operatorname{Tr}\{\ln(I - A) + A\}]$$

$$\leq \exp[\text{const } \|A\|^2_{HS}],$$

where $A = C_N^{1/2} v C_N^{1/2}$, where $\|A\|_{HS}$ denotes the Hilbert-Schmidt norm of A, and where the inequality holds whenever $\|A\|_{HS} < 1$.

Since $\chi_\Delta - P_\Delta$ annihilates constant functions in $L_2(\Delta)$, since $(-\Delta_N + \lambda)^{-1}$ is compact on $L_2(\Delta)$, and since $0 \leq \chi_\Delta - P_\Delta \leq I$,

$$\|A\|^2_{HS} \leq |\ln \lambda|^{-2}\xi^{(2)2} \operatorname{Tr}(C_N(\chi_\Delta - P_\Delta)C_N) = |\ln \lambda|^{-2}\xi^{(2)2} \sum_{j \neq 0} (E_j + \lambda)^{-2},$$

where $E_j = \text{const } |j|^2, j \in Z^2$, are the nonzero eigenvalues of $-\Delta_N$ on $L_2(\Delta)$. The sum $\sum_{j \neq 0} E_j^{-2}$ converges, so for λ sufficiently small, $\|A\|_{HS} < 1$. This completes the proof.

16.3 Symmetry Unbroken, $d = 2$

Consider vector valued fields ϕ, taking values $\phi(x) \in \mathscr{X}$, where $\mathscr{X}$ equals R^n or S^{n-1}, or where $\mathscr{X}$ is a Lie algebra or Lie group. The action functional $\mathscr{A}$ is defined on the space $\mathscr{X}$, or on fields with values in $\mathscr{X}$. Typically $\mathscr{A}$ is invariant under a symmetry group G acting on $\mathscr{X}$. In Section 5.5, for example, we considered an Ising type interaction for a spin field $\sigma(x)$ taking values in S^1, and invariant under the group $U(1)$ of rotations of S^1. Called the rotator or XY model, this interaction has been proposed to describe surface phenomena and melting. Similarly, an Ising type interaction with spins taking values in S^2 is called the Heisenberg or XYZ model, and has been used as a qualitative model of a ferromagnet. The vector ϕ^4 interaction occurs in particle physics as well, where it is called the Higgs field.

The qualitative theory of phase transitions is more complicated for these vector valued models. Let $\mathscr{M}_{cl}$ be the space of configurations ϕ which minimize $\mathscr{A}$. Normally these configurations are constant configurations, $\phi(x) = $ const, and so $\mathscr{M}_{cl}$ is identified with a subset of $\mathscr{X}$. The mean field picture predicts symmetry breaking phase transitions and multiple phases labeled by points $\phi_{cl} \in \mathscr{M}_{cl}$, for low temperature. For example, the S^1 Ising interaction is minimized by a one parameter family of configurations $\phi = (\cos\theta, \sin\theta)$ for $\theta = \theta(x) = $ const $\in [0, 2\pi)$, i.e. $\mathscr{M}_{cl} = S^1$. (In contrast, for the ordinary, or S^0, Ising model, $\mathscr{M}_{cl} = \{\pm 1\} = S^0$ is discrete.)

For fields defined on low dimensional spaces, i.e., $\phi(x)$, with $x \in R^d$ and d small, the predicted mean field phase behavior occurs only at zero temperature, and not at low but positive temperatures. Let d_{cr} be the largest dimension, for a given interaction, having mean field phase behavior only at zero temperature. In the $n = 1$ component $P(\phi)_d$ model and Ising models the mean field picture applies for dimension $d > 1$. In fact the $P(\phi)_1$ models were shown to be equivalent to quantum mechanics of one degree of freedom and to have a unique ground state, following the ideas and methods of Section 3.3. Hence the scalar ($n = 1$ component) models have a critical dimension $d_{cr} = 1$. For $d > d_{cr}$, first order phase transitions occur for sufficiently low temperature.

We show here and in Section 16.4 that $d_{cr} = 2$ for S^{n-1} Ising models, $2 \le n$ components. For $d = 2$ the equilibrium state of the rotator (S^1) model is unique. By showing differentiability of the pressure in some Banach space of potentials (see [Bricmont, Fontaine, and Landau, 1977]) the uniqueness relative to all possible equilibrium states of the same temperature is established. The equilibrium state is symmetric under the action of the rotation group G by construction. By uniqueness, this state is necessarily a pure phase. Thus, in the language of physics, the symmetry group is unbroken. The uniqueness of the equilibrium state (in the above sense) rules out a jump discontinuity in any thermodynamic function and thus a first order phase transition. As explained in Section 5.5, a higher order phase transition and degeneracy of states other than the equilibrium state is not excluded.

A general theory on a mathematical level to determine d_{cr} does not exist. The examples which have been studied suggest the importance of dual (or Fourier transform) variables to describe elementary excitations of the ground state. In scalar $P(\phi)$ or Ising models, the phase boundaries provide these variables. Because the action of an individual phase boundary grows proportionally to β and the size of the boundary, these boundaries are suppressed by an exponentially small activity $O(\exp(-\beta \text{ size}))$. Thus they form a dilute gas for β large, which characterizes a disordered phase of the $P(\phi)$ or Ising system. For the rotator model, the dual variables are vortices and vortex–antivortex dipoles; cf. Section 5.5.

Here we establish the Hohenberg–Mermin–Wagner theorem in a form given by [McBryan and Spencer, 1977]. It implies that $d_{cr} \geq 2$ for S^{n-1} Ising models with $2 \leq n$ components. For simplicity, we restrict attention to the rotator model. Let

$$H = -\sum_{nn} \sigma_i \cdot \sigma_j, \qquad (16.3.1)$$

where the sum extends over nearest neighbor lattice pairs (i, j) and where $\sigma_i \in S^1$. Thus we can write $\sigma = (\cos\theta, \sin\theta)$, and H becomes

$$H = -\sum_{nn} \cos(\theta_i - \theta_j). \qquad (16.3.2)$$

Theorem 16.3.1. *Let $0 < \varepsilon$. There exists $\beta(\varepsilon) < \infty$ such that for $\beta(\varepsilon) < \beta$,*

$$0 \leq \langle \sigma_k \cdot \sigma_l \rangle \leq |k - l|^{-(1-\varepsilon)/(2\pi\beta)}. \qquad (16.3.3)$$

Remark. In view of Theorem 16.1.1, the decay (16.3.3) is strongly suggestive of a unique ground state and absence of a first order phase transition. However, (16.3.3) proves less, and we refer to [Bricmont, Fontaine, and Landau, 1977] for the proof of uniqueness, as well as for continuum $P(\phi)$ models. The case of an arbitrary Lie symmetry group G was considered in [Dobrushin and Shlossman, 1975], where they prove invariance of the equilibrium state under G for bounded spins or with other technical assumptions. See [Ruelle, 1969] for the quantum Heisenberg model.

PROOF. Positivity follows from Corollary 4.7.2. To establish the upper bound we use the representation

$$\langle \sigma_k \cdot \sigma_l \rangle = \text{Re } Z^{-1} \int_{-\pi}^{\pi} \cdots \int_{-\pi}^{\pi} \exp\left\{\beta \sum_{ij} \cos(\theta_i - \theta_j)\right\}$$
$$\times \exp\{i(\theta_k - \theta_l)\} \prod_i d\theta_i. \qquad (16.3.4)$$

Here we assume a finite lattice and establish decay (16.3.3) uniformly in the size of the lattice. We do not discuss convergence of the infinite volume limit.

Using periodicity and analyticity of the integrand in (16.3.4) as a function of $\theta_1, \theta_2, \ldots$, we make the translation of variables

$$\theta_j \rightarrow \theta_j + ia_j,$$

where a_j is a real constant chosen later. In other words, we use Cauchy's formula: Each θ_j integral vanishes if taken around a closed contour in the θ plane. See Figure 16.1. The contributions from the two side segments of Figure 16.1 cancel, due to periodicity. Thus the contribution from the bottom segment equals the contribution from the top, with the sign reversed. The exponent $e^{-\beta H}$ is transformed as

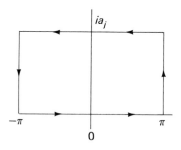

Figure 16.1. θ_j integration in (16.3.4).

$$\cos(\theta_i - \theta_j) \to \cos(\theta_i - \theta_j) \cosh(a_i - a_j) - i \sin(\theta_i - \theta_j) \sinh(a_i - a_j). \quad (16.3.5)$$

Using $|e^{ix}| = 1$ for x real, we find

$$\langle \boldsymbol{\sigma}_k \cdot \boldsymbol{\sigma}_l \rangle \leq \exp[-(a_k - a_l)] \, Z^{-1} \int \exp\left[\beta \sum \cos(\theta_i - \theta_j) \cosh(a_i - a_j)\right] \prod_i d\theta_i$$

$$= \exp[-(a_k - a_l)] \, Z^{-1} \int \exp\left[\beta \sum \cos(\theta_i - \theta_j) (1 + \cosh(a_i - a_j) - 1)\right] \prod_i d\theta_i$$

$$\leq \exp[-(a_k - a_l)] \exp\left[\beta \sum_{ij} (\cosh(a_i - a_j) - 1)\right]. \quad (16.3.6)$$

We now choose

$$a_j = \beta^{-1}[C(j, k) - C(j, l)] = \beta^{-1}\langle \delta_j, (-\Delta)^{-1}(\delta_k - \delta_l)\rangle, \quad (16.3.7)$$

where $C(i, j) = C(i - j)$ is the kernel of the lattice Green's function $(-\Delta)^{-1}$. It follows from (16.3.7) that a_j is bounded uniformly in j, and in fact

$$|a_i| \leq \text{const } \beta^{-1}. \quad (16.3.8)$$

Thus

$$\beta \sum (\cosh(a_i - a_j) - 1) \leq \frac{\beta}{2}(1 + O(\beta^{-2})) \sum_{i, j} (a_i - a_j)^2$$

$$= \left(\frac{\beta}{2} + O(\beta^{-1})\right)\langle a, -\Delta a \rangle$$

$$= \left(\frac{\beta}{2} + O(\beta^{-1})\right)\beta^{-1}(a_k - a_l)$$

$$= \tfrac{1}{2}(a_k - a_l) + O(\beta^{-2})(a_k - a_l).$$

By (16.3.6), for $\beta > \beta(\varepsilon)$ sufficiently large,

$$\langle \sigma_k \cdot \sigma_l \rangle \le \exp[-\tfrac{1}{2}(a_k - a_l)(1 + \varepsilon)]. \tag{16.3.9}$$

Note that by (16.3.7),

$$0 \le a_k - a_l = 2\beta^{-1}(C(0) - C(k - l)). \tag{16.3.10}$$

Here positivity follows from the fact that $C(k)$, as a positive definite function, achieves its maximum value at the origin. The asymptotic behavior of the lattice $d = 2$ Green's function is

$$C(0) - C(k) \sim \frac{1}{2\pi} \ln|k|, \qquad |k| \to \infty. \tag{16.3.11}$$

Inserting (16.3.10–11) into (16.3.9) yields the theorem.

By the correlation inequalities of Corollary 4.7.2,

$$0 \le \langle \sigma_k \cdot \sigma_l \rangle - \langle \sigma_k \rangle \cdot \langle \sigma_l \rangle.$$

Thus by translation invariance of $\langle \cdot \rangle$,

$$0 \le \langle \sigma_k \rangle^2 \le \lim_{|k-l| \to \infty} \langle \sigma_k \cdot \sigma_l \rangle = 0.$$

We thus have proved

Corollary 16.3.2. *With β sufficiently large,*

$$\langle \sigma_k \rangle = 0 \tag{16.3.12}$$

Remark. We can remove the assumption on β in the theorem, if we are satisfied with a smaller rate of decay. In particular, we verify (16.3.12) for all β. We follow the proof above, except we set

$$a_j = \varepsilon(1 + \beta)^{-1}[C(j, k) - C(j, l)] \tag{16.3.13}$$

in place of (16.3.7). Choose $0 < \varepsilon < 1$ to optimize the bound. Thereby follows

Theorem 16.3.3. *There exists a constant $0 < c < 1$ such that for all β,*

$$0 \le \langle \sigma_k \cdot \sigma_l \rangle \le |k - l|^{-c/(1 + \beta)}, \tag{16.3.14}$$

and for all β

$$\langle \sigma_k \rangle = 0. \tag{16.3.15}$$

16.4 Symmetry Broken, $3 \le d$

In this section we show that symmetry breaking and first order phase transitions occur for vector models in $3 \le d$ dimensions. This complements the result of the previous section that symmetry breaking does not occur for such systems for $d = 2$. These methods also apply for continuum $(\phi^2)^2$ fields in $d = 3$ dimensions.

In order to avoid certain technical problems, we study a lattice model with periodic boundary conditions. We assume the existence of the infinite volume limit, and we do not show the equivalence of these boundary conditions to the Dirichlet conditions studied elsewhere in the book.

Consider the Hamiltonian (on a periodic lattice)

$$H(\Lambda) = - \sum_{\substack{|i-j|=1 \\ i,\,j\in\Lambda}} \phi_i \cdot \phi_j - \sum_{j\in\Lambda} \mathbf{h} \cdot \phi_j, \tag{16.4.1}$$

and with a single spin distribution $d\mu_i(\phi)$ which decreases faster than any Gaussian. Thus for all $a < \infty$, assume

$$\int e^{a|\phi|^2}\, d\mu_i(\phi) < \infty, \tag{16.4.2}$$

where $d\mu_i(\phi)$ is a positive and SO(n)-invariant measure on R^n. We assume unit lattice spacing, so Fourier transforms are defined for momentum components $p_\alpha \in [-\pi, \pi]$. Examples of single spin measures are

$$d\mu_i(\phi) = \exp[-P(|\phi|)]\, d\phi, \tag{16.4.3}$$

$$d\mu_i(\phi) = \delta(|\phi|^2 - 1)\, d\phi. \tag{16.4.4}$$

$$d\mu_\Lambda = Z^{-1} \exp[-\beta H(\Lambda)] \prod_{i\in\Lambda} d\mu_i(\phi_i). \tag{16.4.5}$$

Since $H(\Lambda)$ defined in (16.4.1), with periodic boundary conditions, is invariant under the SO(n) symmetry, the models we consider satisfy

$$\langle \phi_l \rangle = 0 \quad \text{if } \mathbf{h} = 0. \tag{16.4.6}$$

The two point function is $\langle \phi_0 \cdot \phi_l \rangle$. In the infinite volume limit, its Fourier transform is

$$S^{\sim}(p) = (2\pi)^{-d/2} \sum_{l\in Z^d} e^{-ip\cdot l} \langle \phi_0 \cdot \phi_l \rangle.$$

The proof of phase transitions by [Fröhlich, Simon, and Spencer, 1976] is based on an infrared (or gradient ϕ) bound; see also [Glimm and Jaffe, 1970a].

Theorem 16.4.1.   *There is a nonnegative constant c such that*

$$0 \le S^{\sim}(p) - (2\pi)^{d/2} c\, \delta(p) \le \frac{n}{4\beta \sum_{\alpha=1}^{d} \sin^2\left(\dfrac{p_\alpha}{2}\right)} \tag{16.4.7}$$

Remark 1.   For a lattice with spacing ε, we obtain $\varepsilon^{-2} \sin^2(\varepsilon p_\alpha/2)$ in place of $\sin^2(p_\alpha/2)$ and $4\varepsilon^{-2} \sum \sin^2(\varepsilon p_\alpha/2) \to p^2$ as $\varepsilon \to 0$.

Remark 2. The eigenvalues of the periodic lattice Laplace operator Δ_p are

$$-4 \sum_{l=1}^{d} \sin^2\left(\frac{p_\alpha}{2}\right), \qquad p_\alpha \in \left|\frac{\pm 2\pi n_\alpha}{L} : 0 \leq n_\alpha \leq \left[\frac{L}{2}\right]\right|, \qquad (16.4.8)$$

where $L = |\Lambda|^{1/d}$ is the integer length of one side of the cubic box and $n_\alpha = 0, 1, 2, \ldots, [L/2]$. The eigenfunctions are $|\Lambda|^{-1/2} \exp(i(p_1 l_1 + p_2 l_2 + \cdots + p_d l_d))$. Thus (16.4.7) is integrable, uniformly for $1 \leq L$, $3 \leq d$.

Corollary 16.4.2. *The vanishing of the two point function*

$$\langle \phi_0 \cdot \phi_l \rangle^T \equiv \langle \phi_0 \cdot \phi_l \rangle - \langle \phi_0 \rangle \cdot \langle \phi_l \rangle$$

as $|l| \to \infty$ is equivalent to the identity

$$\langle \phi_l \rangle^2 = c. \qquad (16.4.9)$$

PROOF. Since (16.4.7) is integrable, its inverse Fourier transform converges to zero, by the Riemann–Lebesgue lemma. Thus

$$\langle \phi_0 \cdot \phi_l \rangle = (2\pi)^{-d/2} \int_{|p_\alpha| \leq \pi} S^\sim(p) e^{il \cdot p} \, dp \to c.$$

Before proving the theorem, however, we use it to establish phase transitions for $3 \leq d$, n-component Ising models.

Theorem 16.4.3. *Let $3 \leq d$, and let β be sufficiently large so that*

$$(2\pi)^{-d/2} n \int_{-\pi}^{\pi} \cdots \int_{-\pi}^{\pi} \left[4 \sum_{\alpha=1}^{d} \sin^2\left(\frac{p_\alpha}{2}\right)\right]^{-1} dp < \beta. \qquad (16.4.10)$$

Assume the single spin distribution (16.4.4), and let $h \equiv 0$. Then

$$\lim_{|l| \to \infty} \langle \phi_0 \cdot \phi_l \rangle^T \neq 0.$$

The infinite volume equilibrium state is not a pure phase.

PROOF. The restriction $3 \leq d$ makes (16.4.10) finite. By Theorem 16.4.1 and (16.4.10),

$$\langle \phi_0^2 \rangle - c = (2\pi)^{-d/2} \int S^\sim(p) \, dp - c$$

$$\leq (2\pi)^{-d/2} \frac{n}{\beta} \int \left[4 \sum_{\alpha=1}^{d} \sin^2\left(\frac{p_\alpha}{2}\right)\right]^{-1} dp < 1. \qquad (16.4.11)$$

However, $\phi_0^2 = 1$ for a rotator (16.4.4). Thus $0 < c$. But $\langle \phi \rangle = 0$ by (16.4.6). The nonvanishing of $\langle \phi_0 \cdot \phi_l \rangle^T$ at $l = \infty$ follows by (16.4.9). The final statement is a consequence of the proof of Proposition 16.1.3, which shows that the equilibrium state is not a pure phase.

Remark. The theorem applies to the ordinary Ising model ($n = 1$). In that case, the Lee–Yang Theorem 4.5.1 assures that for $0 < h$, $\lim_{|l| \to \infty} \langle \phi_0 \, \phi_l \rangle^T = 0$. By Corollary 16.4.2, $\langle \phi \rangle^2 = c$. Let

$$1 - \bar{c} = (2\pi)^{-d/2} \left(\frac{n}{\beta} \right) \int \left[4 \sum_{\alpha=1}^{d} \sin^2 \left(\frac{p_\alpha}{2} \right) \right]^{-1} dp,$$

so that (16.4.11) can be written as $1 - c \le 1 - \bar{c}$, and hence $\bar{c} \le c$. Thus c is bounded away from zero uniformly in h for $0 < h$, and

$$\lim_{h \to 0} \langle \phi_l \rangle_h \neq 0.$$

Thus the Ising model (as was shown in Section 5.4) has *spontaneous magnetization* at low temperature. The Lee–Yang theorem is also known for $n = 2$, 3 component rotators (cf. [Dunlop and Newman, 1975], [Dunlop, 1979a, b]), and the above argument yields

Corollary 16.4.4. *For dimension $d \ge 3$, β sufficiently large, and $n = 1, 2, 3$ component rotators, spontaneous magnetization occurs:*

$$\lim_{h^1 \to 0} \langle \phi_l \rangle_{h^1} \neq 0. \tag{16.4.12}$$

We now return to the proof of (16.4.7). We use the notation of Section 9.5 for lattice gradients, and follow [Fröhlich and Spencer, 1977].

Lemma 16.4.5. *Let $\mathbf{h}_\alpha \in R^n$, and let ∂ denote the forward finite difference quotient. Let $\mathbf{f}_\alpha \in l_2(Z^n)$, $f = \{\mathbf{f}_\alpha\} \in l_2(Z^{nd})$. Then*

$$\left\langle \exp \left(\sum_{\alpha=1}^{d} \phi \cdot (\partial_\alpha \mathbf{f}_\alpha) \right) \right\rangle \le \exp((2\beta)^{-1} \| f \|_{l_2}^2), \tag{16.4.13}$$

where

$$\| f \|_{l_2}^2 = \sum_{l, \alpha} \mathbf{f}_\alpha(l)^2.$$

PROOF OF THEOREM 16.4.1, ASSUMING LEMMA 16.4.5. Subtract 1 from both sides of (16.4.13). Substitute εf for f, multiply by ε^{-2}, and let $\varepsilon \to 0$. Since $\langle \ \rangle$ is invariant under lattice translations, the expectation $\langle \phi(\partial_\alpha f_\alpha) \rangle = 0$. Thus we obtain

$$\left\langle \left(\phi \cdot \sum_{\alpha=1}^{d} \partial_\alpha \mathbf{f}_\alpha \right)^2 \right\rangle \le \beta^{-1} \| f \|_{l_2}^2. \tag{16.4.14}$$

With ∂^* the negative of the backward lattice difference quotient, we choose $\mathbf{f}_\alpha = (|\Lambda|^{-1/2} \partial^* (-\Delta_p)^{-1/2} e^{ip \cdot l}) v_r$, where v_r is a unit basis vector in the spin space R^n. Using (16.4.8), and summing over the n choices of basis vector v_r, we obtain the bound

$$4 \sum_{\alpha=1}^{d} \sin^2 \left(\frac{p_\alpha}{2} \right) S^\sim(p) \le \frac{n}{\beta}. \tag{16.4.15}$$

Since $\langle \phi_0 \cdot \phi_l \rangle$ is a positive definite function, its Fourier transform $S^{\sim}(p)$ is a positive measure. By this fact in combination with (16.4.15), the singularity at $p = 0$ is $(2\pi)^{d/2} c \delta(p)$ for some nonnegative constant c. Thus (16.4.7) follows from division by $4 \sum_{\alpha=1}^{d} \sin^2(p_\alpha/2)$.

PROOF OF LEMMA 16.4.5.　We prove the lemma for a finite periodic lattice (torus) Λ. As we are assuming convergence of the infinite volume limit, it follows that the lemma holds for the lattice Z^d. We use a multiple reflection bound. Let

$$\phi(\partial_\alpha f) = (\partial_\alpha^* \phi)(f) = \sum_{l \in \Lambda} f(l)(-\phi_l + \phi_{l-e_\alpha}).$$

Then

$$I \equiv \left\langle \exp\left(\sum_{\alpha=1}^{d} \phi(\partial_\alpha f_\alpha) \right) \exp\left(-\frac{1}{2\beta} \|f\|_{l_2}^2 \right) \right\rangle$$

$$= \frac{\int \exp\left(-\sum_{l,\alpha} \tfrac{1}{2}\beta(\phi_l - \phi_{l-e_\alpha} + \beta^{-1} f(l))^2 \right) \prod d\mu_i}{\int \exp\left(-\sum_{l,\alpha} \tfrac{1}{2}\beta(\phi_l - \phi_{l-e_\alpha})^2 \right) \prod d\mu_i} \tag{16.4.16}$$

Here we include the (space-independent) linear (magnetic) term in (16.4.1) in $d\mu_i$. The desired inequality is $I \le 1$.

The expectation $\langle \ \rangle$ in (16.4.16) is defined by the measure (16.4.5). As a consequence of Theorems 7.10.3 and 10.4.3 it satisfies reflection positivity. To visualize the reflection property, we imbed Λ in R^{d+1} as in Section 7.10. We now choose a hyperplane Π which bisects the torus Λ, as in Figure 16.2. The spins ϕ_l have been divided into four subsets: $\phi^\pm$ and $\sigma^\pm$. The spins σ^+ couple to spins σ^- across the hyperplane Π. Likewise they may couple to ϕ^+ spins. The spins ϕ^+ and ϕ^- do not couple across Π. The spins $\{\phi^+, \sigma^+\}$ are all the spins in $\Lambda_+ = \Lambda \cap \Pi_+$, etc.

Let A be a function of spins in Λ_- and B a function of spins in Λ_+. After taking the partial expectation $\langle \cdot \rangle$ in the $\phi^\pm$ variables, we obtain a function of the form

$$\exp(\beta \sum \sigma_l^+ \cdot \sigma_{l'}^-) F(\sigma^+) G(\sigma^-) \prod_{l, l'} d\sigma_l^+ d\sigma_{l'}^- ,$$

where by including part of the single spin measures in F, G we write explicitly the Lebesgue measure $d\sigma^\pm$.

If $A = \overline{\theta B}$, then $F = \bar{G}$ and by reflection positivity

$$0 \le \langle \overline{\theta B} \, B \rangle = \int \exp(\beta \sum \sigma_l^+ \cdot \sigma_{l'}^-) G(\sigma^+) G(\sigma^-) \prod_{l, l'} d\sigma_l^+ d\sigma_{l'}^-. \tag{16.4.17}$$

We introduce the Fourier transform variables p dual to $(\sigma_l^+ - \sigma_{l'}^-)$. Thus we rewrite (16.4.17), using the notation b to denote the bond (l, l'), and with general $A \ne \overline{\theta B}$, as

$$\langle AB \rangle = \text{const} \int \exp\left(-\frac{\beta}{2} \sum (\sigma_l^+ - \sigma_{l'}^-)^2 \right) F(\sigma^+) G(\sigma^-) d\sigma^+ d\sigma^-$$

$$= \text{const} \int \exp\left(-\frac{1}{2\beta} \sum p_b^2 \right) \tilde{F}(-p) \tilde{G}(p) \prod_b dp_b. \tag{16.4.18}$$

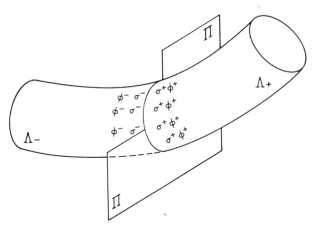

Figure 16.2. The torus Λ bisected by the hyperplane Π. Shown here is one of the intersections of Π with Λ. The bonds connecting the σ^+ and σ^- spins have been cut by Π. By translation invariance, Π can be rotated by an angle which respects the lattice symmetry.

Note that translation $\sigma_l^+ - \sigma_{l'}^- \to \sigma_l^+ - \sigma_{l'}^- + g(l, l')$ of the exponent in (16.4.18) becomes multiplication by $e^{ipg(l,\, l')}$ in the Fourier transform. Thus the reflection positive inner product (16.4.18) yields

$$|\langle AB\rangle| \leq (\langle \overline{\theta A}\, A\rangle \langle \overline{\theta B}\, B\rangle)^{1/2}. \tag{16.4.19}$$

Applying (16.4.19) to bound (16.4.16), we eliminate the $f(l)$ factors coupling to bonds $b = (l, l - e_\alpha)$ which cross Π. Furthermore we obtain, for the remaining functions f, functions which are θ-reflection-symmetric.

Proceeding in this manner, we use all choices of Π to eliminate all $f(l)$ factors from (16.4.16). Thus $I \leq 1$ as desired.

The ϕ^4 Critical Point

17.1 Elementary Considerations

To fix the notation, we consider the interaction

$$V(\phi) = \lambda\phi^4 + \sigma\phi^2 - \mu\phi \qquad (17.1.1)$$

with λ, σ, and μ real and $0 < \lambda$. By the Lee–Yang theorem, there is no phase transition for $\mu \neq 0$. The high temperature series expansions (Chapter 18) show that there is no phase transition for $\mu = 0$ and σ sufficiently large. By Section 16.2, there is a phase transition for $\mu = 0$ and σ sufficiently negative. In the latter region one expects exactly two phases, and a unique value of σ, $\sigma = \sigma_c$, which separates the one and two phase regions. Throughout this chapter we define σ_c as the infinum of the values of σ for which (17.1.1) has a unique phase and exponential decay of correlations. (Thus H has a gap in its spectrum, which separates 0, the spectrum of the vacuum Ω, from the rest of the spectrum. We call this gap a mass $m > 0$.)

Correlation inequalities are used here to analyze the critical point. Other useful methods to study critical behaviour are infrared bounds (Section 16.5) and exactly soluble models ([McCoy and Wu, 1973], [Wu, McCoy, Tracy, and Barouch, 1976]). Series expansion methods are used in numerical analyses of critical behavior. Since the critical point is a singularity on the boundary of the region of convergence, these methods are slowly convergent and difficult to apply. On a more formal level, the renormalization group is used to describe critical behavior.

Theorem 17.1.1. *For the field constructed from* (17.1.1) *with* $0 \leq \mu$ *and with Dirichlet boundary conditions, the Schwinger functions*

$$\langle \phi(x_1) \cdots \phi(x_n) \rangle \tag{17.1.2}$$

are monotone increasing in μ *and* $-\sigma$.

PROOF. Because of convergence of the finite volume theories, it is sufficient to give the proof for a finite volume expectation $\langle \cdot \rangle_\Lambda$. With the customary abbreviation

$$\langle \phi(x_1) \cdots \phi(x_n) \rangle = \langle 1, \ldots, n \rangle,$$

we have

$$-\frac{d}{d\sigma} \langle 1, \ldots, n \rangle_\Lambda$$

$$= \int_\Lambda [\langle 1, \ldots, n, :\phi^2(x): \rangle_\Lambda - \langle 1, \ldots, n \rangle_\Lambda \langle :\phi^2(x): \rangle_\Lambda] \, dx, \tag{17.1.3}$$

which is positive by the second Griffiths inequality. Note that the (infinite) constant in $:\phi^2:$ drops out of (17.1.3) as in (10.2.4). The same proof applies to μ, but not to higher polynomials in ϕ.

For $\mu \neq 0$, the magnetization $M = M(\sigma, \mu)$ is correctly defined as

$$M(\sigma, \mu) = \langle \phi \rangle,$$

because the $\mu \neq 0$ theories are pure phases. This definition is incorrect for $\mu = 0$, at least in the two phase region. In fact, because we have used boundary conditions symmetric under the $\phi \to -\phi$ isomorphism, $\langle \phi \rangle = 0$ for $\mu = 0$ and arbitrary σ. Assuming that the $\mu = 0$ theories have at most two phases, the correct definition of $\mu = 0$ magnetization is

$$M(\sigma) = \pm \left(\lim_{|x-y| \to \infty} \langle \phi(x)\phi(y) \rangle \right)^{1/2}; \tag{17.1.4}$$

cf. Section 16.1. Again assuming at most two phases, we define the mass $m(\sigma)$ as the exponential decay rate of

$$\langle \phi(x)\phi(y) \rangle - M(\sigma)^2 \sim e^{-m(\sigma)|x-y|}. \tag{17.1.5}$$

Corollary 17.1.2. *For* $0 < \mu$, $M(\sigma)$ *is monotonically decreasing in* σ; *for* $\sigma_c < \sigma$, $M(\sigma) = 0$. *Also* $m(\sigma)$ *is monotonically increasing in* σ *for* $\sigma_c \leq \sigma$.

PROOF. This is a direct consequence of the theorem, and of the definitions of $M(\sigma)$, $m(\sigma)$, and σ_c.

17.2 The Absence of Even Bound States

In a single phase, even ϕ^4 model, i.e. for $\sigma_c < \sigma$, we show that the Hamiltonian, restricted to $\mathcal{H}_{\text{even}}$, has no spectrum in the interval $(0, 2m)$. Here $\mathcal{H}_{\text{even}}$ is the subspace of $\mathcal{H}$ invariant under the $\phi \to -\phi$ isomorphism.

Two-particle bound states, which we presume to be even states, thus do not exist. We remark that $\mathscr{H}_{\text{even}}$ is spanned by the projection $\wedge$ into $\mathscr{H}$ of the Euclidean vectors Ω, $\phi(f_1) \cdots \phi(f_n)\Omega$, $n = 2, 4, \ldots$, where suppt f_j is contained in $t > 0$. For $A = \{x_1, \ldots, x_r\}$, let $\phi_A \equiv \phi(x_1) \cdots \phi(x_r)$.

Theorem 17.2.1. *Consider a ϕ^4 field or Ising model with zero external field and $\sigma_c < \sigma$, and let A and B have an even number of elements. Then*

$$\langle \phi_A \phi_B \rangle - \langle \phi_A \rangle \langle \phi_B \rangle$$
$$\leq \sum_{\substack{A_1 \subset A, \, A_1 \text{ odd} \\ B_1 \subset B, \, B_1 \text{ odd}}} \langle \phi_{A_1} \phi_{B_1} \rangle \langle \phi_{A-A_1} \phi_{B-B_1} \rangle.$$

PROOF. We use Corollary 4.3.3. The inequalities are preserved under removal of the lattice and volume cutoffs. Since A, B are even, and $A \backslash A_1$, $B \backslash B_1$ are odd, it follows that A_1, B_1 are odd.

Corollary 17.2.2. *Under the hypothesis of Theorem 17.2.1, there are no even bound states with energy below the two particle threshold.*

PROOF. Let Ω be the vacuum state in $\mathscr{H}$, unique because it is assumed that $\sigma_c < \sigma$. We write $x = x_1, \ldots, x_d$ as

$$x = (\mathbf{x}, t)$$

with $\mathbf{x} \in R^{d-1}$. In particular with θ: $(\mathbf{x}, t) = (\mathbf{x}, -t)$ and

$$A + s = \{(\mathbf{x}, t + s) : (\mathbf{x}, t) \in A\},$$

then

$$\langle (\phi_{\theta A})^\wedge, \, e^{-sH}(\phi_B)^\wedge \rangle_{\mathscr{H}} = \langle \phi_A \phi_{B+s} \rangle$$

is valid when the times in A precede the times in B. In particular we choose A to have only negative times $t \leq 0$, so B, chosen as

$$B = \{\mathbf{x}, -t) : (\mathbf{x}, t) \in A\}$$

has only positive times. With this choice of A and B, and with P_Ω the projection operator on $\mathscr{H}$ onto the vacuum state, we recognize

$$\langle \phi_{A-s} \phi_{B+s} \rangle - \langle \phi_{A-s} \rangle \langle \phi_{B+s} \rangle = \| e^{-sH}(I - P_\Omega) \phi_A^\wedge \Omega \|^2$$

so that Theorem 17.2.1 gives a bound on the decay rates which occur in e^{-sH} on the subspace $(I - P_\Omega)\mathscr{H}$. For A_1 odd, $\phi_{A_1}^\wedge \Omega$ is perpendicular to the vacuum $(\langle \Omega, \phi_{A_1}^\wedge \Omega \rangle = \langle \phi_{A_1} \rangle = 0)$, and so $\langle \phi_{A_1-s} \phi_{B_1-s} \rangle$ has an exponential decay rate at least m, by definition of the mass of the theory. Thus

$$\langle \phi_{A_1-s} \phi_{B_1+s} \rangle \leq C_{A_1, B_1} e^{-2ms}$$

for some constant C_{A_1, B_1} depending on A_1 and B_1. The same bound holds for $\langle \phi_{(A-A_1)-s} \phi_{(B-B_1)+s} \rangle$, and so by Theorem 17.2.1,

$$\| e^{-sH}(I - P_\Omega) \phi_A^\wedge \Omega \|^2 \leq \text{const } e^{-4ms}.$$

Thus, there are no even states, except Ω, with energy below $2m$, hence in particular no even bound states in this energy range.

17.3 A Bound on the Coupling Constant λ_{phys}

Define the dimensionless ϕ^4 coupling constant by

$$\lambda_{\text{phys}} = -m^{d-4}\chi^{-4}\int \langle \phi(x_1) \cdots \phi(x_4)\rangle^T \, dx_1 \, dx_2 \, dx_3,$$

where

$$\chi = \int \langle \phi(x_1)\phi(x_2)\rangle^T \, dx_1 = \int_0^\infty \frac{d\rho(a)}{a}$$

and $d\rho(a)$ is the Lehmann spectral measure, with $a = \text{mass}^2$. By Theorem 4.1.1, $0 \leq \chi$. For a massive, single phase, even ϕ^4 interaction, $0 \leq \lambda_{\text{phys}}$ by Corollary 4.3.3. We now assume in addition that the proper field strength renormalization has been performed; in the case of an isolated particle of mass m, this means $d\rho(a) = \delta(a - m^2) \, da$ in a neighborhood of $m^2 = a$. We then prove an upper bound on λ_{phys}.

Theorem 17.3.1 [Glimm and Jaffe, 1975a]. *Under the above assumptions,*

$$0 \leq \lambda_{\text{phys}} \leq \text{const},$$

where the dimensionless constant is independent of all parameters (e.g., λ, σ).

OUTLINE OF PROOF. For details, see the original paper. We use the basic Griffiths inequality (4.1.1) to derive (writing 1 for $\phi(x_1)$, etc.)

$$0 \leq \langle 1234\rangle - \langle 12\rangle\langle 34\rangle = \langle 1234\rangle^T + \langle 13\rangle\langle 24\rangle + \langle 14\rangle\langle 23\rangle.$$

By Corollary 4.3.3, $\langle 1234\rangle^T \leq 0$ and thus

$$0 \leq -\langle 1234\rangle^T \leq \langle 13\rangle\langle 24\rangle + \langle 14\rangle\langle 23\rangle. \tag{17.3.1}$$

After symmetrization over the choices of variables,

$$-\langle 1234\rangle^T \leq (\langle 13\rangle\langle 24\rangle + \langle 14\rangle\langle 23\rangle)^{1/3}(\langle 12\rangle\langle 34\rangle + \langle 13\rangle\langle 24\rangle)^{1/3}$$
$$\times (\langle 14\rangle\langle 23\rangle + \langle 12\rangle\langle 34\rangle)^{1/3}. \tag{17.3.2}$$

From elementary properties of the Green's function $(-\Delta + a)^{-1}(x, y)$, we find

$$\langle xy\rangle = \int_0^\infty (-\Delta + a)^{-1}(x, y) \, d\rho(a)$$

$$\leq \text{const } \chi|x - y|^{-d} \exp(-m|x - y|/2).$$

Inserting this in our bound (17.3.2) for $-\langle 1234\rangle_T$ gives

$$\lambda_{\text{phys}} \leq \text{const } m^{-4}\chi^{-2}.$$

Using the above assumption of proper field strength renormalization,

$$\chi = \int_{m^2}^\infty \frac{d\rho(a)}{a} \geq m^{-2},$$

we obtain $\lambda_{\text{phys}} \leq \text{const}$ as claimed.

Observe that the final bound does not depend on m, and hence also holds in the limit $m \to 0$. Hence the critical point (which for $d < 4$ should be an infrared stable fixed point of the renormalization group) occurs for finite λ_{phys}.

We now define a somewhat more general ϕ^4 coupling constant λ_l; we show that the above analysis also bounds λ_l. Particle amputation is defined by multiplying each (external) variable of the four point function by

$$m^2 - p^2|_{p=0} = m^2.$$

Propagator amputation, as above, multiplies by χ^{-1}. Define

$$0 \le \lambda_l = -m^{d+4} \int \langle 1234 \rangle^T dx_1 dx_2 dx_3 (m^{-2}\chi^{-1})^l.$$

In λ_l, l legs have propagator amputation and $4 - l$ legs have particle amputation. Above we showed

$$(m^2\chi)^{-1} \le 1,$$

so that

$$\lambda_4 \le \cdots \le \lambda_1 \le \lambda_0.$$

It was shown above that

$$\lambda_{\text{phys}} \equiv \lambda_4 \le \lambda_2 = m^4 \chi^2 g = -m^d \int \frac{G^{(4)}}{\chi^2} \le \text{const}.$$

We now see that $\lambda_l / \lambda_{l+1} = m^2 \chi$ is bounded if conventional scaling ideas hold. In particular, we have

Theorem 17.3.2 [Glimm and Jaffe, 1980]. *Assume for a single phase ϕ_d^4 model that*

$$F(x) \equiv m^{-d+2} \langle \phi(0)\phi(x/m) \rangle$$

$$\le \begin{cases} |O(1)||x|^{-(d-1)/2} e^{-|x|} & \text{if } |x| \ge 1, \\ |O(1)||x|^{-d+2-\eta} & \text{if } |x| \le 1, \end{cases}$$

with $\eta \le 1$ (cf. Section 17.4) and with universal constants $O(1)$. Then

$$m^2 \chi \le \text{const}$$

and

$$0 \le \lambda_4 \le \lambda_3 \le \lambda_2 \le \lambda_0 \le \text{const}$$

with a universal constant.

Remark. Either the λ_l are all nonzero or they are all zero. In the latter case, and only then, the theory is a generalized free field. The final statement follows from [Newman, 1975b].

We next establish the continuity of λ_{phys} under a limit process. We assume that m is bounded away from zero in the limit process. This allows a (positive mass) scaling limit at the critical point, and removal of ultra-violet cutoffs, for noncritical couplings. By scaling, we assume m is fixed at 1.

To illustrate the idea, we see that λ_l/λ_{l+1} is convergent by the Lebesgue bounded convergence theorem. Thus it is sufficient to consider λ_0. We assume the hypotheses of Theorem 17.3.2. Then each λ_l is continuous in a limit process for which the mass m is fixed and not zero, and for which the two and four point functions $S^{(2)}(x, 0)$ and $S_T^{(4)}(x_1, \ldots, x_4)$ are pointwise convergent almost everywhere.

To establish this fact we use (17.3.2) as above, for any permutation $\{i_1, \ldots, i_4\}$ of $\{1, \ldots, 4\}$. After a permutation and translation of variables, we can suppose that $i_v = v$, $x_1 = 0$ and $|x_2 - x_3| \le |x_3 - x_4|$. Also we may take $m = 1$. Then

$$\int F(x_1 - x_2)F(x_3 - x_4)\, dx_2\, dx_3\, dx_4$$

$$\le \int_{|x_2 - x_3| \le |x_3 - x_4|} F(x_3 - x_4)F(x_2)\, d(x_2 - x_3)\, d(x_3 - x_4)\, dx_2$$

$$\le \int x^d F(x)\, dx \int F(x)\, dx < \infty.$$

Continuity follows from the Lebesgue bounded convergence theorem.

17.4 Existence of Particles and a Bound on $dm^2/d\sigma$

Here we consider a canonical, single phase ϕ^4 model (canonical means without field strength renormalization). We establish

$$\frac{dm^2(\sigma)}{d\sigma} \le Z(\sigma) \tag{17.4.1}$$

for $\sigma_c < \sigma$, from which our next result follows by approximation methods. Here $Z(\sigma)$ is the field strength renormalization constant, defined in (17.4.2).

Theorem 17.4.1. (See Glimm and Jaffe, 1977a). *For almost every value of m, particles exist, i.e., $Z \ne 0$.*

PROOF OF (17.4.1). Consider $\Gamma(p) = -S(p)^{-1}$, where $S(p)$ is the Fourier transform of $\langle \phi(x)\phi(0) \rangle$ and p is a Euclidean momentum. Note that

$$S(p) = \int \langle \phi(x)\phi(0) \rangle e^{-ipx}\, dx$$

$$= \frac{Z}{p^2 + m^2} + \int_{m^2 + 0}^{\infty} \frac{d\rho(a)}{p^2 + a}. \tag{17.4.2}$$

The assumed canonical condition is $Z + \int_{m^2+0}^{\infty} d\rho(a) = 1$ and

$$Z^{-1} = -\left(\frac{d\Gamma}{dp^2}\right)_{p^2 = -m^2}. \tag{17.4.3}$$

Since $\Gamma = 0$ on the one particle curve $p^2 = -m^2(\sigma)$, $\nabla\Gamma$ must be orthogonal to the vector $(dm^2/d\sigma, 1)$ in the $-p^2$, σ space. Thus for $p^2 = -m^2$,

$$0 = -\frac{\partial\Gamma}{\partial p^2}\frac{dm^2}{d\sigma} + \frac{\partial\Gamma}{\partial\sigma} = Z^{-1}\frac{dm^2}{d\sigma} + \frac{\partial\Gamma}{\partial\sigma}.$$

The desired inequality follows from

Theorem 17.4.2. *Under the above assumptions,*

$$-1 \le \left(\frac{\partial\Gamma}{\partial\sigma}\right)_{p^2 = -m^2} \le 0. \tag{17.4.4}$$

PROOF.

$$-\frac{dS(-ip)}{d\sigma} = \frac{1}{2}\iint [\langle x0zz\rangle - \langle x0\rangle\langle zz\rangle]\, dz\, e^{-px}\, dx,$$

and for p real,

$$0 \le \iint \langle xz\rangle\langle yz\rangle e^{-p(x-z)}e^{-pz}\, dx\, dz = S(-p)^2.$$

Thus by Corollary 4.3.3 and (17.3.1),

$$0 \le \frac{dS(ip)^{-1}}{d\sigma} \le 1.$$

However for $\mathbf{p} = 0$, $S(-ip)^{-1}|_{p^2=m^2} = -\Gamma(p)|_{p^2=-m^2}$ with p real, so (17.4.4) is proved.

17.5 Existence of the ϕ^4 Critical Point

We show that the correlation length becomes infinite as $\sigma \downarrow \sigma_c$, following [Baker, 1975]; see also [J. Rosen, 1980] and [McBryan and J. Rosen, 1976]. For simplicity, we only consider here the case of a lattice field, and we define (for the purpose of this section only)

$$m(\sigma) = \lim_{|x-y|\to\infty} -\frac{\ln\langle\phi(x)\phi(y)\rangle}{|x-y|}. \tag{17.5.1}$$

For $\sigma > \sigma_c$, $m(\sigma)$ is the mass (energy of the lowest nonvacuum state), and for $\sigma \le \sigma_c$ it is zero.

Theorem 17.5.1. *The mass $m(\sigma)$ in (17.5.1) is continuous as a function of σ. In particular, the mass defined in (17.1.5) tends to zero as $\sigma \downarrow \sigma_c$.*

First we define a pseudomass $m^\sim = m^\sim(\sigma)$, for σ in a bounded interval, $a \le \sigma \le b$. Let $\langle \cdot \rangle_{\sigma, \Lambda}$ denote the finite volume expectation with domain $\Lambda \subset R^d$. Let

$$A = 2 \left(\sup_{\sigma \in [a, b], \, \Lambda \subset R^d, \, x, \, y \in \Lambda} \langle \phi(x)\phi(y) \rangle_{\sigma, \Lambda} \right). \tag{17.5.2}$$

The supremum is finite, and is achieved by $\Lambda = R^d$, $\sigma = a$, $x = y$. Let

$$m^\sim = m^\sim(x, y, \sigma, \Lambda)$$

be the unique solution of the equation

$$A \frac{e^{-m^\sim|x-y|}}{1 + (m^\sim |x - y|)^\alpha} = \langle \phi(x)\phi(y) \rangle_{\sigma, \Lambda}. \tag{17.5.3}$$

Here α is a constant chosen so that

$$d - 1 \le \alpha, \qquad d/2 < \alpha. \tag{17.5.4}$$

Note that for $0 < x$,

$$\frac{d}{dx} \frac{e^{-x}}{1 + |x|^\alpha} < 0,$$

which guarantees the existence of $m^\sim$ for $x \ne y$. Now let

$$m^\sim(\sigma, \Lambda) = \inf_{x \ne y \in \Lambda} m^\sim(x, y, \sigma, \Lambda),$$

$$m^\sim(\sigma) = \inf_\Lambda m^\sim(\sigma, \Lambda) = \lim_{\Lambda \uparrow R^d} m^\sim(\sigma, \Lambda).$$

Lemma 17.5.2. $m^\sim(\sigma, \Lambda)$ *is continuous in σ and strictly positive for Λ bounded and connected. Also*

$$0 \le m^\sim(\sigma) \le m(\sigma) \le \text{const } m^\sim(\sigma),$$

$$0 = m^\sim(\sigma) \quad \text{for } \sigma < \sigma_c.$$

PROOF. $m^\sim(\sigma, \Lambda)$ is strictly positive because $\langle \phi(x)\phi(y) \rangle_{\sigma, \Lambda}$ is. To see the latter, expand all factors of the form

$$\exp[\phi(x_i)\phi(x_j)]$$

in the unnormalized expectation $\langle \cdot \rangle_{\sigma, \Lambda}$. (This expansion of the gradient terms occurs also in the cluster expansion and the proof of Griffiths' inequalities; see Chapter 18.) All such terms are positive, and for Λ connected, at least one is nonzero.

The bound $m^\sim(\sigma) \le m(\sigma) + \varepsilon$, for arbitrary $\varepsilon > 0$, follows from the inequalities

$$e^{-(m+\varepsilon)|x-y|} \le (2A)^{-1} \langle \phi(x)\phi(y) \rangle_\sigma$$

$$\le A^{-1} \langle \phi(x)\phi(y) \rangle_{\sigma, \Lambda}$$

$$\le \frac{e^{-m^\sim|x-y|}}{1 + (m^\sim |x - y|)^\alpha} \le e^{-m^\sim|x-y|}.$$

Here x, y are chosen to make the first inequality hold; with x, y given, Λ is chosen to make the second inequality valid. Thus $m^\sim \leq m$. The bound in the reverse direction follows from the fact that (17.5.1), with the help of the transfer matrix, gives an exponential decay in all correlations, with a rate $e^{-m \, \text{dist}}$, where dist is a distance of lattice hyperplane separation between correlation sites. (See also Section 17.1.) Thus

$$\langle \phi(x)\phi(y)\rangle_\Lambda \leq \langle \phi(x)\phi(y)\rangle \leq Ae^{-m \, \text{dist}},$$

where dist $\geq |x - y|/\sqrt{d}$.

PROOF OF THEOREM 17.5.1. We show that $m^\sim(\sigma, \Lambda)^{2\alpha+1}$ is Lipschitz continuous in σ, with a constant uniform in σ and Λ, for σ in the bounded interval $[a, b]$. From this statement, the theorem follows, by Lemma 17.5.2. Since $m^\sim(\sigma, \Lambda)^{2\alpha+1}$ is the lower envelope of a finite family of functions, we prove the Lipschitz continuity for each such function on the interval for which it coincides with $m^{\sim 2\alpha+1}$. Thus we choose $x_0 \neq y_0$ so that

$$m^\sim(x_0, y_0, \sigma, \Lambda) = m^\sim(\sigma, \Lambda).$$

The definition of $m^\sim$ gives the identity

$$m^\sim|x_0 - y_0| - \ln A + \ln(1 + (m^\sim|x_0 - y_0|)^\alpha) = -\ln\langle\phi(x_0)\phi(y_0)\rangle_{\sigma, \Lambda}.$$

Differentiation with respect to σ gives the equality below:

$$|x_0 - y_0|\frac{dm^\sim}{d\sigma} \leq |x_0 - y_0|\frac{dm^\sim}{d\sigma}\left(1 + \frac{\alpha(m^\sim|x_0 - y_0|)^{\alpha-1}}{1 + (m^\sim|x_0 - y_0|)^\alpha}\right)$$

$$= \sum_{z \in \Lambda} \frac{\langle\phi(x_0)\phi(y_0)\phi^2(z)\rangle - \langle\phi_0)\phi(y_0)\rangle\langle\phi^2(z)\rangle}{\langle\phi(x_0)\phi(y_0)\rangle}$$

$$\leq 2A + 2 \sum_{\substack{z \in \Lambda \\ z \neq x_0, y_0}} \frac{\langle\phi(x_0)\phi(z)\rangle\langle\phi(y_0)\phi(z)\rangle}{\langle\phi(x_0)\phi(y_0)\rangle}.$$

In the last inequality, we have used Lebowitz's inequality (Corollary 4.3.3) to bound the four point function by a product of two-point functions, and the definition of A to bound the two terms $z = x_0$ and $z = y_0$. Substituting the definition of $m^\sim$ (which majorizes each factor of the numerator and is an identity for the denominator) and using the inequality $e^{-a} \leq 1$ for $0 \leq a$,

$$|x_0 - y_0|\frac{dm^\sim}{d\sigma} \leq 2A + 2 \sum_{\substack{z \in \Lambda \\ z \neq x_0, y_0}} \frac{1 + (m^\sim|x_0 - y_0|)^\alpha}{(1 + (m^\sim|x_0 - z|)^\alpha)(1 + (m^\sim|y_0 - z|)^\alpha)}$$

$$\leq 2A + 2m^{\sim -2\alpha} \text{const} |x_0 - y_0|^\alpha \sum_{\substack{z \in \Lambda \\ z \neq x_0, y_0}} \frac{1}{|x_0 - z|^\alpha |y_0 - z|^\alpha}$$

$$\leq 2A + 2m^{\sim -2\alpha} \text{const} |x_0 - y_0|^{d-\alpha}.$$

With $d - \alpha - 1 \leq 0$, we conclude that

$$m^{\sim 2\alpha}\frac{dm^\sim}{d\sigma} \leq \text{const}$$

and the proof is complete.

17.6 Continuity of $d\mu$ at the Critical Point

For the $\lambda\phi^4 + \sigma\phi^2$ model, we show that the Schwinger functions are continuous in σ on the closed interval $\sigma_c \leq \sigma < \infty$. In particular they are continuous as $\sigma \downarrow \sigma_c$. It follows from the monotonicity of $S_\Lambda^{(n)}$ in Λ and σ, as established in Section 17.5.1, that the infinite volume Schwinger functions $S^{(n)}$ are semicontinuous from above and monotonic in σ and $-\mu$, $\mu \geq 0$. Because of the possibility of phase transitions, this general argument gives only one sided continuity. However, two sided continuity for $\sigma_c < \sigma$ follows from the existence of the derivative in Theorem 17.6.1 below.

For σ (or μ) sufficiently large, we fall into the region of convergent cluster expansions (cf. Chapter 18). The definition of $S^{(n)}$ by decreasing σ, $-\mu$ monotonically from such values is known as *weak coupling boundary conditions* (cf. [Glimm and Jaffe, 1975b]).

Theorem 17.6.1. *For $\sigma_c < \sigma$ (with σ_c defined in Section 17.1) the derivatives*

$$\frac{\partial S_n(x)}{\partial \sigma} \tag{17.6.1}$$

exist.

PROOF. The proof proceeds through the Lebowitz inequalities of Corollary 4.3.3, as in Section 17.5 or [Glimm and Jaffe, 1975b]. The derivatives (17.6.1) are bounded from above by sums of products of the two-point functions S_2.

Remark. The result extends to the truncated Schwinger functions defined in Section 14.1 and to the vertex functions Γ; see [Glimm and Jaffe, 1975b]. The derivatives are typically divergent at $\sigma = \sigma_c$, and the rate of divergence is governed by some critical exponent. The derivatives $\partial S_n/\partial \sigma$ are themselves (partially) truncated Schwinger functions, with one $:\phi^2:$ vertex. Because of monotonicity (Theorem 17.1.1), the derivative $\partial S_n/\partial \sigma$ is absolutely continuous in σ. Thus

$$S_n|_{\sigma = \sigma_0} = S_n|_{\sigma = \sigma_1} - \int_{\sigma_0}^{\sigma_1} \frac{\partial S_n}{\partial \sigma}\, d\sigma,$$

and so the derivative $\partial S_n/\partial \sigma$ can be used to study the asymptotic behavior of S_n as $\sigma \downarrow \sigma_c$. When the derivative can be bounded by a correlation inequality, a differential inequality can be obtained. The solution of this inequality gives a rigorous upper bound on some critical exponent. An extension of this point of view leads to the Callan–Symanzik equation [Domb and Green, 1976, v.6] and renormalization group methods.

17.7 Critical Exponents

The thermodynamic functions and correlation functions are most difficult to study near critical points. The leading asymptotic behavior is normally of a power law type, and thus is determined by a few key parameters: exponents and coefficients. It is expected that the exponents are *universal* in the sense that they coincide for broad classes of related interactions (e.g., for ϕ^4 and Ising models, defined for various lattices or—in the ϕ^4 case—as a continuum theory). However, the exponents depend on the dimension d of space or space-time, as well as the number n of components of the vector ϕ. Since the exponents are difficult to determine either numerically or experimentally, exact theoretical relations (inequalities and proposed identities) have received considerable attention. A systematic discussion of these exponents is contained in [Stanley, 1971]. Here we only illustrate the ideas by deriving a few standard and not so standard inequalities from Griffiths' and Lebowitz' inequalities. Because we use Lebowitz' inequalities below, we are restricted to the case $n = 1$.

Assume the canonical normalization and definition of $S(p)$ of (17.4.2), and let

$$\chi = S^{(2)}(0), \qquad \varepsilon = \sigma - \sigma_c. \qquad (17.7.1)$$

Then the critical exponents v, γ, η, and ζ are defined by

$$m \sim \varepsilon^v, \qquad \chi \sim \varepsilon^{-\gamma},$$

$$\langle \phi(x)\phi(0) \rangle |_{\sigma = \sigma_c} \sim x^{-d+2-\eta}, \qquad Z \sim \varepsilon^\zeta.$$

Table 17.1. Values of the critical exponent v, based on exact or numerical calculations, theoretical estimates, and experiments.

d \ n	$n = 1$ Ising model	$n = 2$ xy model	$n = \infty$ Gaussian model
$d = 1$	∞ (exact)		$\frac{1}{2}$ (exact)
$d = 2$	1 (exact)	∞ (theoretical and experimental)	$\frac{1}{2}$ (exact)
$d = 3$	.63 (numerical)	.67 (numerical)	$\frac{1}{2}$ (exact)
$d = 4$	$\frac{1}{2}$ (theoretical)	$\frac{1}{2}$ (theoretical)	$\frac{1}{2}$ (exact)

Theorem 17.7.1. *The Gaussian (mean field) values for v, γ, η, and ζ are*

$$v_{cl} = \tfrac{1}{2}, \quad \gamma_{cl} = 1, \quad \eta_{cl} = 0, \quad \zeta_{cl} = 0.$$

For a single-component ϕ^4 field theory, each exponent v, γ, η, ζ is greater than or equal to its classical value.

PROOF. The calculation of the Gaussian values is elementary. For example, in the Gaussian case, $\sigma_c = 0$, $\varepsilon = \sigma = \frac{1}{2}m^2$, $Z = 1$, $\rho(a)\,da = 0$, and $S(p) = 1/(p^2 + m^2)$. By definition, $0 \leq Z \leq 1$, so $0 = \zeta_{cl} \leq \zeta$. Similarly $0 = \eta_{cl} \leq \eta$ follows from the Lehmann spectral formula (6.2.9).

The bound $dm^2/d\sigma \leq Z$ of Section 17.4 implies for critical exponents:

$$1 \leq \left(2 - \frac{\zeta}{v}\right)v \leq (2 - \eta)v, \qquad (17.7.2)$$

and as a special case $\frac{1}{2} \leq v$. The canonical bound on γ follows from

$$0 \leq \frac{d\chi^{-1}}{d\sigma} \leq 1,$$

established in Section 17.4.

Let us return to the bound $\lambda_2 \leq \text{const}$ of Theorem 17.3.2. As a consequence, $\lambda_2(\varepsilon)$ must have a nonnegative exponent. For $d = 3$,

$$\lambda_2 \sim \varepsilon^{3v + 2\gamma - (2\Delta + \gamma)}$$
$$= \varepsilon^{3v + \gamma - 2\Delta},$$

where γ is the exponent for the susceptibility and Δ is the "gap" exponent relating the four point to the two point function. We conclude

$$0 \leq 3v + \gamma - 2\Delta.$$

On the other hand, if $\lambda_2 \neq 0$ at $\sigma = \sigma_c$, it is necessary that a "hyperscaling" relation hold,

$$3v + \gamma - 2\Delta = 0.$$

There are two frameworks within which v and the hyperscaling relations have been calculated: high temperature series and Borel summation. High temperature series have been used by Wortis et al. in the Ising model case. These methods were also used by Baker and Kincaid [1980] in the strong coupling (Ising model: [J. Rosen, 1977], [Caginalp, 1980a,b], [Constantinescu, 1980], [Constantinescu and Storter, 1980]) region for the $\lambda\phi_3^4$ model. Borel resummation of the continuum $\lambda\phi_3^4$ perturbation series of Section 9.4 has been used by LeGuillou and Zinn–Justin [1977]. Neither method has mathematically justified error bounds. The results for $3v + \gamma - 2\Delta$ (or equivalent) are

H.T.:	$0.039^{+0.02}_{-0.03}$	Wortis et al.,
H.T.:	0.028 ± 0.003	Baker, Kincaid,
Borel:	0.000 ± 0.003	LeGuillou, Zinn–Justin.

Thus the high temperature (H.T.) series suggests a breakdown of hyperscaling. It is possible to argue that the difference in these calculations can be attributed to the exponent v. In particular, we have

$$\text{H.T.:} \qquad v = 0.638^{+0.002}_{-0.001},$$

$$\text{Borel:} \qquad v = 0.6300 \pm 0.0008.$$

Thus

$$3(v_{\text{H.T.}} - v_{\text{Borel}}) = 0.024^{+0.006}_{-0.004}$$

accounts for the discrepancy in hyperscaling.

These results indicate that there are errors in at least one of the following: (1) universality—Ising $\approx \phi^4$; (2) hyperscaling—$0 = 3v + \gamma - 2\Delta$; (3) high temperature series estimated error bounds; (4) Borel summation estimated error bounds. New terms in the high temperature series calculations for both the spin $\frac{1}{2}$ and high spin Ising models [Nickel, 1980] suggest that the error estimates above may be too optimistic. See also [Bender, Cooper, Guralnik, Roskies, and Sharp, 1981]. The analysis suggests that it is important to include corrections to scaling behavior in the calculations, as well as scaling behavior itself. In fact, the present discrepancies between universality and hyperscaling may disappear as calculations improve.

17.8 $\eta \leq 1$

Here we study the exponent η in more detail. More precisely, we show that sufficiently rapid polynomial decay of $\langle \phi(x)\phi(y) \rangle$ implies exponential decay and thus that $\sigma_c < \sigma$. For comparison note that

$$\eta = \begin{cases} 1 & \text{if } d = 1 & \text{(closed form calculation),} \\ 0.25 & \text{if } d = 2 & \text{(closed form calculation),} \\ 0.041 & \text{if } d = 3 & \text{(numerical high temperature series} \\ & & \text{calculation or Borel summation),} \\ 0 & \text{if } d \geq 4 & \text{(renormalization group).} \end{cases}$$

Theorem 17.8.1. *Consider a ϕ^4 lattice field or Ising model, and assume*

$$\lim_{|x| \to \infty} \langle \phi(0)\phi(x) \rangle |x|^{d-1} = 0. \tag{17.8.1}$$

Then there exists $m > 0$ such that

$$\langle \phi(0)\phi(x) \rangle \leq O(1)e^{-m|x|}, \qquad |x| \to \infty.$$

Thus $\eta \leq 1$. For a ϕ^4 continuum field, $\eta \leq 2$.

Remark 1. By the second Griffiths inequality (4.1.11),

$$0 \leq \langle \phi(x)\phi(y) \rangle - \langle \phi(x) \rangle \langle \phi(y) \rangle,$$

so the assumption (17.8.1) ensures that $\langle \phi(x) \rangle = 0$. By Theorem 16.1.1, $\langle \cdot \rangle$ defines a pure phase.

Remark 2. For $d = 1$, the theorem shows that vanishing of $\langle \phi(x)\phi(y) \rangle$ as $x - y \to \infty$ ensures exponential decrease. In the $d = 1$ Ising model, the function $\langle \phi(i)\phi(j) \rangle$ vanishes at $i - j = \infty$ unless $T = 0$. For $T = 0$, $\langle \phi(i)\phi(j) \rangle \equiv 1$.

Remark 3. The bound $\eta \leq 2$ was established by Glimm and Jaffe [1977a]. The bound $\eta \leq 1$ is due to Dobrushin 1979]. We follow [Simon 1980]; see also [Aizenmann and Simon 1980], [Lieb 1980], [Rivasseau 1980].

PROOF OF THEOREM 17.8.1 FOR LATTICE FIELDS AND THE ISING MODEL.

Since $\langle \phi \rangle = 0$, we can use the ϕ^4 inequality of Corollary 4.3.3. This takes the form

$$0 \leq \langle \phi(i)\phi(j)\phi(k)\phi(l) \rangle - \langle \phi(i)\phi(j) \rangle \langle \phi(k)\phi(l) \rangle$$

$$\leq \langle \phi(i)\phi(k) \rangle \langle \phi(j)\phi(l) \rangle + \langle \phi(i)\phi(l) \rangle \langle \phi(j)\phi(k) \rangle. \tag{17.8.2}$$

Furthermore, consider the interpolating expectations $\langle \ \rangle_s$ defined by the measures

$$d\mu_s = Z_s^{-1} \exp\left[-s\beta \sum_\Gamma \phi(k)\phi(l)\right] d\mu. \tag{17.8.3}$$

For simplicity we choose unit lattice spacing. Here $d\mu$ is the measure for $\langle \ \rangle$; $0 \leq s \leq 1$; Z_s normalizes $d\mu_s$; and Γ is a finite set of nearest neighbor bonds (k, l) which we now specify. Let B_r denote a ball of radius r centred at the origin, and let $\Gamma = \Gamma_r$ denote the bonds which intersect ∂B_r. Then $d\mu = d\mu_{s=0}$, and $d\mu_{s=1}$ is a measure which factorizes

$$d\mu_1 = d\mu^{\text{int}} \otimes d\mu^{\text{ext}}$$

with $d\mu^{\text{int}}$ depending only on $\phi(i)$, $i \in$ Interior B_r, and $d\mu^{\text{ext}}$ depending only on $\phi(i)$, $i \in$ Exterior B_r. All the measures $d\mu_s$ are ferromagnetic, are even, and satisfy the ϕ^4 inequality (17.8.2). Furthermore, if $r < |i|$, then

$$\langle \phi(0)\phi(i) \rangle_1 = \langle \phi(0) \rangle_1 \langle \phi(i) \rangle_1 = 0.$$

Thus we can write

$$\langle \phi(0)\phi(i) \rangle = -\int_0^1 \left(\frac{d}{ds} \langle \phi(0)\phi(i) \rangle_s\right) ds$$

$$= \int_0^1 \beta \sum_\Gamma (\langle \phi(0)\phi(i)\phi(k)\phi(l) \rangle_s$$

$$- \langle \phi(0)\phi(i) \rangle_s \langle \phi(k)\phi(l) \rangle_s) ds$$

$$\leq \int_0^1 \beta \sum_{(k, l) \in \Gamma \text{ or } (l, k) \in \Gamma} \langle \phi(0)\phi(k) \rangle_s \langle \phi(l)\phi(i) \rangle_s \, ds. \tag{17.8.4}$$

In the last inequality we use (17.8.2). Again by (17.8.2), $\langle \phi(0)\phi(k)\rangle_s$ is monotonic decreasing in s. Hence

$$\langle \phi(0)\phi(i)\rangle \leq \beta \sum_\Gamma \langle \phi(0)\phi(k)\rangle\langle \phi(l)\phi(i)\rangle$$

$$\leq \beta\left(\sum_\Gamma \langle \phi(0)\phi(k)\rangle\right) \sup_\Gamma \langle \phi(l)\phi(i)\rangle. \tag{17.8.5}$$

Using the bound (17.8.1) and the fact that $|\Gamma| \leq \text{const } r^{d-1}$, we have for r sufficiently large

$$\alpha^{-1} \equiv \beta \sum_\Gamma \langle \phi(0)\phi(k)\rangle < 1.$$

With this choice of α and r, (17.8.5) implies

$$\sup\{\langle \phi(0)\phi(i)\rangle : D \leq |i|\} \leq \alpha \sup\{\langle \phi(0)\phi(i)\rangle : D - r \leq |i|\}.$$

We now proceed by induction. After $|i|/(r+1)$ steps, we have

$$\langle \phi(0)\phi(i)\rangle \leq \alpha^{-|i|/(r+1)} \sup_{k \neq l} \langle \phi(k)\phi(l)\rangle$$

$$\leq c\alpha^{-|i|/(r+1)}.$$

Hence the theorem is proved, with $m = (r+1)^{-1} \ln \alpha$.

17.9 The Scaling Limit

The thermodynamic functions are not the only quantities with a simple power law leading asymptotic behavior at the critical point. The entire field theory is presumed to have an asymptotic limit at the critical point. This limit, called the scaling limit, is defined by an infinite scale (change of length scale) transformation, as described in Section 6.6. The limit, if it exists, is again a Euclidean field theory. Existence of the scaling limit (convergence through a convergent subsequence of scale transformations) has been reduced to a uniform bound

$$S^{(2)}(f \times g) \leq |f|_{\mathscr{S}} \, |g|_{\mathscr{S}} \tag{17.9.1}$$

on the two point function [Glimm and Jaffe, 1974d].

Theorem 17.9.1. *Let $\langle \cdot \rangle_j$ denote a sequence of single phase ϕ^4 lattice or continuum field theories which satisfy a bound of the form (17.9.1), where $|\cdot|_{\mathscr{S}}$ is a Schwartz norm, independent of j. Then a subsequence of these field theories converges as $j \to \infty$. If the lattice spacing goes to zero with this limit, then the limit theory satisfies the Osterwalder–Schrader axioms, with the possible exception of a unique vacuum and (when the approximate theories are lattice theories) of Euclidean covariance.*

17.10 The Conjecture $\Gamma^{(6)} \leq 0$

The unamputated six point vertex function is defined by

$$\Gamma^{(6)}(xxxyyy) = (\langle xxxyyy \rangle^T + \int \langle xxxz \rangle^T \Gamma(zz')\langle z'yyy \rangle^T \, dz \, dz'$$

$$+ 9 \int \langle xxyz \rangle^T \Gamma(zz')\langle z'xyy \rangle^T \, dz \, dz'. \qquad (17.10.1)$$

Here $\Gamma = -S_2^{-1}$, where the inverse is the operator inverse of the kernel $S_2(x, y)$. The conjecture

$$\Gamma^{(6)}(xxxyyy) \leq 0 \qquad (17.10.2)$$

has a number of interesting consequences: e.g., the absence of three-particle bound states in the propagator, the existence of the scaling limit, and certain bounds on critical exponents (see [Glimm and Jaffe, 1975c, 1976b]).

There is some evidence for (17.10.2) in single phase, even ϕ^4 models. For example it is true in perturbation theory (i.e., for $\sigma \gg 0$ or high temperature), and there is a heuristic argument that it holds near σ_c. However, some good new idea is needed to prove (17.10.2).

In this section we illustrate some uses of (17.10.2). For example, we have

Theorem 17.10.1. *If* (17.10.2) *holds, then*

$$0 \leq \Gamma(x) \leq e^{-3m|x|}, \qquad |x| \to \infty. \qquad (17.10.3)$$

Remark. The bound (17.10.3) excludes spectrum in $\Gamma(x)$ in the interval $(0, 3m)$, and hence spectrum in $d\sigma(a)$ in the interval $(m, 3m)$. Thus no three-particle bound states occur in the propagator, i.e., in the states spanned by $\phi(x)\Omega$.

OUTLINE OF PROOF. We use the integration by parts formula [Glimm and Jaffe, 1975c]

$$\int \phi(x)A(\phi) \, d\mu(\phi) = \langle \phi(x)A \rangle$$

$$= \int dy \, S(x - y)\left[\left\langle \frac{\delta A}{\delta \phi(y)} \right\rangle - \langle V'(y)(I - P_1)A \rangle \right]. \qquad (17.10.4)$$

Here $V = \lambda \int :\phi^4: dx$ is the interaction, and

$$P_1 A = \int \phi(z)\Gamma(z - z')\langle \phi(z')A \rangle \, dz \, dz'.$$

From (17.10.4), it follows that for $x \neq y$.

$$\Gamma(x - y) = \langle V'(x)(I - P_1)V'(y) \rangle = 16\lambda^2 \langle \phi^3(x)(I - P_1)\phi^3(y) \rangle. \qquad (17.10.5)$$

Expanding (17.10.5), and using (17.10.1),

$$(16)^{-1}\lambda^{-2}\Gamma(x-y) = 6\langle xy\rangle^3 + 9\langle xxyy\rangle^T\langle xy\rangle$$

$$-9\int \langle xxyz\rangle^T\Gamma(zz')\langle z'xyy\rangle^T \, dz \, dz' + \Gamma^{(6)}(xxxyyy). \quad (17.10.6)$$

The first term in (17.10.6) is $O(e^{-m|x-y|})^3$ for $|x-y| \to \infty$. The second term is negative. The third term also has a three particle decay, which can be established using the absence of two-particle bound states in $\langle xxyz\rangle^T$; see [Glimm and Jaffe, 1975c]. Thus (17.10.6) results in

$$\Gamma(x-y) \le e^{-3m|x-y|}, \qquad |x-y| \to \infty.$$

The positivity of $\Gamma(x-y)$ follows from the fact that it is the Fourier transform of a Herglotz function. This completes the outline of the proof.

We finish this section with another consequence of (17.10.1).

Theorem 17.10.2. [Glimm and Jaffe, 1976b]. *Assume $d \ge 6$, $g_0(\delta) = \lambda\delta^{4-d} \le$ const (finite charge renormalization), and assume (17.10.1). If the $\delta \to 0$ limit of the $\lambda\phi_d^4$ lattice field theory is Euclidean-invariant, then the limit is a free field.*

Remark. This result is a weak form of the idea that renormalization is necessary (that is, unrenormalized theories are incorrect). It does not contribute particularly to the question of whether the renormalized ϕ^4, Yukawa, and quantum electrodynamic fields are correct (nontrivial) in $d = 4$ dimensions.

CHAPTER 18

The Cluster Expansion

18.1 Introduction

The cluster expansion [Glimm, Jaffe and Spencer, 1973, 1974] allows a detailed study of the properties of quantum fields. Consequences of the cluster expansion, in addition to the proof of existence of the infinite volume limit, are the detailed properties of the spectrum: the multiplicity of the ground state, the existence of isolated particle spectrum, the existence or absence of bound states, the completeness of scattering states at low energies, analyticity in the coupling constants, Borel summability, etc. Applications have been indicated in Chapter 14 to the study of particles and in Chapters 5 and 16 to the study of phase transitions.

The cluster expansion converges whenever the quantum field is sufficiently far from the critical point(s), i.e., close to Gaussian. Included in the convergence region (for a more complicated expansion than that presented here) are the multiple phase theories whose pure phase components are each nearly Gaussian. The cluster expansion provides the tool to analyze the mean field picture of Chapter 5, in regions of parameter space far from critical points.

The expansion is related to the virial and cluster expansions of statistical mechanics; see Chapter 2. The relevant formula from statistical mechanics is the following expansion for the density of the Gibbs ensemble:

$$\prod_{i<j} e^{-\beta V(x_i - x_j)} = \prod_{i<j} [1 + (e^{-\beta V(x_i - x_j)} - 1)]$$

$$= \sum_{\Gamma} \prod_{(i,\,j) \in \Gamma} (e^{-\beta V(x_i - x_j)} - 1). \tag{18.1.1}$$

Here Γ is a set of distinct unordered pairs (i, j), i.e. Mayer graphs, and the sum extends over all such graphs. This formula expresses the interaction between the distinct i and j particles $(i \neq j)$ particles, $e^{-\beta V(x_i - x_j)}$, as a sum of a zero-interaction term, 1, for which there is no coupling between the particles, and a perturbation $e^{-\beta V(x_i - x_j)} - 1$ which is small at high temperatures $kT = \beta^{-1}$. Heuristically, the role of the Gibbs density in the $P(\phi)_2$ field theory is replaced by the measure

$$e^{-\int [\mathscr{A}_0(x) + \lambda P(\phi(x))] \, dx} \prod_{x \in R^2} d\phi(x) \qquad (18.1.2)$$

Here

$$\mathscr{A}_0(x) = \tfrac{1}{2}[\nabla \phi(x)^2 + m_0^2 \phi(x)^2], \qquad (18.1.3)$$

and the formal expression

$$e^{-\int \mathscr{A}_0(x) \, dx} \prod_{x \in R^2} d\phi(x) \qquad (18.1.4)$$

denotes the Gaussian measure $d\phi_{C_\varnothing}$ on $\mathscr{S}'(R^2)$ with mean zero and covariance $C_\varnothing$.

In (18.1.2), the coupling between distinct points comes entirely from the $\nabla \phi$ term in $\mathscr{A}_0(x)$, and so $e^{-(1/2) \int (\nabla \phi)^2}$ in (18.1.2) plays the role of $e^{-\beta V}$ in (18.1.1). Our cluster expansion is constructed in the spirit of (18.1.1). In a strict analog of the completely decoupled theory, the Laplacian in

$$\int \mathscr{A}_0(x) \, dx = \frac{1}{2} \int \phi(x)(-\Delta + m_0^2)\phi(x) \, dx$$

is replaced by zero. The resulting ultralocal theory is very singular relative to the theory defined by (18.1.2). We reduce and control the singularity of the difference between the coupled and decoupled theories in two separate steps. As the first step we modify the above ultralocal strategy by introducing a lattice structure into R^2 and into the expansion generalizing (18.1.1). This expansion decouples only on lines of the lattice. In this way we reduce the singularity of the decoupled measure with respect to the undecoupled one. We do not introduce the lattice into the formulation of Theorems 18.1.1–2, and so the resulting expansion is an expansion for the continuum infinite volume $P(\phi)_2$ theory, and not for a lattice approximation to this theory. Let Γ denote a set of lattice lines joining nearest neighbor lattice points in Z^2, let Δ_Γ be the Laplace operator with Dirichlet boundary conditions on Γ, and let

$$C_\Gamma = (-\Delta_\Gamma + m_0^2)^{-1}. \qquad (18.1.5)$$

Then $d\phi_{C_\Gamma}$ plays the role of the decoupled measure, with decoupling along the curve Γ. In summary, the lattice structure gives discrete variables in the sum and product

$$\sum_\Gamma \prod_{(i, j) \in \Gamma}$$

in (18.1.1), even when this formula is applied to the continuum $P(\phi)_2$ model.

The second step regularizes the differences corresponding to

$$e^{-\beta V(x_i - x_j)} - 1$$

in (18.1.1). The difference between two Gaussian measures can be expressed as

$$d\phi_{C_1} - d\phi_{C_2} = \int_0^1 \frac{d}{ds} d\phi_{C(s)},$$

where

$$C(s) = sC_1 + (1 - s)C_2.$$

Then $(d/ds)\, d\phi_{C(s)}$ can be evaluated by (9.1.33), i.e. by integration by parts.

For small values of λ/m_0^2, we prove an exponential cluster property on the Schwinger functions. We prove the cluster property in a finite volume with bounds independent of this volume. From these results, it follows easily that the Schwinger functions converge in the infinite volume limit, and that the infinite volume Schwinger functions satisfy an exponential cluster property. (They are also independent of the boundary conditions.) Using the Osterwalder–Schrader reconstruction theorem, the infinite volume $P(\phi)_2$ field theory is constructed from the Schwinger functions, and in this theory the Wightman axioms are satisfied and the physical mass is strictly positive. We also show that the Schwinger functions are analytic in λ, for λ in the bounded sector

$$0 < |\lambda| < \varepsilon, \qquad -\pi/2 < \arg \lambda < \pi/2. \tag{18.1.6}$$

The finite volume Schwinger functions are by definition the moments of the measure $d\mu_\Lambda$ of (11.2.1):

$$S_\Lambda(x_1, \ldots, x_n) = \int \phi(x_1) \cdots \phi(x_n)\, d\mu_\Lambda. \tag{18.1.7}$$

For convenience, however, we replace the Dirichlet covariance $C_{\partial\Lambda}$ in (11.2.1) by the free covariance $C_\varnothing$ so that the finite volume cutoff Λ occurs only in the interaction $V = V(\Lambda)$. Then S_Λ is defined as a tempered distribution in $\mathscr{S}'(R^{2n})$. In addition to the monomials in (18.1.7), it is useful to integrate products of Wick-ordered polynomials, namely

$$A = \int :\phi(x_1)^{n_1}: \cdots :\phi(x_j)^{n_j}: w(x_1, \ldots, x_j)\, dx. \tag{18.1.8}$$

We assume $w \in \mathscr{S}(R^{2j})$, although weaker bounds would suffice. We define suppt A to be the intersection of all closed subsets $C \subset R^2$ with

$$\text{suppt } w \subset C \times \cdots \times C \quad (j \text{ factors}). \tag{18.1.9}$$

Theorem 18.1.1. *Let λ belong to the closure of the half circle (18.1.6), and let ε/m_0^2 be sufficiently small. Let A and B be functions on $\mathscr{S}'$ of the form*

(18.1.8). *Let d be the width of a strip in R^2 separating suppt A and suppt B. There is a constant $M = M_{A,B}$ and a positive constant m independent of A and B such that*

$$\left| \int AB \, d\mu_\Lambda - \int A \, d\mu_\Lambda \int B \, d\mu_\Lambda \right| \leq M_{A,B} e^{-md}, \qquad (18.1.10)$$

uniformly in Λ, as $|\Lambda| \to \infty$. Furthermore, M is independent of translations in either A or B.

Theorem 18.1.2. *Let λ belong to the closure of the half circle (18.1.6), and let ε/m_0^2 be sufficiently small. For A of the form (18.1.8), $|\int A \, d\mu_\Lambda|$ is bounded uniformly in λ and in Λ as $\Lambda \to \infty$.*

Corollary 18.1.3. *Under the hypotheses of Theorem 18.1.1, the infinite volume limit*

$$\int A \, d\mu = \lim_{\Lambda \uparrow R^2} \int A \, d\mu_\Lambda \qquad (18.1.11)$$

exists and satisfies

$$\left| \int AB \, d\mu - \int A \, d\mu \int B \, d\mu \right| \leq M_{AB} e^{-md}. \qquad (18.1.12)$$

PROOF. Apply the theorem to the interaction

$$V_\alpha \equiv V(\Lambda) + \alpha V(\Delta), \qquad 0 \leq \alpha \leq 1,$$

to add a unit square $\Delta \subset R^2 \backslash \Lambda$ to the interaction region Λ. Thus

$$\left| \frac{d}{d\alpha} \int A \, d\mu_\alpha \right| = \left| \int AV(\Delta) \, d\mu_\alpha - \int A \, d\mu_\alpha \int V(\Delta) \, d\mu_\alpha \right|$$

$$\leq M_A e^{-md},$$

where $d = \text{dist}(\text{suppt } A, \Delta)$. Using this exponential decay, sum over Δ covering $\Lambda' \backslash \Lambda$ to establish that $\int A \, d\mu_\Lambda$ is a Cauchy sequence as $\Lambda \uparrow R^2$. The uniform bound (18.1.10) then also holds for the infinite volume expectations (8.1.11).

From analyticity of the finite volume Schwinger functions, from Vitali's theorem, and from the convergence, as $\Lambda \to \infty$, for λ real and small, we have

Corollary 18.1.4. *The Schwinger functions are analytic in λ in (18.1.6) for ε/m_0^2 small.*

18.2 The Cluster Expansion

The proofs of Theorems 18.1.1 and 18.1.2 are based on an expansion which we now derive. Let $\mathscr{B}$ be a set of line segments in R^2. We are interested in two examples: Either $\mathscr{B} = (Z^2)^*$, the set of all lattice lines (bonds) joining nearest neighbor lattice sites in Z^2, or $\mathcal{B} = (Z^2)^* \backslash \Gamma$ with Γ a finite subset of $(Z^2)^*$. We identify a subset $\Gamma \subset \mathcal{B}$ with the subset

$$\Gamma = \bigcup_{b \in \Gamma} b \subset R^2.$$

The subsets $\Gamma \subset \mathcal{B}$ label terms in our expansion. The term labeled by Γ is decoupled across

$$\Gamma^c = \mathcal{B} \backslash \Gamma \qquad (18.2.1)$$

and formed by choosing a Gaussian measure with Dirichlet covariance on Γ^c. Thus the lines b in Γ^c are called Dirichlet lines. For the lines $b \in \Gamma$ there are differences between coupled and decoupled measures as in (18.1.1). These differences are expressed in terms of derivatives by the fundamental theorem of calculus. Thus the $b \in \Gamma$ are called derivative lines.

The covariance operators we consider are convex combinations of the operators C_Γ. For each $b \in \mathcal{B}$, we introduce a parameter $s_b \in [0, 1]$ to measure the strength of the coupling across b. The parameter value $s_b = 0$ corresponds to full zero Dirichlet data on b, and hence to zero coupling across b, while $s_b = 1$ corresponds to full coupling across b. With

$$s = (s_b)_{b \in \mathcal{B}}, \qquad (18.2.2)$$

we define the multiparameter family of covariance operators

$$C(s) = \sum_{\Gamma \subset \mathcal{B}} \prod_{b \in \Gamma} s_b \prod_{b \in \Gamma^c} (1 - s_b) C_{\Gamma^c}. \qquad (18.2.3)$$

Since the coefficients in (18.2.3) are the terms in the expansion of

$$1 = \prod_{b \in \mathcal{B}} 1 = \prod_{b \in \mathcal{B}} [s_b + (1 - s_b)],$$

(18.2.3) is a convex sum, as asserted. The free covariance is now denoted

$$C_\varnothing = C(1, 1, \ldots) = (-\Delta + m_0^2)^{-1},$$

and the completely decoupled covariance is

$$C_\mathcal{B} = C(0, 0, \ldots) = (-\Delta_\mathcal{B} + m_0^2)^{-1}.$$

The Schwinger functions and the partition function are, in a natural way, functions of s, and we use the notation

$$Z(s)S_s(x) = Z(\Lambda, s)S_{\Lambda, s}(x) = \int \prod_i \phi(x_i) e^{-\lambda V(\Lambda)} \, d\phi_s,$$

$$Z(s) = Z(\Lambda, s) = \int e^{-\lambda V(\Lambda)} \, d\phi_s, \qquad (18.2.4)$$

where

$$d\phi_s = d\phi_{C(s)} \quad \text{and} \quad V(\Lambda) = \int_\Lambda \lambda : P(\phi(x)): dx.$$

The goal of the cluster expansion is to express coupled quantities ($s_b \equiv 1$) in terms of decoupled (and hence finite volume) quantities. The decoupled quantities have $s_b = 0$ for many $b \in \mathcal{B}$, and in order to formulate them, it is convenient to define

$$s(\Gamma) = \{s(\Gamma)_b\}_{b \in \mathcal{B}}$$

by the formula

$$s(\Gamma)_b = \begin{cases} s_b & \text{if } b \in \Gamma, \\ 0 & \text{if } b \notin \Gamma. \end{cases} \tag{18.2.5}$$

For Γ finite, $s(\Gamma)$ specifies Dirichlet boundary conditions at large distances (on Γ^c), while s can be thought of as giving general boundary conditions on Γ^c, and agreeing with $s(\Gamma)$ on Γ. Thus the following definition is a way of saying that F is independent of the boundary conditions at ∞.

Definition 18.2.1. A function $F(s)$ is called regular at infinity if for each s,

$$F(s) = \lim_{\{\Gamma \uparrow \mathcal{B}: \Gamma \text{ finite}\}} F(s(\Gamma)). \tag{18.2.6}$$

Proposition 18.2.1. *The functions* (18.2.4) *are regular at infinity. Here* S *converges in* $\mathcal{S}'$, *and* $\mathcal{B} \subseteq (Z^2)^*$.

Because Λ is fixed and bounded, the limit (18.2.6) is elementary. The proposition follows from (9.1.33), Lemma 8.5.2, and Theorem 8.6.2.

The first step in the cluster expansion is to apply the fundamental theorem of calculus to the finite number of nonzero parameters in $F(s(\Gamma))$. Define

$$\partial^\Gamma = \prod_{b \in \Gamma} \frac{d}{ds_b}, \tag{18.2.7}$$

and for two coordinate values s and σ define order coordinatewise, so that

$$\sigma \leq s \quad \Leftrightarrow \quad \sigma_b \leq s_b \quad \forall b \in \mathcal{B}.$$

Proposition 18.2.2. *Let* $F(s)$ *be smooth and regular at infinity. Then*

$$F(s) = \sum_{\{\Gamma \subset \mathcal{B}: \Gamma \text{ is finite}\}} \int_{0 \leq \sigma \leq s(\Gamma)} \partial^\Gamma F(\sigma(\Gamma)) \, d\sigma. \tag{18.2.8}$$

PROOF. Let $G(s)$ denote the right side of (18.2.8). We assert that $F(s(B)) = G(s(B))$, for B any finite subset of $\mathcal{B}$. Now $G(s(B))$ is just the sum in (18.2.8), restricted to sets $\Gamma \subset B$. Since $F(s)$ is regular at infinity, convergence of the sum in (18.2.8) follows from the assertion. Also (18.2.8) follows by limits in the assertion, as $B \uparrow \mathcal{B}$.

For a function $f(s_b)$ of a single variable, let

$$(\delta^b f)(s_b) = f(s_b) - f(0) = \int_0^{s_b} \partial^b f(\sigma_b) \, d\sigma_b,$$

$$(E_0^b f)(s_b) = f(0).$$

Then $I = E_0^b + \delta^b$ (fundamental theorem of calculus), and so

$$I = \prod_{b \in B} (E_0^b + \delta^b) = \sum_{\Gamma \subset B} E_0^{B \backslash \Gamma} \, \delta^\Gamma, \tag{18.2.9}$$

where

$$\delta^\Gamma = \prod_{b \in \Gamma} \delta^b, \qquad E_0^{B \backslash \Gamma} = \prod_{b \in B \backslash \Gamma} E_0^b.$$

It is easy to see that (18.2.9) yields the desired identity $F(s(B)) = G(s(B))$.

The next step in the cluster expansion is factorization and partial resummation of (18.2.8). Write

$$R^2 \backslash \Gamma^c = X_1 \cup X_2 \cup \cdots \cup X_r, \tag{18.2.10}$$

so that each X_i is a union of connected components such that $X_i \cap X_j = \varnothing$ for $i \neq j$.

Definition 18.2.2. A function $F(\Lambda, s)$ decouples at $s = 0$ provided

$$F(\Lambda, s(\Gamma)) = \prod_{i=1}^{r} F(\Lambda \cap X_i, s(\Gamma \cap X_i)) \tag{18.2.11}$$

whenever (18.2.10) holds.

Proposition 18.2.3. *Given* (18.2.10), *the measures* $d\phi_{s(\Gamma)}$ *and* $e^{-\lambda V(\Lambda)} \, d\phi_{s(\Gamma)}$ *each factor into a product of* r *measures, and the* ith *factor measure is defined on* $\mathscr{S}'(X_i)$.

PROOF. Since $C(s(\Gamma))$ has zero Dirichlet data on Γ^c, $C(s(\Gamma))$ is the direct sum of the operators

$$C(s(\Gamma)) \mid L_2(X_i).$$

The factorization of $d\phi_{s(\Gamma)}$ follows from this fact, and can be seen from (9.1.16). Because $: P(\phi(x)):$ is a local function, $e^{-\lambda V(\Lambda)}$ also factors, and so does $e^{-\lambda V} \, d\phi_{s(\Gamma)}$.

Corollary 18.2.4. *The functions* ZS *and* Z *of* (18.2.4) *decouple at* $s = 0$.

The cluster expansion represents F as a sum of products of contributions from connected graphs. We start with some set X_0 of interest (e.g. $X_0 = \{x_1, \ldots, x_n\}$ in (18.2.4)). The graphs which do not meet X_0 are resummed, giving a single factor of F in an exterior region.

We now carry out this resummation in more detail. Substitute (18.2.11) in the expansion (18.2.8). This yields

$$F(\Lambda, s) = \sum_{\Gamma} \prod_{i=1}^{r} \int_0^{s(\Gamma_i)} \partial^{\Gamma_i} F(\Lambda \cap X_i, \sigma(\Gamma_i)) \, d\sigma, \tag{18.2.12}$$

where $\Gamma_i = \Gamma \cap X_i$. If the X_i are connected, this is the sum over products of connected graphs. Now we choose X_1 in (18.2.10) to be the union of all components meeting X_0, and let X_2 be the union of the remaining components. The resummation consists of holding X_1 and Γ_1 fixed, while summing over all choices of Γ_2. Explicitly,

$$F(\Lambda, s) = \sum_{X_1, \Gamma_1} \int_0^{s(\Gamma_1)} \partial^{\Gamma_1} F(\Lambda \cap X_1, \sigma(\Gamma_1))\, d\sigma$$

$$\times \sum_{\Gamma_2} \int_0^{s(\Gamma_2)} \partial^{\Gamma_2} F(\Lambda \cap X_2, \sigma(\Gamma_2))\, d\sigma.$$

The Γ_2 sum runs over all finite sets Γ_2 of bonds in $\mathscr{B}\backslash X_1^-$. For this reason, the Γ_2 sum can be evaluated by (18.2.8) as $F(\Lambda \cap X_2, s(\mathscr{B}\backslash X_1^-))$. Setting $X = X_1^-$ and writing Γ for Γ_1, the expansion has the form

$$F(\Lambda, s) = \sum_X K(X_0, X) F(\Lambda\backslash X, s(\mathscr{B}\backslash X)),$$

$$\hspace{8cm} (18.2.13)$$

$$K(X_0, X) = \sum_\Gamma \int_0^{s(\Gamma)} \partial^\Gamma F(\Lambda \cap X, \sigma(\Gamma))\, d\sigma.$$

In these sums, X ranges over finite unions of closed lattice squares containing X_0 while Γ ranges over finite subsets of $\mathscr{B}$ such that

(i)   each component of $X\backslash\Gamma^c$ meets X_0,
(ii) $\Gamma \subset \text{Int } X$.

If no such Γ exists for a given X, then $K(X_0, X) \equiv 0$.

Theorem 18.2.5. *Let X_0 be bounded, and let F be smooth, be regular at infinity, and decouple at $s = 0$. Then the cluster expansion (18.2.13) holds.*

EXAMPLE 1: Let $\mathscr{B} = (Z^2)^*$ be the set of all lattice lines, and let $X_0 = \{x_1, \ldots, x_n\}$. We substitute $ZS = F$ (see (18.2.4)). Thus observe that for $s = 1$

$$F(\Lambda\backslash X, s(\mathscr{B}\backslash X)) = \int e^{-\lambda V(\Lambda\backslash X)}\, d\phi_{s(\mathscr{B}\backslash X)}$$

$$= Z(\Lambda\backslash X, s(\mathscr{B}\backslash X)) \equiv Z_{\partial X}(\Lambda\backslash X), \hspace{1cm} (18.2.14)$$

since a change in the data within X does not affect the integral. Thus if we divide by Z, (18.2.13) yields

$$S_\Lambda(x) = \sum_{X, \Gamma} \int \partial^\Gamma \int \prod_i \phi(x_i) e^{-\lambda V(\Lambda \cap X)}\, d\phi_{s(\Gamma)}\, ds(\Gamma) \frac{Z_{\partial X}(\Lambda\backslash X)}{Z(\Lambda)}. \hspace{0.5cm} (18.2.15)$$

If we write $X\backslash\Gamma^c$ as a union of connected components, the integral in (18.2.15) can be factored, as in (18.2.12).

EXAMPLE 2: Let $\Gamma_1 \subset (Z^2)^*$ and $\Gamma_2 = \Gamma_1\backslash b_1$ be finite sets of lattice bonds, where b_1 is the first element of Γ_1 in some lexicographic order of $(Z^2)^*$. We use the cluster expansion (in all bonds $b \neq b_1$) to study the difference $Z_{\Gamma_1} - Z_{\Gamma_2}$,

where

$$Z_\Gamma = \int e^{-\lambda V(\Lambda)} \, d\phi_{C_\Gamma},$$

and so we take $\mathcal{B} = (Z^2)^* \backslash b_1$, $X_0 = b_1$. Z is now a function of the pair (s, s_{b_1}). We define

$$F(\Lambda, s) = \begin{cases} |Z(\Lambda, s(\Gamma_2^c), 0) - Z(\Lambda, s(\Gamma_2^c), 1) & \text{if } b_1 \subset \Lambda, \\ |Z(\Lambda, s(\Gamma_2^c), 0) & \text{if } b_1 \cap \Lambda = \varnothing. \end{cases}$$

This definition is arranged so that

$$F(\Lambda, s = 1) = Z_{\Gamma_1} - Z_{\Gamma_2} \quad \text{for } b_1 \subset \Lambda.$$

Also F is smooth, is regular at infinity, and decouples at $s = 0$. (The last property depends on the exclusion of b_1 from $\mathcal{B}$.) We note that F is independent of the variables s_b for $b \in \Gamma_2$, and so $\partial^b F = 0$ for $b \in \Gamma_2$. Thus in all nonzero terms, Γ_2 consists of Dirichlet lines in the cluster expansion for F, and in (18.2.13), we may impose the further restriction on $\sum_\Gamma$:

(iii) $\Gamma \cap \Gamma_2 = \varnothing.$

From (18.2.13), we have

$$\begin{aligned} F(\Lambda, s = 1) &= Z_{\Gamma_1}(\Lambda) - Z_{\Gamma_2}(\Lambda) \\ &= \sum_X K(b_1, \Gamma_1, X) F(\Lambda \backslash X, s(\mathcal{B} \backslash X)) \\ &= \sum_X K(b_1, \Gamma_1, X) Z_{\Gamma_1 \cup X^*}(\Lambda \backslash X). \end{aligned}$$

Here X^* is the set of lattice lines in X, and the last line is justified by the fact that the second factor above is independent of s_b, $b \in \text{Int } X$. We multiply and divide by

$$Z_{\partial \Delta}(\Delta)^{|\Lambda \cap X|},$$

where Δ is a single lattice square. Since

$$Z_{\partial \Delta}(\Delta)^{|\Lambda \cap X|} Z_{\Gamma_1 \cup X^*}(\Lambda \backslash X) = Z_{\Gamma_1 \cup X^*}(\Lambda),$$

we obtain (with a new K)

$$Z_{\Gamma_1} = Z_{\Gamma_1 \backslash b_1} + \sum_X K(b_1, \Gamma_1, X) Z_{\Gamma_1 \cup X^*}(\Lambda), \tag{18.2.16}$$

and now

$$K(b_1, \Gamma_1, X) = Z_{\partial \Delta}(\Delta)^{-|\Lambda \cap Z|} \int \sum_\Gamma \partial^{\Gamma \cup b_1} Z(\Lambda \cap X, s(\Gamma \cup b_1)) \, ds(\Gamma \cup b_1). \tag{18.2.17}$$

These equations have a structure intermediate between the Kirkwood–Salsburg and the Mayer–Montroll equations. They will be studied in Section 18.5 in order to obtain bounds on Z_Γ / Z, i.e. the second factor in (18.2.15).

18.3 Clustering and Analyticity

We prove the cluster property and analyticity—the main results of this chapter—in this section, using as hypothesis convergence of the cluster expansion, (18.2.15). Let $T(x, \Lambda, X, \Gamma)$ denote the X, Γ term in (18.2.15), so that

$$S_\Lambda(x) = \sum_{X, \Gamma} T(x, \Lambda, X, \Gamma). \tag{18.3.1}$$

For each test function $w \in \mathscr{S}(R^{2n})$, the series

$$\int S_\Lambda(x)w(x) \, dx = \sum_{X, \Gamma} \langle w, T \rangle \tag{18.3.2}$$

converges absolutely. The rate of convergence is governed by $|X|$, the area of X. With $K > 0$, we prove that

$$\sum_{\{X, \Gamma : |X| \geq D\}} |\langle w, T \rangle| \leq |w| e^{-K(D-n)}. \tag{18.3.3}$$

where $|w|$ is some (n-dependent) $\mathscr{S}$ norm on w.

Theorem 18.3.1. *Let $K > 0$ be given. Let m_0 be large and ε be small (depending on K), and let λ belong to the closure of (18.1.6). There is an $\mathscr{S}$ norm $|w|$ such that (18.3.3) holds uniformly in λ, m_0, and D, and $|w|$ is invariant under translations in any of its variables.*

The proof allows integrands A of the form (18.1.8) to be substituted in (18.2.15). Thus Theorem 18.1.2 and Corollary 18.1.3 follow from the case $K = 1$, $D = 1$ of (18.3.3).

Theorem 18.1.1 also follows from the convergence of the cluster expansion. The proof for the two point function with an even interaction is conceptually easier, but still illustrates the main idea, so we consider that case first.

PROOF OF THEOREM 18.1.1 FOR $n = 2$, P EVEN, ASSUMING THEOREM 18.3.1. Because P, the interaction polynomial in (18.1.2), is even, $\phi(x) \to -\phi(x)$ is a symmetry of the theory. For Gaussian integrals defined by the factorizing measure $e^{-\lambda V} \, d\phi_{C(s(\Gamma))}$ (cf. Proposition 18.2.3), more is true:

$$\phi(x) \to \sigma(x)\phi(x) \tag{18.3.4}$$

is a symmetry, where

$$\sigma(x) = \pm 1, \qquad \sigma(x) = \text{const} \quad \text{on each } X_i.$$

Because of the symmetry $\phi \to -\phi$, $S_\Lambda(x_1, \ldots, x_n) = 0$ for n odd. Similarly,

$$\int \prod \phi(x_i) e^{-\lambda V(\Lambda \cap X)} \, d\phi_{C(s(\Gamma))} = 0$$

unless each of the connected components $X_1, \ldots, X_r$ contains an even number of x_i's. Since $\partial^\Gamma 0 = 0$, the same restriction $T(x, \Lambda, X, \Gamma) = 0$ applies unless each X_j has an even number of x_i's. Since each X_j has at least one x_i by (i) of Section 18.2, each X_j must have at least two x_i's. Thus for $n = 2$, there is only one X_j. In other words $X \backslash \Gamma^c$ is connected. Let d be defined by Theorem 18.1.1, and let $w = w_1 \otimes w_2$. Because X is connected and supp $w_i \cap X \neq \varnothing$, we have $d \leq |X| + 1$. Thus by (18.3.3),

$$\left| \int S_\Lambda(x_1, x_2) w_1(x_1) w_2(x_2) \, dx \right| \leq |w| e^{-K(D-2)} \leq M_w e^{-Kd}.$$

Since the one point functions $S_\Lambda(x_1)$ vanish for P even, this bound completes the proof of Theorem 18.1.1.

PROOF OF THEOREM 18.1.1 (GENERAL CASE) ASSUMING THEOREM 18.3.1. The idea, as before, is to reduce the cluster expansion to terms involving only one connected component $X_1 = X$. The terms with two or more connected components are to vanish, because of some symmetry. Since the general case (P not even) does not have such a symmetry, we follow [Ginibre, 1971] and introduce a new theory containing an artificial symmetry.

To construct the new theory, let $d\phi_{\mathcal{C}*}^*$ denote an isomorphic copy of the measure $d\phi_C$ ($C^* \cong C$) defined on an isomorphic copy $\mathcal{S}'^*$ of $\mathcal{S}'$. The new theory has free measure $d\phi_C \times d\phi_{\mathcal{C}*}^*$, covariance

$$C \otimes I + I \otimes C^* = C^\sim,$$

and normalized physical measure

$$Z^{\sim -1} e^{-V(\Lambda)} e^{-V(\Lambda)*} \, d\phi_C \times d\phi_{\mathcal{C}*}^* = d\mu^\sim$$

and a field

$$\phi^\sim = \phi \otimes I + I \otimes \phi^*.$$

This theory is invariant (even) under the symmetry $\phi \leftrightarrow \phi^*$, which interchanges the two factors.

We now apply the cluster expansion to the expression

$$Z^\sim \int (A - A^*)(B - B^*) \, d\mu^\sim.$$

The covariance operators which arise in this expansion have the form

$$C(s)^\sim = C(s) \otimes I + I \otimes C(s)^*,$$

so that the symmetry $\phi \leftrightarrow \phi^*$ is preserved in the Gaussian measure of each term of the expansion. (However, the expansion is modified by choosing $\mathcal{B} = (Z^*) \backslash \Gamma$, where Γ is the set of lattice lines in two connected sets, one containing supp $A = $ supp$(A - A^*)$, and the other containing supp B. Thus for each component X_i, $X_i \supset$ supp A or $X_i \cap$ supp $A = \varnothing$, and the same for supp B. With this restriction, there are at most two components. Since the n in (18.3.3) is a bound on the number of components, as one can check, $n = 2$ in (18.3.3).) Now consider a term in (18.2.15) with components $X_1, X_2, \ldots$, satisfying

$$\text{supp } A \subset X_i, \quad \text{supp } B \subset X_j, \quad i \neq j. \tag{18.3.5}$$

For this term, the symmetry $\phi \leftrightarrow \phi^*$ can be applied separately in each X_i. However $A - A^*$ is odd for the X_i symmetry, so terms satisfying (18.3.5) must vanish. The nonvanishing terms, which violate (18.3.5) contain a component $X_1 = X$ with $d \leq O(|X|)$, and so for $d \geq 4$,

$$\left| \int (A - A^*)(B - B^*) \, d\mu^{\sim} \right| \leq M_{AB} e^{-md}.$$

Expanding the integrand on the left yields four terms, and in each term the $d\mu$ integral factors. We evaluate the result as

$$2 \left| \int AB \, d\mu_{\wedge, c} - \int A \, d\mu_{\wedge, c} \int B \, d\mu_{\wedge, c} \right|,$$

and Theorem 18.1.1 follows.

18.4 Convergence: The Main Ideas

There are two main ideas in this chapter. The first is the formula (18.2.15) giving the cluster expansion for the Schwinger functions. The second is the estimates which lead to the convergence of this expansion, uniformly as $\Lambda \to \infty$. In this section we state these estimates as a series of three propositions, and using them, give the proof of Theorem 18.3.1. The simpler estimates are then proved; the harder ones are postponed for later sections. The most difficult of these estimates is contained in Proposition 18.5.3. Since it is in some sense the core of the chapter, we discuss the ideas involved in its proof at the end of this section.

The proof of convergence depends on estimates of the following types:

(a) Combinatoric estimates to count the number of terms in the expansion, and especially to count or bound the number of terms of some special type.
(b) Single particle estimates, to bound kernels of covariance operators and their derivatives $\partial^\Gamma C(s)$.
(c) Estimates on function space integrals. Typically estimates of type (a) and (b) will be components of an estimate of type (c).

Returning to specifics, (18.3.3) or (18.2.15) is a sum in which each term is a product of two factors: a ratio of partition functions times a function space integral. The first proposition is purely combinatoric—type (a)—and it counts the number of terms with $|X| =$ area X held fixed. The second and third propositions bound the two factors making up a term in (18.2.15). These latter two propositions are hybrids, since their proof uses estimates of types (a), (b), and (c).

Proposition 18.4.1. *There is a purely geometric constant K_1 such that the number of terms in (18.2.15–16) with a fixed value of $|X|$ is bounded by $e^{K_1|X|}$.*

Proposition 18.4.2.   *There is a constant K_2, independent of λ in the closure of (18.1.6), and of Λ and m_0, such that for ε small and m_0 large,*

$$\left| \frac{Z_{\partial X}(\Lambda \backslash X)}{Z(\Lambda)} \right| \leq e^{K_2 |X|}.$$

Proposition 18.4.3.   *There is a constant K_3 and a norm $|w|$ on test functions such that for any $K > 0$, for any Λ, and for λ in the closure of (18.1.6) with m_0 large,*

$$\left| \left\langle \int \partial^\Gamma \int \prod_{i=1}^n \phi(x_i) e^{-\lambda V(\Lambda)} \, d\phi_{C(s(\Gamma))} \, ds(\Gamma), w \right\rangle \right| \leq e^{-K|\Gamma| + K_3 |\Lambda|} |w|.$$

(The m_0 bound depends on K, and $|w|$ is invariant under translation in any of the variables of w.)

Remark.   Wick polynomials, as in (18.1.8), are also allowed in the integrand above.

PROOF OF THEOREM 18.3.1.   We replace Λ by $\Lambda \cap X$ in Proposition 18.4.3. For the X in (18.2.15), we have $X = \bigcup_{i=1}^r X_i^-$ with $r \leq n$ and X_i connected. Moreover

$$\Gamma \subset \bigcup_{i=1}^r \text{Int } X_i^-,$$

and so "many" of the lattice lines in X_i^- belong to Γ. In fact, since $X_i^- \backslash \Gamma^c = X_i$ is connected,

$$|X_i| - 1 \leq 2|\Gamma \cap \text{Int } X_i^-| \quad \text{and} \quad |X| - n \leq 2|\Gamma|. \qquad (18.4.1)$$

Thus we replace the upper bound in Proposition 18.4.3 by

$$e^{-K(|X| - n)} |w|,$$

with a new choice of K and $|w|$. Theorem 18.3.1 follows directly from this bound combined with Propositions 18.4.1 and 18.4.2.

PROOF OF PROPOSITION 18.4.1.   We consider (18.2.15). First we bound the number of ways of choosing the component X_i containing a particular x_j, $1 \leq j \leq n$. We identify each lattice square Δ with a lattice point located at its center, and we identify each lattice line b lying between two squares Δ and Δ' with the line joining the center of Δ to the center of Δ'. Hence we must count the number of ways of drawing a connected graph with nearest neighbor bonds. We assert that each such graph may be constructed by starting at x_j and drawing an oriented path formed of unit line segments which traverses each segment at most twice. In fact if we regard the lattice sites as islands and the line segments as bridges, this simply follows from the solution to the Königsberg bridge problem. See [Blackett, 1967, p. 159]. The number of connected paths of length l formed from lattice line segments and starting at x_j is at most 4^l. Since $l \leq 8|X_i|$, the number of choices for X_i is at most $O(1) 2^{16|X_i|}$. Since the number of choices for Γ is at most $4^{|X|}$, the number of choices for pairs X, Γ is at most

$$O(1) 2^{|X|} 4^{|X|} \prod_i 2^{16|X_i|} = O(1) 2^{19|X|},$$

since $2^{|X|}$ bounds the number of choices of $|X_i|$ with $|X|$ given. This completes the proof. The case (18.2.16) is similar.

Proposition 18.4.2 is proved in Section 18.5 using equations related to the Kirkwood–Salsburg equations. We only remark here that the change in partition function involves both a change in the interaction region and a change in covariance ($\Lambda \backslash X \to \Lambda$ and $C_{\partial X} \to C$).

Discussion of Proposition 18.4.3. Each derivative d/ds_b in ∂^Γ differentiates either the measure or the integrand. Derivatives of the measure are evaluated by (9.1.34), and produce s-dependent kernels $C'(s)$ in the integrand. Repeated differentiation yields

(a)   a sum of terms coming from the iterated functional derivatives $\partial^2/\partial\phi^2$ in (9.1.34),
(b)   a sum over terms arising from possible repeated derivatives $\partial^{\Gamma_i} C(s)$ of each $C'(s)$ introduced by (9.1.34).

Each derivative d/ds_b also produces convergence in one of two ways. Differentiation of the measure introduces a kernel C', and

(c) C' and C are small in the sense that

(c$_1$) $\|C'(x, y)\|_{L_p} \le O(m_0^{-\varepsilon})$,

(c$_2$) $0 \le C'(x, y) = \partial^b C(x, y) \le e^{-m_0(\text{dist}(x, b) + \text{dist}(y, b))}$,

(c$_3$) $0 \le C(x, y) \le e^{-m_0|x-y|}$, $|x - y| \ge 1$.

Repeated differentiation of $C(s)$ produces the further convergence

(d) $\partial^\Gamma C$ is small in the sense that

(d$_1$) $\|\partial^\Gamma C(x, y)\|_{L_p} \le O(m_0^{-\varepsilon|\Gamma|})$,

(d$_2$) $0 \le \partial^\Gamma C(x, y) \le e^{-m_0 d}$,

where d is the length of the shortest path connecting x and y and passing through each $b \in \Gamma$. In fact from the Weiner integral representation for $(-\Delta + m_0^2)^{-1}$, $\partial^l C(s)$ is the integral over Wiener paths from x to y passing through each $b \in \Gamma$; see Section 18.6.

We use (c$_2$) to control (a). In fact, by (c$_2$), only x and y near b contribute to the Laplacian $C'(s) \cdot \Delta_\phi$ of (9.1.34). By the nature of the expansion, there is at most one Laplacian Δ_ϕ per bond $b \in \mathcal{B}$, and hence a bounded power Δ_ϕ locally. Evaluation of Δ^n leads to $K(n) < \infty$ terms, and this evaluation, repeated over disjoint local regions, leads to $e^{O(\text{vol})}$ terms. We use (d$_2$) to control (b). The basic convergence, $m_0^{-\varepsilon|\Gamma|}$, comes from (d$_1$).

After performing these steps, there remains a function space integral of the form $\int Re^{-\lambda V}\,d\phi$. Then

$$\left|\int Re^{-\lambda V(\Lambda)}\,d\phi\right| \le \left(\int R^2\,d\phi\right)^{1/2}\left(\int e^{-2\,\mathrm{Re}\,\lambda V(\Lambda)}\,d\phi\right)^{1/2},$$

and each factor on the right is $e^{O(\mathrm{vol})}$. Now (c_3) produces the decoupling of distant local regions in R^2.

18.5 An Equation of Kirkwood–Salsburg Type

Equations (18.2.16–17) can be rewritten as a Banach space equation

$$\rho = Z(\Lambda)1 + \mathscr{K}\rho \tag{18.5.1}$$

with a unique solution

$$\rho = (I - \mathscr{K})^{-1}Z(\Lambda)1$$

satisfying the bound

$$|\rho| \le |(I - \mathscr{K})^{-1}|\,|Z(\Lambda)| \le 4|Z(\Lambda)|. \tag{18.5.2}$$

This bound is essentially Proposition 18.4.2, as we shall see.

Let $\mathscr{X}$ be the Banach space of functions f defined on finite subsets $\Gamma \subset (Z^2)^*$. For $f \in \mathscr{X}$ let $(f_n)_{n \ge 0}$ denote the restriction of f to n-element subsets. The norm on $\mathscr{X}$ is defined to be

$$|f| = \sup_{\{n,\,\Gamma:|\Gamma|=n\}} 2^{-n}|f_n(\Gamma)|. \tag{18.5.3}$$

We define $\rho_\Lambda \equiv (\rho_{\Lambda,n})_{n \ge 0}$ to be the function

$$\Gamma \to Z_\Gamma(\Lambda).$$

Theorem 18.5.1. *Let $|\lambda| \le \varepsilon$, $\mathrm{Re}\,\lambda \ge 0$, where $\varepsilon > 0$ is small, and let m_0 be sufficiently large. Then $Z(\Lambda) \ne 0$, and ρ_Λ defined above is the unique solution in $\mathscr{X}$ of Equation (18.5.1). Moreover ρ_Λ satisfies the bounds (18.5.2).*

By the dominated convergence theorem, $Z_{\partial\Delta}(\Delta) \to 1$ as $\varepsilon \to 0$. Thus for small ε,

$$\tfrac{1}{2} \le |Z_{\partial\Delta}(\Delta)| \le 2.$$

This restriction fixes the ε in Theorem (18.5.1), and in fact is the only restriction on ε in this chapter.

PROOF OF PROPOSITION 18.4.2, ASSUMING THEOREM 18.5.1.

$$\left|\frac{Z_{\partial X}(\Lambda\backslash X)}{Z(\Lambda)}\right| = \left|\frac{Z_{X^*}(\Lambda)}{Z(\Lambda)}\right||Z_{\partial\Delta}(\Delta)|^{-|\Lambda \cap X|}$$
$$\le 2^{|X^*|}\|(1 - \mathscr{K})^{-1}\|\,|Z_{\partial\Delta}(\Delta)|^{-|\Lambda \cap X|}$$
$$\le e^{K_2|X|}.$$

PROOF OF THEOREM 18.5.1.   Let $1 = (1, 0, 0, \ldots) \in \mathscr{X}$, and define $\mathscr{K}$ by the equations

$$(\mathscr{K}f)_n(\Gamma) = f_{n-1}(\Gamma \backslash b_1) + \sum_{m=1}^{\infty} \sum_{X, |X| = m} K(b_1, \Gamma, X) f_{|X^* \cup \Gamma|}(X^* \cup \Gamma) \quad (18.5.4)$$

for $n \geq 2$, where the sum over X (as in (18.2.16) and (18.2.17)) ranges over connected unions of lattice squares such that $b_1 \in X$. For $n = 1$, we omit the first term on the right of (18.5.4), and for $n = 0$, $(\mathscr{K}f)_0 \equiv 0$.

We now show that $\mathscr{K}$ is a contraction, $|\mathscr{K}| \leq \frac{3}{4}$, in order to obtain (18.5.2). It suffices to show that

$$\frac{1}{2} + \sup_{\Gamma \subset \mathscr{B}} 2^{-|\Gamma|} \sum_{m=1}^{\infty} \sum_{\{X : |X| = m\}} |K(b_1, \Gamma, X)| \, 2^{|\Gamma \cup X^*|} \leq \frac{3}{4}. \quad (18.5.5)$$

The proof of (18.5.5) is similar to the proof of Theorem 18.3.1 given in Section 18.4. In particular it depends on Propositions 18.4.1 and 18.4.3. By Proposition 18.4.1 there are at most $e^{K_1|X|}$ terms for each fixed value of $|X|$. From (18.2.17) each term has at least one derivative ∂^{b_1}. Using this fact and (18.4.1), the bound from Propositions 18.4.1 and 18.4.3 is

$$|K(b_1, \Gamma, X)| \leq e^{-K(|X|+1)} e^{K_3|X|}.$$

For K sufficiently large (18.5.5) follows. We complete the proof by showing $Z(\Lambda) \neq 0$. By the dominated convergence theorem, $Z_{\partial\Lambda}(\Delta) \to 1$ as $\varepsilon \to 0$. For ε sufficiently small, $|Z_{\Lambda^*}| = |Z_{\partial\Lambda}(\Delta)|^{|\Lambda|} \neq 0$. Thus $\rho_\Lambda \neq 0$, and by (18.5.2), $Z(\Lambda) \neq 0$.

18.6 Covariance Operators

The basic facts for the kernels $C_\emptyset(x, y)$ and $C_\Gamma(x, y)$ were derived in Chapter 7. Here we show that the differentiated covariance operator $\partial^\gamma C$ satisfies stronger bounds, including the exponential decay $e^{-O(d)}$ of Section 18.4, formula (d_2). Let $dW^T_{xy}(\omega)$ be the conditional Wiener measure on paths $\omega(\tau)$ starting at x at $\tau = 0$ and ending at y at $\tau = T$. Let $\chi^T_\Gamma(\omega)$ be the characteristic function of paths not crossing Γ for $0 \leq \tau \leq T$; see Section 7.8. Then C_Γ has a Wiener integral expression, using χ^T_Γ (see (7.8.3)), and similarly

$$C(s)(x, y) = \int_0^\infty dT \, e^{-m_0^2 T} \int \prod_{b \in \mathscr{B}} (s_b + (1 - s_b) \chi^T_b) \, dW^T_{xy}.$$

Consequently

$$(\partial^\gamma C(s))(x, y) = \int_0^\infty dT \, e^{-m_0^2 T'}$$

$$\cdot \int \prod_{b \in \gamma} (1 - \chi^T_b) \prod_{b \in \mathscr{B} \backslash \gamma} [s_b + (1 - s_b) \chi^T_b] \, dW^T_{xy}. \quad (18.6.1)$$

We need the improved bounds on $\partial^\gamma C$ for two reasons. The first is to localize x and y, with γ given. For this purpose, let $j = (j_1, j_2)$ be the pair of lattice points closest to x and y respectively. Then

$$d(j, \gamma) = \sup_{b \in \gamma} \{\text{dist}(\Delta_{j_1}, b) + \text{dist}(\Delta_{j_2}, b)\} \qquad (18.6.2)$$

is sufficient, as a crude lower bound on d.

We now explain the second use of bounds on $\partial^\gamma C$. Let $\mathscr{P}(\Gamma)$ be the set of all partitions π of the set of lattice line segments Γ. In Proposition 18.4.3, we are called on to bound $\partial^\Gamma \int F \, d\phi_s$, which by Leibnitz's rule and by (9.1.33) is just

$$\partial^\Gamma \int F \, d\phi_s = \sum_{\pi \in \mathscr{P}(\Gamma)} \int \left(\prod_{\gamma \in \pi} \tfrac{1}{2} \partial^\gamma C \cdot \Delta_\phi \right) F \, d\phi_s . \qquad (18.6.3)$$

The second use of the bounds on $\partial^\gamma C$ is to control $\sum_{\pi \in \mathscr{P}(\Gamma)}$. As in Chapter 7, we also find a factor $m_0^{-O(|\gamma|)}$, which yields the overall convergence of the expansion.

Proposition 18.6.1. *Let $1 \le q \le \infty$, and let m_0 be sufficiently large. There are constants $K_4(q, \gamma)$ and $K_5(q)$, independent of m_0, such that*

$$\|\partial^\gamma C\|_{L_q(\Delta_{j_1} \times \Delta_{j_2})} \le K_4(q, \gamma) m_0^{-|\gamma|/2q} \exp\left(-\frac{m_0 d(j, \gamma)}{2} \right), \qquad (18.6.4)$$

$$\sum_{\pi \in \mathscr{P}(\Gamma)} \prod_{\gamma \in \pi} K_4(q, \gamma) \le e^{K_5(q)|\Gamma|} . \qquad (18.6.5)$$

PROOF. We use the Wiener integral representation (18.6.1) for $\partial^\gamma C$. The proof consists of estimates on the Wiener measure of paths $\omega(\tau)$ which cross the lattice lines $b \in \gamma$ in some definite order, together with combinatoric arguments to count the number of ways the lines $b \in \gamma$ can be so ordered.

Let $L(\gamma)$ be the set of all possible linear orderings of the lines $b \in \gamma$, and for $l \in L(\gamma)$, let $\mathscr{W}(l)$ be the set of Wiener paths which cross all lines $b \in \gamma$, and whose order of first crossing is l. Then

$$0 \le \partial^\gamma C(s) \le \int_0^\infty e^{-m_0^2 T} \int \prod_{b \in \gamma} (1 - \chi_b^T(\omega)) \, dW_{x, y}^T \, dT = \partial^\gamma C_\varnothing , \qquad (18.6.6)$$

and

$$\partial^\gamma C_\varnothing(x, y) = \sum_{l \in L(\gamma)} \int_0^\infty e^{-m_0^2 T} \int_{\mathscr{W}(l)} dW_{x, y}^T \, dT . \qquad (18.6.7)$$

Let $b_1, b_2, \ldots$, be the elements of γ, as ordered by l. Let b_2' be the first of the b's not touching $b_1 = b_1'$, let b_3' be the first of the b's after b_2' and not touching b_2', etc. Set

$$a_j = \text{dist}(b_{j+1}', b_j'), \qquad 1 \le j \le J,$$

and define

$$|l| = \sum_{j=1}^J a_j .$$

If there is no such b_2', we set $|l| = 0$ by convention.

With these definitions, we bound the $l \in L(\gamma)$ term in (18.6.7), i.e.

$$K(l, x, y) \equiv \int_0^\infty e^{-m_0^2 T} \int_{\mathcal{W}(l)} dW_{xy}^T \, dT.$$

We use induction on J, i.e. the number of b_j', and the strong Markov property. The strong Markov property [McKean, 1969, p. 10] states that the first passage time

$$\tau_1 = \inf\{t \geq 0 : \omega(t) \in b_1\}$$

is a measurable function of the path ω, and that the process

$$s \to \omega(s - \tau_1)$$

is a Wiener process conditioned to begin on b_1.

Let l' be the linear ordering of $\{b_1', \ldots, b_J'\}$ defined by l, and let l_1' be the ordering of $\{b_2', \ldots, b_J'\}$ defined by l. Then

$$K(l, x, y) \leq K(l', x, y) \equiv \int_0^\infty e^{-m_0^2 T} \int_{\mathcal{W}(l')} dW_{xy}^T \, dT \qquad (18.6.8a)$$

Let

$$\mathcal{W}(l', t_1) = \{\omega \in \mathcal{W}(l') : \tau_1(\omega) \leq t_1\},$$

$$v(l', t_1) = \int_{\mathcal{W}(l', t_1)} dW_{x, y}^T.$$

Then $v(l', t_1)$ is a monotonically increasing function of t_1 and

$$\int_{\mathcal{W}(l', t_1)} dW_{x, y}^T = v(l', \infty) = v(l', \infty) - v(l', 0)$$

$$= \int_0^\infty dv(l', t_1).$$

Thus substitution into (18.6.8a) yields

$$K(l, x, y) \leq \int_0^\infty \int_0^\infty \exp(-m_0^2 T) \, dv(l', t_1) \, dT. \qquad (18.6.8b)$$

By the strong Markov property, the integral over the part of the path for $t \geq \tau_1$ can be written as a new Wiener measure, conditioned to begin on b_1. To make this explicit, we define

$$\mathcal{W}(t_1) = \{\omega \in \mathcal{W} : \tau_1(\omega) \leq t_1\},$$

$$v(t_1, x) = \int_{\mathcal{W}(t_1)} dW_x.$$

Here $\mathcal{W}$ is the set of Wiener paths $\omega(\cdot)$ with $\omega(0) = x$, and dW_x is the Wiener measure on $\mathcal{W}$. Also define

$$\xi = \xi(\omega) = \omega(\tau_1(\omega)) \in b_1$$

to be the point of first entry of the path into b_1. It is a measurable function of ω, since τ_1 is. Then the strong Markov property is the identity

$$dv(l', t_1) = \int_{\mathcal{W}(l')} dW_{\xi(\omega), y}^{T - t_1} \, dv(t_1, x).$$

From this identity,

$$dv(l', t_1) \leq dv(t_1, x) \sup_{\xi \in b_1} \int_{\mathcal{W}(l')} dW_{\xi, y}^{T - t_1}.$$

Substitution of this inequality and the identities $T = T - \tau_1 + \tau_1$ and $dT = dT - \tau_1$ into (18.6.8b) gives

$$K(l, x, y) \leq \int_0^\infty \exp(-m_0^2 t_1) \, dv(t_1, x) \sup_{\xi \in b_1} \int_0^\infty \exp(-m_0^2 s) \int_{\mathcal{W}(l_1')} dW_{\xi, y}^s \, ds.$$

$$(18.6.8c)$$

The first factor, namely

$$\int \exp(-m_0^2 \tau_1(\omega)) \, dW_x,$$

decreases exponentially in the distance $d(x, b_1)$. In fact, this factor is dominated by

$$\int \exp(-m_0^2 \sigma) \, dW_x,$$

where σ is the first passage time for the path to cross an infinite strip separating x and b_1, of width $d(x, b_1)$. The integral $\int \exp(-m_0^2 \sigma) \, dW_x$ reduces to a one dimensional Wiener integral, and is evaluated in [McKean, 1969, p. 27] as $e^{-m_0 d}$. This shows that for $J = 0$ and 1, $K(l, x, y)$ has the desired exponential decay.

Now proceed by induction on J. Then the second factor above decreases exponentially in $|l_1'|$. Combining these two bounds,

$$K(l, x, y) \leq K_6^J \, e^{-m_0 |l|}$$

for $|l| \geq 1$. For $|l| = 0$, we use the fact that

$$0 \leq \frac{d}{ds_b} (s_b + (1 - s_b)\chi_b^T) = 1 - \chi_b^T \leq 1$$

to infer

$$0 \leq \partial^\gamma C \leq C_\varnothing,$$

and reduce to estimates of Section 7.2. There is an entirely similar estimate based on the distance $d(j, \gamma)$ of (18.6.2), and taking geometric means of these two bounds yields for $2\delta < 1$,

$$\|\partial^\lambda C\|_{L_q(\Delta_{j_1} \times \Delta_{j_2})} \leq \sum_{l \in L(\gamma)} K_7^{|\gamma|} e^{-m_0 |l|/(2 + 2\delta)} e^{-m_0 \, d(j, \gamma)/(2 - \delta)}. \quad (18.6.9)$$

for m_0 large. If $|l| \geq 1$ for all $l \in L(\gamma)$, then we can include a factor $m_0^{-|\gamma|}$ on the right side of (18.6.9), by increasing δ. If $|l| < 1$ for some l, then $|l| = 0$, and in this case $|\gamma| \leq 4$. With $|\gamma| \leq 4$ and $d(j, \gamma) \geq 1$, we can still include the factor $m_0^{-|\gamma|}$ in (18.6.9) by increasing δ. Finally, for $|\gamma| \leq 4$ and $d(j, \gamma) = 0$, the factor $m_0^{-|\gamma|/2q} \geq m_0^{-2/q}$ in (18.6.4) comes from scaling, as in Proposition 7.9.1.

We define

$$K_4(q, \gamma) = \text{const} \sum_{l \in L(\gamma)} K_7^{|\gamma|} e^{-m_0 |l|/(2 + 2\delta)}. \quad (18.6.10)$$

With this definition, (18.6.4) follows; in the case $d(j, \gamma) = 0$ and $|l| = 0$ for some l, we use the bound of Proposition 7.9.4 to establish (18.6.4).

We complete the proof by establishing (18.6.5) as a separate proposition.

Proposition 18.6.2. *For m_0 sufficiently large,*

$$\sum_{\pi \in \mathscr{P}(\Gamma)} \prod_{\gamma \in \pi} \sum_{l \in L(\gamma)} e^{-m_0|l|/3} \le e^{K_8|\Gamma|}. \tag{18.6.11}$$

PROOF. Let $\mathscr{L}(\Gamma)$ be the set of linear orderings defined on subsets of Γ. Thus $L(\Gamma) \subsetneqq \mathscr{L}(\Gamma)$. As before, we define $|l|$ for $l \in \mathscr{L}(\Gamma)$. We assert that the number of $l \in \mathscr{L}(\Gamma)$ with $|l| \le r$ is bounded by

$$|\Gamma| e^{K_9(r+1)} \tag{18.6.12}$$

Using (18.6.12), we complete the proof. Let $A_l = \exp(-m_0|l|/3)$. Expanding $\sum \prod \sum A_l$ in (18.6.11), we get a sum of terms of the form

$$A_{l_1} A_{l_2} \cdots A_{l_j},$$

where the l_j are distinct elements of $\mathscr{L}(\Gamma)$. Adding all terms of this form, we bound (18.6.11) by

$$\sum \prod \sum A_l \le \sum_{l \in \mathscr{L}(\Gamma)} A_{l_1} \cdots A_{l_j} = \prod_{l \in \mathscr{L}(\Gamma)} (1 + A_l)$$

$$\le \prod_{l \in \mathscr{L}(\Gamma)} \exp A_l = \exp \sum_{l \in \mathscr{L}(\Gamma)} A_l$$

$$\le \exp(O(1)|\Gamma|).$$

Here, in the last expression, we have used the bound (18.6.12) to estimate $\sum_{l \in \mathscr{L}(\Gamma)} A_l$ and we choose m_0 sufficiently large.

Next we establish (18.6.12). Suppose the integer parts $[a_i]$ of the distances a_i are given. We choose $b_1 = b'_1$ in $|\Gamma|$ ways, and we choose the b's between b'_1 and b'_2 in $O(1)$ ways, since they all must overlap b_1. Next b'_2 is chosen in $O(1)[a_i]$ ways, namely from the lattice line segments b with

$$[a_1] \le \text{dist}(b, b'_1) < [a_1] + 1.$$

Continuing in this fashion, we choose all the b's in

$$|\Gamma| \prod_i O(1)[a_i] \le |\Gamma| e^{O(1) \sum [a_i]} \le |\Gamma| e^{O(1)r}$$

ways. Finally we count the number of choices of the $[a_i]$. This is the number of ways of choosing integers $r_i \ge 1$ with $\sum r_i \le r$, namely 2^r. In fact, suppose $\sum r_i = r$, and we distribute the r units in $\sum r_i$ as follows: The first 1 goes into a_1 (no choice). The second 1 goes to a_1 or a_2 (one binary choice). If the jth 1 goes to a_i, the $j + 1$st goes to a_i or a_{i+1} (one binary choice). Thus there are $r - 1$ binary choices, or 2^{r-1} ways to choose r_i with $\sum r_i = r \ge 1$. Summing $j = \sum r_i$ gives $\sum_{j=1}^{r} 2^{j-1} = 2^j - 1$. Finally, we get one more choice from $|l| = 0$ (no a_i's).

18.7 Convergence: The Proof Completed

PROOF OF PROPOSITION 18.4.3. Without loss of generality, the kernel w is localized, i.e. has its support contained in a product of lattice squares, and in this case we take $|w| = |w|_{L_2}$. The expression we want to estimate is

$$\left\langle \int \partial^{\Gamma} \int \prod_{i=1}^{n} \phi(x_i) e^{-\lambda V(\Lambda)} \, d\phi_{s(\Gamma)} \, ds(\Gamma), \, w \right\rangle. \tag{18.7.1}$$

Let $\mathscr{P}(\Gamma)$ be the set of all partitions π of Γ. By Leibnitz's rule and (9.1.34), (18.7.1) equals

$$\left\langle \int \sum_{\pi \in \mathscr{P}(\Gamma)} \int \left(\prod_{\gamma \in \pi} \tfrac{1}{2} \partial^{\gamma} C \cdot \Delta_{\phi} \right) \prod_{i} \phi(x_i) e^{-\lambda V(\Gamma)} \, d\phi_{s(\Gamma)} \, ds(\Gamma), \, w \right\rangle \tag{18.7.2}$$

where $C = C(s(\Gamma))$ and $\partial^{\gamma} C \cdot \Delta_{\phi} \equiv \Delta_{\partial^{\gamma} C}$; cf. (9.1.10).

For $j \in Z^2$, let Δ_j denote (1) the lattice square containing j, (2) the characteristic function of this lattice square, and (3) the multiplication operator by this function Δ_j. Using Δ_j in the third of these three meanings, we define

$$\partial^{\gamma} C(j_{\gamma}) = \Delta_{j_{1, \gamma}} \, \partial^{\gamma} C \, \Delta_{j_{2, \gamma}},$$

where $j_{\gamma} = (j_{1, \gamma}, j_{2, \gamma}) \in Z^4$, so that the two derivatives in $\partial^{\gamma} C(j_{\gamma}) \cdot \Delta_{\phi}$ are localized in Δ_{j_1} and Δ_{j_2} respectively, and

$$\partial^{\gamma} C = \sum_{j_{\gamma}} \partial^{\gamma} C(j_{\gamma}).$$

We substitute this identity into (18.7.2) and expand.

The resulting sum is now indexed by localizations $\{j_{\gamma}\}$ and partitions $\pi \in \mathscr{P}(\Gamma)$. For a given term let $M = M(\pi, \{j_{\gamma}\})$ be the number of terms resulting from the differentiations Δ_{ϕ} in (18.7.2). By Theorem 8.5.5, Proposition 10.3.1, and Corollary 10.3.2, each of the resulting terms can be estimated by

$$\|w'\|_{L_p} < \|w\|_{L_2} \left\| \prod_{\gamma \in \pi} \partial^{\gamma} C(j) \right\|_{L_q}$$

$$\leq \|w\|_{L_2} m_0^{-|\Gamma|/2q} \prod_{\gamma \in \pi} K_4(q, \gamma) e^{-m_0 d(j_{\gamma}, \gamma)/2}.$$

Now using (18.6.5) to control the sum over $\pi \in \mathscr{P}(\Gamma)$, we can bound (18.7.2) by

$$\|w\|_{L_2} e^{K_5 |\Gamma|} m_0^{-|\Gamma|/2q} \sum_{\{j_{\gamma}\}} \max_{\pi \in \mathscr{P}(\Gamma)} M \prod_{\gamma \in \pi} e^{-m_0 d(j_{\gamma}, \gamma)/2} \prod_{\Delta} N(\Delta)!.$$

The proof of Proposition 18.4.3 follows from two lemmas which control M and the sum over $\{j_{\gamma}\}$ respectively. Let $M(\Delta)$ be the number of elements in the set $\{j_{i, \gamma} : \Delta_{j_{i, \gamma}} = \Delta, i = 1 \text{ or } 2, \gamma \in \pi\}$.

Lemma 18.7.1. *There exists a constant K_{10}, independent of m_0, such that*

$$M \leq e^{K_{10}|\Gamma|} \prod_{\Delta} (M(\Delta)!)^p$$

and

$$\prod_\Delta N(\Delta)! \le e^{K_{10}|\Gamma|} \prod_\Delta (M(\Delta)!)^p.$$

Lemma 18.7.2. *Given* $\pi \in \mathscr{P}(\Gamma)$ *and* $r > 0$, *there exists a constant* K_{11}, *independent of* m_0, *such that*

$$\sum_{\{j_\gamma\}} \prod_{\gamma \in \pi} e^{-m_0 \, d(j_\gamma, \, \gamma)/2} \prod_\Delta M(\Delta)!^r \le e^{K_{11}|\Gamma|}. \tag{18.7.3}$$

PROOF OF LEMMA 18.7.1.   Let $N_0(\Delta)$ be the number of x_i, $1 \le i \le n$, which are localized in Δ. The number of terms resulting from $M(\Delta)$ differentiations in Δ is bounded by

$$(N_0(\Delta) + 1)(N_0(\Delta) + p + 1) \cdots (N_0(\Delta) + p(M(\Delta) - 1) + 1).$$

Since $N_0(\Delta) \le \sum_\Delta N_0(\Delta) = n$, we have M, the total number of terms resulting from all $\partial/\partial\phi(y)$ differentiations, bounded by

$$M \le \prod_\Delta p^{pM(\Delta)}(N_0(\Delta) + 1 + pM(\Delta))^{N_0(\Delta)+1} M(\Delta)!^p,$$

using the inequalities

$$(a + b)! \le (a + b)^a(b!) \quad \text{and} \quad (ab)! \le a^{ab}(b!)^a.$$

Furthermore with $N(\Delta)$, as defined in Section 18.5, the number of legs in Δ, after differentiation, we have

$$N(\Delta) \le N_0(\Delta) + (p - 1)M(\Delta),$$

and so $N(\Delta)!$ is bounded as above.

PROOF OF LEMMA 18.7.2.   The sum $\sum_{\{j_\gamma\}}$ is controlled by the exponentially decreasing distance factor, so it is sufficient to show

$$\prod_\Delta M(\Delta)!^r \le \prod_\Delta e^{\text{const } |\gamma|} e^{\text{const } \Sigma_\gamma \, d(j, \gamma)}$$

with constants independent of m_0, γ, $\{j_\gamma\}$ and π.

Recall that $d(j, \gamma)$, defined by (18.6.2), contains the distance from $j_{1, \gamma}$ and $j_{2, \gamma}$ to some $b \in \gamma$. Thus for fixed Δ, there are at most $O(1)r^2$ values of γ within a fixed partition π such that

$$\Delta_{j_{v, \gamma}} = \Delta, \qquad v = 1 \text{ or } 2, \tag{18.7.4}$$

and $d(j, \gamma) \le r$. By definition there are $M(\Delta)$ γ's which satisfy (18.7.4). The most distant half $(= M(\Delta)/2)$ of these γ's must also satisfy

$$M(\Delta)^{1/2} \le \text{const } d(j, \gamma) + \text{const},$$

because the γ's are nonoverlapping. Hence

$$M(\Delta)^{3/2} \le \text{const} \sum_\gamma (\{d(j, \gamma) : \Delta_{j_{v, \gamma}} = \Delta\} + \text{const}),$$

and so the proof is completed by the inequality

$$\prod_{\Delta} M(\Delta)!^r \le \exp\{r \sum_{\gamma} M(\Delta) \ln M(\Delta)\}$$

$$\le \exp\left(O\left(\sum_{\gamma} \{M(\Delta)^{1+\delta} : M(\Delta) > 0\}\right)\right)$$

$$\le \exp\left\{O\left(\sum_{\gamma} d(j, \gamma)\right)\right\} \exp(O(|\Gamma|)).$$

The methods of this chapter have been extended to allow other interactions and other (connected, irreducible, etc.) Schwinger functions, and to include phase transitions.

References include: [Glimm, Jaffe, and Spencer 1973, 1974], [Spencer, 1974b] [Glimm, Jaffe, and Spencer, 1976a], and [Spencer, 1975]. See also Chapters 14 and 20.

CHAPTER 19

From Path Integrals to Quantum Mechanics

19.1 Reconstruction of Quantum Fields

The main goal of this chapter is the verification of the Wightman and Haag–Kastler axioms, including the proof of Theorems 6.1.5–6. We assume that we are given a measure $d\mu$ on $\mathscr{D}'(R^d)$ which satisfies the axioms OS0–3 given in Chapter 6. In that chapter we constructed Euclidean field operators ϕ acting on $\mathscr{E} = L_2(\mathscr{D}', d\mu)$ and we constructed the Hilbert space $\mathscr{H}$ of quantum mechanical states as the completion of $\mathscr{E}_+/\mathscr{N}$.

Let Y denote the Banach space obtained by completing $C_0^\infty(R^d)$ in the norm

$$\|f\| = \|f\|_{L_1} + \|f\|_{L_p}. \tag{19.1.1}$$

Because the generating function $S\{f\}$ of the measure $d\mu$ is assumed to be continuous in the L_1 and L_p norms in (6.1.5), we have

Proposition 19.1.1. *Assume OS0–1. Then $S\{f\}$ extends by continuity to an entire analytic function on Y, and the moments S_n of $d\mu$ extend to continuous, multilinear functionals on $Y \times \cdots \times Y$.*

PROOF. We use Vitali's theorem to establish analyticity of $S\{f\}$ for $f \in Y$: A sequence of functions on a compact set $K \subset C^n$ which is uniformly bounded and pointwise convergent, converges to a function analytic on K. The continuity of S_n on $Y \times \cdots \times Y$ follows from the Cauchy integral theorem, as in the proof of Proposition 6.1.4.

Remark. In view of the proposition, the measure $d\mu$ is concentrated on $\mathscr{S}'(R^d)$. Thus all integrals will be taken over $\mathscr{S}'$ in the remainder of the chapter.

We now define $Y(0, t) \subset Y$ as the subset of functions $f \in Y$ supported in the interval $(0, t)$. Let $M(f) \equiv \|f\|_{L_1} + \|f\|_{L_p}^p$. Recall that $\mathcal{S}(0, t)$ denotes the subset of $\mathcal{S}$ with functions f supported in the time interval $(0, t)$. Also $\mathcal{S}_+ = \mathcal{S}(0, \infty)$. In Chapter 6 we defined the canonical projection $^\wedge$ from $\mathscr{E}_+$ to $\mathscr{H}$, namely $^\wedge : \mathscr{E}_+ \to \mathscr{E}_+/\mathscr{N} \subset \mathscr{H}$ such that $^\wedge$ transfers certain operators S on $\mathscr{E}_+$ to become operators $S^\wedge$ on $\mathscr{H}$ by the formula (6.1.12).

Proposition 19.1.2. *Assume OS0–3, and let $f \in Y(0, t)_{\text{real}}$. Then $^\wedge$ maps the functions $\phi(f)^r$ and $e^{\phi(f)}$ into operators on $\mathscr{H}$ with domain $e^{-tH}(\mathscr{E}_+ \cap L_\infty)^\wedge$. Furthermore*

$$\|(e^{\phi(f)})^\wedge e^{-tH}\| \le e^{M(2f)/2}, \tag{19.1.2}$$

$$\|(\phi(f)^r)^\wedge e^{-tH}\| \le (c\|f\|)^r r!^q, \tag{19.1.3}$$

with $c < \infty$ and $q = (p-1)/p$.

Remark. As a special case of (19.1.3),

$$\left| \int \phi(f)^r \, d\mu \right| = |\langle \Omega, (\phi(f)^r)^\wedge \Omega \rangle| \le (c\|f\|)^r r!^q.$$

PROOF. The operators $(\phi^r)^\wedge$ and $(e^\phi)^\wedge$ are densely defined by Propositions 10.5.4 and 6.1.4. To establish (19.1.2), let $k = e^{\phi(f)}$ and calculate M_n defined in (10.5.4). Since the supports of the time translates $f_{(2j-1)t}$ and $(\theta f)_{(2j-1)t}$ are all disjoint (for fixed t and $j = 1, 2, \ldots$), it follows by (6.1.5) that $M_n \le e^{M(2f)/2}$. Then (19.1.2) is a consequence of Theorem 10.5.5.

To prove (19.1.3), we define, for $w_j, z_j \in C$,

$$g = -i \sum_{j=1}^{n} (w_j f_{(2j-1)t} + z_j(\theta f)_{(2j-1)t}).$$

Then

$$M_n^{4n} = \left(\prod_{j=1}^{n} \frac{d^{2r}}{dw_j^{2r}} \frac{d^{2r}}{dz_j^{2r}} \right) S\{g\} \Big|_{g=0}. \tag{19.1.4}$$

We bound (19.1.4) using the Cauchy integral formula

$$M_n^{4n} = \left(\frac{(2r)!}{2\pi i} \right)^{2n} \oint S\{g\} \left(\prod_{j=1}^{n} \frac{dw_j}{w_j^{2r+1}} \frac{dz_j}{z_j^{2r+1}} \right),$$

with each integral taken over a circle of radius $4\varepsilon^{-1}$ centered at the origin. The upper bound for $M(g)$ on this circle is $2nM(4\varepsilon^{-1}f)$, since the terms in g arising from different (or time-reflected) values of j have disjoint supports. Thus

$$M_n \le \exp\left(\frac{M(4\varepsilon^{-1}f)}{2} \right) \left(\frac{\varepsilon}{4} \right)^r (2r)!^{1/2}$$

$$\le \exp\left(\frac{M(4\varepsilon^{-1}f)}{2} \right) \varepsilon^r r!.$$

To minimize M_n, first replace ε by $\varepsilon|f|$, yielding $M_n \le \exp(c\varepsilon^{-1})(\varepsilon|f|)^r r!$. Then choose $\varepsilon = (cp/r)^{1/p}$ to obtain (19.1.3).

Recall from (10.5.13) the definition $\mathscr{H}_\delta = e^{-\delta H}\mathscr{H}$.

Proposition 19.1.3. *There exists a unique bilinear form* $\Phi(h)$ *defined on the domain* $\mathcal{H}_\delta \times \mathcal{H}_\delta$ *(for any* $\delta > 0$*) and satisfying (for* $f = h \otimes \alpha$*)*

$$\phi(h \otimes \alpha)^\wedge e^{-tH} = \int e^{-sH} \Phi(h) e^{-(t-s)H} \alpha(s)\, ds \qquad (19.1.5)$$

on $\mathcal{H}_\delta \otimes \mathcal{H}_\delta$. *Furthermore, with* $\|h\| = \|h\|_{L_1(R^{d-1})} + \|h\|_{L_p(R^{d-1})}$,

$$\|e^{-\delta H}\Phi(h)e^{-\delta H}\| \le K_\delta \|h\|,$$
$$\|e^{-\delta H}\phi(f)^\wedge e^{-(t+\delta)H}\| \le K_\delta \|h\| \|\alpha\|_{L_1}. \qquad (19.1.6)$$

PROOF.   By Corollary 10.5.6, $\phi(h \otimes \alpha_\tau)^\wedge$ is a form on $\mathcal{H}_\delta \times \mathcal{H}_\delta$, which is C_∞ in τ for $0 \le \tau < \delta/4$, $\delta' \ge t + \delta/2$. By (10.5.14) and (19.1.3),

$$\left\| e^{-\delta H}\phi\left(h \otimes \frac{d^n}{d\tau^n}\alpha_\tau\right)^\wedge e^{-\delta' H}\right\| \le K_{n,\delta} \|h\|(\|\alpha\|_{L_1} + \|\alpha\|_{L_p}). \qquad (19.1.7)$$

We appeal to the fact that a distribution in s, all of whose derivatives have a fixed order, must be a C^∞ function of s. Thus on $\mathcal{H}_\delta \times \mathcal{H}_{\delta'}$, there exists a bilinear form $\Phi(h)$ which satisfies

$$\phi(h \otimes a)^\wedge = \int_0^t e^{-sH}\Phi(h)e^{sH}\alpha(s)\, ds$$

and such that for all $\psi \in \mathcal{H}_\delta$, $\psi' \in \mathcal{H}_{\delta'}$

$$F(s) = \langle e^{-sH}\psi, \Phi(h)e^{sH}\psi'\rangle$$

is a C^∞ function.

We now improve (19.1.7) to eliminate $\|\alpha\|_{L_p}$. With $n = 1$, and $\alpha = \chi_{(t_1, t_2)}$ the characteristic function of $(t_1, t_2) \subset (0, t)$, $t \le 1$, we bound

$$\|e^{-\delta H}\phi(h \otimes (\delta_{t_1} - \delta_{t_2}))^\wedge e^{-\delta' H}\| \le 2\tilde{K}_{1,\delta}\|h\|.$$

We integrate t_2 from s_1 to $s_2 = s_1 + o(\delta)$. Since

$$(s_2 - s_1)\delta_{t_1} = \chi_{(s_1, s_2)} + \int_{s_1}^{s_2}(\delta_{t_1} - \delta_{t_2})\, dt_2,$$

we have

$$\|e^{-\delta H}\phi(h \otimes \delta_{t_1})^\wedge e^{-\delta' H}\| \le (K_{0,\delta}o(\delta^{-1}) + 2K_{1,\delta})\|h\|$$

Now integration over t_1 shows that

$$\|e^{-\delta H}\phi(h \otimes \alpha)^\wedge e^{-\delta' H}\| \le K_\delta \|h\| \|\alpha\|_{L_1}.$$

Proposition 19.1.4. *Assume* OS0–3, *and let* $f = h \otimes \chi_{0, t}$, *where* $h \in \mathcal{D}(R^{d-1})$ *is real and where* $\chi_{0, t}$ *is the characteristic function of the interval* $(0, t)$. *Then*

$$\int \phi(f)^2\, d\mu = o(t), \qquad t \to 0. \qquad (19.1.8)$$

PROOF. By Proposition 19.1.2,

$$\int \phi(f)^2 \, d\mu \le \text{const}(\|f\|_{L_1}^2 + \|f\|_{L_p}^2).$$

For $f = h \otimes \chi_{0,t}$,

$$\|f\|_{L_p}^2 = O(t^{2/p}),$$

and so for $p < 2$ the proof is complete. However $p \le 2$, by the statement of OS1, Section 6.1, and for $p = 2$, the statement of OS1 contains an extra regularity condition: the two point function $S_2(x - y) \in L_1^{loc}(R^d)$. In this case

$$\int \phi(f)^2 \, d\mu = \int S_2(x - y) f(x) f(y) \, dx \, dy.$$

Changing to sum and difference variables

$$\xi = \tfrac{1}{2}(x + y), \qquad \eta = \tfrac{1}{2}(x - y),$$

we have

$$\left| \int \phi(f)^2 \, d\mu \right| \le \text{const } t \int_{R^{d-1} \times R^{d-1} \times [-t,\, t]} \left| S_2(2\eta) h(\xi + \eta) h(\xi - \eta) \right| \, d\xi \, d\eta$$

$$\le o(t) \quad \text{as } t \to 0.$$

19.2 The Feynman–Kac Formula

In this section we derive the Feynman–Kac formula on $\mathcal{H}$ suitable for studying perturbations of H by the bilinear form $\Phi(h)$, for $h \in \mathcal{S}(R^{d-1})_{\text{real}}$. We find that

$$\left(e^{-\phi(h \otimes \chi_{0,t})}\right)^{\wedge} e^{-tH} = e^{-tH(h)}. \tag{19.2.1}$$

Here $H(h)$ is a self-adjoint operator, bounded from below, which satisfies

$$H(h) = H + \Phi(h) \tag{19.2.2}$$

as an equation for bilinear forms. Also for h real, we find that

$$\pm \Phi(h) \le \text{const} \|h\| (H + I).$$

Formally $\Phi(k) = \phi(h \otimes \delta)^{\wedge}$.

Theorem 19.2.1. Let $S\{f\}$ satisfy OS0–3. Then the left side of (19.2.1) is a semigroup $S(t)$ with a self-adjoint generator $H(h)$ satisfying

$$-c(\|h\| + 2^{p-1}\|h\|^p) \le H(h). \tag{19.2.3}$$

Furthermore, (19.2.2) is satisfied on the domain $\mathcal{H}_\delta \times \mathcal{H}_\delta$ for any $\delta > 0$.

PROOF. Let $S(t)$ denote the left side of (19.2.1). By Proposition 19.1.2, $S(t)$ is a bounded operator, and $\|S(t)\| \le \exp(tc(\|h\| + 2^{p-1}\|h\|^p))$. Also $S(t + s) = S(t)S(s)$ and $S(t) = S(t)^*$ follow from the definition (19.2.1). We next extablish weak differentiability of $S(t)$ at $t \to 0$ on the dense domain $\mathcal{H} \otimes \mathcal{H}$, where $\mathcal{H} = e^{-\delta H}(\mathscr{E}_+ \cap L_\infty)$. Weak (and strong) continuity of $S(t)$ follow, and hence the existence of a self-adjoint generator $H(g)$ satisfying (19.2.3). Write $f = -h \otimes \chi_{0,t}$, so that on $\mathcal{H} \times \mathcal{H}$

$$
\begin{aligned}
t^{-1}(S(t) - I) &= t^{-1}(e^{-tH} - I) + t^{-1}\phi(f)^\wedge e^{-tH} \\
&\quad + t^{-1}(e^{\phi(f)} - I - \phi(f))^\wedge e^{-tH}.
\end{aligned}
\tag{19.2.4}
$$

The first term on the right of (19.2.4) tends to $-H$. By Proposition 19.1.3, the second term converges to $-\Phi(h)$. The third term converges to zero, which we see as follows: Let $A \in \mathscr{E}_+ \cap L_\infty$. Then using

$$
e^x - 1 - x \le \text{const } (x^2 + x^N) + \sum_{j > N/2}^{\infty} \frac{x^{2j}}{(2j)!},
$$

we have

$$
\begin{aligned}
&\left| t^{-1}\langle e^{-\delta H} A^\wedge, (e^{\phi(f)} - I - \phi(f))^\wedge e^{-(\delta+t)H} A^\wedge \rangle \right| \\
&\quad \le t^{-1} \|A\|_{L_\infty}^2 \int |e^{\phi(f)} - 1 - \phi(f)| \, d\mu \\
&\quad \le \text{const } t^{-1} \|A\|_{L_\infty}^2 \Bigg[\int \phi(f)^2 \, d\mu + \int \phi(f)^N \, d\mu \\
&\qquad\qquad\qquad\qquad\qquad + \sum_{j>N/2}^{\infty} \frac{1}{(2j)!} \int \phi(f)^{2j} \, d\mu \Bigg]
\end{aligned}
\tag{19.2.5}
$$

for any even $N \ge 2$. The first term on the right of (19.2.5) tends to zero (as $t \to 0$) by Proposition 19.1.4. Choose $N > p$, so the remaining terms in (19.2.5) tend to zero (as $t \to 0$) by (19.1.3) and the fact that $\|f\| = \|h\|(t + t^{1/p}) \le O(t^{1/p})$. This completes the proof of weak differentiability and the identity (19.2.2) on $\mathcal{H} \times \mathcal{H}$. By Proposition 19.1.3, $e^{-\delta H}\Phi(h)e^{-\delta H}$ is bounded. Hence (19.2.2) extends by continuity to $\mathcal{H}_\delta \times \mathcal{H}_\delta$.

Corollary 19.2.2. *The forms $\Phi(h)$ and $H(h)$ extend by continuity to the domain $\mathscr{D}(H^{1/2}) \times \mathscr{D}(H^{1/2})$, and they satisfy (19.2.2–3). Also*

$$
\|(H+I)^{-1/2}\Phi(h)(H+I)^{-1/2}\| \le \text{const}\|h\|.
\tag{19.2.6}
$$

PROOF. On $\mathcal{H}_\delta \times \mathcal{H}_\delta$ we infer from (19.2.2–3) that for $1 \le c(\|h\| + 2^{p-1}\|h\|^p)$,

$$
\pm \Phi(h) \le c(\|h\| + 2^{p-1}\|h\|^p)(H+I),
$$

or

$$
\pm \Phi(h) = \pm 2\|h\|\Phi\left(\frac{h}{2\|h\|}\right) \le 2c\|h\|(H+I).
$$

This bound now extends by continuity, as claimed.

19.3 Self-Adjoint Fields

We define real time field operators $\Phi(f)$ and establish their properties as operators on $\mathscr{H}$. On $\mathscr{D}(H^{1/2})$ the bilinear form

$$\Phi(h, t) = e^{itH}\Phi(h)e^{-itH} \tag{19.3.1}$$

is continuous in t. Thus we define the Minkowski field

$$\phi_M(f) = \int \Phi(f^{(t)}, t)\, dt \tag{19.3.2}$$

where $f^{(t)}(\mathbf{x}) \equiv f(\mathbf{x}, t) \in \mathscr{S}(R^{d-1})$ for $f \in \mathscr{S}(R^d)$.

Theorem 19.3.1. *Assume $S\{f\}$ satisfies OS0–3. Then for $f \in \mathscr{S}(R^d)_{\text{real}}$, the bilinear form (19.3.2) uniquely determines a self-adjoint operator $\phi_M(f)$ on $\mathscr{H}$, which is essentially self-adjoint on any core for H. Moreover*

$$\|(H + I)^{-1/2}\phi_M(f)(H + I)^{-1/2}\| \leq \text{const} \int \|f^{(t)}\|\, dt, \tag{19.3.3}$$

$$\phi_M(f): \mathscr{D}(H^n) \to \mathscr{D}(H^{n-1}), \tag{19.3.4}$$

and

$$[iH, \phi_M(f)] = \phi_M\left(\frac{\partial f}{\partial t}\right) \quad \text{on } \mathscr{D}(H^2). \tag{19.3.5}$$

PROOF. The bound (19.3.3) for $\phi_M(f)$ as a bilinear form follows from Corollary 19.2.3. The relation (19.3.5) for $\phi_M(f)$ as a bilinear form follows from (19.3.3). The remaining assertions follow by Theorems 19.4.1–3 below.

Corollary 19.3.2. *The Wightman functions*

$$W_n(f_1, \ldots, f_n) \equiv \langle \Omega, \phi_M(f_1) \cdots \phi_M(f_n)\Omega \rangle \tag{19.3.6}$$

exist and are tempered distributions in $\mathscr{S}'(R^{nd})$.

PROOF. Since $H\Omega = 0$, $\Omega \in \bigcap_n \mathscr{D}(H^n)$, i.e., Ω is a C^∞ vector for H. By (19.3.4) and by induction on j, $\phi_M(f_1) \cdots \phi_M(f_j)\Omega$ is a C^∞ vector for H also. In particular (19.3.6) is defined, and is bounded by a product of norms as in (19.3.3), to complete the proof.

Let $\mathscr{D} \subset \mathscr{H}$ denote the finite linear span of vectors of the form $\phi_M(f_1) \cdots \phi_M(f_n)\Omega$, $f_j \in \mathscr{S}(R^d)_{\text{real}}$.

Theorem 19.3.3. *$\mathscr{D}$ is dense in $\mathscr{H}$.*

PROOF. Recall that $\mathscr{H}$ is spanned by vectors of the form $^\wedge(e^{i\phi(f)}) = (e^{i\phi(f)})^\wedge\Omega$ for $f \in \mathscr{S}_+$. Using Proposition 19.1.2, it is no loss of generality to restrict attention to $f = \sum_{j=1}^{n} h_j \otimes \alpha_j$, with $f_j = h_j \otimes \alpha_j \in \mathcal{S}(0, t)$. We first choose α_j such that $0 < t_1 < t_2 < \cdots < t_n < t$ for $t_j \in \text{supp} \, \alpha_j$, and define for real s

$$\theta(s) = \int e^{ist_1 H} \phi_M(f_1) e^{-is(t_1 - t_2)H} \phi_M(f_2) \cdots \phi_M(f_n)\Omega \in \mathcal{D}. \qquad (19.3.7)$$

By Corollary 10.5.6 and Theorem 19.3.1, $\theta(s)$ is the boundary value of a function of s, analytic for $\text{Im} \, s > 0$. If χ is orthogonal to $\mathcal{D}$, then $\langle \chi, \theta(s) \rangle = 0$ for s real, and hence $\langle \chi, \theta(s = i) \rangle = 0$, i.e.

$$0 = \int \langle \chi, e^{-t_1 H} \phi(h_1 \; 0)^\wedge e^{-|t_1 - t_2|H} \phi(h, 0)^\wedge \cdots \rangle \prod_{j=1}^{n} \alpha_j(t_j) \, dt_j. \qquad (19.3.8)$$

By the continuity of (19.3.9) in α_j which follows by Proposition 19.1.2, (19.3.8) remains valid if we translate the supports of α_j to overlap. Thus

$$0 = \langle \chi, (\phi(f)^n{}^\wedge \Omega \rangle.$$

Again, using (19.1.3), we may sum the exponential series to establish $\langle \chi, (e^{i\phi(f)})^\wedge \Omega \rangle = 0$. Thus $\chi = 0$, and $\mathcal{D}$ is dense.

Corollary 19.3.4. ϕ_M *is essentially self-adjoint on the domain* $\mathcal{D}$.

PROOF. We established in the theorem that $\mathcal{D}$ is dense in $\mathcal{H}$. Note that $e^{itH} : \mathcal{D} \to \mathcal{D}$, and $\mathcal{D} \subset (\bigcap_n \mathcal{D}(H^n))$ by (19.3.4). Thus $\mathcal{D}$ is a core for H and, by Theorem 19.3.1, a core for $\phi_M(f)$.

19.4 Commutators

We now prove four technical results concerning commutators and self-adjointness of operators. Let $0 \le H = H^*$ be a positive, self-adjoint operator, and let $\mathcal{D} \subset \bigcap_n \mathcal{D}(H^n)$ be a core of C^∞ vectors for H. Let $R(\lambda) = (H + (\lambda + 1)I)^{-1}$, so that $R \equiv R(0) = (H + I)^{-1}$. Let A be a bilinear form defined on the domain $\mathcal{D} \times \mathcal{D}$, and let $\delta(A) = [iH, A]$ again be a bilinear form on $\mathcal{D} \times \mathcal{D}$.

Theorem 19.4.1. *Let* $R^{1/2}\delta(A)R^{1/2}$ *be bounded. Then for any positive integer* n, $R^{n/2}AR^{n/2}$ *is bounded if and only if* AR^n *is bounded. Also* $\|R^{n/2}AR^{n/2} - AR^n\| \le n\|R^{1/2}\delta(A)R^{1/2}\|$.

PROOF. It is sufficient to study the difference

$$AR^n - R^{n/2}AR^{n/2} = [A, R^{n/2}]R^{n/2}$$

$$= \sum_{j=0}^{n-1} R^{j/2}[A, R^{1/2}]R^{(2n-j-1)/2}.$$

The factors $R^{j/2}$ and $R^{(2n-j-1)/2}$ are bounded by I, while the commutator is studied through the Cauchy integral formula [Kato, 1966, p. 282]

$$R^{1/2} = \pi^{-1} \int_0^\infty \lambda^{-1/2} R(\lambda) \, d\lambda.$$

Thus

$$[A, R^{1/2}] = -i\pi^{-1} \int_0^\infty R(\lambda)\delta(A)R(\lambda)\lambda^{-1/2} \, d\lambda,$$

which is bounded in norm by $\|R^{1/2}\delta(A)R^{1/2}\|$, using

$$\pi^{-1} \int_0^\infty (\lambda + 1)^{-1} \lambda^{-1/2} \, d\lambda = 1.$$

Theorem 19.4.2. *If $R^{1/2}\delta(A)R^{1/2}$ and $R^{n/2}AR^{n/2}$ are bounded, then A uniquely determines an operator (also denoted A) with domain $\mathcal{D}$. The operator A is symmetric if the form A is real.*

PROOF. This follows from Theorem 19.4.1 and the Riesz representation theorem.

Theorem 19.4.3. *Let A be a symmetric operator defined on $\mathcal{D}$, and suppose that $R^{1/2}\delta(A)R^{1/2}$ and AR^n are bounded for some $n \geq 1$. Then A is essentially self-adjoint on any core for H^n.*

PROOF. Since AR^n is bounded, $\mathcal{D}(A^-) \supset \mathcal{D}(H^n)$ and it is sufficient to prove essential self-adjointness on $\mathcal{D}(H^n)$. Let $\theta \in \mathcal{D}(A^*)$, $\chi \in \mathcal{D}(H^n)$. Then $R(\lambda)^n\theta \in \mathcal{D}(H^n) \subset \mathcal{D}(A)$, and

$$\begin{aligned}
\langle \chi, A\lambda^n R(\lambda)^n\theta \rangle &= \lambda^n \langle R(\lambda)^n A\chi, \theta \rangle \\
&= \langle A\lambda^n R(\lambda)^n\chi, \theta \rangle + \langle [\lambda^n R(\lambda)^n, A]\chi, \theta \rangle \qquad (19.4.1) \\
&= \langle \chi, \lambda^n R(\lambda)^n A^*\theta \rangle + \langle [\lambda^n R(\lambda)^n, A]\chi, \theta \rangle.
\end{aligned}$$

We take the limit $\lambda \to \infty$. Since $\lambda^n R(\lambda)^n \to I$ strongly, we evaluate the first term using

$$\lambda^n R(\lambda)^n A^*\theta \to A^*\theta.$$

The commutator in the second term is bounded uniformly as $\lambda \to \infty$ because

$$\begin{aligned}
\|[A, \lambda^n R(\lambda)^n]\| &\leq \sum_{r=0}^{n-1} \lambda^n \|R(\lambda)^r [A, R(\lambda)]R(\lambda)^{n-r-1}\| \\
&\leq \sum_{r=0}^{n-1} \lambda^n \|R(\lambda)^{r+1}\delta(A)R(\lambda)^{n-r}\| \\
&\leq O(1).
\end{aligned}$$

Thus by (19.4.1)

$$\lim_\lambda A\lambda^n R(\lambda)^n\theta = A^*\theta + \lim_\lambda [\lambda^n R(\lambda)^n, A]^*\theta. \qquad (19.4.2)$$

We complete the proof by showing that the last term in (19.4.2) has a limit equal to zero. Thus $\theta \in \mathcal{D}(A^-)$ and A is self-adjoint.

Because of the uniform bound on the norm above, it is sufficient to prove that $[A, \lambda^n R(\lambda)^n]$ converges to zero on the dense set $\mathcal{D}(H^n)$. Let $\psi \in \mathcal{D}(H^n)$. Then

$$[\lambda^n R(\lambda)^n, A]^*\psi = A\lambda^n R(\lambda)^n\psi - \lambda^n R(\lambda)^n A\psi.$$

The second term on right converges to $-A\psi$. The first term, written as

$$AR^n\lambda^nR(\lambda)^n(H+I)^n\psi$$

converges to $A\psi$. This completes the proof.

Theorem 19.4.4. *Let* A, B, $\delta(A)$, *and* $\delta(B)$ *all satisfy the hypotheses of Theorem 19.4.3 with* $n = 1$. *Let* AB *and* BA *be defined on* $\mathscr{D}$, *and* $[A, B]\mathscr{D} = 0$. *Then* A^- *and* B^- *commute.*

PROOF. Specifically we are assuming $R^{1/2}\delta(X)R^{1/2}$ and XR bounded for $X = A$, B, $\delta(A)$, and $\delta(B)$. As an approximation to A, define

$$A_\lambda = \lambda R(\lambda)^{1/2}AR(\lambda)^{1/2}.$$

We assert that $A_\lambda \to A$ strongly on the domain $\mathscr{D}$. Since $\mathscr{D} \subset \bigcap_n\mathscr{D}(H^n)$, any $\theta \in \mathscr{D}$ has the form $\theta = R\psi$. Then

$$A_\lambda\theta = A_\lambda R\psi = AR\lambda R(\lambda)\psi + \lambda[R(\lambda)^{1/2}, A]RR(\lambda)\psi.$$

Since AR is bounded and $\lambda R(\lambda) \to I$ strongly, the first term above converges to $AR\psi = A\theta$. The second is analyzed as in the proof of Theorem 19.4.1. Namely,

$$\lambda^{1/2}[R(\lambda)^{1/2}, A] = \lambda^{1/2}i\pi^{-1}\int_0^\infty R(\mu)\delta(A)R(\mu)\mu^{-1/2}\,d\mu.$$

Thus

$$\left\|\lambda^{1/2}[R(\lambda)^{1/2}, A]R^{1/2}\right\| \leq \lambda^{1/2}\pi^{-1}\int_0^\infty \mu^{-1/2}(\lambda + \mu + 1)^{-3/2}\,d\mu$$

$$\to 0 \quad \text{as } \lambda \to \infty,$$

as asserted.

By Theorem 19.4.3, $\mathscr{D}$ is a core for A, so $e^{iA_\lambda} \to e^{iA^-}$ strongly. We now show $[e^{iA_\lambda}, e^{iB_\lambda}] \to 0$. First we establish the identity

$$[A_\lambda, B_\lambda] = -i\lambda^2 R(\lambda)^{3/2}[\delta(A)R(\lambda)B - \delta(B)R(\lambda)A]R(\lambda)^{1/2}. \tag{19.4.3}$$

The calculations used to derive this identity are performed on the domain $R(\lambda)^{-1/2}\mathscr{D}$. This domain is dense because $\mathscr{D}$ is a core for $R(\lambda)^{-1/2}$ (e.g. by Theorem 19.4.3 with $A = R(\lambda)^{-1/2}$ and $n = 1$). On this domain we use the identity $R(\lambda)^{3/2}[A, B]R(\lambda)^{1/2} = 0$ as part of the derivation of (19.4.3). From (19.4.3), and the boundedness of $R^{1/2}\delta(A)R^{1/2}$, AR, etc., we infer that

$$\|R[A_\lambda, B_\lambda]R\| \leq O(\lambda^{-1/2}). \tag{19.4.4}$$

Next we claim that

$$\|(H + I)e^{itA_\lambda}R\| \leq e^{K|t|}, \tag{19.4.5}$$

where K is a constant independent of λ. To establish (19.4.5), we integrate the differential inequality

$$\frac{dF(t, \mu)}{dt} \leq KF(t, \mu), \tag{19.4.6}$$

where

$$F(t, \mu) \equiv \mu^2 Re^{-itA_\lambda} R(\mu)(H + I)^2 R(\mu) e^{itA_\lambda} R.$$

To derive (19.4.6), write

$$\frac{dF(t, \mu)}{dt} = \mu^2 Re^{-itA_\lambda}[R(\mu)(H + I)^2 R(\mu), iA_\lambda] e^{itA_\lambda} R.$$

Then

$$[R(\mu)(H + I)^2 R(\mu), iA_\lambda] = R(\mu)^2(H + I)\delta(A_\lambda)R + R(\mu)^2(H + I)(-i\delta(A_\lambda))R(\mu)$$
$$+ \text{Hermitian conjugate}$$
$$= R(\mu)^2(H + I)\delta(A_\lambda)R(H + \mu + I)(H + I)R(\mu)$$
$$+ R(\mu)(H + I)\{R(\mu)(-i\delta(A_\lambda))R\}(H + I)R(\mu)$$
$$+ \text{Hermitian conjugate}.$$

Next we commute $\delta(A_\lambda)$ with the factor $(H + \mu + I)$, introducing a second commutator. This gives for the first term above

$$R(\mu)(H + I)\{\delta(A_\lambda)R + R(\mu)i\delta^2(A_\lambda)RR(\mu)\}(H + I)R(\mu).$$

Because

$$\|\delta(A_\lambda)R\| + \|R^{1/2}\delta^2(A_\lambda)R^{1/2}\| \leq \text{const}$$

by hypothesis, (19.4.6) is established. Note that the constant K does not depend on λ, μ, or t. Thus (19.4.6) yields on integration

$$F(t, \mu) \leq e^{K|t|}\mu^2 R(\mu)^2 \leq e^{K|t|}. \tag{19.4.7}$$

Since $\mu R(\mu)$ is monotone increasing in μ, $F(t, \mu)$ also is monotone increasing in μ, and the limit $F(t, \mu) \uparrow F(t)$ exists [Kato, 1966, p. 459]. Thus the range of $e^{itA_\lambda}R$ lies in the domain of $H + I$, and (19.4.5) holds. Similar estimates hold with B replacing A.

Finally, we study the identity between bounded operators:

$$-R[e^{iA_\lambda}, e^{iB_\lambda}]R = \int_0^1 ds \int_0^1 dt\ RQ(-s, -t)^*[A_\lambda, B_\lambda]Q(1 - s, 1 - t)R, \tag{19.4.8}$$

where $Q(s, t) \equiv e^{itA_\lambda}e^{isB_\lambda}$ is unitary. By (19.4.4), (19.4.5), we infer $\|R[e^{iA_\lambda}, e^{iB_\lambda}]R\| \leq O(\lambda^{-1/2})$. The left side of (19.4.8) converges strongly as $\lambda \to \infty$ to $R[e^{iA^-}, e^{iB^-}]R$, while the bound converges to zero. Thus A^- and B^- commute.

Note that (19.4.8) follows from the following identity for bounded C, D:

$$[e^C, e^D] = \int_0^1 ds\left(\frac{d}{ds} e^{sC} e^D e^{(1-s)C}\right) = \int_0^1 ds\ e^{sC}[C, e^D]e^{(1-s)C},$$

so

$$[e^C, e^D] = \int_0^1 ds \int_0^1 dt\ e^{sC}e^{tD}[C, D]e^{(1-t)D}e^{(1-s)C}.$$

19.5 Lorentz Covariance

The main result of this section is the Lorentz covariance of ϕ_M and the Lorentz invariance of Ω. We also prove analyticity of the Schwinger functions at noncoinciding points.

Theorem 19.5.1. *Let $d\mu$ satisfy OS0–3. Then there exists a strongly continuous, unitary representation $U(g)$ of the inhomogeneous Lorentz group $\mathscr{L}$ on $\mathscr{H}$ such that*

$$U(g)\Omega = \Omega,$$

$$U(g)\phi_M(f)U(g)^{-1} = \phi_M(g^{-1}f) \tag{19.5.1}$$

for all $g \in \mathscr{L}$. In terms of $\phi_M(x)$,

$$U(g)\phi_M(x)U(g)^{-1} = \phi_M(gx). \tag{19.5.2}$$

Proposition 19.5.2. *Let $d\mu$ satisfy OS0–3, and let $g \to V(g)$ be a strongly continuous unitary representation of a group $\mathscr{G}$ on $\mathscr{E}$ such that*

$$V(g)1 = 1, \quad V(g)\mathscr{E}_+ \subset \mathscr{E}_+,$$

$$\theta V(g) = V(g)\theta, \quad T(t)V(g) = V(g)T(t). \tag{19.5.3}$$

Then $U(g)$ defined by

$$U(g)A^\wedge = (V(g)A)^\wedge, \quad g \in \mathscr{G},$$

is a continuous unitary representation of $\mathscr{G}$ on $\mathscr{H}$ such that

$$U(g)\Omega = \Omega, \quad e^{itH}U(g) = U(g)e^{itH}. \tag{19.5.4}$$

PROOF. As in the proof of Theorem 6.1.3, $V(g)$ maps $\mathscr{E}_+$ and $\mathscr{N}$ into themselves, so $U(g)$ is defined on the domain $\mathscr{E}_+^\wedge$. Furthermore, $U(g)$ is unitary, since $V(g)$ commutes with θ. In fact

$$\langle U(g)A^\wedge, B^\wedge \rangle_{\mathscr{H}} = \langle \theta V(g)A, B \rangle_{\mathscr{E}} = \langle V(g)\theta A, B \rangle_{\mathscr{E}} = \langle A, V(g)^{-1}B \rangle_{\mathscr{E}}$$

$$= \langle A^\wedge, {}^\wedge V(g^{-1})B \rangle_{\mathscr{H}} = \langle A^\wedge, U(g^{-1})B^\wedge \rangle_{\mathscr{H}},$$

so $U(g)^* = U(g)^{-1} = U(g^{-1})$. Thus $U(g)$ extends to a representation of $\mathscr{G}$ on all of $\mathscr{H}$. Since $V(g)$ commutes with $T(t)$, $U(g)$ commutes with e^{-tH} and hence with e^{itH}. Strong continuity of $U(g)$ follows from the strong continuity of $V(g)$, while $U(g)\Omega = \Omega$ follows from $V(g)1 = 1$.

We now study the distributions

$$W_n(x_1, \ldots, x_n) = \langle \Omega, \phi_M(x_1) \cdots \phi_M(x_n)\Omega \rangle,$$

$$W_n(\underline{h}, \underline{t}) = \langle \Omega, \phi_M(h_1, t_1) \cdots \phi_M(h_n, t_n)\Omega \rangle, \tag{19.5.5}$$

which are densities of the Wightman functions (19.3.6). Here for brevity we use the notation $h = \{h_1, \ldots, h_n\}$, $\underline{t} = \{t_1, \ldots, t_n\}$.

Proposition 19.5.3. *The $W_n(\underline{h}, t)$ are boundary values in $\mathscr{S}'(R_n)$ of analytic functions $W_n(\underline{h}, z)$. Here $z_j = t_j + is_j$, and the $W(\underline{h}, z)$ are analytic in the region $s_{j+1} - s_j > 0$, $j = 1, 2, \ldots, n - 1$. Furthermore, for $t_j = 0$ and $s_1 < s_2 < \cdots < s_n$,*

$$W_n(\underline{h}, i\underline{s}) = S_n(\underline{h}, \underline{s}) = \int \phi(h_1, s_1) \cdots \phi(h_n, s_n)\, d\mu. \qquad (19.5.6)$$

PROOF. The bound of Theorem 19.3.1 shows that $W_n(\underline{h}, z)$ is analytic for $s_{j+1} - s_j > 0$, $j = 1, 2, \ldots, n - 1$. The bound of Corollary 19.2.2 shows that $W_n(\underline{h}, \underline{z}) \to W_n(\underline{h}, \underline{t})$ in $\mathscr{S}'(R^n)$ as $s_{j+1} - s_j \downarrow 0$. For $t_j = 0$ and $s_{j+1} - s_j > 0$, the $W_n(\underline{h}, \underline{z})$ agree with the Schwinger functions, by the definitions of Section 19.3. Note the Schwinger functions are already known to be analytic in $s_{j+1} - s_j$ by Corollary 10.5.6.

PROOF OF THEOREM 19.5.1. The covariance under space-time translations and space rotations follows from Proposition 19.5.2. The field ϕ transforms covariantly by definition. To complete the proof we construct the Lorentz boost transformations. Let us consider a pure Lorentz rotation Λ_α by the hyperbolic angle α in the $(t, x_1) \equiv (t, x)$ plane. The infinitesimal operator for this rotation on the Wightman functions is

$$L_n = \sum_{j=1}^{n} \left(t_j \frac{\partial}{\partial x_j} + x_j \frac{\partial}{\partial t_j} \right)$$

We show $W_n(L_n F) = 0$ for all $F \in \mathscr{S}(R^{nd})$. In particular

$$\frac{d}{d\alpha} W_n(\Lambda_\alpha F) = W_n(L_n \Lambda_\alpha F) = 0,$$

so each W_n is Lorentz-invariant. It follows that there exists a unitary group $U(\Lambda_\alpha)$ on $\mathscr{H}$ which implements Λ_α. The Lorentz group multiplication law for the $U(\Lambda_\alpha)$ follows from the Lorentz group multiplication law for the Λ_α.

The Euclidean invariance of the Schwinger function S_n, in infinitesimal form, states that

$$0 = \sum_{j=1}^{n} \left(s_j \frac{\partial}{\partial x_j} - x_j \frac{\partial}{\partial s_j} \right) S_n(\mathbf{x}, \underline{s}). \qquad (19.5.7)$$

We analytically continue (19.5.7) to complex $s_j = \varepsilon_j - it_j$ with $\varepsilon_{j+1} - \varepsilon_j > 0$, i.e. within the domain of analyticity of S_n. For complex s we rewrite (19.5.7) as

$$0 = \sum_{j=1}^{n} (\varepsilon_j - it_j) \left[\frac{\partial}{\partial x_j} - x_j \frac{\partial}{\partial(-it_j)} \right] S_n(\mathbf{x}, \underline{\varepsilon} - i\underline{t}). \qquad (19.5.8)$$

We now take $\varepsilon_{j+1} - \varepsilon_j \to 0$ and find $L_n W_n = 0$. With test functions, this states $W_n(L_n F) = 0$, as desired.

Corollary 19.5.4. *The energy-momentum spectrum lies in the forward cone, $|\mathbf{P}| \leq H$. Here $\mathbf{P}$ is the momentum operator, the generator of space translation on $\mathscr{H}$.*

Theorem 19.5.5. *Assume OS0–3. Let $\delta > 0$. There exists a bilinear form $\phi_M(x)$ on $\mathcal{H}_\delta \times \mathcal{H}_\delta$ such that*

$$e^{-\delta H} \phi_M(x) e^{-\delta H} \tag{19.5.9}$$

is a bounded operator, analytic in x for $|\operatorname{Im} x| < \delta$, and such that

$$\phi_M(f) = \int \phi_M(x) f(x)\, dx. \tag{19.5.10}$$

PROOF. Since $\mathbf{P}$ and H commute, we infer from Corollary 19.5.4 that the series

$$e^{itH - i\mathbf{x}\cdot\mathbf{P} - \delta H} = \sum_{n, m = 0}^{\infty} \frac{(itH)^n}{n!} \frac{(-i\mathbf{x}\cdot\mathbf{P})^m}{m!} e^{-\delta H}$$

converges in norm $|t| + |\mathbf{x}| < \delta$. Using this fact together with (19.3.3), it follows that for $|t| + |\mathbf{x}| < \delta$,

$$e^{-\delta H} \phi_M(f_{t,\mathbf{x}}) e^{-\delta H} = F(x)$$

is a real analytic function of x, with

$$\|\partial_x^n F(x)\| \le K(\varepsilon)\varepsilon^{-|n|} n! \int \|f^{(t)}\|\, dt, \tag{19.5.11}$$

for $\varepsilon < \delta$ and x, t real but otherwise unrestricted. Thus as in (19.1.7), $F(0)$ is the integral of a bounded C^∞ function $F(0) = \int G(x) f(x)\, dx$, where $G(x) = e^{-\delta H} \phi_M(x) e^{-\delta H}$ defines $\phi_M(x)$. Repeating the argument leading to (19.5.11), we obtain

$$\|\partial_x^n e^{-\delta H} \phi_M e^{-\delta H}\| \le K(\varepsilon, \delta)\varepsilon^{-|n|} n!, \tag{19.5.12}$$

where

$$K(\varepsilon, \delta) = \|e^{-(\delta - \varepsilon)H} \phi_M(x) e^{-(\delta - \varepsilon)H}\|.$$

Since $K(\varepsilon, \delta)$ is independent of x (for x real), the claimed analyticity follows by (19.5.12).

Corollary 19.5.6. *The Schwinger functions $S_n(x_1, \ldots, x_n)$ are real analytic functions of $x_1, \ldots, x_n$ at noncoinciding points (i.e., $x_i \ne x_j$ for all $i \ne j$).*

PROOF. Since the S_n are symmetric under permutation of $x_1, \ldots, x_n$, we may assume that $t_1 \le t_2 \le \cdots \le t_n$. If some times are equal, but no two points coincide, then a small Euclidean rotation produces all unequal times. The corollary now follows from the analyticity of (19.5.9).

Proposition 19.5.7. *Let x, y be given, with $\mathbf{x} - \mathbf{y} \ne 0$. Let B be a subset of R^d of points $(z_0, \mathbf{z})$ such that $\mathbf{z}$ projected on the line $\mathbf{x} - \mathbf{y}$ does not lie between $\mathbf{x}$ and $\mathbf{y}$. Let $f_1, \ldots, f_n \in \mathcal{S}$ be supported in B. Then $S_{n+2}(f_1, \ldots, f_n, x, y)$ is analytic in $x - y$ for $\mathbf{x} - \mathbf{y}$ real and $|x_0 - y_0| < |\mathbf{x} - \mathbf{y}|$.*

PROOF. Perform a Euclidean rotation so that $\mathbf{x} - \mathbf{y}$ is the new time axis. By construction B contains no points in the time interval between $\mathbf{x}$ and $\mathbf{y}$. The proof now follows the proof of Theorem 19.5.5.

Note that in the $\xi = x_0 - y_0$ plane, the proposition allows us to connect the half planes Re $\xi > 0$ and Re $\xi < 0$ of analyticity by a slit $|\text{Im } \xi| < |\mathbf{x} - \mathbf{y}|$ on the imaginary ξ axis.

19.6 Locality

Theorem 19.6.1. *Assume OS0–3, and let* f, $g \in \mathscr{S}_{\text{real}}$ *have spacelike separated supports. Then three forms of locality hold:*

(i) $[e^{i\phi_M(f)}, e^{i\phi_M(g)}] = 0.$

(ii) $[\phi_M(f), \phi_M(g)]\mathscr{D} = 0.$

(iii) $W_{n+2}(f_1, \ldots, f, g, \ldots, f_n) = W_{n+2}(f_1, \ldots, g, f, \ldots, f_n)$ *for all* n *and all* $f_j \in \mathscr{S}.$

Given $T > 0$ and $\mathbf{z} \neq 0$, $\mathbf{z} \in R^{d-1}$, let $B = B(T, \mathbf{z})$ be the subset of R^d of points $(t, \mathbf{x})$ with $t \geq T$, $\mathbf{x} \cdot \mathbf{z} \geq \mathbf{z}^2$. Geometrically, B is a quarter plane in the two-dimensional space spanned by t and $\mathbf{z}$, and is unrestricted in the two space dimensions orthogonal to $\mathbf{z}$. Let

$$\mathscr{D}_B = \text{span}\{(\phi(f)^n)^\wedge : f \in \mathscr{S}(B)\}.$$

Proposition 19.6.2. $\mathscr{D}_B$ *is a core of* C^∞ *vectors for* H.

PROOF. Every vector in $\mathscr{D}_B$ is in the range of e^{-TH} and $e^{-tH}\mathscr{D}_B \subset \mathscr{D}_B$. Thus we need only show that $\mathscr{D}_B$ is dense. Let $\chi \perp \mathscr{D}_B$, and define

$$F(t, \mathbf{x}) = \langle \chi, e^{-tH + i\mathbf{x}\cdot\mathbf{P}}\phi(g)^n\rangle \equiv \langle \chi, \psi\rangle,$$

where $g \in \mathscr{S}_+ \cap C_0^\infty$. For t and $\mathbf{x} \cdot \mathbf{z}$ sufficiently large, $\psi \in \mathscr{D}_B$, and $F(t, \mathbf{x}) = 0$. Clearly F is real analytic in t for $t > 0$, and by Corollary 19.5.4, F is real analytic in $\mathbf{x}$. Thus $F \equiv 0$, and in particular $F(0, 0) = 0$. But the estimates (19.1.1) allow summation of the exponential series $e^{\phi(g)^\wedge}$ and thus ensure $\chi \perp e^{\phi(g)^\wedge}$. However, these vectors span $\mathscr{H}$. Thus $\chi = 0$, and $\mathscr{D}_B$ is dense.

Proposition 19.6.3. *Let* f, $g \in C_0^\infty$ *have spacelike separated supports. Then for some* B *as above,*

$$[\phi_M(f), \phi_M(g)]\mathscr{D}_B = 0.$$

PROOF. Study the Schwinger functions

$$S_{n+2}(\theta f_1, \ldots, \theta f_r, x, y, f_{r+1}, \ldots, f_n) = S_{n+2}(\theta f_1, \ldots, \theta f_r, y, x, f_{r+1}, \ldots, f_n), \quad (19.6.1)$$

with f_j supported in $B(T, \mathbf{z})$. Choose $\mathbf{z}$ sufficiently large and so that no point in $B \cup \theta B$ lies in the strip bounded by two hyperplanes normal to $\mathbf{x} - \mathbf{y}$ and passing through x and y respectively. Then Proposition 19.5.7 applies, and (19.6.1) is analytic in $x - y$ for $\mathbf{x} - \mathbf{y}$ real and $|x_0 - y_0| < |\mathbf{x} - \mathbf{y}|$. We evaluate (19.6.1) at pure imaginary $x_0 = it$, $y_0 = is$. (Note that we may choose one B for all $x \in \text{suppt } f$, $y \in \text{suppt } g$.) Then multiply by $f(x)g(y)$ and integrate over x, y. Thus the analytic continuation of (19.6.3) is the statement

$$\langle \theta_1, [\phi_M(f), \phi_M(g)]\theta_2 \rangle_{\mathscr{H}} = 0,$$

where θ_1, $\theta_2 \in \mathscr{D}_B$. Note that the restriction $|t - s| < |\mathbf{x} - \mathbf{y}|$ is just the condition that f, g have spacelike separated supports.

PROOF OF THEOREM 19.6.1. By Proposition 19.6.2 and Theorems 19.4.1, 19.4.3, $\mathscr{D}_B$ is a core for $\phi_M(f)$, $\phi_M(g)$ and in the domain of their product. We let f, $g \in C_{0,\text{real}}^\infty$, and apply Proposition 19.6.3. By Theorem 19.4.4 (with $\mathscr{D}_B$ the domain of definition) part (i) of Theorem 19.6.1 holds. Since the vectors in $\mathscr{D}$ are C^∞ for all products of field operators, parts (ii) and (iii) extend by continuity from f, $g \in C_0^\infty$ to f, $g \in \mathscr{S}$, from which part (i) follows by a second application of Theorem 19.4.4.

We remark that at this point we have proved Theorems 6.1.5–6, with the exception of uniqueness of the vacuum (Wightman axioms) and irreducibility (Haag–Kastler axioms). These properties are analyzed below.

19.7 Uniqueness of the Vacuum

We recall the following condition on $d\mu(\phi)$, introduced in Section 6.1.

OS4 (Ergodicity). The time translation subgroup $T(t)$ of $\mathscr{L}$ acts ergodically on the measure space $\{\mathscr{D}'(R^d)_{\text{real}}, d\mu\}$.

Theorem 19.7.1. *Let $S\{f\}$ satisfy OS0–3. Then OS4 is satisfied if and only if Ω is the unique vector (up to scalar multiples) in $\mathscr{H}$ which is invariant under time translation e^{itH}.*

Remark. Ergodicity of $d\mu$ is equivalent to the statement that 1 is the unique invariant vector for the unitary group $T(t)$ acting on the Hilbert space $\mathscr{E} = L_2(\mathscr{S}', d\mu)$. This in turn is equivalent to the cluster property

$$0 = \lim_{t \to \infty} t^{-1} \int_0^t [\langle A T(s)B \rangle - \langle A \rangle \langle B \rangle] \, ds, \qquad (19.7.1)$$

for A, B in a dense subspace of $\mathscr{E}$. Here $\langle \cdot \rangle$ denotes $\int \cdot \, d\mu$. In particular, exponential clustering of the Schwinger functions ensures ergodicity of $d\mu$.

PROOF. For a self-adjoint contraction semigroup e^{-tH} acting on a Hilbert space $\mathscr{H}$,

$$\text{st lim}_{t \to \infty} t^{-1} \int_0^t e^{-sH} \, ds = P_{\text{inv}}$$

is the projection onto the subspace of invariant vectors. (Similarly, for a unitary group $T(t)$, $P_{\text{inv}} = \text{st lim } t^{-1} \int^t T(s) \, ds$.) Thus (19.7.1) is equivalent to the statement that 1 spans the invariant subspace of $T(t)$ in $\mathscr{E}$. Let A, B in (19.7.1) be finite linear combinations of functions $e^{i\phi(f)}$, with $f \in C_{0,\,\mathrm{real}}^{\infty}$.

It is no loss in generality to require $A \in \mathscr{E}_-$, $B \in \mathscr{E}_+$, because of the time translation $T(s)$ in (19.7.1). Then (19.7.1) is equivalent to the cluster property

$$0 = \lim t^{-1} \int_0^t \left[\langle (\theta A)^\wedge, e^{-tH}B^\wedge \rangle_{\mathscr{H}} - \langle (\theta A)^\wedge, \Omega \rangle_{\mathscr{H}} \langle \Omega, B^\wedge \rangle_{\mathscr{H}} \right] ds, \quad (19.7.2)$$

and thus to the uniqueness of Ω as a ground state for H.

We remark that the proof of Theorem 6.1.5 is now complete. Furthermore, assuming OS4, we now prove that the global von Neumann algebra generated by $\bigcup_B \mathfrak{A}(B)$ is irreducible and complete the proof of Theorem 6.1.6. To establish irreducibility, suppose that A commutes with all observables. Since each algebra $\mathfrak{A}(B)$ is closed under taking of adjoints, we see that A^* commutes with all observables also. Thus $\mathrm{Re}\, A = (A^* + A)/2$ and $\mathrm{Im}\, A = (A^* - A)/2i$ commute with all observables, and without loss of generality we assume that A is self-adjoint. Thus for $C = C^* \in \mathfrak{A}(B)$, we have $C(t) = U(t)CU(t)^* \in \mathfrak{A}(B_t)$ and

$$\langle A\Omega, U(t)C\Omega \rangle = \langle \Omega, AC(t)\Omega \rangle = \langle C(t)\Omega, A\Omega \rangle$$
$$= \langle U(t)C\Omega, A\Omega \rangle = \langle A\Omega, U(t)C\Omega \rangle^-.$$

Thus $\langle A\Omega, U(t)C\Omega \rangle$ is real, and so its Fourier transform with respect to t is an even function of the dual variable ω. However, by positivity of the energy, the Fourier transform is supported in the half line $\omega = \text{energy} \geq 0$. Thus $A\Omega$ is orthogonal to strictly positive energy states of form $C\Omega$. However, the states $C\Omega$ are dense in $\mathscr{H}$ by Theorem 19.3.3 and the fact that vectors in $\mathscr{D}$ can be obtained by differentiating exponentials $\exp(\phi_M(f_1)) \cdots \exp(\phi_M(f_n)) \Omega$ in $C\Omega$. Thus $A\Omega$ is a zero energy state. However by the axiom OS4, Ω is the unique zero energy state, up to scalar multiples. Thus $A\Omega = \lambda\Omega$, and hence for $\chi, \psi \in \mathfrak{A}(B)\Omega$, $\langle \chi, A\psi \rangle = \langle \chi, \psi \rangle \lambda$. Thus $A = \lambda I$.

The standard methods of constructing quantum fields lead naturally to a measure $d\mu$ which satisfies OS0–3 but not necessarily OS4. To deal with this possibility and to provide the technical tools for the study of mixed states, we examine more carefully the ergodicity properties of $d\mu$.

Definition 19.7.2. Let $\mathscr{E}_1$ denote the subspace of $\mathscr{E}$ which is invariant under the time translation group $T(t)$, $t \in R^1$. Likewise, let $\mathscr{E}_E \subset \mathscr{E}$ denote the subspace of $\mathscr{E}$ invariant under the full Euclidean group. Let $\mathscr{E}_\infty$ denote the subspace of $\mathscr{E}$ of functions which are measurable "at ∞." In other words, $\mathscr{E}_\infty = \cap\{\mathscr{E}(\sim B): B \text{ bounded}\}$. Let $\mathscr{E}_\pm$ denote the positive and negative time subspaces of $\mathscr{E}$, as in Section 6.1.

Proposition 19.7.3. *Let $d\mu$ satisfy OS2. Then*

$$\mathscr{E}_1 \subset \mathscr{E}_\infty \cap \mathscr{E}_+ \cap \mathscr{E}_-.$$

PROOF. Let $A \in \mathscr{E}_1$. Then $A = \lim A_n$, where $\|A_n\| \leq \|A\|$ and A_n is a function of the field $\phi(x)$ for $|x| \leq n$. Such a sequence A_n can be constructed by orthogonal projection onto $\mathscr{E}_n = \mathscr{E}(|x| \leq n)$, i.e. by conditional expectations, and convergence follows from the limit $E_n \uparrow I$, where E_n is the projection onto the subspace $\mathscr{E}_n$ of $\mathscr{E}$ consisting of all L_2 functions of $\phi(x)$, $|x| \leq n$. Then strong convergence of $T(t)A_n$ follows as $n \to \infty$, since for every $t \in R^1$,

$$\|T(t)A_n - A\| = \|A_n - T(-t)A\| = \|A_n - A\| \to 0.$$

Choosing $t = 2n$, n, and $-n$, respectively, we see that $A \in \mathscr{E}_\infty$, $\mathscr{E}_+$, and $\mathscr{E}_-$, respectively.

Proposition 19.7.4. *Let $d\mu$ satisfy OS2 and OS3. Then the elements of $\mathscr{E}_1$ are invariant under reflections θ about the $t = 0$ hyperplane.*

PROOF. We may assume $0 \leq A$. Let $\langle\ \rangle_{\mathscr{E}}$ and $\langle\ \rangle_{\mathscr{H}}$ denote inner products in $\mathscr{E}$ and $\mathscr{H}$, respectively. By Proposition 19.7.3, A, $\theta A \in \mathscr{E}_+ \cap \mathscr{E}_-$. Thus

$$\langle \theta A, A \rangle_{\mathscr{E}} \leq \langle \theta A, A \rangle_{\mathscr{E}}^{1/2} \langle A, A \rangle_{\mathscr{E}}^{1/2} = \langle A^\wedge, (\theta A)^\wedge \rangle_{\mathscr{H}}^{1/2} \langle (\theta A)^\wedge, A^\wedge \rangle_{\mathscr{H}}^{1/2}$$

by the Schwarz inequality in $\mathscr{E}$. This chain of inequalities can be continued, using the Schwarz inequality in $\mathscr{H}$, as follows:

$$|\langle A^\wedge, (\theta A)^\wedge \rangle_{\mathscr{H}}| \leq \langle A^\wedge, A^\wedge \rangle_{\mathscr{H}}^{1/2} \langle (\theta A)^\wedge, (\theta A)^\wedge \rangle_{\mathscr{H}}^{1/2}$$

$$= |\langle \theta A, A \rangle_{\mathscr{E}}| = \langle \theta A, A \rangle_{\mathscr{E}}.$$

Thus all the inequalities are equalities, the first of which implies that $\theta A = A$.

Theorem 19.7.5. *Let $d\mu$ satisfy OS2 and OS3. Then $\mathscr{E}_1 = \mathscr{E}_E$.*

Corollary 19.7.6. *$d\mu$ as above is ergodic under the full Euclidean group if and only if it is R^1-ergodic.*

PROOF OF THEOREM 19.7.5. Write $x = (\mathbf{x}, t)$, $\mathbf{x} = x_1$, and let $T(e^{i\theta})$ denote a rotation by the angle θ in the t, x_1 plane. Then

$$T(x_1) = \lim_{t \to \infty} T(-t)T(e^{-ix_1/t})T(t).$$

With $A \in \mathscr{E}_1$, let $A_t = A - T(e^{-ix_1/t})A$. Then $\|A_t\| \to 0$, because $T(e^{i\theta})$ is strongly continuous in θ. Thus $T(-t)A_t \to 0$ also, and so

$$T(x_1)A = \lim_{t \to \infty} T(-t)T(e^{-ix_1/t})A$$

$$= \lim_{t \to \infty} T(-t)A - T(-t)A_t = A.$$

Thus A is invariant under translations in the x_1 direction, and similarly under all translations in R^d. By Proposition 19.7.4, A is invariant under reflections about an arbitrary hyperplane. Since translations and reflections generate the Euclidean group, $A \in \mathscr{E}_E$.

The bounded functions $\mathscr{E}_E \cap L_\infty$ define a direct integral decomposition of the measure $d\mu$. Let $\mathscr{Z}$ denote the spectrum of $\mathscr{E}_E \cap L_\infty$. Then there is a positive measure $d\zeta$ on $\mathscr{Z}$ of total mass one, and for a.e. $\zeta \in \mathscr{Z}$ a measure $d\mu_\zeta$ on $\mathscr{S}'(R^d)$, such that

$$d\mu = \int_{\zeta \in \mathscr{Z}} d\mu_\zeta$$

and if

$$S_\zeta\{f\} = \int e^{i\phi(f)} \, d\mu_\zeta,$$

then

$$S\{f\} = \int_{\zeta \in \mathscr{Z}} S_\zeta\{f\} \, d\zeta.$$

Theorem 19.7.7. *Let $d\mu$ satisfy OS0–3. Then for a.e. ζ, $d\mu_\zeta$ satisfies OS0–4.*

PROOF. We identify $\mathscr{E}_E \cap L_\infty$ with the functions $L_\infty \, (\mathscr{Z}, d\zeta)$ of ζ. By definition,

$$\int AZ \, d\mu_\zeta = \left(\int AZ \, d\mu \right)(\zeta),$$

where $Z \in \mathscr{E}_E \cap L_\infty$ and $\int AZ \, d\mu$ is regarded as a function of ζ. The invariance of $d\mu_\mathscr{Z}$ then follows from the invariance of Z and the invariance of $d\mu$. Similarly, reflection positivity follows from the positivity of $d\zeta$ and from the reflection positivity of the inner product:

$$\mathscr{E}_+ \ni A, B \quad \Rightarrow \quad \int (\theta A)^- BZ \, d\mu = \int \theta(Z^{1/2}A)^- (Z^{1/2}B) \, d\mu$$

for $Z \in \mathscr{E}_E \cap L_\infty$, $Z \geq 0$. Here we use the facts that $Z^{1/2}$ and $\theta Z^{1/2} \in \mathscr{E}_+ \cap L_\infty$, by Propositions 19.7.3,4. The ergodicity of $d\mu_\zeta$ for a.e. ζ follows from the general theory of direct integral decompositions, and we refer the reader to [Dixmier, 1957] for details.

We now show that the regularity axiom OS1 holds in a.e. pure phase ζ, using the method of multiple reflections. It is sufficient to show that

$$\left| \int Ze^{i\langle f, \phi \rangle} \, d\mu \right| \leq \left(\int Z \, d\mu \right) \exp(\|f\|_{L_1} + \|f\|_{L_p}^2) \tag{19.7.3}$$

for any projection $Z \in \mathscr{E}_E$, in order to prove (19.7.3) for a.e. pure phase. It is sufficient to take f to be supported in a time strip, say $0 \leq t \leq T$. By the Schwarz inequality in $\mathscr{H}$,

$$\left| \int Ze^{i\langle f, \phi \rangle} \, d\mu \right| = |\langle Z^{1/2 \wedge}, (Z^{1/2}e^{i\langle f, \phi \rangle})^\wedge \rangle|$$

$$\leq \left(\int Z \, d\mu \right)^{1/2} \left(\int Ze^{i\langle f, \phi \rangle}(e^{i\langle \theta f, \phi \rangle})^- \, d\mu \right)^{1/2}.$$

Now f and θf can be combined into a single test function, and after a time translation, the process can be repeated in the second factor. After n steps, we get a test function of the form $g = \sum_{j=1}^{2^n} f_j$, and the f_j have support in disjoint time intervals. For these f_j,

$$\|g\|_{\mathcal{L}_p}^p = \|\sum f_j\|_{\mathcal{L}_p}^p = \sum \|f_j\|_{\mathcal{L}_p}^p = 2^n \|f\|_{\mathcal{L}_p}^p.$$

We substitute OS1 and let $n \to \infty$ to complete the proof.

CHAPTER 20

Further Directions

To complete the main text and provide the reader with a guide to the bibliography and literature, we survey here some of the more important developments of constructive quantum field theory which were omitted from the earlier chapters.

20.1 The ϕ_3^4 Model

The constructive field theory program gives a ϕ^4 model in $d = 3$ dimensions. Here an infinite mass renormalization is required. We state the basic existence theorem: Let $d\phi$ denote the Gaussian measure on $\mathscr{D}'(R^3)$ with mean 0 and covariance $(-\Delta + 1)^{-1}$. Let ϕ_κ denote the cutoff field, either a lattice with spacing κ^{-1} or a convolution cutoff $\phi_\kappa \equiv \phi * \delta_\kappa$ as in Section 8.1. Let

$$V(\Lambda, \kappa) = \int_\Lambda [\lambda\phi^4 + a(\kappa, \lambda)\phi^2] \, d^3x,$$

and

$$S_{\Lambda, \kappa}(f) = \int e^{\phi(f)} \, d\mu_{\Lambda, \kappa},$$

where

$$d\mu_{\Lambda, \kappa} = Z(\Lambda, \kappa)^{-1} \exp[-V(\Lambda, \kappa)] \, d\phi,$$
$$Z(\Lambda, \kappa) = \int \exp[-V(\Lambda, \kappa)] \, d\phi.$$

Theorem 20.1.1. *There exist constants α, $\beta \geq 0$ such that for all $0 < \lambda$, $\sigma \in R$, and with*

$$a(\kappa) = -\alpha\lambda\kappa + \beta\lambda^2 \ln \kappa + \sigma,$$

it follows that

$$S(f) = \lim_{\Lambda \uparrow R^3} \lim_{\kappa \to \infty} S_{\Lambda, \kappa}(f)$$

exists and satisfies the axioms OS0–3.

Theorem 20.1.2. *There exist finite numbers $\sigma_{\pm}(\lambda)$ as follows: For $\sigma > \sigma_{+}(\lambda)$, $S(f)$ satisfies OS4 (and yields a field theory with a unique vacuum vector for the Hamiltonian H). For $\sigma < \sigma_{-}(\lambda)$, $S(f)$ does not satisfy OS4. (At least two vacuum vectors exist.)*

The fundamental step in the existence proof, namely the construction of $d\mu(\Lambda, \kappa = \infty)$, was established in [Glimm and Jaffe, 1972b]. The existence of the $\Lambda \uparrow R^3$ limit was carried out for $\sigma > \sigma_{+}$ by [Feldman and Osterwalder, 1976] and [Magnen and Sénéor, 1976a] using cluster expansions. Convergence as the lattice spacing tends to zero, for Λ fixed, was proved by [Park, 1977] and establishes correlation inequalities. The infinite volume limit for all σ was constructed by [Seiler and Simon, 1976]. The proof that multiple vacua (i.e., phase transitions) exist was given by an extension of the methods described in Section 16.4 to continuum theories [Fröhlich, Simon, and Spencer, 1976]. The particle spectrum of the $\sigma \gg 1$ model was studied by [Burnap, 1977], who showed the existence of an isolated one particle state. An alternative construction of the finite volume ϕ_3^4 model has been given by [Benfatto et al., 1980], and is based on ideas related to the renormalization group; cf. [Wilson and Kogut, 1974], [Ma, 1976].

20.2 Borel Summability

The Schwinger functions for $\lambda P(\phi)_2$ models are nonanalytic in λ at $\lambda = 0$, since stability is lost when $\lambda < 0$. The model is, however, analytic in the pie-shaped sector with opening angle $\theta > \pi$ illustrated in Figure 20.1. The Schwinger functions moreover have one sided (right) derivatives of all orders at zero (i.e., for $\lambda \in [0, \varepsilon]$), and Dimock [1974, 6] has shown that these derivatives agree with the infinite volume perturbation theory. The scattering matrix elements also are asymptotic to perturbation theory [Osterwalder and Sénéor, 1976], [Eckmann, Epstein, and Fröhlich, 1976]).

In fact, the ϕ^4 Schwinger functions can be recovered from the perturbation series coefficients by Borel summation [Eckmann, Magnen, and Sénéor, 1975]. This means that while the r point Schwinger function

$$S_r(\lambda) \sim \sum_{n=0}^{\infty} a_n \lambda^n \tag{20.2.1}$$

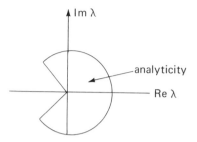

Figure 20.1. Known analyticity domain in complex λ for $\lambda P(\phi)_2$ quantum fields.

has a divergent expansion in ϕ_2^4 models because $|a_n| > O(n!)$, nevertheless the Borel transform

$$h(t) \equiv \sum_{n=0}^{\infty} \frac{a_n}{n!} t^n \tag{20.2.2}$$

has a convergent expansion near $t = 0$. Furthermore, $h(t)$ has an analytic continuation for $t > 0$ for which the integral

$$S_n(\lambda) \equiv \int_0^{\infty} e^{-t} h(\lambda t)\, dt \tag{20.2.3}$$

exists. A recent reference on the Borel transform is [Sokal, 1980]. One shows that $S_n(\lambda)$ equals the ϕ_2^4 Schwinger function constructed in Chapter 11. The mass $m(\lambda)$ is also Borel summable [Eckmann and Epstein, 1979b].

The results on Borel summability of the $\lambda \phi^4$ Schwinger functions extend to $d = 3$ [Magnen and Sénéor, 1977]. The scattering matrix elements are also asymptotic [Constantinescu, 1977]. It is an open problem whether resummation techniques can be used for $(\lambda \phi^4 - \phi^2)_2$ quantum fields.

20.3 Euclidean Fermi Fields

The Euclidean scalar fields ϕ discussed throughout this section form the path space for the Feynman–Kac formula

$$\langle A^{\wedge}, e^{-t(H+V)} A^{\wedge} \rangle = \int (\overline{\theta A})(\phi) A(\phi) e^{-\int_0^t V(\phi(s))\, ds}\, d\mu(\phi),$$

or

$$e^{-t(H+V)} = \left(e^{-\int_0^t V(\phi(s))\, ds} T(t) \right)^{\wedge}$$

in the notation of Section 6.1. The generalization to Fermi fields of this formula is given in [Osterwalder and Schrader, 1972a, 1973a,b]; it appears to have been new, even on a formal level. We explain the main features of this generalization.

In the first place, the real time field ψ and its adjoint $\bar{\psi}$ are replaced by independent, anticommuting Euclidean fields ψ_1 and ψ_2. As a consequence, time zero Euclidean fields are constructed in a different space from $\mathcal{H}$.

Secondly, since the Euclidean fields anticommute (take their values in a Grassmann algebra), the inner product in Euclidean space is not given by integration with respect to a positive measure $d\mu$. Rather, the inner product is defined as a positive state (expectation) ρ on the algebra of field operators generated by $\psi_1(f)$, $\psi_2(g)$. For a polynomial $P(\psi_1, \psi_2)$, $\rho(P^*P) \geq 0$. Such a state defines an integral.

Thirdly, typical interaction energy densities, such as $\bar\psi\psi$, $\bar\psi\gamma^5\psi$, $\sum_n \bar\psi\gamma^\mu\psi A_\mu$, are real for ordinary Fermi fields. These are represented in Euclidean space by $\psi_2\psi_1$, $\psi_2\gamma^5\psi_1$, $\sum_n \psi_2\gamma^\mu\psi_1 A_\mu$, etc. As a consequence, the Euclidean action function is not real.

Nevertheless, it is possible to generalize the axioms OS0–4 to this case, and to establish a Feynman–Kac formula

$\exp[-t(H + V(\psi, \bar\psi, \phi))]$

$$= \left(\exp\left[-\int_0^t V(\psi_1(s), \psi_2(s), \phi(s))\, ds\right] T(t)\right)^\wedge. \quad (20.3.1)$$

The Hamiltonian $H + V$ defined by (20.3.1) acts on the Hilbert space $\mathcal{H}$ of the quantum theory. When $V(\psi, \bar\psi, \phi)$ is formally real, the definition (20.3.1) yields a symmetric $H + V$, agreeing with the standard canonical construction.

20.4 Yukawa Interactions

Interactions between fermions and bosons have also been studied, in particular the Yukawa$_2$ interactions

$$\lambda\bar\psi\psi\phi \quad \text{and} \quad \lambda\bar\phi\gamma^5\psi\phi,$$

scalar and pseudoscalar, respectively. These models require a logarithmically divergent mass renormalization in $d = 2$. A survey giving details of and references to the original construction of these models by Hamiltonian methods can be found in [Glimm and Jaffe, 1971b]. Euclidean methods have also been developed for these models and lead to an improved treatment of the Yukawa$_2$ fields. The finite volume estimates were given in [Seiler, 1975], and the construction of the infinite volume limit (using uniform bounds and convergent subsequences) is in [McBryan, 1975a,b,c] and [Seiler and Simon, 1975a,b and 1976]. Convergence of the weak coupling cluster expansion ([Magnen and Sénéor, 1976b], Cooper and Rosen, 1977]) and Borel summability [Renouard, 1977, 79] has also been established. For $d = 3$, Magnen and Sénéor [1980] established existence and Borel summability.

In the low temperature (strong coupling) region the pseudoscalar theory has a phase transition. The proof is based on a cluster expansion for this region [Bałaban and Gawedzki, 1980]. The fermion expectation can be taken in the Yukawa model, for fixed boson paths, yielding an effective even potential $V(\phi)$. For $\lambda \gg 0$, the potential V has two minima and the $\phi \to -\phi$ symmetry is broken. The scalar Yukawa model is believed to have a phase transition at low temperatures also. In the region of convergence of the cluster expansions the Wightman axioms are verified. For arbitrary coupling, only the Haag–Kastler axioms have been verified; see [Schrader, 1972] and [McBryan and Park, 1975].

20.5 Low Temperature Expansions and Phase Transitions

The cluster expansion of Chapter 18 is analogous to high temperature expansions in statistical physics, valid far from critical points in the single phase region. The expansion basically reduces the study of an infinite volume theory to a finite volume. The expansion converges when the theory is close to Gaussian: for example, $\lambda\phi^4 + \phi^2$ models with $\lambda \approx 0$.

Such expansions can also be developed for multiphase quantum field models, e.g., for the two-dimensional $\lambda\phi^4 - \phi^2$ models, $\lambda \approx 0$. These are the low temperature expansions, valid in the region of phase transitions. One must take into account the probability for fluctuations about configurations near one ground state to make a transition to another ground state. In other words one must improve the estimates such as those in Section 16.2 to show that phase boundaries occur with small probability $\Pr(\Gamma) < \exp(-\lambda^{-1/2}|\Gamma|)$; the goal is to obtain an asymptotic expansion of the Schwinger functions with $\lambda \ll 1$. For $\lambda\phi^4 - \phi^2$ models this program has been carried out [Glimm, Jaffe, and Spencer, 1976a], and showed exponential decay of the pure phase truncated Schwinger functions. These methods also yield low temperature expansions in statistical mechanics [Schor, 1978b].

The multiphase analysis has been extended in quantum field theory by [Gawedzki, 1978a] to establish coexistence of three phases in equilibrium (for particular values of λ, σ) in the $\lambda\phi^6 + \sigma\phi^4$ model—as illustrated by the triple point of Figure 4.1.

The low temperature expansion methods have also been used to study the spectrum of multiphase ϕ^4 models ([Imbrie, 1980], and [Koch, 1981]).

The basic idea of the low temperature expansion can be illustrated with the polynomial $V(\phi) = \lambda\phi^4 - 2\phi^2$. It can be reparametrized by expanding about each of the two minima $\phi = \pm\lambda^{-1/2}$. Thus

$$V(\phi) = V_+(\psi_+) = \lambda\psi_+^4 + 4\lambda^{1/2}\psi_+^3 + 2\psi_+^2 - \lambda^{-1}$$
$$= V_-(\psi_-) = \lambda\psi_-^4 - 4\lambda^{1/2}\psi_-^3 + 2\psi_-^2 - \lambda^{-1}.$$

By imposing boundary conditions which ensure either $\psi_+ = 0$ at $|x| = \infty$, or else $\psi_- = 0$, we make an asymptotic expansion in powers of $\lambda^{1/2}$, starting from $V_+(\psi_+)$ or $V_-(\psi_-)$, and using ψ_+ or ψ_- as variable in place of ϕ.

This expansion is generated by integration by parts, as in (8.4.3). Then we perform a second expansion to show that the probability of starting with $\psi_+ \approx 0$ and making a transition to $\psi_- \approx 0$ is $O(\exp(-\lambda^{-1}))$, at least for small λ, as in the proof in Section 16.2 of phase transitions for the models. Here, however, we give explicit boundary conditions which yield two different field theories, the two "phases" of the model. These expansions are defined and are shown to be convergent in [Glimm, Jaffe, and Spencer, 1976a]. They have been extended to give a complete description of the phase diagram for a $P(\phi)_2$ model in the mean field region [Imbrie, 1980b, 1981].

20.6 Debye Screening and the Sine–Gordon Transformation

The statistical physics of a classical Coulomb gas reduces to the study of the grand canonical partition function Ξ_{Coulomb}, defined as follows. Let us introduce a lattice $Z^d(\delta, \Lambda)$ with fixed lattice spacing δ and finite volume Λ, in dimension d. We use the lattice in order to avoid short distance singularities in the Coulomb potential, which we choose to be $C(i, j) = -\Delta^{-1}(i, j)$, where Δ is the $Z^d(\delta, \Lambda)$ lattice Laplacian. Define the n particle canonical distribution for charges $q_k = \pm e$ located at i_k by

$$\mu_{\text{canonical},\, n} \equiv n!^{-1} \exp\left[-\frac{\beta}{2} \sum_{\substack{k \neq l \\ 1 \leq k,\, l \leq n}} q_k C(i_k,\, i_l) q_l \right]. \tag{20.6.1}$$

Then define

$$\Xi_{\text{Coulomb}} = \sum_{n=0}^{\infty} z^n \, \delta^{nd} \sum_{\substack{i_k \in \Lambda \\ q_k = \pm e \\ k = 1,\, 2,\, \dots,\, n}} \mu_{\text{canonical},\, n}, \tag{20.6.2}$$

where for convenience we let $i \in \Lambda$ denote $i \in Z^d(\delta, \Lambda) \cap \Lambda$. Note that for $d = 3$, $C(i, j) \sim (4\pi |i - j|)^{-1}$ as $|i - j| \to \infty$, which is the standard Coulomb potential. We assume even (e.g., Dirichlet) boundary conditions on $\partial\Lambda$.

The problem is to determine the long range behavior of the correlation between two test charges located at i and j as $|i - j| \to \infty$. This correlation is defined by

$$\langle q_i q_j \rangle = \lim_{\Lambda \uparrow Z^d} \Xi^{-1} \sum_{n,\, i_k,\, q_k} \delta^{nd} q_i q_j z^n \mu_{\text{canonical},\, n}. \tag{20.6.3a}$$

Note that by symmetry $\langle q_i \rangle = 0$. One might expect that $\langle q_i q_j \rangle$ would decay asymptotically as a multiple of the potential $C(i, j) = O(|i - j|^{-d+2})$. In the case $|z| \ll 1$, $\beta e^2 \ll 1$ (i.e., for a "high temperature dilute plasma"), it is known that $\langle q_i q_j \rangle$ does not behave in this way. Rather,

$$|\langle q_i q_j \rangle| \leq O\left(\exp\left(-\frac{|i - j|}{\xi} \right) \right), \qquad |i - j| \to \infty, \tag{20.6.3b}$$

where ξ is finite. The infimum of ξ satisfying an exponential bound of the form (20.6.3b) is by definition the Debye screening length ξ_D.

The picture of Debye is that the test charge q_i in the equilibrium distribution becomes surrounded by a cloud of opposite charges. These tend to neutralize q_i and prevent it from interacting with the second test charge q_j. The effect of the screening is to replace the long range Coulomb potential $C(i, j)$ by an exponentially small (short range) potential of the form $\exp(-|i - j|/\xi_D)$. The original justification of this picture involved mean field arguments of a formal nature applied to $\langle q_i q_j \rangle$.

Recently, Brydges and Federbush have used the techniques of constructive quantum field theory, in particular the cluster expansion of Chapter 18 as well as low temperature expansions, to establish the correctness of the picture for the expectation (20.6.3). Their investigation is based on the sine–Gordon transformation, which we now explain.

Consider the canonical partition function for a Euclidean field theory with interaction $V(\phi) = \gamma \cos \alpha\phi$. The classical equation of motion (in the real time continuum model) is

$$-\Box\phi = \alpha\gamma \sin \alpha\phi, \tag{20.6.4}$$

and for this reason it has been called the sine–Gordon equation. Let $d\phi$ denote the lattice Gaussian measure (for lattice $Z^d(\delta, \Lambda)$) with mean zero and covariance $C = -\Delta^{-1}$. The canonical sine–Gordon partition function is defined with appropriate constants α, γ by

$$Z_{\text{sine–Gordon}} = \int \exp\left[2z \sum_{j \in \Lambda} \delta^d : \cos \beta^{1/2}e\phi_j:\right] d\phi. \tag{20.6.5}$$

This is the partition function for a lattice field theory.

Proposition 20.6.1 [Stratonovich, 1957], [Edwards, 1959], [Edwards and Lenard, 1962]. *With Ξ defined by (20.6.2) and Z defined by (20.6.5),*

$$\Xi_{\text{Coulomb}} = Z_{\text{sine–Gordon}}. \tag{20.6.6}$$

PROOF. Let $q_k, j_k, k = 1, \ldots, n$ be specified charges and positions of n particles. Define

$$\phi(f) = \delta^d \sum_i \phi_i f_i$$

and

$$f_i^{(n)} = \beta^{1/2} \sum_{k=1}^{n} q_k \delta_{i, j_k}.$$

By the definition of Gaussian measures, it follows that

$$\int e^{i\phi(f^{(n)})} d\phi = \exp\left[-\frac{1}{2} \sum_{i, j} \delta^{2d}f_i^{(n)}C(i, j)f_j^{(n)}\right]$$

$$= \exp[-\tfrac{1}{2}n\beta e^2 \, \delta^{2d}C(0, 0)]\mu_{\text{canonical}, n}; \tag{20.6.7}$$

cf. (9.1.16). Next use the identity

$$\sum_{\substack{q_k = \pm e \\ k = 1, 2, \ldots, n}} \sum_{\substack{j_k \in \Lambda \\ k = 1, 2, \ldots, n}} \delta^{nd} e^{i\phi(f^{(n)})} = \left[\delta^d \sum_{j \in \Lambda} 2 \cos(\beta^{1/2} e\phi_j) \right]^n. \qquad (20.6.8)$$

By (20.6.7–8),

$$\sum_{\substack{j_k \in \Lambda \\ q_k = \pm e \\ k = 1, 2, \ldots, n}} \mu_{\text{canonical}, n} = \exp(\tfrac{1}{2} n \beta e^2 \, \delta^{2d}) \left[\delta^d \sum_j 2 \cos(\beta^{1/2} e\phi_j) \right]^n$$

$$= \; : \left[\delta^d \sum_j 2 \cos(\beta^{1/2} e\phi_j) \right]^n : . \qquad (20.6.9)$$

In deriving the last equality, we have used the identity for a Wick-ordered exponential, namely

$$e^{\pm i\alpha\phi_j} = e^{-(1/2)\alpha^2 C(0,\, 0)} \; : e^{\pm i\alpha\phi_j} :, \qquad (6.3.10)$$

which entails

$$[\cos(\alpha\phi_j)]^n = e^{-(1/2)\alpha^2 n C(0,\, 0)} \; : [\cos(\alpha\phi_j)]^n : .$$

Inserting (20.6.9) in the definition (20.6.2) yields the proposition.

The difficult work now goes into studying the lattice sine–Gordon field theory. On a formal level, the identity $\Xi_C = Z_{sG}$ exhibits the origin of the screening length. Assuming the cosine function in (20.6.5) can be expanded as a series in the small parameter $\beta^{1/2} e$, the quadratic term in ϕ, namely

$$- 2z e^2 \beta \left(\tfrac{1}{2} \, \delta^d \sum_j \phi_j^2 \right), \qquad (20.6.10)$$

can be incorporated as a mass perturbation of the measure $d\phi$, according to the relation (9.1.25). The remaining terms in the cosine give corrections to this mass. Thus we expect that the inverse screening length ξ_D^{-1} has an asymptotic expansion

$$m_D = \xi_D^{-1} \sim (2z\beta e^2)^{1/2} \left[1 + \sum_{n + m > 1} a_{nm} z^n (\beta e^2)^m \right], \qquad (20.6.11)$$

with the coefficients a_{nm} given by perturbation theory. The leading term $m_{\text{mean field}}^{-1} = \xi_{\text{mean field}} = (2z\beta e^2)^{-1/2}$ is the mean field theory value of the screening length. The basic result of [Brydges, 1978], [Brydges and Federbush, 1980, 1] is stated in terms of lattice (or continuum) observables $q(f) = \int q(x) f(x) \, dx, f \in C_0^\infty$, and $A = \prod_{i=1}^N q(f_i)$.

Theorem 20.6.2. *For β, z sufficiently small, the infinite volume state*

$$\langle A \rangle = \lim_{\Lambda \uparrow \infty} \langle A \rangle$$

exists and clusters exponentially. For any $\varepsilon > 0$, with β, z sufficiently small (depending on ε),

$$|\langle AB \rangle - \langle A \rangle\langle B \rangle| \leq C_A C_B e^{-md},$$

where d is the distance between the supports of A, B. Here $m = (1 - \varepsilon) m_{\text{mean field}} = (1 - \varepsilon)(2z\beta e^2)^{1/2}$.

20.7 Dipoles Don't Screen

We consider a gas of dipoles with Coulomb interaction. As in the previous section, we consider a d dimensional lattice gas, to avoid short distance singularities. A lattice dipole $\mathbf{D}$ is a pair of charges (q_i, q_j) on nearest neighbor lattice sites (i, j) with the same magnitude e, and with opposite signs for their charges. Let $\mathbf{D}$ denote a vector of length $2\delta e$, directed from the negative to the positive charge in the dipole pair; this is the dipole moment. Label $\mathbf{D} = \mathbf{D}_b$ by the bond b connecting (i, j).

The interaction energy of a pair of dipoles $\mathbf{D}_b$, $\mathbf{D}_{b'}$ on bonds b, b' has the form

$$\langle \mathbf{D}_b, V\mathbf{D}_{b'} \rangle = \sum_{\alpha, \beta = 1}^{d} D_{b, \alpha} V_{\alpha\beta}(b, b') D_{b', \beta}, \tag{20.7.1}$$

where $V_{\alpha\beta}(b, b')$ is a $d \times d$ matrix determined by the constituent charges of the dipoles interacting in pairs with a Coulomb interaction, but excluding interactions within each dipole. Asymptotically, for b, b' separated from one another by a large vector $\mathbf{r}_{bb'} = \mathbf{r}$ of length r,

$$V_{\alpha\beta}(b, b') \sim \begin{cases} (d-2)\Omega(d-1)^{-1}r^{-d}\left(\delta_{\alpha\beta} - (d-1)\dfrac{r_\alpha r_\beta}{r^2}\right) & \text{if } d > 2, \\[3mm] (2\pi)^{-1}r^{-2}\left(2\dfrac{r_\alpha r_\beta}{r^2} - \delta_{\alpha\beta}\right) & \text{if } d = 2. \end{cases} \tag{20.7.2}$$

The dipole potential is not absolutely integrable, but integrated over a sphere it averages to zero.

The grand canonical partition function is

$$\Xi_{\text{dipole}} = \sum_{n=0}^{\infty} z^n \, \delta^{nd} \sum_{\substack{b_k \in \Lambda \\ D_k = \pm \\ k = 1, 2, ..., n}} \mu_{\text{canonical}, n}, \tag{20.7.3}$$

where the sum over $D_k = \pm$ denotes a sum over the two directions of a dipole on a particular bond b_k, and where

$$\mu_{\text{canonical}, n} = (n!)^{-1} \exp\left[-\frac{\beta}{2} \sum_{\substack{k \neq l \\ 1 \leq k, l \leq n}} \langle \mathbf{D}_{b_k}, V\mathbf{D}_{b_l} \rangle\right]. \tag{20.7.4}$$

We define the dipole pair correlation by

$$\langle \mathbf{D}_b \mathbf{D}_{b'} \rangle = \lim_{\Lambda \uparrow \mathbb{Z}^d} \Xi_{\text{dipole}}^{-1} \sum_{n=0}^{\infty} z^n \, \delta^{nd} \sum_{\substack{b_k \\ D_k = \pm \\ k = 1, 2, ..., n}} \mathbf{D}_b \mathbf{D}_{b'} \mu_{\text{canonical}, n}. \tag{20.7.5}$$

Unlike the Coulomb gas, we do not expect the dilute dipole gas to screen. The argument of Section 20.6 can be repeated to show

$$\Xi_{\text{dipole}} = Z_{V = 2z \cos(\beta^{1/2} e |\nabla \phi|)}$$

$$= \int \exp\left[2z \sum_{j \in \Lambda} \delta^d : \cos(\beta^{1/2} e |\nabla \phi|_j): \right] d\phi. \qquad (20.7.6)$$

Here $(\nabla \phi)^2$ replaces ϕ^2 in (20.6.5). The representation (20.7.6) shows that for $|z| \ll 1$, $\beta e^2 \ll 1$, expanding the cosine gives a quadratic term

$$-2ze^2\beta\left(\tfrac{1}{2} \delta^d \sum_j (\nabla \phi)_j^2\right). \qquad (20.7.7)$$

This does not give a mass, but rather gives a coefficient $(1 + 2ze^2\beta)$ to the kinetic energy. Hence (20.7.7) leads to the mean field prediction that (20.7.5) has the following behavior for small z, $e^2\beta$:

$$\langle \mathbf{D}_b \mathbf{D}_{b'} \rangle \sim \varepsilon(z, e^2\beta)^{-1} \langle \mathbf{D}_b, V\mathbf{D}_{b'} \rangle + O(r^{-6}), \qquad (20.7.8)$$

where the dielectric constant ε has the form

$$\varepsilon = 1 + ze^2\beta + \sum_{n+m>1} a_{nm} z^n(\beta e^2)^m.$$

In the language of field theory renormalization, $\varepsilon = Z^{-1}$, where Z is a field strength renormalization. Because of the effective long range tail for the dipole forces, the expansion techniques are harder to apply than for the dilute Coulomb gas. Block spin methods have been used successfully to study the free energy of a dipole gas [Glimm and Jaffe, 1977b]. Park [1979] and Fröhlich and Spencer [1981a] have established the lack of screening in this model. Renormalization group methods have been applied to the related $(\nabla\phi)^4$ [Gawędzki and Kupiainen, 1980]; cf. also [Bricmont, Fontaine, Lebowitz, and Spencer, 1980, 1], [Bricmont, Fontaine, Lebowitz, Lieb and Spencer, 1981].

In spite of the lack of screening, one expects the dipole gas to undergo a phase transition in $d \geq 2$, from a dissociated dipole phase to a condensed phase; cf. Figure 20.2. The most difficult case to study is $d = 2$; since this transition is mathematically related to the roughening transition (Section 20.8) and the rotator transition (Section 5.5), it is of interest to prove its existence (cf. [Fröhlich and Spencer, 1981b]).

(a)   (b)

Figure 20.2. (a) Dissociated dipoles, (b) condensed dipole phase.

20.8 Solitons

In presenting and analyzing the mean field picture in the book, we have generally expanded about configurations $\phi = \text{const}$ which are absolute minima of the action a. There may also be other time-independent solutions to the classical field equations. A simple example occurs in the $d = 2$, ϕ^4 model with interaction $\lambda(\phi^2 - a)^2$. Then the time-independent solutions

$$\phi = \pm\sqrt{a}\,\tanh((2\lambda a)^{1/2}x_1) \tag{20.8.1}$$

are the soliton and antisoliton. This solution is real only for $a \geq 0$. In that case, the soliton is also believed to influence the particle spectrum. While soliton states are not believed to occur in the space $\mathscr{H}$ (since they connect two distinct vacuum states in distinct representations), pairs of solitons (which are approximations to classical solutions) are believed to give rise to a particle spectrum in the two phase region. The classical state (see Figure 20.3(b)) suggests the existence of a soliton–antisoliton bound state in the spectrum of the quantum theory.

Furthermore, one could imagine constructing a superselection sector containing the solition sector but not the vacuum state. This has been achieved for certain ϕ_2^4, $(\phi^2)_2^2$, and sine–Gordon$_2$ models ([Fröhlich, 1976b], [Bellisard, Fröhlich and Gidas, 1978], and [Gidas, 1979a]). We mention that the $d = 2$ continuum sine–Gordon theory and has been constructed [Fröhlich and Seiler, 1976].

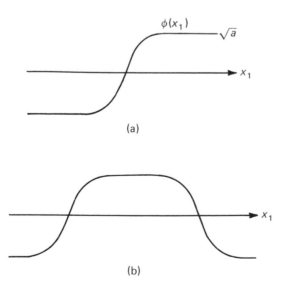

Figure 20.3. (a) Soliton classical solution. (b) Two soliton approximate solution for ϕ^4 model.

The roughening phase transition is an interesting and related problem. Consider a $d = 3$ Ising model with the infinite volume limit of the boundary conditions $+$ for $x_1 > 0$ and $-$ for $x_1 < 0$, as in Figure 20.4. It is known that for $T < (T_{cr})_{\mathrm{Ising}_2} < (T_{cr})_{\mathrm{Ising}_3}$, a sharp interface exists. Also, for $(T_{cr})_{\mathrm{Ising}_3} < T$ we have a translation-invariant state (no interface). It is of interest to understand whether the roughening (disappearance) of the interface occurs for $T < (T_{cr})_{\mathrm{Ising}_3}$.

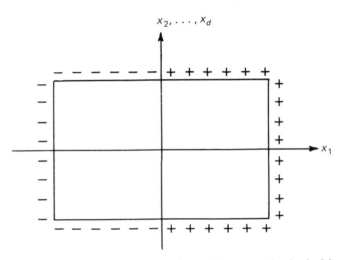

Figure 20.4. Boundary conditions which give rise to phase separation in the Ising₃ model, for $T < (T_{cr})_{\mathrm{Ising}_2}$.

20.9 Gauge Theories

The only gauge theory considered in this book has been electrodynamics. There are three aspects of gauge theories which have been pursued from a mathematical point of view: the classical and semiclassical approximations, the formulation of axioms, and the construction of the simplest continuum models as the lattice spacing $\delta \to 0$.

The action for a pure gauge theory is

$$\mathscr{A} = \tfrac{1}{4}\|F\|^2,$$

where

$$F_{ij} = \partial_i A_j - \partial_j A_i + [A_i, A_j]$$

takes its values in the Lie Algebra of the guage group G. The Euclidean classical Yang–Mills equations,

$$\sum D_\mu F_{\mu\nu} = 0 = \sum (\partial_\mu F_{\mu\nu} + [A_\mu, F_{\mu\nu}]),$$

have a finite action solution, the instanton, discovered by [Belavin,

Polyakov, Schwartz, and Tyupkin, 1976],

$$A_\mu = \frac{x^2}{x^2 + \lambda} g \, \partial_\mu g^{-1}, \tag{20.9.1}$$

where $g(x) = x_0 I + \mathbf{x} \cdot \boldsymbol{\sigma}$ and $\boldsymbol{\sigma}$ are the Pauli matrices. It has the property $F = *F$. In fact, all classical, finite action solutions satisfying $F = \pm *F$ have been constructed [Atiyah, Drinfeld, Hitchin, and Manin, 1978]. The instanton has been useful in the formal understanding of the semiclassical approximation to quantum gauge theories; cf. [Coleman, 1977, 1979]. Axioms for guage fields have been formulated by [Strocchi, 1978].

The mathematical construction of quantized gauge theories is now beginning. A lattice cutoff model [Wilson, 1974] is useful. The Lie algebra valued component A_i of the connection in coordinate direction i is replaced by the group element $\exp(\delta A_i) = \gamma_b$, assigned to the lattice bond b. Osterwalder and Seiler [1977] formulated and established the reflection positivity axiom for lattice models, with an action function

$$\mathscr{A} = \sum_p \text{tr}(\gamma_{b_1} \cdots \gamma_{b_4}) \tag{20.9.2}$$

with $b_1, \ldots, b_4$ the bonds bounding an elementary lattice square p. They used cluster expansions to analyze the infinite volume limit. For the gauge group $G = Z_2$, the Lee–Yang theorem applies [Dunlop, 1981]. Phase transitions for the Z_2 gauge theory have been analyzed by Balian Drouffe and Itzykson [1975]. See [Creutz, Jacobs, and Rebbi, 1979], [Drouffe, 1980] and [Creutz, 1980a] for Z_n theories. For nonabelian models, see [Migdal, 1976], [Kadanoff, 1977], ['t Hooft, 1978, 1980, 1981], [Glimm and Jaffe, 1979], [Creutz, 1980b], [Fröhlich, 1980–81], [Itzykson, 1980], [Mack, 1980], [Wilson, 1980], and [Seiler, 1981].

The continuum limit has been studied only in $d = 2$ dimensions. There a pure gauge theory is trivial, but the Higgs model with gauge group U(1) and action

$$\mathscr{A} = \tfrac{1}{4}\|F\|^2 + \tfrac{1}{2}\|D\phi\|^2 + \frac{\lambda}{8}\|(|\phi|^2 - 1)\|^2 \tag{20.9.3}$$

has been constructed by [Brydges, Fröhlich, and Seiler, 1979, 1980a,b]. They have established the Osterwalder–Schrader axioms for gauge-independent quantities, but so far no detailed spectral information has been established.

Because the ultraviolet behavior of gauge theories (especially the renormalizability for $d \leq 4$) depends crucially on the gauge invariance of the action, it is essential that ultraviolet cutoffs preserve gauge invariance. As an alternative to Wilson's lattice gauge covariant cutoff, Singer [1977] has advocated a mathematical construction based on gauge covariant continuum concepts, including a gauge covariant ζ function regularization. This point of view was used to analyze the Gribov ambiguity [Singer, 1978], which states broadly that the perturbative definition of the gauge field measure $d\mu$ of (6.6.5) may be incomplete on a formal level. Asorey and Mitter [1981] have begun the nonperturbative definition of $d\mu$ in this framework by establishing the existence of the regularized measure; their regularization uses powers of the covariant spatial Laplacian. See also [Narasimhan and Ramadas, 1979].

20.10 The Higgs Model and Superconductivity

In the 1950s, Ginzburg and Landau proposed the $d = 3$ action (20.9.3) to describe superconductors. In particular, the complex scalar field Φ is interpreted as a Schrödinger wave function for electron (Cooper) pairs moving in the magnetic field $\mathbf{F} = \text{curl } \mathbf{A}$. The field is also known as the "order parameter" of the Ginzburg–Landau theory. The coupling constant $\lambda < 1$ corresponds to type I superconductors, while $\lambda > 1$ corresponds to type II. For the former, the magnetic field is completely expelled from the superconductor (the Meissner effect). For the latter, the magnetic field partially penetrates the superconductor in integer multiples of a basic flux unit. In fact classical time-independent solutions to the equations exist, which are independent of one spatial coordinate direction. These solutions describe tubes of magnetic flux, which in the units of (20.9.3) satisfy $\int \mathbf{F} \cdot d\mathbf{x} = 2\pi N$ for integer N. A microscopic justification of the Ginzburg–Landau equations arises from the theory of Bardeen, Cooper, and Schrieffer (BCS); see [Huebener, 1979] and [Fetter and Walecka, 1971].

In the neighborhood of a flux tube, the order parameter Φ displays a vortex-like behavior (with the vortex center the axis of the flux tube). For this reason, the flux tubes are also known as "vortices". Then $\Phi \approx 0$ characterizes the normal region of the sample (nonzero magnetic flux) and $|\Phi| \approx 1$ the superconducting region (flux expulsion, Meissner effect). In fact, a triangular lattice of flux tubes in type II superconductors was predicted by Abrikosov and has been observed experimentally. Vortex-like solutions to the equations are possible because the action (20.9.3) requires $|\Phi| \to 1$ as $|\mathbf{x}| \to \infty$. This leads to the interpretation of the vortex number as the winding number of the function $\Phi/|\Phi|$ (considered as a map into S^1, away from the zeros of Φ). In fact smooth vortex solutions to the equations exist and are characterized classically by two length scales, a penetration depth for the magnetic field (inverse photon mass) and a correlation length for vortices (inverse Higgs particle mass); see [Jaffe and Taubes, 1980].

The lattice statistical mechanics of this model (or of a closely related model) has been studied; see [Isreal and Nappi, 1979a, b], [Guth, 1980], and [Fröhlich and Spencer, 1981b]. The quantum field theory of the action (20.9.3) for $d = 2$ (i.e. for the $d = 3$ problem when considered independent of one coordinate) is believed to have $\langle \Phi \rangle = 0$ (no symmetry breaking) but to exhibit the classical length scales (subject to quantum corrections). See [Callan, Dashen, and Gross, 1977] and [Coleman, 1977]. This partial "Higgs effect" is not yet understood at a mathematical level. For $d \geq 3$ field theories one expects mass generation to be accompanied by symmetry breaking, $\langle \Phi \rangle \neq 0$.

Bibliography

Abers, E. and Lee, B. W. (1973). Gauge theories, *Phys. Rept.* **9C**, 1–141.

Abraham, D. (1978). n-point functions for the rectangular Ising ferromagnet, *Comm. Math. Phys.* **60**, 205–213.

Agmon, S. (1965). *Lectures on Elliptic Boundary Value Problems*, Princeton: Van Nostrand.

Aguilar, J. and Combes, J. M. (1971). A class of analytic perturbations for one-body Schrödinger Hamiltonians, *Comm. Math. Phys.* **22**, 269–279.

Aizenman, M., Goldstein, S. and Lebowitz, J. L. (1978). Conditional equilibrium and the equivalence of microcanonical and grandcanonical ensembles in the thermodynamic limit, *Comm. Math. Phys.* **62**, 279–302.

Aizenman, M. and Simon, B. (1980). Local Ward identities and the decay of correlations in ferromagnets, *Comm. Math. Phys.* **77**, 137–144.

Albeverio, S., Gallavotti, G. and Høegh-Krohn, R. (1979a). Some results for exponential interaction in two or more dimensions, *Comm. Math. Phys.* **70**, 187–192.

Albeverio, S. and Høegh-Krohn, R. (1973). Uniqueness of the physical vacuum and the Wightman functions in the infinite volume limit for some nonpolynomial interactions, *Comm. Math. Phys.* **30**, 171–200.

Albeverio, S. and Høegh-Krohn, R. (1979). Uniqueness and the global Markov property for Euclidean fields. The case of trigonometric interactions, *Comm. Math. Phys.* **68**, 95–128.

de Alfaro, V., Fubini, S. and Furlan, G. (1976). A new classical solution of the Yang-Mills field equations, *Phys. Lett. B* **65**, 163–166.

Araki, H. (1960). Hamiltonian formalism and the canonical commutation relations in quantum field theory, *J. Math. Phys.* **1**, 492–504.

Araki, H. (1963). A lattice of Von Neumann algebras associated with the quantum theory of a free Bose field, *J. Math. Phys.* **4**, 1343–1362.

Araki, H. (1964a). On the algebra of all local observables, *Prog. Theor. Phys.* **32**, 844–854.

Araki, H. (1964b). The type of Von Neumann algebra associated with the free field, *Prog. Theor. Phys.* **32**, 956–965.

Araki, H. (1964c). Von Neumann algebras of local observables for free scalar field, *J. Math. Phys.* **5**, 1–13.

Araki, H. ed. (1975). *Mathematical Problems in Theoretical Physics*, (Kyoto, 1975), New York: Springer-Verlag.

Asano, T. (1970a). Lee-Yang theorem and the Griffiths inequality for the anisotropic Heisenberg ferromagnet, *Phys. Rev. Lett.* **24**, 1409–1411.

Asano, T. (1970b). Theorems on the partition functions of the Heisenberg ferromagnets, *J. Phys. Soc. Jap.* **29**, 350–359.

Ashcroft, N. and Mermin, D. (1976). *Solid State Physics*, New York: Holt, Rinehart and Winston.

Asorey, M. and Mitter, P. K. (1981). Regularized, continuum Yang-Mills process and Feynman-Kac functional integral, *Comm. Math. Phys.*, to appear.

Atiyah, M. F., Drinfeld, V. G., Hitchin, N. J. and Manin, Yu. I. (1978). Construction of instantons, *Phys. Lett.* **65A**, 185–187.

Atiyah, M. F., Hitchin, N. J. and Singer, I. M. (1978). Self-duality in four-dimensional Riemannian geometry, *Proc. Roy. Soc. London Series A* **362**, 421–457.

Baker, G. (1975). Self-interacting boson quantum field theory and the thermodynamic limit in d dimensions, *J. Math. Phys.* **16**, 1324–1346.

Baker, G. (1977). Analysis of hyperscaling in the Ising model by the high-temperature series method, *Phys. Rev.* **B15**, 1553–1559.

Baker, G. and Kincaid, J. M. (1981). The continuous-spin Ising model, $g_0:\phi^4:_d$ field theory, and the renormalization group, *J. Stat. Phys.* **24**, 469–528.

Balaban, T. and Gawedzki, K. (1981). A low temperature expansion for the pseudo-scalar Yukawa model of quantum fields in two space-time dimensions, *Ann. l'Inst. Henri Poincaré*, to appear.

Balian, R., Drouffe, J. and Itzykson, C. (1975). Gauge fields on a lattice I, II, III. *Phys. Rev. D* **10**, 3376–3395; *D* **11**, 2008–2013; *D* **11**, 2104–2119.

Balslev, E. and Combes, J. M. (1971). Spectral properties of many-body Schrödinger operators with dilation analytic interactions, *Comm. Math. Phys.* **22**, 280–294.

Battle, G. A. and Rosen, L. (1980). The FKG inequality for the Yukawa₂ quantum field theory, *J. Stat. Phys.* **22**, 123–192.

Baumel, R. T. (1979). On spontaneously broken symmetry in the $P(\phi)_2$ model quantum field theory, *Princeton University Thesis*.

Baym, G. (1969). *Lectures on Quantum Mechanics*, New York: Benjamin.

Becchi, C., Rouet, A. and Stora, R. (1975). Renormalization of the abelian Higgs–Kibble model, *Comm. Math. Phys.* **42**, 127–162.

Becchi, C., Rouet, A. and Stora, R. (1976). Renormalization of guage theories, *Ann. Phys.* **98**, 287–321.

Belavin, A. A., Polyakov, A. M., Schwartz, A. S. and Tyupkin, Yu. S. (1975). Pseudoparticle solutions of the Yang-Mills equations, *Phys. Lett.* **B59**, 85–87.

Béllisard, J., Fröhlich, J. and Gidas, B. (1978). Soliton mass and surface tension in the $(\lambda \phi^4)_2$ quantum field theory, *Comm. Math. Phys.* **60**, 37–72.

Bender, C., Cooper, F., Guralnik, G., Roskies, R. and Sharp, D. (1981). Numerical computation of the renormalized effective potential in the strong coupling limit. *Phys. Rev. D*, to appear.

Bender, C. M. and Wu, T. T. (1969). Anharmonic oscillator, *Phys. Rev.* **184**, 1231–1260.

Benfatto, G., Cassandro, M., Gallavotti, G., Nicolò, F., Oliveri, E., Presutti, E. and Scacciatelli, E. (1978). Some probabilistic techniques in field theory, *Comm. Math. Phys.* **59**, 143–166.

Benfatto, G., Cassandro, M., Gallavotti, G., Nicolò, F., Olivieri, E., Presutti, E. and Scacciatelli, E. (1980). On the ultraviolet stability in the Euclidean scalar field theories, *Comm. Math. Phys.* **71**, 95–130.

Benfatto, G., Gallavotti, G. and Nicolò, F. (1980). Elliptic equations and Gaussian processes, *J. Funct. Anal.* **36**, 343–400.

Berezin, F. A. (1966). *The Method of Second Quantization*, New York: Academic Press.

Berezinskii, V. L. (1971). Destruction of long-range order in one-dimensional and two-dimensional systems having a continuous symmetry group I. Classical systems, *Soviet Phys. JETP* **32**, 493–500.

Bers, L., John, F. and Schechter, M. (1964). *Partial Differential Equations*, New York: Interscience Publishers.

Bisognano, J. and Wichmann, E. (1975). On the duality condition for a Hermitean scalar field, *J. Math. Phys.* **16**, 985–1007.

Bjorken, J. and Drell, S. (1964). *Relativistic quantum mechanics*, New York: McGraw-Hill.

Bjorken, J. and Drell, S. (1965). *Relativistic quantum fields*, New York: McGraw-Hill.

Bleher, P. M. and Sinai, Ya. G. (1973). Investigation of the critical point in models of the type of Dyson's hierarchical models, *Comm. Math. Phys.* **33**, 23–42.

Bleher, P. M. and Sinai, Ya. G. (1975). Critical indices for Dyson's asymptotically hierarchical models, *Comm. Math. Phys.* **45**, 247–278.

Bogoliubov, N. N., Logunov, A. A. and Todorov, R. T. (1975). *Introduction to Axiomatic Quantum Field Theory*, Reading: Benjamin, Translation and revision of original publication in 1969.

Bogoliubov, N. N. and Parasiuk, O. S. (1957). Über die Multiplikation der Kausalfunktionen in der Quantentheorie der Felder, *Acta Math.* **97**, 227–266.

Bogoliubov, N. N. and Shirkov, D. V. (1959). *Introduction to the Theory of Quantized Fields*, New York: Interscience.

Borchers, H. J. (1962). On structure of the algebra of field operators, *Nuovo Cimento* **24**, 214–236.

Borchers, H. J. (1966). Energy and momentum as observables in quantum field theory, *Comm. Math. Phys.* **2**, 49–54.

Borchers, H. J. and Yngvason, J. (1976). Necessary and sufficient conditions for integral representations of Wightman functionals at Schwinger points, *Comm. Math. Phys.* **47**, 197–213.

Bratteli, O. (1972). Conservation of estimates in quantum field theory, *Comm. Pure and Appl. Math.* **25**, 759–779.

Bratteli, O. and Robinson, D. W. (1981). *Operator algebras and quantum statistical mechanics II. Equilibrium States. Models in Quantum Statistical Mechanics*, New York: Springer Verlag.

Bricmont, J. (1976). Correlation inequalities for two-component fields, *Ann. Soc. Sr. Brussels* **90**, 245–252.

Bricmont, J., Fontaine, J. R. and Landau, L. J. (1977). On the uniqueness of the equilibrium state for plane rotators, *Comm. Math. Phys.* **56**, 281–296.

Bricmont, J., Fontaine, J. R. and Landau, L. J. (1979). Absence of symmetry breakdown and uniqueness of the vacuum for multicomponent field theories, *Comm. Math. Phys.* **64**, 49–72.

Bricmont, J., Fontaine, J. R., Lebowitz, J. L. and Spencer, T. (1980–1). Lattice systems with a continuous symmetry. I, II, *Comm. Math. Phys.* **78**, 281–302; 363–372.

Bricmont, J., Fontaine, J. R., Lebowitz, J. L., Lieb, E. and Spencer, T. (1981). Lattice systems with a continuous symmetry III. *Comm. Math. Phys.* **78**, 545–566.

Bricmont, J. Lebowitz, J. L. and Pfister, C. (1979). On the equivalence of boundary conditions, *J. Stat. Phys.* **21**, 573–582.

Bricmont, J., Lebowitz, J. L. and Pfister, C. E. (1980). Low temperature expansion for continuous spin-Ising models, *Comm. Math. Phys.* **78**, 117–136.

Bros, J. (1970). Some analyticity properties implied by the two particle structure of Green's functions in general quantum field theory *In: Analytic Methods in Mathematical Physics*, R. Gilbert and R. Newton, eds., New York: Gordon and Breach.

Bros, J., Epstein, H. and Glaser, V. (1967). On the connection between analyticity and Lorentz covariance of Wightman functions, *Comm. Math. Phys.* **6**, 77–100.

Bros, J. and Iagolnitzer, D. (1973). Causality and local analyticity: mathematical study, *Ann. l'Inst. Henri Poincaré* **18**, 147–184.

Bros, J. and LaSalle, M. (1977). Analyticity properties and many body structure in general quantum field theory. III, *Comm. Math. Phys.* **54**, 33–62.

Brydges, D. (1975). Boundedness below for fermion model theories. I, *J. Math. Phys.* **16**, 1649–1661.

Brydges, D. (1976a). Cluster expansions for fermion fields by the time-dependent Hamiltonian approach, *J. Math. Phys.* **17**, 1118–1124.

Brydges, D. C. (1976b). Boundedness below for fermion model theories. Part II. The linear lower bound, *Comm. Math. Phys.* **47**, 1–24.

Brydges, D. (1978). A rigorous approach to Debye screening in dilute classical Coulomb systems, *Comm. Math. Phys.* **58**, 313–350.

Brydges, D. and Federbush, P. (1974). A semi-Euclidean approach to boson-fermion model theories, *J. Math. Phys.* **15**, 730–732.

Brydges, D. and Federbush, P. (1976). The cluster expansion in statistical mechanics, *Comm. Math. Phys.* **49**, 233–246.

Brydges, D. and Federbush, P. (1977). The cluster expansion for potentials with exponential fall-off, *Comm. Math. Phys.* **53**, 19–30.

Brydges, D. and Federbush, P. (1978a). A lower bound for the mass of a random Gaussian lattice. *Comm. Math. Phys.* **62**, 79–82.

Brydges, D. and Federbush, P. (1978b). A new form of the Mayer expansion in classical statistical mechanics, *J. Math. Phys.* **19**, 2064–2067.

Brydges, D. and Federbush, P. (1980). Debye screening, *Comm. Math. Phys.* **73**, 197–246.

Brydges, D. and Federbush, P. (1981). Debye screening in classical Coulomb systems, In: *1980 Erice summer school lectures*, G. Velo and A. Wightman, eds. To appear.

Brydges, D., Fröhlich, J. and Seiler, E. (1979). On the construction of quantized gauge fields, I. General results, *Ann. Phys.* **121**, 227–284.

Brydges, D., Fröhlich, J. and Seiler, E. (1980). Construction of quantized guage fields. II. Convergence of the lattice approximation, *Comm. Math. Phys.* **71**, 159–205.

Brydges, D., Fröhlich, J. and Seiler, E. (1981). On the construction of quantized gauge theories III, *Comm. Math. Phys.*, **79**, 353–399.

Buchholz, D. and Fredenhagen, K. (1980). Clustering, charge-screening and the mass-spectrum in local quantum field theory *In: Mathematical Problems in Theoretical Physics* K. Osterwalder, ed., Berlin: Springer-Verlag.

Buchvostov, A. P. and Lipatov, L. N. (1977). High orders of the perturbation theory in scalar electrodynamics, *Phys. Lett. B* **70**, 48–50.

Burnap, C. (1976). *The particle structure of Boson quantum field theory models*, Harvard University Thesis.

Burnap, C. (1977). Isolated one particle states in boson quantum field theory models, *Ann. Phys.* **104**, 184–196.

Caginalp, G. (1980a). The ϕ^4 lattice field theory as an asymptotic expansion about the Ising limit, *Ann. Phys.* **124**, 189–207.

Caginalp, G. (1980b). Thermodynamic properties of the ϕ^4 lattice field theory near the Ising limit, *Ann. Phys.*, **126**, 500–511.

Callen, C., Dashen, R. and Gross, D. (1977). A mechanism for quark confinement, *Phys. Lett.*, *B* **66**, 375–381.

Camp, W. J., Saul, D. M., Van Dyke, J. P. and Wortis, M. (1976). Series analysis of corrections to scaling for the spin-pair correlations of the spin-s Ising model: Confluent singularities, universality, and hyperscaling, *Phys. Rev.* **B14**, 3990–4001.

Cannon, J. (1974). Continuous sample paths in quantum field theory, *Comm. Math. Phys.* **35**, 215–234.

Cannon, J. and Jaffe, A. (1970). Lorentz covariance of the $\lambda(\phi^4)_2$ quantum field theory, *Comm. Math. Phys.* **17**, 261–321.

Cartier, P. (1974). Unpubished.

Chiu, S. T. and Weeks, J. D. (1978). Dynamics of the roughening transition, *Phys. Rev. Lett.* **40**, 733–736.

Chretien, M. and Deser, S., eds. (1966). *Particle Symmetries and Axiomatic Field Theory*, (Brandeis, 1965), New York: Gordon and Breach.

Coleman, S. (1972). Scaling anomalies *In: Developments in High Energy Physics* R. Gatto, ed., New York: Academic Press.

Coleman, S. (1973a). There are no Goldstone bosons in two dimensions, *Comm. Math. Phys.* **31**, 259–264.

Coleman, S. (1973b). Dilations *In: Properties of the Fundamental Interactions* A. Zichichi, ed., Bologna: Bologna Press.

Coleman, S. (1975a). Quantum Sine-Gordon equation as the massive Thirring model, *Phys. Rev. D* **11**, 2088–2097.

Coleman, S. (1975b). Secret symmetry: an introduction to spontaneous symmetry breakdown and gauge fields *In: Laws of Hadronic Matter* A. Zichichi, ed., New York: Academic Press.

Coleman, S. (1977). Classical lumps and their quantum descendants *In: New Phenomena in Subnuclear Physics*, A. Zichichi, ed., New York: Plenum Press.

Coleman, S. (1979). The uses of instantons *In: The Whys of Subnuclear Physics* A. Zichichi, ed., New York: Plenum Press.

Collett, P. and Eckmann, J.–P. (1978). *A Renormalization Group Analysis of the Hierarchical Model in Statistical Mechanics*, New York: Springer-Verlag.

Combes, J. M., Schrader, R. and Seiler, R. (1978). Classical bounds and limits for energy distributions of Hamiltonian operators in electromagnetic fields, *Ann. Phys.* **111**, 1–18.

Combescure, M. and Dunlop, F. (1979). N-particle-irreducible functions in Euclidean quantum field theory, *Ann. Phys.* **122**, 102–150.

Constantinescu, F. (1977). Nontriviality of the scattering matrix for weakly coupled ϕ_3^4 models, *Ann. Phys.* **108**, 37–48.

Constantinescu, F. (1980). *Expansion of the double-well model near the Ising model*, Institüt für angewandte Mathematik Johann Wolfgang Goethe Universität Preprint.

Constantinescu, F. and Stroter, B. (1980). The Ising limit of the double-well model, *J. Math. Phys.* **21**, 881–890.

Cooper, A. and Rosen, L. (1977). The weakly coupled Yukawa$_2$ field theory: Cluster expansion and Wightman axioms, *Trans. Amer. Math. Soc.* **234**, 1–88.

Creutz, M. (1980a). Phase diagram for coupled spin gauge theories, *Phys. Rev.* **D21**, 1006–1012.

Creutz, M. (1980b). Monte Carlo study of quantized SU(2) gauge theory, *Phys. Rev. D* **21**, 2308–2315.

Creutz, M., Jacobs, L. and Rebbi, C. (1979). Monte Carlo study of abelian lattice gauge theories, *Phys. Rev. D* **20**, 1915–1922.

Dashen, R., Hasslacher, B. and Neveu, A. (1974). Nonperturbative methods and extended hadron models in field theory, I, II, III, *Phys. Rev.* **D10**, 4114–4141.

Dashen, R., Hasslacher, B. and Neveu, A. (1975). Particle spectrum in model field theories from semiclassical functional integral techniques, *Phys. Rev.* **D11**, 3424–3450.

De Angelis, G. F. and de Falco, D. (1977). Correlation inequalities for lattice gauge fields, *Lett. Nuovo Cimento* **18**, 536–538.

De Angelis, G. E., de Falco, D. and Guerra, F. (1977a). Scalar quantum electrodynamics on the lattice as classical statistical mechanics, *Comm. Math. Phys.* **57**, 201–212.

De Angelis, G. E., de Falco, D. and Guerra, F. (1977b). Lattice gauge models in the strong coupling regime, *Lett. Nuovo Cimento* **19**, 55–58.

DeDominicis, C. and Martin, P. (1964). Stationary entropy principle and renormalization in normal and superfluid systems I, II. *J. Math. Phys.* **5**, 14–30, 31–59.

Dell'Antonio, G., Doplicher, S. and Jona-Lasinio, G. (1978). *Mathematical problems in theoretical physics*, (Rome, 1977), New York: Springer-Verlag.

DeWitt, C. and Stora, R., eds. (1971). *Statistical Mechanics and Quantum Field Theory*, (Les Houches, 1970), New York: Gordon and Breach.

Dimock, J. (1972a). Estimates, renormalized currents and field equations for the Yukawa field theory, *Ann. Phys.* **72**, 177–242.

Dimock, J. (1972b). Spectrum of local Hamiltonians in the Yukuwa field theory, *J. Math. Phys.* **13**, 477–481.

Dimock, J. (1974). Asymptotic perturbation expansion in the $P(\phi)_2$ quantum field theory, *Comm. Math. Phys.* **35**, 347–356.

Dimock, J. (1976). The $P(\phi)_2$ Green's functions: asymptotic perturbation expansion, *Helv. Phys. Acta* **49**, 199–216.

Dimock, J. (1977). The non-relativistic limit of $P(\phi)_2$ quantum field theories: Two particle phenomena, *Comm. Math. Phys.* **57**, 51–66.

Dimock, J. (1980). Algebras of local observables on a manifold, *Comm. Math. Phys.* **77**, 219–228.

Dimock, J. and Eckmann, J.-P. (1976). On the bound state in weakly coupled $\lambda(\phi^6 - \phi^4)_2$, *Comm. Math. Phys.* **51**, 41–54.

Dimock, J. and Eckmann, J.-P. (1977). Spectral properties and bound state scattering for weakly coupled $P(\phi)_2$ models, *Ann. Phys.* **103**, 289–314.

Dimock, J. and Glimm, J. (1974). Measures on Schwarz distribution space and applications to $P(\phi)_2$ field theories, *Adv. Math.* **12**, 58–83.

Dixmier, J. (1957). *Les Algèbres d'Opérateurs dans l'Espace Hilbertien (Algèbres de von Neumann)*, Paris: Gauthiers-Villars.

Dobrushin, R. L. (1965). Existence of a phase transition in the two-dimensional and three-dimensional Ising models, *Soviet Phys. Doklady* **10**, 111–113.

Dobrushin, R. L. (1979). Talk presented at the Conference on Random Fields, Esztergom, Hungary.

Dobrushin, R. L. and Minlos, R. (1973). Construction of one dimensional quantum field via a continuous Markov field, *Funct. Anal. Appl.* **7**, 324–325.

Dobrushin, R. L. and Shlosman, S. B. (1975). Absence of breakdown of continuous symmetry in two dimensional models of statistical physics, *Comm. Math. Phys.* **42**, 31–40.

Domb, C. and Green, M. (1972–). *Phase Transitions and Critical Phenomena*, Vol. 1–6, New York: Academic Press.

Donald, M. (1981). The classical field limit of $P(\phi)_2$ quantum field theory, *Comm. Math. Phys.*, to appear.

Doplicher, S., Haag, R. and Roberts, J. (1969). Fields, observables, and gauge transformations. I, II, *Comm. Math. Phys.* **13**, 1–23; **15**, 173–200.

Doplicher, S., Haag, R. and Roberts, J. (1971). Local observables and particle statistics. I, *Comm. Math. Phys.* **23**, 199–230.

Doplicher, S., Haag, R. and Roberts, J. (1974). Local observables and particle statistics. II, *Comm. Math. Phys.* **35**, 49–85.

Doplicher, S., Kadison, R. V., Kastler, D. and Robinson, D. W. (1967). Asymptotically abelian systems, *Comm. Math. Phys.* **6**, 101–120.

Drechsler, W. and Mayer, M. E. (1977). *Fibre Bundle Techniques in Gauge Theories: Lectures in Mathematical Physics at the University of Texas at Austin*, New York: Springer-Verlag.

Driessler, W. (1977). On the type of local algebras in quantum field theory, *Comm. Math. Phys.* **53**, 295–297.

Driessler, W. (1979). Duality and absence of locally generated superselection sectors for CCR-type algebras, *Comm. Math. Phys.* **70**, 213–220.

Driessler, W. and Fröhlich, J. (1977). The reconstruction of local observable algebras from the Euclidean Green's functions of relativistic quantum field theory, *Ann. l'Inst. Henri Poincarè* **27**, 221–236.

Drinfeld, V. G. and Manin, Yu. I. (1978). A description of instantons, *Comm. Math. Phys.* **63**, 177–192.

Drouffe, J. M. (1980). Series analysis in four-dimensional Z_n lattice guage systems, *Nucl. Phys.* **B 170[FS1]**, 91–97.

Duneau, M., Iagolnitzer, D. and Souillard, B. (1973). Properties of truncated correlation functions and analyticity properties for classical lattices and continuous systems, *Comm. Math. Phys.* **31**, 191–208.

Duneau, M., Iagolnitzer, D. and Souillard, B. (1974). Strong cluster properties for classical systems with finite range interaction, *Comm. Math. Phys.* **35**, 307–320.

Duneau, M., Iagolnitzer, D. and Souillard, B. (1975). Decay of correlations for infinite-range interactions, *J. Math. Phys.* **16**, 1662–1666.

Dunlop, F. (1976). Correlation inequalities for multicomponent rotators, *Comm. Math. Phys.* **49**, 247–256.

Dunlop, F. (1977). Zeros of partition functions via correlation inequalities, *J. Stat. Phys.* **17**, 215–228.

Dunlop, F. (1979a). Analyticity of the pressure for Heisenberg and plane rotator models, *Comm. Math. Phys.* **69**, 81–88.

Dunlop, F. (1979b). Zeros of the partition function and Gaussian inequalities for the plane rotator model, *J. Stat. Phys.* **21**, 561–572.

Dunlop, F. (1981). Zeros of the partition function for some generalized Ising models, *To be published in: Rigorous Results in Statistical Mechanics and Quantum Field Theory*. J. Fritz and D. Szasz, eds.

Dunlop, F. and Newman, C. (1975). Multicomponent field theories and classical rotators, *Comm. Math. Phys.* **44**, 223–235.

Durhuus, B. and Fröhlich, J. (1980). A connection between v-dimensional Yang-Mills theory and $(v - 1)$-dimensional, non-linear σ-models, *Comm. Math. Phys.* **75**, 103–153.

Dyson, F. (1969a). Existence of a phase transition in a one-dimensional Ising ferromagnet, *Comm. Math. Phys.* **12**, 91–107.

Dyson, F. J. (1969b). Nonexistence of spontaneous magnetization in a one-dimensional Ising ferromagnet, *Comm. Math. Phys.* **12**, 212–215.

Dyson, F. J. (1971). An Ising ferromagnet with discontinuous long-range order, *Comm. Math. Phys.* **21**, 269–283.

Dyson, F. and Lenard, A. (1967–8). Stability of matter I, II, *J. Math. Phys.* **8**, 423–434; **9**, 698–711.

Dyson, F., Lieb, E. H. and Simon, B. (1978). Phase transitions in quantum spin systems with isotropic and nonisotropic interactions, *J. Stat. Phys.* **18**, 335–383.

Eckmann, J.-P. (1970). A model with persistent vacuum, *Comm. Math. Phys.* **18**, 247–264.

Eckmann, J.-P. (1972). Representation of the CCR in the $(\phi^4)_3$ model: independence of space cut-off, *Comm. Math. Phys.* **25**, 1–61.

Eckmann, J.-P. (1977). Remarks on the classical limit of quantum field theories, *Lett. Math. Phys.* **1**, 387–394.

Eckmann, J.-P. and Epstein, H. (1979a). Time-ordered products and Schwinger functions, *Comm. Math. Phys.* **64**, 95–130.

Eckmann, J.-P. and Epstein, H. (1979b). Borel summability of the mass and the S-matrix in ϕ^4 models, *Comm. Math. Phys.* **68**, 245–258.

Eckmann, J.-P., Epstein, H. and Fröhlich, J. (1976). Asymptotic perturbation expansion for the S-matrix and the definition of time ordered functions in relativistic quantum field models, *Ann. l'Inst. Henri Poincaré* **25**, 1–34.

Eckmann, J.-P., Magnen, J. and Sénéor, R. (1975). Decay properties and Borel summability for the Schwinger functions in $P(\phi)_2$ theories, *Comm. Math. Phys.* **39**, 251–271.

Eckmann, J.-P. and Osterwalder, K. (1971). On the uniqueness of the Hamiltonian and of the representation of the CCR for the quartic boson interaction in three dimensions, *Helv. Phys. Acta* **44**, 884–909.

Eckmann, J.-P. and Osterwalder, K. (1973). On application of Tomita's theory of modular Hilbert algebras: Duality for free bose fields, *J. Funct. Anal.* **13**, 1–12.

Edwards, S. F. (1959). The statistical thermodynamics of a gas with long and short-range forces, *Phil. Mag.* **4**, 1171–1182.

Edwards, S. F. and Lenard, A. (1962). Exact statistical mechanics of a one-dimensional system with Coulomb forces. II. The method of functional integration, *J. Math. Phys.* **3**, 778–792.

Ellis, R. S., Monroe, J. L. and Newman, C. (1976). The GHS and other correlation inequalities for a class of even ferromagnets, *Comm. Math. Phys.* **46**, 167–182.

Ellis, R. S. and Newman, C. (1976). Quantum mechanical soft springs and reverse correlation inequalities, *J. Math. Phys.* **17**, 1682–1683.

Ellis, R. S. and Newman, C. (1978). Necessary and sufficient conditions for the GHS inequality with applications to analysis and probability, *Trans. Amer. Math. Soc.* **237**, 83–99.

Enss, V. (1978). Asymptotic completeness for quantum mechanical potential scattering, *Comm. Math. Phys.* **61**, 258–291.

Epstein, H. (1966). Some analytic properties of scattering amplitudes *In: Axiomatic Field Theory* M. Chretien and S. Deser, eds., New York: Gordon and Breach.

Epstein, H. and Glaser, V. (1972). Renormalization of non polynomial Lagrangians in Jaffe's class, *Comm. Math. Phys.* **27**, 181–194.

Epstein, H. and Glaser, V. (1973). The role of locality in perturbation theory, *Ann. l'Inst. Henri Poincaré* **19**, 211–295.

Epstein, H. and Glaser, V. (1976). Adiabatic limit in perturbation theory *In: Renormalization Theory* G. Velo, A. S. Wightman, eds., Dordrecht: D. Reidel.

Epstein, H., Glaser, V. and Jaffe, A. (1965). Nonpositivity of the energy density in quantized field theories, *Nuovo Cimento*, **36**, 1016–1022.

Erdélyi, A., Magnus, W., Oberhettinger, F. and Tricomi, F. G. (1953). *Higher Transcendal Functions*, Vol. I, II, III, New York: McGraw-Hill.

Ezawa, H. and Swieca, J. A. (1967). Spontaneous breakdown of symmetries and Zero-mass states, *Comm. Math. Phys.* **5**, 330–336.

Fabrey, J. (1970). Exponential representations of the canonical commutation relations, *Comm. Math. Phys.* **19**, 1–30.

Faddeev, L. D. and Slavnov, A. A. (1980). *Gauge fields: introduction to quantum Scattering Theory*, Moscow: Works of the Steklov Mathematical Institute, vol. 69.

Faddeev, L. D. and Slavnov, A. A. (1980). *Gauge fields: introduction to quantum theory*, Reading, Mass.: Benjamin/Cummings Publishing Co.

Faris, W. and Lavine, R. (1974). Commutators and self-adjointness of Hamiltonian operators, *Comm. Math. Phys.* **35**, 39–48.

Federbush, P. (1969). A partially alternative derivation of a result of Nelson, *J. Math. Phys.* **10**, 50–52.

Federbush, P. (1971a). Renormalization of some one-space dimensional quantum field theories by unitary transformation, *Ann. Phys.* **68**, 94–97.

Federbush, P. (1971b). Unitary renormalization of $[\phi^4]_{2+1}$, *Comm. Math. Phys.* **21**, 261–268.

Federbush, P. (1973). Positivity for some generalized Yukawa models in one space dimension, *J. Math. Phys.* **14**, 1532–1542.

Federbush, P. (1975). A new approach to the stability of matter problem. I, II, *J. Math. Phys.* **16**, 347–351; **16**, 706–709.

Federbush, P. (1976). The semi-Euclidean approach in statistical mechanics. I, II, *J. Math. Phys.* **17**, 200–203; 204–207.

Federbush, P. (1981). A zero mass cluster expansion, *Comm. Math. Phys.*, to appear.

Federbush, P. and Gidas, B. (1971). Renormalization of the one-space dimensional Yukawa$_2$ model by unitary transformation, *Ann. Phys.* **68**, 98–101.

Feldman, J. (1973). A relativistic Feynman-Kac formula, *Nuclear Physics B* **52**, 608–614.

Feldman, J. (1974a). On the absence of bound states in the $\lambda\phi_2^4$ quantum field model without symmetry breaking, *Canad. Jour. Phys.* **52**, 1583–1587.

Feldman, J. (1974b). The $\lambda\phi_3^4$ field theory in a finite volume, *Comm. Math. Phys.* **37**, 93–120.

Feldman, J. and Osterwalder, K. (1976). The Wightman axioms and the mass gap for weakly coupled $(\phi^4)_3$ quantum field theories, *Ann. of Phys.* **97**, 80–135.

Feldman, J. and Raczka, R. (1977). The relativistic field equations of the $\lambda\phi_3^4$ quantum field theory, *Ann. Phys.* **10**, 212–229.

Fetter, A. and Walecka, J. (1971). *Quantum Theory of Many-Particle Systems*, New York: McGraw-Hill.

Feynman, R. P. and Hibbs, A. P. (1965). *Quantum Mechanics and Path Integrals*, New York: McGraw-Hill.

Fisher, M. (1969). Rigorous inequalities for critical point correlation exponents, *Phys. Rev.* **180**, 594–600.

Fortuin, C., Kastelyn, P. and Ginibre, J. (1971). Correlation inequalities on some partially ordered sets, *Comm. Math. Phys.* **22**, 89–103.

Fredenhagen, K. (1981). On the existence of antiparticles. *Comm. Math. Phys.* **79**, 141–151.

Friedman, H. (1962). *Ionic Solution Theory Based on Cluster Expansion Methods*, New York: Interscience.

Friedrichs, K. (1965). *Perturbation of Spectra in Hilbert Space*, Providence: American Mathematical Society.

Fröhlich, J. (1973). On the infrared problem in a model of scalar electrons and massless, scalar bosons, *Ann. l'Inst. Henri Poincaré* **A19**, 100–103.

Fröhlich, J. (1974a). Verification of axioms for Euclidean and relativistic fields and Haag's theorem in a class of $P(\phi)_2$ models, *Ann. l'Inst. H. Poincaré* **21**, 271–317.

Fröhlich, J. (1974b). Schwinger functions and their generating functionals, I. *Helv. Phys. Acta* **47**, 265–306.

Fröhlich, J. (1975a). The quantized "Sine-Gordon" equation with a non-vanishing mass term in two space-time dimensions, *Phys. Rev. Lett.* **34**, 833–836.

Fröhlich, J. (1975b). The reconstruction of quantum fields from Euclidean Green's functions at arbitrary temperatures in models of a self-interacting Bose field in two space-time dimension, *Helv. Phys. Acta* **48**, 355–363.

Fröhlich, J. (1976a). Classical and quantum statistical mechanics in one and two dimensions: Two component Yukawa and Coulomb systems, *Comm. Math. Phys.* **47**, 233–268.

Fröhlich, J. (1976b). New super selection sectors (soliton-states) in two dimensional Bose quantum field models, *Comm. Math. Phys.* **47**, 269–310.

Fröhlich, J. (1976c). Phase transitions, Goldstone Boson, and topological super-selection rules, *Acta. Phys. Austriaca, Suppl. XV*, 133–269.

388   Bibliography

Fröhlich, J. (1976d). The pure phases, the irreducible quantum fields, and dynamical symmetry breaking in Symanzik-Nelson quantum field theories, *Ann. Phys.* **97**, 1–54.

Fröhlich, J. (1977a). Application of commutator theorems to the integration of representations of Lie algebras and commutation relations, *Comm. Math. Phys.* **54**, 135–150.

Fröhlich, J. (1977b). Schwinger functions and their generating functionals. II. Markovian and generalized path space measures on L^1, *Adv. Math.* **23**, 119–180.

Fröhlich, J. (1979). The charged sectors of quantum electrodynamics in a framework of local observables, *Comm. Math. Phys.* **66**, 223–265.

Fröhlich, J. (1980). Some results and comments on quantized gauge fields *In: Recent Developments in Gauge Theories* (Cargèse, 1979). G. 'tHooft et al eds., New York: Plenum Press.

Fröhlich, J. (1981). Some comments on the crossover between strong and weak coupling in SU(2) pure Yang-Mills theory. In: *Proceedings of Les Houches workshop "Common Trends in Particle and Condensed Matter Physics"* Feb. 18–29, 1980. To appear.

Fröhlich, J., Israel, R., Lieb, E. and Simon, B. (1978). Phase transitions and reflection positivity, I. General theory and long-range lattice models, *Comm. Math. Phys.* **62**, 1–34.

Fröhlich, J., Israel, R. B., Lieb, E. H. and Simon, B. (1980). Phase transitions and reflection positivity, II. Lattice systems with short-range and Coulomb interactions, *J. Stat. Phys.* **22**, 297–347.

Fröhlich, J. and Lieb, E. (1978). Phase transitions in anisotropic spin systems, *Comm. Math. Phys.* **60**, 233–267.

Fröhlich, J., Morchio, G. and Strocchi, F. (1979a). Charged sectors and scattering states in quantum electrodynamics, *Ann. Phys.* (*N.Y.*), **119**, 241–284.

Fröhlich, J., Morchio, G. and Strocchi, F. (1979b). Infrared problem and spontaneous breaking of the Lorentz group in QED, *Phys. Lett.* **89B**, 61–64.

Fröhlich, J. and Osterwalder, K. (1974). Is there a Euclidean field theory for fermions? *Helv. Phys. Acta* **47**, 781–805.

Fröhlich, J. and Park, Y. M. (1977). Remarks on exponential interactions and the quantum sine-Gordon equation in two space-time dimensions, *Helv. Phys. Acta.* **50**, 315–329.

Fröhlich, J. and Park, Y. M. (1978). Correlation inequalities and thermodynamic limit for classical and quantum continuous systems, *Comm. Math. Phys.* **59**, 235–266.

Fröhlich, J. and Park, Y. M. (1980). Correlation inequalities and the thermodynamic limit for classical and quantum continuous systems, II. Bose-Einstein and Fermi-Dirac statistics, *J. Stat. Phys.* **23**, 701–753.

Fröhlich, J. and Seiler, E. (1976). The massive Thirring-Schwinger model (QED_2) convergence of perturbation theory and particle structure, *Helv. Phys. Acta* **49**, 889–924.

Fröhlich, J. and Simon, B. (1977). Pure states for general $P(\phi)_2$ theories: Construction, regularity and variational equality, *Ann. Math.* **105**, 493–526.

Fröhlich, J., Simon, B. and Spencer, T. (1976). Infrared bounds, phase transitions, and continuous symmetry breaking, *Comm. Math. Phys.* **50**, 79–85.

Fröhlich, J. and Spencer, T. (1977). Phase transitions in statistical mechanics and quantum field theory *In: New Developments in Quantum Field Theory*, M. Levy and P. Mitter, eds., New York: Plenum Press.

Fröhlich, J. and Spencer, T. (1981a). On the statistical mechanics of classical Coulomb and dipole gases, *J. Stat. Phys.* **24**, 617–701.

Fröhlich, J. and Spencer, T. (1981b). The Kosterlitz–Thouless transition in two-dimensional abelian spin systems and the Coulomb gas, *Comm. Math. Phys.*, to appear.

Fröhlich, J. and Spencer, T. (1981c). Phase diagrams and critical properties of (classical) Coulomb systems. In: *Proceedings of the Erice 1980 Summer School*, A. Wightman and G. Velo, eds. (To appear.)

Galindo, A. (1962). On a class of perturbations in quantum field theory, *Proc. Nat. Acad. Sci. (U.S.A.)* **48**, 1128–1134.

Gallavotti, G. and Knops, H. (1975). The hierarchical model and the renormalization group, *Rivista Nuovo Cimento* **5**, 341–368.

Gallavotti, G. and Martin-Löf, A. (1972). Surface tension in the Ising model, *Comm. Math. Phys.* **25**, 87–126.

Gallavotti, G. and Miracle-Sole, S. (1972). Equilibrium states of the Ising model in the two phase region, *Phys. Rev.* **5B**, 2555–2559.

Garber, W., Ruijsenaars, S., Seiler, E. and Burns, D. (1979). On the finite action solutions of the nonlinear σ-model, *Ann. Phys.* **119**, 305–325.

Gawedzki, K. (1978a). Existence of three phases for a $P(\phi)_2$ model of quantum field, *Comm. Math. Phys.* **59**, 117–142.

Gawedzki, K. (1978b). On confinement of fermions in strongly coupled lattice gauge theory, *Comm. Math. Phys.* **63**, 31–47.

Gawedzki, K. and Kupiainen, A. (1980). A rigorous block spin approach to massless lattice theories, *Comm. Math. Phys.* **77**, 31–64.

Gelfand, I. and Shilov, G. E. (1964–8). *Generalized Functions* Vols. I-III (English Translation), New York: Academic Press.

Gelfand, I. and Vilenkin, N. (1964). *Generalized Functions*, Vol. 4 (English Translation), New York: Academic Press.

Gidas, B. (1974a). Properties of the $(\phi^4)_{1+1}$ interaction Hamiltonian, *J. Math. Phys.* **15**, 861–866.

Gidas, B. (1974b). On the self adjointness of the Lorentz generator for $(\phi^4)_{1+1}$, *J. Math. Phys.* **15**, 867–869.

Gidas, B. (1979). The Glimm-Jaffe-Spencer expansion for the classical boundary conditions and coexistence of phases in the $\lambda\phi_2^4$ Euclidean (quantum) field theory, *Ann. Phys.* **118**, 18–83.

Ginibre, J. (1969). Existence of phase transitions for quantum lattice systems, *Comm. Math. Phys.* **14**, 205–234.

Ginibre, J. (1970). General formulation of Griffiths' inequalities, *Comm. Math. Phys.* **16**, 310–328.

Ginibre, J. (1971). Some applications of functional integration in statistical mechanics *In: Statistical Mechanics and Quantum Field Theory*, C. DeWitt and R. Stora, eds., New York: Gordon and Breach.

Ginibre, J. and Velo, G. (1979). The classical field limit of scattering theory for non-relativistic many-Boson systems I, II, *Comm. Math. Phys.* **66**, 37–76; **68**, 45–68.

Glaser, V. (1974). On the equivalence of the Euclidean and Wightman formulation of field theory, *Comm. Math. Phys.* **37**, 257–272.

Glimm, J. (1967). The Yukawa coupling of quantum fields in two dimensions. I, II, *Comm. Math. Phys.* **5**, 343–386; **6**, 61–76.

Glimm, J. (1968a). Boson fields with non-linear self-interaction in two dimensions, *Comm. Math. Phys.* **8**, 12–25.

Glimm, J. (1968b). Boson fields with the ϕ^4 interaction in three dimensions, *Comm. Math. Phys.* **10**, 1–47.

Glimm, J. and Jaffe, A. (1968a). A $\lambda(\phi^4)_2$ quantum field theory without cutoffs. I, *Phys. Rev.* **176**, 1945–1951.

Glimm, J. and Jaffe, A. (1968b). A Yukawa interaction in infinite volume, *Comm. Math. Phys.* **11**, 9–18.

Glimm, J. and Jaffe, A. (1969a). An infinite renormalization of the Hamiltonian is necessary, *J. Math. Phys.* **10**, 2213–2214.

Glimm, J. and Jaffe, A. (1969b). Singular perturbations of self-adjoint operators, *Comm. Pure Appl. Math.* **22**, 401–414.

Glimm, J. and Jaffe, A. (1970a). Energy-momentum spectrum and vacuum expectation values in quantum field theory, *J. Math. Phys.* **11**, 3335–3338.

Glimm, J. and Jaffe, A. (1970b). The $\lambda(\phi^4)_2$ quantum field theory without cutoffs. II. The field operators and the approximate vacuum, *Ann. Math.* **91**, 362–401.

Glimm, J. and Jaffe, A. (1970c). The $\lambda(\phi^4)_2$ quantum field theory without cutoffs. III. The physical vacuum, *Acta Math.* **125**, 204–267.

Glimm, J. and Jaffe, A. (1970d). Self adjointness of the Yukawa$_2$ Hamiltonian, *Ann. Phys.* **60**, 321–383.

Glimm, J. and Jaffe, A. (1971a). The energy momentum spectrum and vacuum expectation values in quantum field theory. II, *Comm. Math. Phys.* **22**, 1–22.

Glimm, J. and Jaffe, A. (1971b). Field theory models *In: Statistical Mechanics and Quantum Field Theory*. C. DeWitt and R. Stora, eds., New York: Gordon and Breach.

Glimm, J. and Jaffe, A. (1971c). Positivity and self adjointness of the $P(\phi)_2$ Hamiltonian, *Comm. Math. Phys.* **22**, 253–258.

Glimm, J. and Jaffe, A. (1971d). The Yukawa$_2$ quantum field theory without cutoffs, *J. Funct. Anal.* **7**, 323–357.

Glimm, J. and Jaffe, A. (1972a). Boson quantum field models *In: Mathematics of Contemporary Physics* R. Streater, ed., New York: Academic Press.

Glimm, J. and Jaffe, A. (1972b). The $(\lambda\phi^4)_2$ quantum field theory without cutoffs. IV. Perturbations of the Hamiltonian, *J. Math. Phys.* **13**, 1568–1584.

Glimm, J. and Jaffe, A. (1973). Positivity of the ϕ_3^4 Hamiltonian, *Fort. Phys.* **21**, 327–376.

Glimm, J. and Jaffe, A. (1974a). Critical point dominance in quantum field models, *Ann. l'Inst. Henri Poincaré* **21**, 27–41.

Glimm, J. and Jaffe, A. (1974b). Entropy principle for vertex functions in quantum field models, *Ann. l'Inst. H. Poincaré* **21**, 1–26.

Glimm, J. and Jaffe, A. (1974c). ϕ_2^4 quantum field model in the single phase region: Differentiability of the mass and bounds on critical exponents, *Phys. Rev.* **10**, 536–539.

Glimm, J. and Jaffe, A. (1974d). A remark on the existence of ϕ_4^4, *Phys. Rev. Lett.* **33**, 440–442.

Glimm, J. and Jaffe, A. (1975a). Absolute bounds on vertices and couplings, *Ann. l'Inst. Henri Poincaré A* **22**, 97–107.

Glimm, J. and Jaffe, A. (1975b). On the approach to the critical point, *Ann. l'Inst. Henri Poincaré* **22**, 13–26.

Glimm, J. and Jaffe, A. (1975c). Three particle structure of ϕ^4 interactions and the scaling limit, *Phys. Rev. D* **11**, 2816–2827.

Glimm, J. and Jaffe, A. (1975d). Particles and bound states and progress toward unitarity and scaling *In: Mathematical Problems in Theoretical Physics* H. Araki, ed., New York: Springer-Verlag.

Glimm, J. and Jaffe, A. (1975e). Two and three body equations in quantum field models, *Comm. Math. Phys.* **44**, 293–320.

Glimm, J. and Jaffe, A. (1975f). ϕ^j bounds in $P(\phi)_2$ quantum field models *In: Mathematical Methods of Quantum Field Theory*, Paris: Centre National de la Recherche Scientifique.

Glimm, J. and Jaffe, A. (1976a). Critical exponents and renormalization in the ϕ^4 scaling limit *In: Quantum Dynamics: Models and Mathematics* L. Streit, ed., New York: Springer-Verlag.

Glimm, J. and Jaffe, A. (1976b). Particles and scaling for lattice fields and Ising models, *Comm. Math. Phys.* **51**, 1–14.

Glimm, J. and Jaffe, A. (1977a). Critical exponents and elementary particles, *Comm. Math. Phys.* **52**, 203–209.

Glimm, J. and Jaffe, A. (1977b). Instantons in a U(1) lattice gauge theory: A Coulomb dipole gas, *Comm. Math. Phys.* **56**, 195–212.

Glimm, J. and Jaffe, A. (1977c). Quark trapping for U(1) lattice gauge fields, *Phys. Lett. B* **66**, 67–69.

Glimm, J. and Jaffe, A. (1978a). Droplet model for quark confinement, *Phys. Rev. D* **18**, 463–467.

Glimm, J. and Jaffe, A. (1978b). Meron pairs and quark confinement, *Phys. Rev. Lett.* **40**, 277–282.

Glimm, J. and Jaffe, A. (1978c). Multiple meron solutions of the classical Yang-Mills equation, *Phys. Lett. B* **73**, 167–170.

Glimm, J. and Jaffe, A. (1979a). Changes, vortices and confinement, *Nucl. Phys. B* **149**, 49–60.

Glimm, J. and Jaffe, A. (1979b). A note on reflection positivity, *Lett. Math. Phys.* **3**, 377–378.

Glimm, J. and Jaffe, A. (1979c). The resummation of one particle lines, *Comm. Math. Phys.* **67**, 267–293.

Glimm, J. and Jaffe, A. (1980). The coupling constant in a ϕ^4 field theory *In: Recent Developments in Gauge Theories* (Cargèse, 1979). G. 't Hooft et al, eds., New York: Plenum Press.

Glimm, J., Jaffe, A. and Spencer, T. (1973). The particle structure of the weakly coupled $P(\phi)_2$ model and other applications of high temperature expansions *In: Constructive Quantum Field Theory* G. Velo and A. S. Wightman, eds., New York: Springer-Verlag.

Glimm, J., Jaffe, A. and Spencer, T. (1974). The Wightman axioms and particle structure in the $P(\phi)_2$ quantum field model, *Ann. Math.* **100**, 585–632.

Glimm, J., Jaffe, A. and Spencer, T. (1975). Phase transitions for ϕ_2^4 quantum fields, *Comm. Math. Phys.* **45**, 203–216.

Glimm, J., Jaffe, A. and Spencer, T. (1976a). A convergent expansion about mean field theory. Parts I, II, *Ann. Phys.* **101**, 610–630; 631–669.

Glimm, J., Jaffe, A. and Spencer, T. (1976b). Existence of phase transitions for quantum fields *In: Mathematical Methods of Quantum Field Theory*, Paris: Editions du Centre National de la Recherche Scientifique.

Goldstone, J. (1961). Field theories with "superconductor" solutions, *Nuovo Cimento*, **19**, 154–164.

Goldstone, J., Salam, A. and Weinberg, S. (1962). Broken symmetries, *Phys. Rev.* **127**, 965–970.

Goodman, R. and Segal, I. eds. (1966). *Proceedings of the Conference on the Mathematical Theory of Elementary Particles* (Dedham, 1965), Cambridge: MIT Press.

Graffi, S., Grecchi, V. and Simon, B. (1970). Borel summability: Application to the harmonic oscillator, *Phys. Lett. B* **32**, 631–634.

Griffiths, R. B. (1964). Peierls proof of spontaneous magnetization of a two-dimensional Ising ferromagnet, *Phys. Rev. A* **136**, 437–439.

Griffiths, R. B. (1967). Correlation in Ising ferromagnets I, II, III, *J. Math. Phys.* **8**, 478–484; 484–489; *Comm. Math. Phys.* **6**, 121–127.

Griffiths, R. B. (1969). Rigorous results for Ising ferromagnets of arbitrary spin, *J. Math. Phys.* **10**, 1559–1565.

Griffiths, R. B. (1970). Phase transitions *In: Statistical Mechanics and Quantum Field Theory*, C. DeWitt and R. Stora, eds., New York: Gordon and Breach.

Griffiths, R. B., Hurst, C. A. and Sherman, S. (1970). Concavity of magnetization of an Ising ferromagnet in a positive external field, *J. Math. Phys.* **11**, 790–795.

Gross, L. (1972). Existence and uniqueness of physical ground states, *J. Funct. Anal.* **10**, 52–109.

Gross, L. (1974). Analytic vectors for representations of the canonical commutation relations and nondegeneracy of ground states, *J. Funct. Anal.* **17**, 104–111.

Gross, L. (1975). Hypercontractivity and logarithmic Sobolev inequalities for the Clifford-Dirichlet form, *Duke Math. J.* **42**, 383–396.

Gruber, C. and Kunz, H. (1971). General properties of polymer systems, *Comm. Math. Phys.* **22**, 133–161.

Gruber, C., Merlini, D. and Greenberg, W. (1973). Spin-$\frac{1}{2}$ lattice system: Duality transformation and correlation functions, *Physica*, **65**, 28–40.

Guerra, F. (1972). Uniqueness of the vacuum energy density and van Hove phenomenon in the infinite volume limit for two-dimensional self-coupled Bose fields, *Phys. Rev. Lett.* **28**, 1213–1215.

Guerra, F., Robinson, D. and Stora, R., eds. (1976). *Les Méthodes Mathématiques de la Théorie Quantique des Champs*, (Marseille, 1976), Paris: Editions du Centre National de la Recherche Scientifique.

Guerra, F., Rosen, L. and Simon, B. (1973a). Nelson's symmetry and the infinite volume behavior of the vacuum in $P(\phi)_2$, *Comm. Math. Phys.* **27**, 10–22.

Guerra, F., Rosen, L. and Simon, B. (1973b). The vacuum energy for $P(\phi)_2$: Infinite volume limit and coupling constant dependence, *Comm. Math. Phys.* **29**, 233–247.

Guerra, F., Rosen, L. and Simon, B. (1975a). Correlation inequalities and the mass gap in $P(\phi)_2$ III. Mass gap for a class of strongly coupled theories with nonzero external field, *Comm. Math. Phys.* **41**, 19–32.

Guerra, F., Rosen, L. and Simon, B. (1975b). The $P(\phi)_2$ Euclidean quantum field theory as classical statistical mechanics, *Ann. Math.* **101**, 111–259.

Guerra, F., Rosen, L. and Simon, B. (1976). Boundary conditions in the $P(\phi)_2$ Euclidean field theory, *Ann. l'Inst. Henri Poincaré* **15**, 231–334.

Guth, A. (1980). Existence proof of a nonconfining phase in four-dimensional U(1) lattice gauge theory, *Phys. Rev. D* **21**, 2291–2307.

Haag, R. (1955). On quantum field theories, *Mat.-Fys. Medd. Kong. Danske Videns. Selskab* **29**, *No. 12*.

Haag, R. (1958). Quantum fields with composite particles and asymptotic conditions, *Phys. Rev.* **112**, 669–673.

Haag, R. (1970). Observables and fields *In: Lectures on Elementary Particles and Quantum Field Theory* S. Deser et al, eds., Cambridge: MIT Press.

Haag, R. and Kastler, D. (1964). An algebraic approach to quantum field theory, *J. Math. Phys.* **5**, 848–861.

Haag, R. and Schroer, B. (1962). Postulates of quantum field theory, *J. Math. Phys.* **3**, 248–256.

Hagedorn, G. (1980). Asymptotic completeness for classes of 2, 3 and 4 particle Schrödinger operators, *Trans. Amer. Math. Soc.* **258**, 1–75.

Hawking, S. W. (1977). Zeta function regularization of path integrals in curved space, *Comm. Math. Phys.* **55**, 133–148.

Healy, J. (1973). New rigorous bounds on coupling constants in field theory, *Phys. Rev. D* **8**, 1904–1914.

Hegerfeldt, G. (1974). From Euclidean to relativistic fields and on the notion of Markoff fields, *Comm. Math. Phys.* **35**, 155–171.

Hegerfeldt, G. (1975). Probability measures on distribution spaces and quantum field theoretical models, *Rep. Math. Phys.* **7**, 403–409.

Heifets, E. P. and Osipov, E. P. (1977a). The energy-momentum spectrum in the $P(\phi)_2$ quantum field theory, *Comm. Math. Phys.* **56**, 161–172.

Heifets, E. P. and Osipov, E. P. (1977b). The energy-momentum spectrum in the Yukawa$_2$ quantum field theory, *Comm. Math. Phys.* **57**, 31–50.

Hepp, K. (1964a). Lorentz invariant analytic S-matrix amplitudes, *Helv. Phys. Acta* **37**, 55–73.

Hepp, K. (1964b). On the analyticity properties of the scattering amplitude in relativistic quantum field theory, *Helv. Phys. Acta* **37**, 639–658.

Hepp, K. (1964c). Spatial cluster decomposition properties of the S-matrix, *Helv. Phys. Acta* **37**, 659–662.

Hepp, K. (1965a). One particle singularities of the S-matrix in quantum field theory, *J. Math. Phys.* **6**, 1762–1767.

Hepp, K. (1965b). On the connection between the LSZ and Wightman quantum field theory, *Comm. Math. Phys.* **1**, 95–111.

Hepp, K. (1966a). On the connection between Wightman and LSZ quantum field theory *In: Axiomatic Field Theory* M. Chretien and S. Deser, eds., New York: Gordon and Breach.

Hepp, K. (1969b). Proof of the Bogoliubov–Parasiuk theorem on renormalization, *Comm. Math. Phys.* **2**, 301–326.

Hepp, K. (1969a). On the quantum mechanical N-body problem, *Helv. Phys. Acta* **42**, 425–458.

Hepp, K. (1969b). Renormalized Hamiltonians for a class of quantum fields with infinite mass and charge renormalization *In: Problems of Theoretical Physics: Essays dedicated to N. N. Bogoliubov*, Moscow: Nauka.

Hepp, K. (1970). *Théorie de la Renormalisation*, New York: Springer-Verlag.

Hepp, K. (1971). Renormalization theory *In: Statistical Mechanics and Quantum Field Theory* C. DeWitt and R. Stora, eds., New York: Gordon and Breach.

Hepp, K. (1974). The classical limit for quantum mechanical correlation functions, *Comm. Math. Phys.* **35**, 265–277.

Hepp, K. and Lieb, E. H. (1973). On the superradiant phase transition for molecules in a quantized radiation field: the Dicke maser model, *Ann. Phys.* **76**, 360–404.

Herbst, I. (1976). On canonical quantum field theories, *J. Math. Phys.* **17**, 1210–1221.

Hess, H., Schrader, R. and Uhlenbrock, D. A. (1977). Domination of semigroups and generalization of Kato's inequality, *Duke Math. J.* **44**, 893–904.

Hida, T. (1970). *Stationary Stochastic Processes*, Princeton, N.J.: Princeton University Press.

Higgs, P. W. (1966). Spontaneous symmetry breakdown without massless bosons, *Phys. Rev.* **145**, 1156–1163.

Høegh-Krohn, R. (1971a). A general class of quantum fields without cutoffs in two space-time dimensions, *Comm. Math. Phys.* **21**, 244–255.

Høegh-Krohn, R. (1971b). On the spectrum of the space cutoff: $P(\phi)$: Hamiltonian in two space-time dimensions, *Comm. Math. Phys.* **21**, 256–260.

Høegh-Krohn, R. (1974). Relativistic quantum statistical mechanics in two-dimensional space-time, *Comm. Math. Phys.* **38**, 195–224.

Høegh-Krohn, R. and Simon, B. (1972). Hypercontractive semigroups and two dimensional self-coupled Bose fields, *J. Funct. Anal.* **9**, 121–180.

Hohenberg, P. (1967). Existence of long-range order in one and two dimensions, *Phys. Rev.* **158**, 383–386.

Holley, R. (1974). Remarks on the FKG inequalities, *Comm. Math. Phys.* **36**, 227–231.

Holsztynski, W. and Slawny, J. (1978). Peierls condition and number of ground states, *Comm. Math. Phys.* **61**, 177–190.

't Hooft, G. (1971a). Renormalization of massless Yang-Mills fields, *Nucl. Phys.* *B* **33**, 173–199.

't Hooft, G. (1971b). Renormalizable Lagrangians for massive Yang-Mills fields, *Nucl. Phys.* *B* **35**, 167–188.

't Hooft, G. (1974). Magnetic monopoles in unified gauge theories, *Nucl. Phys.* *B* **79**, 276–284.

't Hooft, G. (1978). On the phase transition towards permanent quark confinement, *Nucl. Phys.* *B* **138**, 1–25.

't Hooft, G. (1980). Which topological features of a gauge theory can be responsible for permament confinement. *In: Recent Developments in Gauge Theories* (Cargèse, 1979) G. 't Hooft et al, eds., New York, Plenum Press.

't Hooft, G. (1981). Topology of the gauge condition and new confinement phases in nonabelian gauge theories. (To appear.)

Huang, K. (1963). *Statistical Mechanics*, New York: Wiley and Sons.

Huebener, R. P. (1979). *Magnetic Flux in Superconductors*, New York: Springer-Verlag.

Iagolnitzer, D. (1978). *The S Matrix*, New York: North-Holland.

Imbrie, J. (1980a). Mass spectrum of the two dimensional $\lambda\phi^4 - \frac{1}{4}\phi^2 - \mu\phi$ quantum field theory, *Comm. Math. Phys.* **78**, 169–200.

Imbrie, J. (1980b). *Cluster expansions and mass spectra for $P(\phi)_2$ models possessing many phases*, Harvard University Thesis.

Imbrie, J. (1981). Phase diagrams and cluster expansions for low temperature $P(\phi)_2$ models. I, II. *Comm. Math. Phys.*, to appear.

Isaacson, D. (1977). The critical behavior of ϕ_1^4, *Comm. Math. Phys.* **53**, 257–275.

Isaacson, D. (1981). The continuum limit of a classical 3 component one-dimensional Heisenberg model is Brownian motion on the sphere. *J. Math. Phys.* (To appear.)

Isaacson, D. and Marchesin, D. (1978). The eigenvalues and eigenfunctions of a spherically symmetric anharmonic oscillator, *Comm. Pure Appl. Math.* **31**, 659–676.

Isaacson, D., Marchesin, D. and Paes-Leme, P. J. (1980). Numerical methods for studying anharmonic oscillator approximations to the ϕ_2^4 quantum field theory, *Int. J. Eng. Soc.* **18**, 341–349.

Israel, R. (1978). *Convexity and the Theory of Lattice Gases*, Princeton: Princeton University Press.

Israel, R. B. and Nappi, C. R. (1979a). Quark confinement in the two-dimensional lattice Higgs-Villian model, *Comm. Math. Phys.* **64**, 177–189.

Israel, R. B. and Nappi, C. R. (1979b). Exponential clustering for long-range integer-spin systems, *Comm. Math. Phys.* **68**, 29–38.

Ito, K. R. (1980). Construction of Euclidean (QED)$_2$ via lattice gauge theory. I, to appear.

Itzykson, C. (1980). Introduction to lattice gauge theories *In: Recent Developments in Gauge Theories* (Cargèse, 1979) G. 't Hooft et al., eds., New York: Plenum Press.

Itzykson, C. and Zuber, J.-B. (1980). *Quantum Field Theory*, New York: McGraw-Hill.

Jaffe, A. (1965a). Divergence of perturbation theory for Bosons, *Comm. Math. Phys.* **1**, 127–149.

Jaffe, A. (1965b). *Dynamics of a cut-off $\lambda\phi^4$ field theory*, Princeton University Thesis.

Jaffe, A. (1976). Problèmes ergodiques dans la théorie quantique des champs, *Soc. Math. Fr.* **40**, 105–112.

Jaffe, A., Lanford, O. and Wightman, A. (1969). A general class of cutoff model field theories, *Comm. Math. Phys.* **15**, 47–68.

Jaffe, A. and McBryan, O. (1974). What constructive field theory says about currents *In: Local Currents and their Applications* D. H. Sharp and A. S. Wightman, eds., Amsterdam: North-Holland.

Jaffe, A. and Powers, R. (1968). Infinite volume limit of a $\lambda\phi^4$ field theory, *Comm. Math. Phys.* **7**, 218–221.

Jaffe, A. and Taubes, C. (1980). *Vortices and Monopoles: Structure of Static Gauge Theories*, Boston: Birkhäuser.

Jauch, J. M. and Rohrlich, F. (1976). *The Theory of Photons and Electrons*, New York: Springer-Verlag.

Jevicki, A. (1980). Statistical mechanics of instantions in quantum chromodynamics, *Phys. Rev. D* **21**, 992–1005.

Jona-Lasinio, G. (1964). Relativistic field theories with symmetry breaking solutions, *Nuovo Cimento (L)* **34**, 1790–1795.

Jona-Lasinio, G. (1975). The renormalization group: A probabilistic view, *Nuovo Cimento B* **26**, 99–119.

Jonsson, T., McBryan, O., Zirrilli, F. and Hubbard, J. (1979). An existence theorem for multimeron solutions to classical Yang-Mills field equations, *Comm. Math. Phys.* **68**, 259–273.

José, J., Kadanoff, L., Kirkpatrick, S. and Nelson, D. (1977). Renormalization, vortices, and symmetry-breaking perturbations in the two-dimensional planar model, *Phys. Rev. B* **16**, 1217–1241.

Jost, R. (1965). *The General Theory of Quantized Fields*, Providence: American Mathematical Society.

Jost, R. ed. (1969). *Local Quantum Theory*, (Varenna, 1968), New York: Academic Press.

Kac, M. (1951). On some connections between probability theory and differential and integral equations *In*: *Proceedings of the Second Berkeley Symposium on Probability and Statistics* J. Neyman, ed., Berkeley: University of California Press.

Kac, M. (1959). *Probability and Related Topics in Physical Sciences*, New York: Interscience Publishers.

Kadanoff, L. (1977). The application of renormalization group techniques to quarks and strings, *Rev. Mod. Phys.* **49**, 267–296.

Kadanoff, L. (1979). Multicritical behavior at the Kosterlitz-Thouless critical point, *Ann. Phys.* **120**, 39–71.

Kadison, R. V. (1965). Transformation of states in operator theory and dynamics, *Topology* **3**, 177–198.

Kashiwara, M. and Kawai, T. (1977). Holonomic systems of linear differential equations and Feynman integrals, *Publ. RIMS Kyoto Univ.* **12** *Supp.*, 131–140.

Kashiwara, M., Kawai, T. and Stapp, H. P. (1979). Micro-analyticity of the *S*-matrix and related functions, *Comm. Math. Phys.* **66**, 95–130.

Kastler, D. (1961). *Introduction a l'Electrodynamique Quantique*, Paris: Dunod.

Kastler, D., Robinson, D. W. and Swieca, A. (1966). Conserved currents and associated symmetries; Goldstone's theorem, *Comm. Math. Phys.* **2**, 108–120.

Kato, T. (1951a). Fundamental properties of Hamiltonian operator of Schrödinger type, *Trans. Am. Math. Soc.* **70**, 195–211.

Kato, T. (1951b). On the existence of solutions of the helium wave equation, *Trans. Amer. Math. Soc.* **70**, 212–218.

Kato, T. (1966). *Perturbation Theory for Linear Operators*, New York/Berlin: Springer-Verlag.

Kato, Y. and Mugibayashi, N. (1963). Regular perturbation and asymptotic limits of operators in quantum field theory, *Prog. in Theor. Phys.* **30**, 103–133.

Kawai, T. and Stapp, H. (1977). Discontinuity formula and Sato's conjecture, *Publ. RIMS Kyoto Univ.*, **12** *Suppl.*, 155–232.

Kelley, D. G. and Sherman, S. (1969). General Griffiths' inequalities on correlations in Ising ferromagnets, *J. Math. Phys.* **9**, 466–484.

Khuri, N. N. (1981). Coupling constant analyticity and the renormalization group. *Phys. Rev. D*, to appear.

Klauder, J. (1973). Field structure through model studies: Aspects of non-renormalizable theories, *Acta Physica Austriaca Suppl.* **11**, 341–387.

Klein, A. (1978). The semigroup characterization of Osterwalder-Schrader path spaces and the construction of Euclidean fields, *J. Funct. Anal.* **27**, 277–291.

Klein, A. and Landau, L. J. (1975a). The $:\phi^2:$ field in the $P(\phi)_2$ model, *Comm. Math. Phys.* **43**, 143–154.

Klein, A. and Landau, L. J. (1975b). Singular perturbations of positivity preserving semigroups via path space techniques, *J. Funct. Anal.* **20**, 44–82.

Koch, H. (1979). Irreducible kernels and bound states in $\lambda P(\phi)_2$ models, *Ann. l'Inst. Henri Poincaré* **31**, 173–234.

Koch, H. (1981). Particles exist in the low temperature ϕ_2^4 model, *Helv. Phys. Acta.* (To appear.)

Kogut, J. (1980). 1/N expansions and the phase diagram of discrete lattice gauge theories with matter fields, *Phys. Rev. D* **21**, 2316–2326.

Kogut, J. and Susskind, L. (1975). Hamiltonian formulation of Wilson's lattice guage theories, *Phys. Rev. D* **11**, 395–408.

Kinoshita, T. (1979). Anomalous magnetic moment of an electron and high precision test of quantum electrodynamics *In: Luminy CNRS colloquium.*

Kosterlitz, J. M. (1974). The critical properties of the two dimensional xy model, *J. Phys. C* **7**, 1046–1060.

Kosterlitz, J. M. (1977). The d-dimensional Coulomb gas and the roughening transition, *J. Phys.* **C10**, 3753–3760.

Kosterlitz, J. M. and Thouless, D. V. (1973). Ordering, metastability and phase transitions in two-dimensional systems, *J. Phys. C.* **6**, 1181–1203.

Kristensen, P., Mejlbo, L. and Poulsen, E. (1965). Tempered distributions in infinitely many dimensions I, *Comm. Math. Phys.* **1**, 175–214.

Kunz, H. and Pfister, C. E. (1976). First order phase transitions in the plane rotor ferromagnetic model in two dimensions, *Comm. Math. Phys.* **46**, 245–251.

Kunz, H., Pfister, C. and Vuillermot, P. (1975). Inequalities for some classical spin vector models, *J. Phys.* **A9**, 1673–1683.

Kunz, H., Pfister, C. and Vuillermot, P. (1976). Correlation inequalities for some classical spin vector models, *Phys. Lett. A* **54**, 428–430.

Kupiainen, A. J. (1980a). On the 1/N expansion, *Comm. Math. Phys.* **73**, 273–294.

Kupiainen, A. J. (1980b). 1/N expansion for a quantum field model, *Comm. Math. Phys.* **74**, 199–222.

Landau, L. and Lifshitz, E. (1969). *Statistical Physics*, Reading: Addison-Wesley.

Lanford, O. (1966). *Construction of quantum fields interacting by a cutoff Yukawa coupling*, Princeton University Thesis.

Lanford, O. (1973). Entropy and equilibrium states in classical statistical mechanics *In: Statistical Mechanics and Mathematical Problems* A. Lenard, ed., New York: Springer-Verlag.

Lanford, O. E. and Ruelle, D. (1969). Observables at infinity and states with short range correlations in statistical mechanics, *Comm. Math. Phys.* **13**, 194–215.

Lassalle, M. (1974). Analyticity properties implied by the many-particle structure of N-point functions in general quantum field theory, I. Convolution of N-point functions associated with a graph, *Comm. Math. Phys.* **36**, 185–226.

Lax, P. and Phillips, R. S. (1967). *Scattering Theory*, New York: Academic Press.

Lebowitz, J. L. (1972a). Bounds on the correlations and analyticity properties of ferromagnetic Ising spin systems, *Comm. Math. Phys.* **28**, 313–321.

Lebowitz, J. L. (1972b). More inequalities for Ising ferromagnets, *Phys. Rev. B* **5**, 2538–2541.

Lebowitz, J. L. (1974). GHS and other inequalities, *Commun. Math. Phys.* **35**, 87–92.

Lebowitz, J. L. (1977). Coexistence of phases in Ising ferromagnets, *J. Stat. Phys.* **16**, 463–476.

Lebowitz, J. L. and Martin-Löf, A. (1972). On the uniqueness of the equilibrium state for Ising spin systems, *Comm. Math. Phys.* **25**, 276–282.

Lebowitz, J. L. and Monroe, J. L. (1972a). Bounds on the correlations and analyticity properties of ferromagnetic Ising spin systems, *Comm. Math. Phys.* **28**, 313–321.

Lebowitz, J. L. and Monroe, J. C. (1972b). Inequalities for higher order Ising spins and for continuum fluids, *Comm. Math. Phys.* **28**, 301–311.

Lebowitz, J. and Penrose, O. (1968). Analytic and clustering properties of thermodynamic functions and distribution functions for classical lattice and continuum systems, *Comm. Math. Phys.* **11**, 99–124.

Lebowitz, J. and Penrose, O. (1973). Decay of correlations, *Phys. Rev. Lett.* **31**, 749–752.

Lee, T. D. and Yang, C. N. (1952). Statistical theory of equations of state and phase transitions II. Lattice gas and Ising model, *Phys. Rev.* **87**, 410–419.

Le Guillou, J. and Zinn-Justin, J. (1977). Critical exponents for the n-vector model in three dimensions from field theory, *Phys. Rev. Lett.* **39**, 95–98.

Lehmann, H., Symanzik, K. and Zimmermann, W. (1955). On the formulation of quantized field theories, *Nuovo Cimento* (*ser. 10*) **1**, 205–225.

Lenard, A. (1973). *Statistical Mechanics and Mathematical Problems*, (Battelle 1971), New York: Springer-Verlag.

Lenard, A. and Newman, C. M. (1974). Infinite volume asymptotics in $P(\phi)_2$ field theory, *Comm. Math. Phys.* **39**, 243–250.

Levine, M. J. and Roskies, R. (1976). Analytic contribution to the g factor of the electron in sixth order, *Phys. Rev. D.* **14**, 2191–2192.

Lévy, M., ed. (1967). *High Energy Electromagnetic Interactions and Field Theory*, (Cargèse, 1964), New York: Gordon and Breach.

Lévy, M. and Mitter, P., eds. (1977). *New Developments in Quantum Field Theory and Statistical Mechanics*, (Cargèse, 1976), New York: Plenum Press.

Lieb, E. (1973). The classical limit of quantum spin systems, *Comm. Math. Phys.* **31**, 327–340.

Lieb, E. (1976). The stability of matter, *Rev. Mod. Phys.* **48**, 553–569.

Lieb, E. (1980). A refinement of Simon's correlation inequality, *Comm. Math. Phys.* **77**, 127–136.

Lieb, E. and Lebowitz, J. L. (1972). The constitution of matter: Existence of thermodynamics for systems composed of electrons and nuclei, *Adv. Math.* **9**, 316–398.

Lieb, E. and Mattis, D. (1965). Exact solution of a many-fermion system and its associated boson field, *J. Math. Phys.* **6**, 304–312.

Lieb, E. and Mattis, D. (1967). *Mathematical Physics in One-Dimension*, New York: Academic Press.

Lieb, E., Mattis, D. and Schultz, T. (1964). Two dimensional Ising model as a soluble problem of many fermions, *Rev. Mod. Phys.* **36**, 856–871.

Lieb, E. and Simon, B. (1974). On solutions to the Hartree Fock problem for atoms and molecules, *J. Chem. Phys.* **61**, 735–736.

Lieb, E. and Simon, B. (1977a). The Hartree-Foch theory for Coulomb systems, *Comm. Math. Phys.* **53**, 185–194.

Lieb, E. and Simon, B. (1977b). The Thomas-Fermi theory of atoms, molecules, and solids, *Adv. in Math.* **23**, 22–116.

Lieb, E. and Sokal, A. (1981). A general Lee-Yang theorem for one-component and multicomponent ferromagnets, *Comm. Math. Phys.* (To appear.)

Lieb, E. and Thirring, W. (1975). Bound for the kinetic energy of fermions which proves the stability of matter, *Phys. Rev. Lett.* **35**, 687–689.

Lipatov, L. N. (1976). Calculation of the Gell-Mann-Low function in scalar theories with strong nonlinearity, *JETP Lett.* **24**, 157–160.

Lipatov, L. N. (1977a). Divergence of the perturbation-theory series and the quasi-classical theory, *Soviet Physics JETP* **45**, 216–223.

Lipatov, L. N. (1977b). Divergence of the perturbation-theory series and pseudo-particles, *JETP Lett.* **25**, 104–107.

Lukaszuk, L. and Martin, A. (1967). Absolute upper bounds for $\pi\pi$ scattering, *Nuovo Cimento A* **52**, 122–145.

Lundeen, S. R. and Pipkin, F. (1981). Measurement of the Lamb shift in hydrogen, $n = 2$, *Phys. Rev. Lett.* **46**, 232–235.

Lüscher, M. (1977a). Absence of spontaneous gauge symmetry breaking in Hamiltonian lattice gauge theories, *Desy Preprint* 77/16.

Lüscher, M. (1977b). Construction of a self-adjoint, strictly positive transfer matrix for Euclidean lattice gauge theories, *Comm. Math. Phys.* **54**, 283–292.

Lüscher, M. (1977c). Dynamical charges in the quantized renormalized massive Thirring model, *Nucl. Phys. B* **117**, 475–492.

Ma, S. (1976). *Modern Theory of Critical Phenomena*, Reading: Benjamin.

MacDermot, A. (1976). *A lattice approximation to the Yukawa$_2$ Euclidean quantum field theory and a correlation inequality*, Cornell University Thesis.

Mack, G. (1980). Properties of lattice guage models at low temperatures *In: Recent Developments in Guage Theories* (Cargèse, 1979) G. 't Hooft et al., eds., New York: Plenum Press.

Mack, G. and Symanzik, K. (1972). Currents, stress tensor and generalized unitarity in conformal invariant quantum field theory, *Comm. Math. Phys.* **27**, 247–281.

Magnen, J. and Sénéor, R. (1976a). The infinite volume limit of the ϕ_3^4 model, *Ann. l'Inst. Henri Poincaré* **24**, 95–159.

Magnen, J. and Sénéor, R. (1976b). The Wightman axioms for the weakly coupled Yukawa model in two dimensions, *Comm. Math. Phys.* **51**, 297–313.

Magnen, J. and Sénéor, R. (1977). Phase space cell expansion and Borel summability for the Euclidean ϕ_3^4 theory, *Comm. Math. Phys.* **56**, 237–276.

Magnen, J. and Sénéor, R. (1980). Yukawa quantum field theory in three dimensions (Y_3) *In: Third International Conference on Collective Phenomena* J. Lebowitz, J. Langer, and W. Glaberson, eds., New York: The New York Academy of Sciences.

Martin, W. T. and Segal, I. eds. (1966). *Mathematical Theory of Elementary Particles*, (Dedham, 1963), Cambridge: MIT Press.

Masson, D. and McClary, W. (1971). On the self adjointness of the $(g(x)\phi^4)_2$ Hamiltonian, *Commun. Math. Phys.* **21**, 71–74.

Matthews, P. T. and Salam, A. (1954). The Green's functions of quantised fields, *Nuovo Cimento Series 9*, **12**, 563–565.

Matthews, P. and Salam, A. (1955). Propagators of quantized field, *Nuovo Cimento Series 10*, **2**, 120–134.

McBryan, O. (1973). Local generators for the Lorentz group in the $P(\phi)_2$ model, *Nuovo Cimento A* **18**, 654–662.

McBryan, O. (1975a). Finite mass renormalizations in the Yukawa$_2$ quantum field theory, *Comm. Math. Phys.* **44**, 237–243.

McBryan, O. (1975b). Higher order estimates for the Yukawa$_2$ quantum field theory, *Comm. Math. Phys.* **42**, 1–7.

McBryan, O. (1975c). Volume dependence of Schwinger functions in the Yukawa$_2$ quantum field theory, *Comm. Math. Phys.* **45**, 279–294.

McBryan, O. (1975d). Self-adjointness of relatively bounded quadratic forms and operators, *J. Funct. Anal.* **19**, 97–103.

McBryan, O. (1975e). Convergence of the vacuum energy density ϕ-bounds and existence of Wightman functions for the Yukawa model *In: Mathematical Methods of Quantum Field Theory*, Paris: Centre National de la Recherche Scientifique.

McBryan, O. (1978). The ϕ_2^4 quantum field as a limit of Sine-Gordon fields, *Comm. Math. Phys.* **61**, 275–284.

McBryan, O. and Park, Y. (1975). Lorentz covariance of the Yukawa$_2$ quantum field theory, *J. Math. Phys.* **16**, 105–110.

McBryan, O. and Rosen, J. (1976). Existence of the critical point in ϕ^4 field theory, *Comm. Math. Phys.* **51**, 97–105.

McBryan, O. and Spencer, T. (1977). On the decay of correlations in SO(n)-symmetric ferromagnets, *Comm. Math. Phys.* **53**, 299–302.

McCoy, B., Tracy, C. and Wu, T. T. (1977). Two dimensional Ising model as an explicitly solvable relativistic field Theory: Explicit formulas for n-point functions, *Phys. Rev. Lett.* **38**, 793–796.

McCoy, B. and Wu, T. T. (1973). *The Two-Dimensional Ising Model*, Cambridge: Harvard University Press.

McCoy, B. M. and Wu, T. T. (1978). Two-dimensional Ising field theory for $T < T_c$: Green's-functions strings in n-point functions, *Phys. Rev. D* **18**, 1253–1258.

McKean, H. P. (1964). Kramers-Wannier duality for the two-dimensional Ising model as an instance of Poisson's summation formula, *J. Math. Phys.* **5**, 775–776.

McKean, H. P. (1969). *Stochastic Integrals*, New York: Academic Press.

Mermin, N. (1967). Absence of ordering in certain classical systems, *J. Math. Phys.* **8**, 1061–1064.

Mermin, N. and Wagner, H. (1966). Absence of ferromagnetism or antiferromagnetism in one- or two-dimensional isotropic Heisenberg models, *Phys. Rev. Letters* **17**, 1133–1136.

Messager, A. and Miracle-Sole, S. (1977). Correlation functions and boundary conditions in the Ising Ferromagnet, *J. Stat. Phys.* **17**, 245–262.

Migdal, A. A. (1975a). Recursion equations in gauge field theories, *Zh. Eksp. Teor. Fiz.* **69**, 810–822. Engl. trans.: *Soviet Physics JETP* **42**, 413–418 (1976).

Migdal, A. A. (1975b). Phase transitions in gauge and spin lattice systems, *Zh. Eksp. Teor. Fiz.* **69**, 1457–1465. Engl. trans.: *Soviet Physics JETP* **42**, 743–746 (1976).

Miller, W. (1972). *Symmetry Groups and their Applications*, New York: Academic Press.

Minlos, R. A. and Sinai, Ya. G. (1970). Investigation of the spectra of stochastic operators arising in lattice models of a gas, *Theor. Math. Phys.* **2**(2), 167–176.

Monroe, J. L. (1975). Correlation inequalities for two-dimensional vector spin systems, *J. Math. Phys.* **16**, 1809–1812.

Moore, M. A., Jasnow, D. and Wortis, M. (1969). Spin-spin correlation function of the three-dimensional Ising ferromagnet above the Curie temperature, *Phys. Rev. Lett.* **22**, 940–943.

Morchio, G. and Strocchi, F. (1980). Infrared singularities, vacuum structure and pure phases in local quantum field theory, *Ann l'Inst. Henri Poincaré* **33**, 251–282.

Nambu, Y. and Jona-Lasino, G. (1961). Dynamical model of elementary particles based on an analogy with superconductivity. I, II, *Phys. Rev.* **122**, 345–358; **124**, 246–254.

Nappi, C. P. (1978). On the scaling limit of the Ising model, *Nuovo Cimento A* **44**, 392–400.

Narasimhan, M. S. and Ramades, T. R. (1979). Geometry of SU(2) gauge fields, *Comm. Math. Phys.* **67**, 121–136.

Nelson, E. (1966). A quartic interaction in two dimensions *In*: *Mathematics Theory of Elementary Particles* R. Goodman and I. Segal, eds., Cambridge: MIT Press.

Nelson, E. (1972). Time ordered operator products of sharp time quadratic forms, *J. Funct. Anal.* **11**, 211–219.

Nelson, E. (1973a). The construction of quantum fields from Markov fields, *J. Funct. Anal.* **12**, 97–112.

Nelson, E. (1973b). The free Markov field, *J. Funct. Anal.* **12**, 211–217.

Nelson, E. (1973c). Probability theory and Euclidean field theory *In*: *Constructive Quantum Field Theory* G. Velo and A. Wightman, eds., New York: Springer-Verlag.

Nelson, E. (1973d). Quantum fields and Markoff fields *In*: *Partial Differential Equations* D. C. Spencer, ed., Providence: American Math. Soc.

Neves di Silva, R. (1981). Three particle bound states in even $\lambda P(\phi)_2$ models. To appear.

Newman, C. (1973). The construction of stationary two dimensional Markoff fields with an application to quantum field theory, *J. Func. Anal.* **14**, 44–61.

Newman, C. (1974). Zeros of the partition function for generalized Ising systems, *Comm. Pure Appl. Math.* **27**, 143–159.

Newman, C. (1975a). Inequalities for Ising models and field theories which obey the Lee-Yang theorem, *Comm. Math. Phys.* **41**, 1–9.

Newman, C. (1975b). Gaussian correlation inequalities for ferromagnets, *Z. für. Wahrscheinlichkeitstheorie* **33**, 75–93.

Newman, C. (1975c). Moment inequalities for ferromagnetic Gibbs distributions, *J. Math. Phys.* **16**, 1956–1959.

Newman, C. (1976a). Classifying general Ising models *In: Mathematics Methods of Quantum Field Theory*, Paris: Editions du Centre National de la Recherche Scientifique.

Newman, C. (1976b). Rigorous results for general Ising ferromagnets, *J. Stat. Phys.* **15**, 399–406.

Newman, C. (1979a). Critical point inequalities and scaling limits, *Comm. Math. Phys.* **66**, 181–196.

Newman, C. (1979b). Short distance scaling and the maximal degree of a field theory, *Phys. Lett. B* **83**, 63–66.

Newman, C. (1980). Normal fluctuations and the FKG inequalities, *Comm. Math. Phys.* **74**, 119–128.

Newman, C. (1981). Critical point dominance in one dimension, *Comm. Math. Phys.*, **79**, 133–140.

Nickel, B. (1980). Hyperscaling and universality in three dimensions. (To appear.)

Nickel, B. (1981). The problem of confluent singularities, *In: Proceedings of the 1980 Cargèse Summer School on Phase Transitions*, London: Plenum Press.

Nicolai, H. (1978). An inequality for Fermion systems, *Comm. Math. Phys.* **59**, 71–78.

Osipov, E. P. (1977). Connection between the spectrum condition and the Lorentz invariance of the Yukawa$_2$ quantum field theory, *Comm. Math. Phys.* **57**, 111–116.

Osipov, E. P. (1979a). The Yukawa$_2$ field theory: linear N_τ bound, locally Fock property. *Ann. l'Inst. Henri Poincaré* **30**, 159–192.

Osipov, E. P. (1979b). The Yukawa$_2$ field theory: the Matthews-Salam formula. *Ann. l'Inst. Henri Poincaré* **30**, 193–206.

Osipov, E. P. (1980). The Yukawa$_2$ quantum field theory: Lorentz invariance, *Ann. Phys.* **125**, 53–66.

Osterwalder, K. (1971). On the Hamiltonian of the cubic boson self-interaction in four dimensional space time, *Fort. Phys.* **19**, 43–113.

Osterwalder, K. (1973a). Duality for free bose fields, *Comm. Math. Phys.* **29**, 1–14.

Osterwalder, K. (1973b). Euclidean Green's functions and Wightman distributions *In: Constructive Quantum Field Theory* G. Velo and A. S. Wightman, eds., New York: Springer-Verlag.

Osterwalder, K. (1980). *Mathematical Problems in Theoretical Physics*, (Lausanne, 1979), New York: Springer.

Osterwalder, K. and Schrader, R. (1972a). Feynman-Kac formula for Euclidean fermi and bose fields, *Phys. Rev. Lett.* **29**, 1423–1425.

Osterwalder, K. and Schrader, R. (1972b). On the uniqueness of the energy density in the infinite volume limit for quantum field models, *Helv. Phys. Acta* **45**, 746–754.

Osterwalder, K. and Schrader, R. (1973a). Euclidean fermi fields and a Feynman-Kac formula for boson-fermion models, *Helv. Phys. Acta* **46**, 277–302.

Osterwalder, K. and Schrader, R. (1973b). Axioms for Euclidean Green's functions. I, *Comm. Math. Phys.* **31**, 83–112.

Osterwalder, K. and Schrader, R. (1975). Axioms for Euclidean Green's functions. II, *Comm. Math. Phys.* **42**, 281–305.

Osterwalder, K. and Seiler, E. (1978). Gauge field theories on the lattice, *Ann. Phys.* **110**, 440–471.

Osterwalder, K. and Séneór, R. (1976). A nontrivial scattering matrix for weakly coupled $P(\phi)_2$ models, *Helv. Phys. Acta* **49**, 525–535.

Ozkaynak, H. (1974). *Euclidean fields for arbitrary spin particles*, Harvard University Thesis.

Paes-Leme, P. (1978). Ornstein-Zernike and analyticity properties for classical lattice spin systems, *Ann. Phys.* **115**, 367–387.

Park, Y. M. (1977). Convergence of lattice approximations and infinite volume limit in the $(\lambda\phi^4 - \sigma\phi^2 - \mu\phi)_3$ field theory, *J. Math. Phys.* **18**, 354–366.

Park, Y. M. (1979). Lack of screening in the continuous dipole systems, *Comm. Math. Phys.* **70**, 161–167.

Peierls, R. (1936). Ising's model of ferromagnetism, *Proc. Cambridge Philos. Soc.* **32**, 477–481.

Percus, J. (1975). Correlation inequalities for Ising spin lattices, *Comm. Math. Phys.* **40**, 283–308.

Percus, J. (1977). One-dimensional Ising model in arbitrary external field, *J. Stat. Phys.* 299–309.

Pirogov, S. A. and Sinai, Ya. G. (1974). Phase transitions of the first kind for small perturbations of the Ising model, *Funct. Anal. Appl.* **8**, 21–25.

Pirogov, S. A. and Sinai, Ya. G. (1975). Phase diagrams of classical lattice systems, *Theor. Mat. Fiz.* **25**, 358–369, (*Eng. translation: Theor. Math. Phys.* **25**, 1185–1192).

Pirogov, S. A. and Sinai, Ya. G. (1976). Phase diagrams of classical lattice systems. Continuation, *Theor. Mat. Fiz.* **26**, 61–76 (*Eng. translation: Theor. Math. Phys.* **26**, 39–49).

Polyakov, A. M. (1975). Compact gauge fields and the infrared catastrophe, *Phys. Lett. B* **59**, 82–84.

Polyakov, A. M. (1977). Quark confinement and topology of gauge theories, *Nucl. Phys. B* **120**, 429–458.

Powers, R. and Stormer, E. (1970). Free states of the canonical anticommutation relations, *Comm. Math. Phys.* **16**, 1–33.

Preston, C. (1974). A generalization of the FKG inequalities, *Comm. Math. Phys.* **36**, 233–241.

Reed, M. and Rosen, L. (1974). Support properties of the free measure for boson fields, *Comm. Math. Phys.* **36**, 123–132.

Reed, M. and Simon, B. (1972-1979). *Methods of Modern Mathematical Physics*, Vols. I, II, III, IV, New York: Academic Press.

Renouard, P. (1977). Analyticité et sommabilitié "de Borel" des fonctions de Schwinger du modèle de Yukawa en dimension $d = 2$, I. Approximation "a volume fini," *Ann. l'Inst. Henri Poincaré* **27**, 237–277.

Renouard, P. (1979). Analyticité et sommabilité "de Borel" des functions de Schwinger du modèle de Yukawa en dimension $d = 2$ II. La "Limite adia-batique," *Ann. l'Inst Henri Poincaré*, **31**, 235–318.

Riedel, E. and Wegner, F. (1972). Tricritical exponents and scaling fields, *Phys. Rev. Lett.* **29**, 349–352.

Rivasseau, V. (1980). Lieb's correlation inequality for plane rotators, *Comm. Math. Phys.* **77**, 145–148.

Roberts, J. (1976). Local cohomology and superselection structure, *Comm. Math. Phys.* **51**, 107–119.

Robinson, D. (1969). A proof of the existence of phase transitions in the anisotropic Heisenberg model, *Comm. Math. Phys.* **14**, 195–204.

Robinson, G. de B. (1961). *Representation of Symmetric Groups*, Toronto: University of Toronto Press.

Rosen, J. (1977). The Ising model limit of ϕ^4 lattice fields, *Proc. AMS* **66**, 114–118.

Rosen, J. (1980). Mass renormalization for the $\lambda\phi^4$ Euclidean lattice field, *Adv. Appl. Math.* **1**, 37–49.

Rosen, L. (1970). A $\lambda\phi^{2n}$ field theory without cutoffs, *Comm. Math. Phys.* **16**, 157–183.

Rosen, L. (1971). The (ϕ^{2n}) quantum field theory: higher order estimates, *Comm. Pure Appl. Math.* **24**, 417–457.

Rosen, L. (1977). Construction of the Yukawa$_2$ field theory with a large external field, *J. Math. Phys.* **18**, 894–897.

Rosen, L. and Simon, B. (1972). The $(\phi^{2n})_2$ field Hamiltonian for complex coupling constant, *Trans. Amer. Math. Soc.* **165**, 365–379.

Ruelle, D. (1962). On the asymptotic condition in quantum field theory, *Helv. Phys. Acta* **35**, 147–163.

Ruelle, D. (1969). *Statistical Mechanics*, New York: Benjamin.

Ruelle, D. (1971). Analyticity of Green's functions of dilute quantum gases, *J. Math. Phys.* **12**, 901–903.

Ruelle, D. (1972a). On the use of "small extremal fields" in the problem of symmetry breakdown in statistical mechanics, *Ann. Phys.* **69**, 364–374.

Ruelle, D. (1972b). Definition of Green's functions for dilute Fermi gases, *Helv. Phys. Acta* **45**, 215–219.

Sato, M., Miwa, T. and Jimbo, M. (1978–80). Holonomic quantum fields, I–V, *Publ. RIMS, Kyoto University* **14**, 223–267; **15**, 201–278; **15**, 577–629; RIMS preprint.

Scadron, M. (1979). *Advanced Quantum Theory*, New York: Springer.

Schor, R. (1978a). The instanton gas for the anharmonic oscillator, *Rockefeller University preprint*.

Schor, R. (1978b). The particle structure in ν-dimensional Ising models at low temperature, *Comm. Math. Phys.* **59**, 219–233.

Schrader, R. (1971). A remark on Yukawa plus boson self interaction in two space time dimensions, *Comm. Math. Phys.* **21**, 164–170.

Schrader, R. (1972). A Yukawa quantum field theory in two spacetime dimensions without cutoffs, *Ann. Phys.* **70**, 412–457.

Schrader, R. (1974a). On the Euclidean version of Haag's theorem in $P(\phi)_2$ theories, *Comm. Math. Phys.* **36**, 133–136; **38**, 81–82.

Schrader, R. (1974b). Local operator products and field equations in $P(\phi)_2$ theories, *Fort. Physik.* **22**, 611–631.

Schrader, R. (1976–7). A possible constructive approach to ϕ_4^4. I, II, III. *Comm. Math. Phys.* **49**, 131–153; **50**, 97–102; *Ann. l'Inst. Henri Poincaré* **26**, 295–301.

Schrader, R. (1977). New correlation inequalities for the Ising model and $P(\phi)$ theories, *Phys. Rev.* **B 15**, 2798–2803.

Schrader, R. (1978). Towards a constructive approach of a gauge invariant, massive $P(\phi)_2$ theory, *Comm. Math. Phys.* **58**, 299–312.

Schrader, R. and Seiler, R. (1978). A uniform lower bound on the renormalized scalar Euclidean functional determinant, *Comm. Math. Phys.* **61**, 169–175.

Schwartz, L. (1950–1). *Theory of Distributions* I, II. Paris: Hermann.

Schweber, S. (1961). *Relativistic Quantum Field Theory*, New York: Harper and Row.

Schwinger, J. (1958). On the Euclidean structure of relativistic field theory, *Proc. N.A.S.* **44**, 956–965.

Schwinger, J. (1959). Euclidean quantum electrodynamics, *Phys. Rev.* **115**, 721–731.

Segal, I. (1963). *Mathematical Problems of Relativistic Physics*, Providence: American Mathematical Society.

Segal, I. (1967). Notes toward the construction of nonlinear relativistic quantum fields. I: The Hamiltonian in two space-time dimensions as the generator of a C^*-automorphism group, *Proc. Nat. Acad. Sci. U.S.A.* **57**, 1178–1183.

Segal, I. (1969). Nonlinear functions of weak processes I, *J. Funct. Anal.* **4**, 404–456.

Segal, I. (1970). Construction of nonlinear local quantum processes: I, *Ann. Math.* **92**, 462–481.

Seiler, E. (1975). Schwinger functions for the Yukawa model in two dimensions with space-time cutoff, *Comm. Math. Phys.* **42**, 163–182.

Seiler, E. (1981). *Guage theories as a problem of constructive field theory and statistical mechanics*, University of Geneva lecture notes.

Seiler, E. and Simon, B. (1975a). Bounds in the Yukawa$_2$ quantum fields theory: Upper bound on the pressure, Hamiltonian bound and linear lower bound, *Comm. Math. Phys.* **45**, 99–114.

Seiler, E. and Simon, B. (1975b). On finite mass renormalizations in the two dimensional Yukawa model, *J. Math. Phys.* **16**, 2289–2293.

Seiler, E. and Simon, B. (1976). Nelson's symmetry and all that in the Yukawa$_2$ and ϕ_3^4 field theories, *Ann. Phys.* **97**, 470–518.

Shlosman, S. B. (1980). Phase transitions for two-dimensional models with isotropic short-range interaction and continuous symmetries, *Comm. Math. Phys.* **71**, 207–212.

Siegert, A. J. F. (1960). Partition functions as averages of functionals of Gaussian random functions, *Physica* **26**, S30–S35.

Sigal, I. (1978). Mathematical foundations of quantum scattering theory for multi-particle systems, *Memoirs of the AMS*, **209**, 1–145.

Simon, B. (1973a). Correlation inequalities and the mass gap in $P(\phi)_2$. I. Domination by the two point function, *Comm. Math. Phys.* **31**, 127–136.

Simon, B. (1973b). Positivity of the Hamiltonian semigroup and the construction of Euclidean region fields, *Helv. Phys. Acta* **46**, 686–696.

Simon, B. (1974). *The $P(\phi)_2$ Euclidean (Quantum) Field Theory*, Princeton: Princeton University Press.

Simon, B. (1975). Correlation inequalities and the mass gap in $P(\phi)_2$ II. Uniqueness of the vacuum in a class of strongly coupled theories, *Ann. Math.* **101**, 260–267.

Simon, B. (1979). *Functional Integration and Quantum Physics*, New York: Academic Press.

Simon, B. (1980). Correlation inequalities and the decay of correlations in ferromagnets, *Comm. Math. Phys.* **77**, 111–126.

Simon, B. and Griffiths, R. (1973). The $(\phi^4)_2$ field theory as a classical Ising model, *Comm. Math. Phys.* **33**, 145–164.

Simon, B. and Hoegh-Krohn, R. (1972). Hypercontractive semigroups and two-dimensional self-coupled bose fields, *J. Func. Anal.* **9**, 121–180.

Sinai, Ya. G. (1981). *The theory of phase transitions: rigorous results*, London: Pergamon.

Singer, I. M. (1977). Unpubished.

Singer, I. M. (1978). Some remarks on the Gribov ambiguity, *Comm. Math. Phys.* **60**, 7–12.

Slawny, J. (1973). Analyticity and uniqueness for spin $\frac{1}{2}$ classical ferromagnetic lattice systems at low temperatures, *Comm. Math. Phys.* **34**, 271–296.

Sokal, A. (1980). An improvement of Watson's theorem on Borel summability, *J. Math. Phys.* **21**, 261–263.

Sokal, A. (1981a). Rigorous proof of the high-temperature Josephson inequality for critical exponents. *J. Stat. Phys.* **25**, 51–56.

Sokal, A. (1981b). More inequalities for critical exponents, *J. Stat. Phys.* **25**, 25–50.

Spencer, T. (1973). Perturbation of the $P(\phi)_2$ quantum field Hamiltonian, *J. Math. Phys.* **14**, 823–828.

Spencer, T. (1974a). The absence of even bound states for $\lambda(\phi^4)_2$, *Comm. Math. Phys.* **39**, 77–79.

Spencer, T. (1974b). The mass gap for the $P(\phi)_2$ quantum field model with a strong external field, *Comm. Math. Phys.* **39**, 63–76.

Spencer, T. (1975). The decay of the Bethe-Salpeter kernel in $P(\phi)_2$ quantum field models, *Comm. Math. Phys.* **44**, 143–164.

Spencer, T. (1980). The Lipatov argument, *Comm. Math. Phys.* **74**, 273–280.

Spencer, T. and Zirilli, F. (1976). Scattering states and bound states in $\lambda P(\phi)_2$, *Comm. Math. Phys.* **49**, 1–16.

Stanley, H. (1971). *Introduction to Phase Transitions and Critical Phenomena*, New York: Oxford University Press.

Stratonovich, R. L. (1957). On a method of calculating quantum distribution functions, *Soviet Physics Doklady* **2**, 416–419.

Streater, R. (1972a). Connection between the spectrum condition and the Lorentz invariance of $P(\phi)_2$, *Comm. Math. Phys.* **26**, 109–120.

Streater, R. ed. (1972b). *Mathematics of Contemporary Physics* (London, 1971), New York: Academic Press.

Streater, R. F. (1975). Outline of axiomatic relativistic quantum field theory, *Reports on Progress in Physics* **38**, 771–846.

Streater, R. and Wightman, A. (1964). *PCT, Spin and Statistics, and All That*, New York: Benjamin.

Streater, R. F. and Wilde, I. F. (1970). Fermion states of a boson field, *Nucl. Phys.* **B24**, 561–575.

Streit, L. ed. (1976). *Quantum Dynamics: Models and Mathematics* (Bielefeld, 1975), New York: Springer-Verlag.

Strocchi, F. (1977). Spontaneous symmetry breaking in the local gauge quantum field theory; the Higgs mechanism, *Comm. Math. Phys.* **56**, 57–78.

Strocchi, F. (1978). Local and covariant gauge quantum field theories. Cluster property, superselection rules, and the infrared problem, *Phys. Rev. D* **17**, 20210–2021.

Strocchi, F. and Wightman, A. S. (1974). Proof of the charge superselection rule in local relativistic quantum field theory, *J. Math. Phys.* **15**, 2198–2224.

Summers, S. (1979). *The phase diagram for a two dimensional bose quantum field model*, Harvard University Thesis.

Summers, S. J. (1980). A new proof of the asymptotic nature of perturbation theory in $P(\phi)_2$ models, *Helv. Phys. Acta* **53**, 1–30.

Summers, S. (1981). On the phase diagram of a $P(\phi)_2$ quantum field model, *Ann. l'Inst. Henri Poincaré*, **34**, 173–229.

Suzuki, M. and Fisher, M. (1971). Zeros of the partition function for the Heisenberg, ferroelectric, and general Ising models, *J. Math. Phys.* **12**, 235–246.

Sylvester, G. (1975). Representations and inequalities for Ising model Ursell functions, *Comm. Math. Phys.* **42**, 209–220.

Sylvester, G. (1976a). *Continuous spin Ising ferromagnets*, MIT Thesis.

Sylvester, G. (1976b). Inequalities for continuous-spin Ising ferromagnets, *J. Stat. Phys.* **15**, 327–341.

Sylvester, G. (1981). Weakly coupled Gibbs measure, *Zeit. für Wahr. Theorie*, to appear.

Symanzik, K. (1960). On the many-particle structure of Green's functions in quantum field theory, *J. Math. Phys.* **1**, 249–273.

Symanzik, K. (1964). A modified model of Euclidean quantum field theory, *New York University Courant Institute of Mathematical Sciences, IMM-NYU 327*.

Symanzik, K. (1966). Euclidean quantum field theory, I. Equations for a scalar model, *J. Math. Phys.* **7**, 510–525.

Symanzik, K. (1969). Euclidean quantum field theory *In: Local Quantum Theory* R. Jost, ed., New York: Academic Press.

Symanzik, K. (1970a). Renormalizable models with simple symmetry breaking. I. Symmetry breaking by a source term, *Comm. Math. Phys.* **16**, 48–80.

Symanzik, K. (1970b). Small distance behavior in field theory and power counting, *Comm. Math. Phys.* **18**, 227–246.

Symanzik, K. (1971). Small distance behaviour analysis and Wilson expansions, *Comm. Math. Phys.* **23**, 49–86.

Symanzik, K. (1973). Infrared singularities and small distance behaviour analysis, *Comm. Math. Phys.* **34**, 7–36.

Symanzik, K. (1975). Renormalization problem in nonrenormalizable massless ϕ^4 theory, *Comm. Math. Phys.* **45**, 79–98.

Taylor, J. C. (1976). *Gauge Theories of Weak Interactions*, New York: Cambridge University Press.

Thirring, W. (1958). *Principles of Quantum Electrodynamics*, New York: Academic Press.

Thompson, C. (1980). *Mathematical Statistical Mechanics*, Princeton: Princeton University Press.

Tracy, C. and McCoy, B. (1973). Neutron scattering and the correlation functions of the Ising model near T_c, *Phys. Rev. Lett.* **31**, 1500–1504.

Tucciarone, A. (1966). A relativistic treatment of the three body problem, *Nuovo Cimento* **41A**, 204–221.

Uhlenbeck, G. and Ford, W. (1963). *Lectures in Statistical Mechanics*, Providence: American Mathematical Society.

Ukawa, A., Windey, P., and Guth, A. (1980). Dual variables for lattice guage theories and phase structure of $Z(N)$ systems. *Phys. Rev. D* **21**, 1013–1036.

van Beijeren, H. (1975). Interface sharpness in the Ising model, *Comm. Math. Phys.* **40**, 1–6.

van Beijeren, H., Gallavotti, G. and Knops, H. (1974). Conservation laws in the hierarchial model, *Physica* **78**, 541–548.

van Beijeren, H. and Sylvester, G. (1978). Phase transitions for continuous spin Ising ferromagnets. *J. Funct. Anal.* **28**, 145–167.

van Dyke, Jr., R. S., Schwinberg, P. B. and Dehmelt, H. G. (1979). Progress of the electron spin anomaly experiment, *Bull. Am. Phys. Soc.*, **24**, 758.

Vasilev, A. N. and Kazanskii, A. K. (1972). Legendre transforms of the generating functionals in quantum field theory, *Teoret. i Mat. Fizika* **12**, 352–369.

Velo, G. and Wightman, A. S. (1973). *Constructive Quantum Field Theory*, (Erice, 1973), Berlin/New York: Springer-Verlag.

Weinberg, S. (1950). High energy behavior in quantum field theory, *Phys. Rev.* **80**, 268–272.

Wightman, A. (1956). Quantum field theory in terms of vacuum expectation values, *Phys. Rev.* **101**, 860–866.

Wightman, A. (1967). Introduction to some aspects of the relativistic dynamics of quantized fields *In: High Energy Electromagnetic Interactions and Field Theory* M. Levy, ed., New York: Gordon and Breach.

Wightman, A. and Gårding, L. (1965). Fields as operator-valued distributions in relativistic quantum theory, *Arkiv för Fysik* **28**, 129–184.

Wigner, E. P. (1959). *Group Theory and Quantum Mechanics*, New York: Academic Press.

Wilde, I. (1974). The free fermion field as a Markov field, *J. Funct. Anal.* **15**, 12–21.

Williams, E. R. and Olsen, P. T. (1979). New measurement of the proton gyromagnetic ratio and a derived value of the fine-structure constant accurate to a part in 10^7, *Phys. Rev. Lett.* **42**, 1575–1579.

Wilson, K. G. (1974). Confinement of quarks, *Phys. Rev.* **D10**, 2445–2459.

Wilson, K. G. (1980). Monte-Carlo calculations for the lattice gauge theory *In: Recent Developments in Gauge Theories* (Cargèse, 1979) G. 't Hooft et al., eds., New York: Plenum Press.

Wilson, K. G. and Kogut, J. (1974). The renormalization group and the ε expansion, *Phys. Rep.* **12C**, 75–200.

Witten, E. (1977). Some exact multi pseudo particle solutions of classical Yang-Mills theory, *Phys. Rev. Lett.* **38**, 121–124.

Wu, T. T., McCoy, B., Tracy, C. and Barouch, E. (1976). Spin-spin correlation functions for the two dimensional Ising model: Exact theory in the scaling region, *Phys. Rev.* **B13**, 316–374.

Yang, C. N. and Lee, T. D. (1952). Statistical theory of equations of state and phase transitions. I. Theory of condensation, *Phys. Rev.* **87**, 404–409.

Yang, C. N. and Mills, R. L. (1954). Conservation of isotopic spin and isotopic gauge invariance, *Phys. Rev.* **96**, 191–195.

Yeh, J. (1973). *Stochastic Processes and the Wiener Integral*, New York: Marcel Dekker.

Zygmund, A. (1959). *Trigonometric Series*, Cambridge: Cambridge University Press.

Index